TOPOLOGICAL GEOMETRY

The New University Mathematics Series

Editor:
Professor E. T. Davies
Department of Mathematics, University of Southampton

This series is intended for readers whose main interest is in mathematics, or who need the methods of mathematics in the study of science and technology. Some of the books will provide a sound treatment of topics essential in any mathematical training, while other, more advanced, volumes will be suitable as preliminary reading for research in the field covered. New titles will be added from time to time.

BROWN and PAGE: *Elements of Functional Analysis*
BURGESS: *Analytical Topology*
COOPER: *Functions of a Real Variable*
CURLE and DAVIES: *Modern Fluid Dynamics* (Vols. 1 and 2)
PORTEOUS: *Topological Geometry*
ROACH: *Green's Functions: Introductory Theory with Applications*
RUND: *The Hamilton-Jacobi Theory in the Calculus of Variations*
SMITH: *Laplace Transform Theory*
SMITH: *Introduction to the Theory of Partial Differential Equations*
SPAIN: *Ordinary Differential Equations*
SPAIN: *Vector Analysis*
SPAIN and SMITH: *Functions of Mathematical Physics*
ZAMANSKY: *Linear Algebra and Analysis*

Topological Geometry

IAN R. PORTEOUS

University of Liverpool

VAN NOSTRAND REINHOLD COMPANY

LONDON

NEW YORK CINCINNATI TORONTO MELBOURNE

VAN NOSTRAND REINHOLD COMPANY LTD
Windsor House, 46 Victoria Street,
London S.W.1

INTERNATIONAL OFFICES

New York Cincinnati Toronto Melbourne

Library of Congress Catalog Card No. 77–77103

First published 1969
Reprinted 1972

Printed in Great Britain by
The Garden City Press Limited,
Letchworth, Hertfordshire, SG6 1JS

FOREWORD

Mathematicians frequently use geometrical examples as aids to the study of more abstract concepts and these examples can be of great interest in their own right. Yet at the present time little of this is to be found in undergraduate textbooks on mathematics. The main reason seems to be the standard division of the subject into several watertight compartments, for teaching purposes. The examples get excluded since their construction is normally algebraic while their greatest illustrative value is in analytic subjects such as advanced calculus or, at a slightly more sophisticated level, topology and differential topology.

Experience gained at Liverpool University over the last few years, in teaching the theory of linear (or, more strictly, affine) approximation along the lines indicated by Prof. J. Dieudonné in his pioneering book *Foundations of Modern Analysis* [14], has shown that an effective course can be constructed which contains equal parts of linear algebra and analysis, with some of the more interesting geometrical examples included as illustrations. The way is then open to a more detailed treatment of the geometry as a Final Honours option in the following year.

This book is the result. It aims to present a careful account, from first principles, of the main theorems on affine approximation and to treat at the same time, and from several points of view, the geometrical examples that so often get forgotten.

The theory of affine approximation is presented as far as possible in a basis-free form to emphasize its geometrical flavour and its linear algebra content and, from a purely practical point of view, to keep notations and proofs simple. The geometrical examples include not only projective spaces and quadrics but also Grassmannians and the orthogonal and unitary groups. Their algebraic treatment is linked not only with a thorough treatment of quadratic and hermitian forms but also with an elementary constructive presentation of some little-known, but increasingly important, geometric algebras, the Clifford algebras. On the topological side they provide natural examples of manifolds and, particularly, smooth manifolds. The various strands of the book are brought together in a final section on Lie groups and Lie algebras.

Acknowledgements

I wish to acknowledge the lively interest of my colleagues and students in the preparation of this book. Particular thanks are due to the students of W3053 (Advanced Calculus) at Columbia, who met the book in embryo, and of BH (Algebra and Geometry), CH (Linear Algebra and Analysis) and DP5 (Lie Groups and Homogeneous Spaces) at Liverpool, who have suffered parts of it more recently. I owe also a considerable debt to Prof. T. J. Willmore, who was closely associated with the earliest drafts of the book and who shared in experiments in teaching some of the more elementary geometrical material to the BH (first-year) class. Various colleagues—including Prof. T. M. Flett, Prof. G. Horrocks and Drs. R. Brown, M. C. R. Butler, M. C. Irwin and S. A. Robertson—have taught the 'Dieudonné course' at Liverpool. Their comments have shaped the presentation of the material in many ways, while Prof. C. T. C. Wall's recent work on linear algebra over rings (for example [57]) has had some influence on the final form of Chapters 11 and 13.

The linear algebra in the first half of the book is fairly standard, as is the treatment of normed linear spaces in Chapter 15 and of topological spaces in Chapter 16. For most of Chapters 9 and 11 my main debt is to Prof. E. Artin's classic [3]. My interest in Clifford algebras and their use in relativity was stimulated by discussions with Dr. R. H. Boyer, tragically killed in Austin, Texas, on August 1st, 1966. Their treatment here is derived from that of M. F. Atiyah, R. Bott and A. Shapiro [4], while the classification of the conjugation anti-involutions in the tables of Clifford algebras (Tables 13–66) is in a Liverpool M.Sc. thesis by A. Hampson. The observation that the Cayley algebra can be derived from one of the Clifford algebras I also owe to Prof. Atiyah. Chapters 18 and 19, on affine approximation, follow closely the route charted by Prof. J. Dieudonné, though the treatment of the Inverse Function Theorem and its geometrical applications is from the Princeton notes of Prof. J. Milnor [42]. The proof of the Fundamental Theorem of Algebra also is Milnor's [44]. The method adopted in Chapter 20 for constructing the Lie algebras of a Lie group was outlined to me by Prof. J. F. Adams.

Finally, thanks are due to Mr. M. E. Matthews, who drew most of the diagrams, and to Miss Gillian Thomson and her colleagues, who produced a very excellent typescript.

References and Symbols

For ease of reference propositions and exercises are numbered consecutively through each chapter, the more important propositions being styled theorems and those which follow directly from their immediate predecessors being styled corollaries. (Don't examine the system too closely—there are many anomalies!)

Implication is often indicated by the symbol $\Rightarrow$ or $\Leftarrow$, the symbol $\Leftrightarrow$ being an abbreviation for 'if, and only if'.

The symbol $\square$ is used to mark the end of a proposition or exercies and such proof or hints at proof as may be given.

Numbers within [ ] are references to the bibliography on pages 435–437. The entries in the bibliography are very varied in character. Some are texts which are readily accessible and which complement the material of this book. Others are given because of their historic interest.

Following the bibliography there is a list of the more important mathematical symbols used in the text, as well as a comprehensive index.

Liverpool, September 1969

IAN R. PORTEOUS

CONTENTS

CHAPTER 0

GUIDE

This short guide is intended to help the reader to find his way about the book.

As we indicated in the Foreword, the book consists of a basic course on affine approximation, which we refer to colloquially as 'the Dieudonné course', linked at various stages with various geometrical examples whose construction is algebraic and to which the topological and differential theorems are applied.

Chapters 1 and 2, on sets, maps and the various number systems serve to fix basic concepts and notations. The Dieudonné course proper starts at Chapter 3. Since the intention is to apply linear algebra in analysis, one has to start by studying linear spaces and linear maps, and this is done here in Chapters 3 to 7 and in the first part of Chapter 8. Next one has to set up the theory of topological spaces and continuous maps. This is done in Chapter 16, this being prefaced, for motivational and for technical reasons, by a short account of normed linear spaces in Chapter 15. The main theorems of linear approximation are then stated and proved in Chapters 18 and 19, paralleling Chapters 8 and 10, respectively, of Prof. Dieudonné's book [47].

The remainder of the book is concerned with the geometry. We risk a brief consideration of the simplest geometrical examples here, leaving the reader to come back and fill in the details when he feels able to do so.

Almost the simplest example of all is the unit circle, S^1, in the plane $\mathbf{R}^2$. This is a smooth curve. It also has a group structure, if one interprets its points as complex numbers of absolute value 1, the group product being multiplication. This group may be identified in an obvious way with the group of rotations of $\mathbf{R}^2$, or indeed of S^1 itself, about the origin.

What about $\mathbf{R}^3$? The situation is now more complicated, but the ingredients are analogous. The complex numbers are replaced, not by a three-dimensional, but by a four-dimensional algebra called the quaternion algebra and identifiable with $\mathbf{R}^4$ just as the complex algebra is identifiable with $\mathbf{R}^2$, and the circle group is replaced by the group of quaternions of absolute value 1, this being identifiable with the unit sphere, S^3, in $\mathbf{R}^4$, the set of points in $\mathbf{R}^4$ at unit distance from 0. As for the group of rotations of $\mathbf{R}^3$, this turns out to be identifiable not with S^3 but with the space obtained by identifying each point of the sphere with the point antipodal to it.

An example of how this model of the group of rotations of $\mathbf{R}^3$ can be used is the following.

Suppose one rotates a solid body continuously about a point. Then the axial rotation required to get from the initial position of the body to the position at any given moment will vary continuously. The initial position may be represented on S^3 by one of the two points representing the identity rotation, say the real quaternion 1 which we may think of as the North pole of S^3. The subsequent motion of the body may then be represented by a continuous path on the sphere. What one can show is that after a rotation through an angle 2π about any axis one arrives at the South pole, the real quaternion -1, After a further full rotation about the same axis one arrives back at 1. There are various vivid illustrations of this, one of the simplest being the soup plate trick, in which the performer rotates a soup plate lying horizontally on his hand through an angle 2π about the vertical line through its centre. His arm is then necessarily somewhat twisted. A further rotation of the plate through 2π about the vertical axis surprisingly brings the arm back to the initial position. The twisting of the arm at any point in time provides a record of a possible path of the plate from its initial to its new position, this being recorded on the sphere by a path from the initial point to the new point. As the arm twists, so the path varies continuously. The reason why it is possible for the arm to return to the initial position after a rotation of the hand through 4π is that it is possible to deform the path of the actual rotation, namely a great circle on S^3, continuously on the sphere to a point. This is not possible in the two-dimensional analogue when the group of rotations is a circle (see Exercise 16.106!). A great circle on S^1 is necessarily S^1 itself, and this is not deformable continuously on itself to a point.

The thing to be noticed here is that the topological (or continuous) features of the model are as essential to its usefulness as the algebraic or geometrical ones.

It is natural to ask, what comes next? For example, which algebras do for higher-dimensional spaces what the complex numbers and the quaternions do for $\mathbf{R}^2$ and $\mathbf{R}^3$, respectively? The answer is provided by the Clifford algebras, and this is the motivation for their study here. Our treatment of quadratic forms and their Clifford algebras in Chapters 9, 10 and 13 is somewhat more general, for we consider there not only the positive-definite quadratic forms necessary for the description of the euclidean (or Pythagorean) distance, but also indefinite forms such as the four-dimensional quadratic form

$$(x,y,z,t) \rightsquigarrow x^2 + y^2 + z^2 - t^2,$$

which arises in the theory of relativity. The Lorentz groups also are introduced.

Chapter 14 contains an alternative answer to the 'what comes next' question.

Analogues of the rotation groups arise in a number of contexts, and Chapter 11 is devoted to their study. One of the principal reasons for the generality given here is to be found towards the end of the chapter on Clifford algebras, Chapter 13. On a first reading it might, in fact, be easier to tackle the early part of Chapter 13 first, before a detailed study of Chapter 11.

Besides the spheres and the rotation groups, there are other examples of considerable interest, such as the projective spaces, their generalization the Grassmannians, and subspaces of them known as the quadrics, defined by quadratic forms. All these are defined and studied in Chapter 8 (the latter part) and Chapter 12.

There remain only two chapters to consider, and both are strongly geometrical in character. In Chapter 17 the topological features of all the examples introduced earlier are studied, while Chapter 20 is devoted to the study of the smoothness of the same examples.

And now a word to the experts about what is not included. In the algebraic direction we stop short of anything involving eigenvalues, while in the analytical direction the exponential function only turns up in an occasional example. The differential geometry is almost wholly concerned with the first differential. Riemannian metrics are nowhere mentioned, nor is there anything on curvature, nor on connections. Finally, in the topological direction there is no discussion of the fundamental group nor of the classification of surfaces. All these topics are, however, more than adequately treated in the existing literature.

CHAPTER 1

MAPS

The language of sets and maps is basic to any presentation of mathematics. Unfortunately, in many elementary school books sets are discussed at length while maps are introduced clumsily, if at all, at a rather late stage in the story. In this chapter, by contrast, maps are introduced as early as possible. Also, by way of a change, more prominence than is usual is given to the von Neumann construction of the set of natural numbers.

Most of the material is standard. Non-standard notations include $f_{\vdash}$ and $f^{\dashv}$, to denote the *forward* and *backward* maps of subsets induced by a map f, and $X!$, to denote the set (and in Chapter 2 the group) of permutations of a set X. The notation ω for the set of natural numbers is that used in [21] and in [34]. An alternative notation in common use is **N**.

Membership

Membership of a *set* is denoted by the symbol $\in$, to be read as an abbreviation for 'belongs to' or 'belonging to' according to its grammatical context. The phrase 'x is a member of X' is denoted by $x \in X$. The phrase 'x is not a member of X' is denoted by $x \notin X$. A member of a set is also said to be an *element* or a *point* of the set. Sets X and Y are *equal*, $X = Y$, if, and only if, each element of X is an element of Y and each element of Y is an element of X. Otherwise the sets are *unequal*, $X \neq Y$. Sets X and Y *intersect* or *overlap* if they have a common member and are *mutually disjoint* if they have no common member.

A set may have no members. It follows at once from the definition of equality for sets that there is only one such set. It is called the *null* or *empty set* or the *number zero* and is denoted by $\emptyset$, or by 0, though the latter symbol, having many other uses, is best avoided when we wish to think of the null set as a set, rather than as a number.

An element of a set may itself be a set. It is, however, not logically permissible to speak of the set of all sets. See Exercise 1.60 (the Russell Paradox).

Sometimes it is possible to list all the members of a set. In such a case the set may be denoted by the list of its members inside $\{\ \}$, the order in which the elements are listed being irrelevant. For example, $\{x\}$ denotes the set whose sole member is the element x, while $\{x,y\}$ denotes the set whose sole members are the elements x and y. Note that $\{y,x\} = \{x,y\}$ and that $\{x,x\} = \{x\}$. The set $\{x\}$ is not the same thing as the element x, though one is often tempted to ignore the distinction for the sake of having simpler notations. For example, let $x = 0$ $(= \emptyset)$. Then $\{0\} \neq 0$, for $\{0\}$ has a member, namely 0, while 0 has no members at all. The set $\{0\}$ will be denoted by 1 and called the *number one* and the set $\{0,1\}$ will be denoted by 2 and called the *number two*.

Maps

Let X and Y be sets. A *map* $f: X \to Y$ associates to each element $x \in X$ a unique element $f(x) \in Y$.

Suppose, for example, that X is a class of students and that Y is the set of desks in the classroom. Then any seating arrangement of the members of X at the desks of Y may be regarded as a map of X to Y (though not as a map of Y to X): to each student there is associated the desk he or she is sitting at. We shall refer to this briefly as a *classroom map*.

Maps $f: X \to Y$ and $f': X' \to Y'$ are said to be *equal* if, and only if, $X' = X$, $Y' = Y$ and, for each $x \in X$, $f'(x) = f(x)$. The sets X and Y are called, respectively, the *domain* and the *target* of the map f. For any $x \in X$, the element $f(x)$ is said to be the *value* of f at x or the *image* of x by f, and we say informally that f *sends* x to $f(x)$. We denote this by $f; x \rightsquigarrow f(x)$ or, if the domain and target of f need mention, by $f: X \to Y$; $x \rightsquigarrow f(x)$.

The arrow $\mapsto$ is used by many authors in place of $\rightsquigarrow$. The arrow $\to$ is also used, but this can lead to confusion when one is discussing maps between sets of sets. For our use of the arrow $\rightarrowtail$, and the term *source* of a map, see page 39. The *image* of a map is defined below, on page 8. The word 'range' has not been used here, either to denote the target or the image of a map. This is because both usages are current. By avoiding the word we avoid confusion.

To any map $f: X \to Y$ there is associated an equation $f(x) = y$. The map f is said to be *surjective* or *a surjection* if, for each $y \in Y$, there is some $x \in X$ such that $f(x) = y$. It is said to be *injective* or *an injection*, if, for each $y \in Y$, there is at most one element $x \in X$, though possibly none, such that $f(x) = y$. The map fails to be surjective if there exists an element $y \in Y$ such that the equation $f(x) = y$ has no *solution* $x \in X$, and fails to be injective if there exist distinct $x, x' \in X$ such that $f(x') = f(x)$. For example, a classroom map fails to be surjective if there is an empty desk and fails to be injective if there is a desk at which there is more than one student. The map f is said to be *constant* if, for all $x, x' \in X, f(x') = f(x)$.

If the map $f: X \to Y$ is both *surjective* and *injective*, it is said to be *bijective* or to be *a bijection*. In this case the equation $f(x) = y$ has a unique solution $x \in X$ for each $y \in Y$. In the classroom *each* desk is occupied by *one* student only.

An injection, or more particularly a bijection, $f: X \to Y$ may be thought of as a labelling device, each element $x \in X$ being labelled by its image $f(x) \in Y$. In an injective seating arrangement each student may, without ambiguity, be referred to by the desk he occupies.

A map $f: X \to X$ is said to be a *transformation* of X, and a bijective map $\alpha: X \to X$ a *permutation* of X.

Example 1.1. Suppose that $f: X \to Y$ is a bijection. Then a second bijection $g: X \to Y$ may be introduced in one of three ways; directly, by stating $g(x)$ for each $x \in X$, or indirectly, either in terms of a permutation $\alpha: X \to X$, with $g(x)$ defined to be $f(\alpha(x))$, or in terms of a permutation $\beta: Y \to Y$, with $g(x)$ defined to be $\beta(f(x))$. These last two possibilities are illustrated by the diagrams

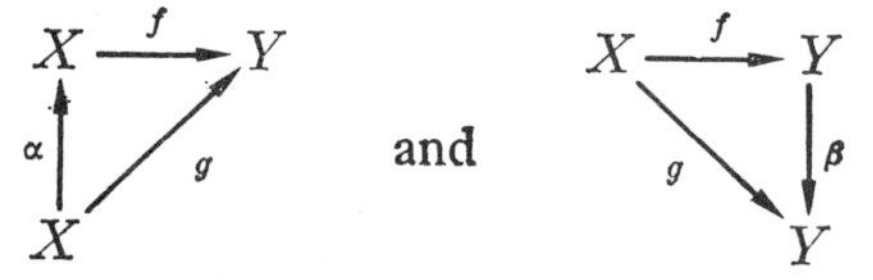

(The maps f and g may be thought of as bijective classroom maps on successive days of the week. The problem is, how to tell the class on Monday the seating arrangement preferred on Tuesday. The example indicates three ways of doing this. For the proof that g defined in either of the last two ways is bijective, see Cor. 1.4 or Prop. 1.6 below.) □

Example 1.1 illustrates the following fundamental concept.

Let $f: X \to Y$ and $g: W \to X$ be maps. Then the map $W \to Y$; $w \rightsquigarrow f(g(w))$ is called the *composite* fg (read 'f *following* g') of f and g. (An alternative notation for fg is $f \circ g$. See also page 30.) We need not restrict ourselves to two maps. If, for example, there is also a map $h: V \to W$, then the map $V \to Y$; $v \rightsquigarrow f(g(h(v)))$ will be called the composite fgh of f, g and h.

Prop. 1.2. For any maps $f: X \to Y$, $g: W \to X$ and $h: V \to W$

$$f(gh) = fgh = (fg)h.$$

Proof For all $v \in V$,

$$(f(gh))(v) = f((gh)(v)) = f(g(h(v)))$$

and

$$((fg)h)(v) = (fg)(h(v)) = f(g(h(v))). \quad \square$$

Prop. 1.3. Let $f: X \to Y$ and $g: W \to X$ be maps. Then

(i) f and g surjective $\Rightarrow$ fg surjective
(ii) f and g injective $\Rightarrow$ fg injective
(iii) fg surjective $\Rightarrow$ f surjective
(iv) fg injective $\Rightarrow$ g injective.

We prove (ii) and (iii), leaving (i) and (iv) as exercises.

Proof of (ii) Let f and g be injective and suppose that $a, b \in W$ are such that $fg(a) = fg(b)$. Since f is injective, $g(a) = g(b)$ and, since g is injective, $a = b$. So, for all $a, b \in W$,

$$fg(a) = fg(b) \Rightarrow a = b.$$

That is, fg is injective.

Proof of (iii) Let fg be surjective and let $y \in Y$. Then there exists $w \in W$ such that $fg(w) = y$. So $y = f(x)$, where $x = g(w)$. So, for all $y \in Y$, there exists $x \in X$ such that $y = f(x)$. That is, f is surjective. $\square$

Cor. 1.4. Let $f: X \to Y$ and $g: W \to X$ be maps. Then

(i) f and g bijective $\Rightarrow$ fg bijective
(ii) fg bijective $\Rightarrow$ f surjective and g injective. $\square$

Bijections may be handled directly. The bijection $1_X: X \to X$; $x \rightsquigarrow x$ is called the *identity map* or *identity permutation* of the set X and a map $f: X \to Y$ is said to be *invertible* if there exists a map $g: Y \to X$ such that $gf = 1_X$ and $fg = 1_Y$.

Prop. 1.5. A map $f: X \to Y$ is invertible if, and only if, it is bijective.
(There are two parts to the proof, corresponding to 'if' and 'only if', respectively.) $\square$

It should be noted that $1_Y f = f = f\, 1_X$, for any map $f: X \to Y$, and that if $g: Y \to X$ is a map such that $gf = 1_X$ and $fg = 1_Y$, then it is the only one. When such a map exists it is called the *inverse* of f and denoted by f^{-1}.

Prop. 1.6. Let $f: X \to Y$ and $g: W \to X$ be invertible. Then f^{-1}, g^{-1} and fg are invertible, $(f^{-1})^{-1} = f$, $(g^{-1})^{-1} = g$ and $(fg)^{-1} = g^{-1}f^{-1}$. $\square$

Let us return for a moment to Example 1.1. The bijections f and g in that example were related by permutations $\alpha : X \longrightarrow X$ and $\beta : Y \longrightarrow Y$ such that $g = f\alpha = \beta f$. It is now clear that in this case α and β exist and are uniquely determined by f and g. In fact $\alpha = f^{-1}g$ and $\beta = g f^{-1}$.

Note in passing that if $h : X \longrightarrow Y$ is a third bijection, then $h f^{-1} = (h g^{-1})(g f^{-1})$, while $f^{-1} h = (f^{-1} g)(g^{-1} h)$, the order in which the permutations are composed in the one case being the reverse of the order in which they are composed in the other case.

Subsets and quotients

If each element of a set W is also an element of a set X, then W is said to be a *subset* of X, this being denoted either by $W \subset X$ or by $X \supset W$. For example, $\{1,2\} \subset \{0,1,2\}$. The injective map $W \longrightarrow X$; $w \rightsquigarrow w$ is called the *inclusion* of W in X.

In practice a subset is often defined in terms of the truth of some proposition. The subset of a set X consisting of all x in X for which some proposition $P(x)$ concerning x is true is normally denoted by $\{x \in X : P(x)\}$, though various abbreviations are in use in special cases. For example, a map $f : X \longrightarrow Y$ defines various subsets of X and Y. The set $\{y \in Y : y = f(x)$, for some $x \in X\}$, also denoted more briefly by $\{f(x) \in Y : x \in X\}$, is a subset of Y called the *image* of f and denoted by $\operatorname{im} f$. It is non-null if X is non-null. It is also useful to have a short name for the set $\{x \in X : f(x) = y\}$, where y is an element of Y. This will be called the *fibre* of f over y. It is a subset of X, possibly null if f is not surjective. The fibres of a map f are sometimes called the *levels* or *contours* of f, especially when the target of f is $\mathbf{R}$, the real numbers (introduced formally in Chapter 2). The set of non-null fibres of f is called the *coimage* of f and is denoted by $\operatorname{coim} f$.

A subset of a set X that is neither null nor the whole of X is said to be a *proper subset* of X.

The elements of a set may themselves be sets or maps. In particular it is quite possible that two elements of a set may themselves be sets which intersect. This is the case with the *set of all subsets* of a set X, Sub X (known also as the *power set* of X for reasons given on pages 10 (Prop 1.11) and 21).

Consider, for example,

$$\text{Sub}\,\{0,1,2\} = \{0,\{0\},\{1\},\{2\},\{0,1\},\{0,2\},\{1,2\},\{0,1,2\}\},$$

where $0 = \emptyset$, $1 = \{0\}$ and $2 = \{0,1\}$ as before. The elements $\{0,1\}$ and $\{0,2\}$ are subsets of $\{0,1,2\}$, which intersect each other. A curious fact about this example is that each *element* of $\{0,1,2\}$ is also a *subset* of $\{0,1,2\}$ (though the converse is not, of course, true). For example, $2 = \{0,1\}$ is both an element of $\{0,1,2\}$ and a subset of $\{0,1,2\}$. We shall return to this later in Prop. 1.34.

Frequently one classifies a set by dividing it into mutually disjoint subsets. A map $f: X \to Y$ will be said to be a *partition* of the set X, and Y to be the *quotient* of X by f, if f is surjective, if each element of Y is a subset of X, and if the fibre of f over any $y \in Y$ is the set y itself.

For example, the map $\{0,1,2\} \to \{\{0,1\},\{2\}\}$ sending 0 and 1 to $\{0,1\}$ and 2 to $\{2\}$ is a partition of $\{0,1,2\}$ with quotient the set $\{\{0,1\},\{2\}\}$ of subsets of $\{0,1,2\}$.

The following properties of partitions and quotients are easily proved.

Prop. 1.7. Let X be a set. A subset $\mathscr{S}$ of $\operatorname{Sub} X$ is a quotient of X if, and only if, the null set is not a member of $\mathscr{S}$, each $x \in X$ belongs to some $A \in \mathscr{S}$ and no $x \in X$ belongs to more than one $A \in \mathscr{S}$. □

Prop. 1.8. A partition f of a set X is uniquely determined by the quotient of X by f. □

Any map $f: X \to Y$ with domain a given set X induces a partition of X, as follows.

Prop. 1.9. Let $f: X \to Y$ be a map. Then $\operatorname{coim} f$, the set of non-null fibres of f, is a quotient of X.

Proof By definition the null set is not a member of $\operatorname{coim} f$. Also, each $x \in X$ belongs to the fibre of f over $f(x)$. Finally, x belongs to no other fibre, since the statement that f belongs to the fibre of f over y implies that $y = f(x)$. □

It is occasionally convenient to have short notations for the various injections and surjections induced by a map $f: X \to Y$. Those we shall use are the following:

f_{inc} for the inclusion of $\operatorname{im} f$ in Y,
f_{par} for the partition of X on to $\operatorname{coim} f$,
f_{sur} for f 'made surjective', namely the map

$$X \to \operatorname{im} f;\ x \rightsquigarrow f(x),$$

f_{inj} for f 'made injective', namely the map

$$\operatorname{coim} f \to Y;\ f_{\text{par}}(x) \rightsquigarrow f(x),$$

and, finally,

f_{bij} for f 'made bijective', namely the map

$$\operatorname{coim} f \to \operatorname{im} f;\ f_{\text{par}}(x) \rightsquigarrow f(x).$$

Clearly,

$$f = f_{\text{inc}} f_{\text{sur}} = f_{\text{inj}} f_{\text{par}} = f_{\text{inc}} f_{\text{bij}} f_{\text{par}}.$$

These maps may be arranged in the diagram

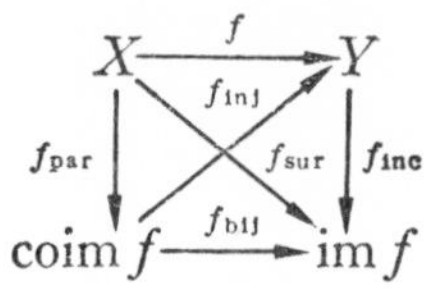

Such a diagram, in which any two routes from one set to another represent the same map, is said to be a *commutative diagram.* For example, the triangular diagrams on page 6 are commutative. For the more usual use of the word 'commutative' see page 15.

The composite $fi: W \to Y$ of a map $f: X \to Y$ and an inclusion $i: W \to X$ is said to be the *restriction* of f to the subset W of X and denoted also by $f \mid W$. The target remains unaltered.

If maps $f: X \to Y$ and $g: Y \to X$ are such that $fg = 1_Y$ then, by Prop. 1.3, f is surjective and g is injective. The injection g is said to be a *section* of the surjection f. It selects for each $y \in Y$ a *single* $x \in X$ such that $f(x) = y$.

It is assumed that any surjection $f: X \to Y$ has a section $g: Y \to X$, this assumption being known as the *axiom of choice.*

Notice that to prove that a map $g: Y \to X$ is the inverse of a map $f: X \to Y$ it is not enough to prove that $fg = 1_Y$. One also has to prove that $gf = 1_X$.

Prop. 1.10. Let $g, g': Y \to X$ be sections of $f: X \to Y$. Then $g = g' \Leftrightarrow \operatorname{im} g = \operatorname{im} g'$.

Proof $\Rightarrow$: Clear.

$\Leftarrow$: Let $y \in Y$. Since $\operatorname{im} g = \operatorname{im} g'$, $g(y) = g'(y')$ for some $y' \in Y$. But $g(y) = g'(y') \Rightarrow fg(y) = fg'(y') \Rightarrow y = y'$. That is, for all $y \in Y$, $g(y) = g'(y)$. So $g = g'$. □

A map $g: B \to X$ is said to be a *section* of the map $f: X \to Y$ *over the subset* B of Y if, and only if, $fg: B \to Y$ is the inclusion of B in Y.

The set of maps $f: X \to Y$, with given sets X and Y as domain and target respectively, is sometimes denoted by Y^X, for a reason which will be given on page 21.

Prop. 1.11. Let X be any set. Then the map

$$2^X \to \operatorname{Sub} X : f \rightsquigarrow \{x \in X : f(x) = 0\}$$

is bijective. □

Many authors prefer 2^X to Sub X as a notation for the set of subsets of X.

The set of permutations $f: X \to X$ of a given set X will be denoted by $X!$. This notation is non-standard. The reason for its choice is given on page 22.

Forwards and backwards

A map $f: X \to Y$ induces maps $f_{\vdash}$ (*f forwards*) and $f^{\dashv}$ (*f backwards*) as follows:

$$f_{\vdash}: \text{Sub } X \to \text{Sub } Y; \quad A \rightsquigarrow \{f(x) : x \in A\}$$
$$f^{\dashv}: \text{Sub } Y \to \text{Sub } X; \quad B \rightsquigarrow \{x : f(x) \in B\}.$$

The set $f_{\vdash}(A)$ is called the *image* of A by f in Y and the set $f^{\dashv}(B)$ is called the *counterimage* or *inverse image* of B by f in X.

The notations $f_{\vdash}$ and $f^{\dashv}$ are non-standard. It is the usual practice to abbreviate $f_{\vdash}(A)$ to $f(A)$ and to write $f^{-1}(B)$ for $f^{\dashv}(B)$, but this can, and does, lead to confusion, since in general $f^{\dashv}$ is *not* the inverse of $f_{\vdash}$. The absence of a notation also makes difficult direct reference to the maps $f_{\vdash}$ and $f^{\dashv}$.

There are several unsatisfactory alternatives for $f_{\vdash}$ and $f^{\dashv}$ in circulation. An alternative that is almost as satisfactory is to denote $f_{\vdash}$ by f_* and $f^{\dashv}$ by f^*, but the 'star' notations are often wanted for other purposes. The positions of the marks $\vdash$ and $\dashv$ or $*$ have not been chosen at random. They conform to an accepted convention that the lower position is used in a notation for an induced map going 'in the same direction' as the original map, while the upper position is used for an induced map going 'in the opposite direction'.

Prop. 1.12. Let $f: X \to Y$ be a map and let A be a subset of X and B a subset of Y. Then

$$A \subset f^{\dashv}(B) \Leftrightarrow f_{\vdash}(A) \subset B. \qquad \square$$

Prop. 1.13. Let $f: X \to Y$ and $g: W \to X$ be maps. Then

$$(fg)_{\vdash} = f_{\vdash}g_{\vdash} : \text{Sub } W \to \text{Sub } Y,$$
$$(fg)^{\dashv} = g^{\dashv}f^{\dashv} : \text{Sub } Y \to \text{Sub } W$$

and $$((fg)^{\dashv})_{\vdash} = (g^{\dashv})_{\vdash}(f^{\dashv})_{\vdash} : \text{Sub Sub } Y \to \text{Sub Sub } W. \qquad \square$$

Prop. 1.14. Let $f: X \to Y$ be any map. Then

$$f^{\dashv}f_{\vdash} = 1_{\text{Sub } X} \Leftrightarrow f \text{ is injective.}$$

Proof Note first that $f^{\dashv}f_{\vdash} = 1_{\text{Sub } X}$ if, and only if, for all $A \in \text{Sub } X$, $f^{\dashv}f_{\vdash}(A) = A$. Now, for any $A \in \text{Sub } X$, $f^{\dashv}f_{\vdash}(A) \supset A$. For let $x \in A$.

Then $f(x) \in f_{\vdash}(A)$; that is, $x \in f^{\dashv}f_{\vdash}(A)$. So $f^{\dashv}f_{\vdash} = 1_{\operatorname{Sub} X}$ if, and only if, for all $A \in \operatorname{Sub} X$, $f^{\dashv}f_{\vdash}(A) \subset A$.

$\Leftarrow$: Suppose that f is injective. By what we have just said it is enough to prove that, for all $A \in \operatorname{Sub} X$, $f^{\dashv}f_{\vdash}(A) \subset A$. So let $x \in f^{\dashv}f_{\vdash}(A)$. Then $f(x) \in f_{\vdash}(A)$. That is, $f(x) = f(a)$, for some element $a \in A$. However, f is injective, so that $x = a$ and $x \in A$.
(Note here how the datum 'f is injective' becomes relevant only in the last line of the proof!)

$\Rightarrow$: Suppose that f is not injective. Then it has to be proved that it is not true that, for all $A \in \operatorname{Sub} X$, $f^{\dashv}f_{\vdash}(A) \subset A$. To do so it is enough to construct, or exhibit, a single subset A of X such that $f^{\dashv}f_{\vdash}(A) \not\subset A$. Now f will fail to be injective only if there exist distinct $a, b \in X$ such that $f(a) = f(b)$. Choose such a, b and let $A = \{a\}$. Then $b \in f^{\dashv}f_{\vdash}(A)$, but $b \notin A$. That is $f^{\dashv}f_{\vdash}(A) \not\subset A$.
(There are points of logic and of presentation to be noted here also!)

This completes the proof. □

Prop. 1.15. Let $f: X \longrightarrow Y$ be any map. Then

$$f_{\vdash}f^{\dashv} = 1_{\operatorname{Sub} Y} \Leftrightarrow f \text{ is surjective.} \qquad \square$$

As an immediate corollary of the last two propositions we also have:

Prop. 1.16. Let $f: X \longrightarrow Y$ be any map. Then

$$f^{\dashv} = (f_{\vdash})^{-1} \Leftrightarrow f \text{ is bijective.} \qquad \square$$

Notice that, for any map $f: X \longrightarrow Y$ and any $x \in X$, $f_{\text{par}}(x) = f^{\dashv}\{f(x)\}$.

Pairs

A surjection $2 \longrightarrow W$ is called an *ordered pair*, or simply a *pair*, any map $2 \longrightarrow X$ being called a *pair of elements* or *2-tuple* of the set X. The standard notation for a pair is (a,b), where $a = (a,b)(0)$ and $b = (a,b)(1)$, the image of the pair being the set $\{a,b\}$. Possibly $a = b$. The elements a and b are called the *components* of (a,b). We shall frequently be perverse and refer to a as the 0th component of (a,b) and to b as the 1st component of (a,b). It is of course more usual to call a and b, respectively, the first and second components of (a,b). Two pairs (a,b) and (a',b') are equal if, and only if, $a = a'$ and $b = b'$.

Let X and Y be sets. The set of pairs

$$X \times Y = \{(x,y) : x \in X, y \in Y\}$$

is called the (*cartesian*) *product* of the pair (X,Y), often referred to loosely as the *product* of X and Y. (The term 'cartesian' refers to R. Descartes, who introduced the set $\mathbf{R} \times \mathbf{R}$ of pairs of real numbers.)

The set X^2 of maps $2 \rightarrow X$ is identified with the set $X \times X$ in the obvious way. (In particular, $\mathbf{R} \times \mathbf{R}$ is usually denoted by $\mathbf{R}^2$.)

Prop. 1.17. Let X and Y be sets. Then

$$X \times Y = \emptyset \Leftrightarrow X = \emptyset \text{ or } Y = \emptyset. \qquad \square$$

Prop. 1.18. Let X, Y, U and V be non-null sets. Then

$$X \times Y = U \times V \Leftrightarrow X = U \text{ and } Y = V. \qquad \square$$

If $f: W \rightarrow X$ and $g: W \rightarrow Y$ are maps with the same source W, the symbol (f,g) is used to denote not only the pair (f,g) but also the map

$$W \rightarrow X \times Y; \quad w \rightsquigarrow (f(w),g(w)).$$

Prop. 1.19. Let $(f,g): W \rightarrow X \times Y$ and $(f',g'): W \rightarrow X \times Y$ be maps. Then

$$(f,g) = (f',g') \Leftrightarrow f = f' \text{ and } g = g'. \qquad \square$$

The maps f and g are called the *components* of (f,g). In particular, the map $1_{X \times Y}$ has components $p: X \times Y \rightarrow X$; $(x,y) \rightsquigarrow x$ and $q: X \times Y \rightarrow Y$; $(x,y) \rightsquigarrow y$, the *projections* of the product $X \times Y$.

Prop. 1.20. Let X and Y be sets, not both null, and let $(p,q) = 1_{X \times Y}$. Then the projection p is surjective $\Leftrightarrow$ Y is non-null, and the projection q is surjective $\Leftrightarrow$ X is non-null. $\square$

Prop. 1.21. Let $f: W \rightarrow X \times Y$ be a map and let $(p,q) = 1_{X \times Y}$. Then $f = (pf, qf)$. $\square$

The *graph* of a map $f: X \rightarrow Y$ is, by definition, the subset

$$\{(x,y) \in X \times Y : y = f(x)\} = \{(x,f(x)) : x \in X\}$$

of $X \times Y$.

Prop. 1.22. Let f and g be maps. Then

$$\text{graph}\, f = \text{graph}\, g \Leftrightarrow f_{\text{sur}} = g_{\text{sur}}. \qquad \square$$

Prop. 1.23. Let $f: X \rightarrow Y$ be a map. Then the map $X \rightarrow X \times Y$; $x \rightsquigarrow (x,f(x))$ is a section of the projection $X \times Y \rightarrow X$; $(x,y) \rightsquigarrow x$, with image graph f. $\square$

Equivalences

A partition of a set X is frequently specified by means of an equivalence on the set. An *equivalence* on X is by definition a subset E of $X \times X$ such that, for all $a, b, c \in X$,

(i) $(a,a) \in E$

(ii) $(a,b) \in E \Rightarrow (b,a) \in E$

and (iii) (a,b) and $(b,c) \in E \Rightarrow (a,c) \in E$, elements a and b of X being said to be *E-equivalent* if $(a,b) \in E$, this being denoted also by $a \sim_E b$, or simply by $a \sim b$ if the equivalence is implied by the context. Frequently, one refers loosely to the equivalence $\sim_E$ or $\sim$ on X.

Prop. 1.24. Any map $f: X \rightarrow Y$ determines an equivalence E on X defined, for all $(a,b) \in X \times X$, by

$$a \sim_E b \Leftrightarrow f(a) = f(b),$$

that is, $a \sim_E b$ if, and only if, a and b belong to the same fibre of f.

Conversely, each equivalence E on X is so determined by a unique partition $f: X \twoheadrightarrow Y$ of X, namely that defined, for all $x \in X$, by

$$f(x) = \{a \in X : a \sim_E x\}. \qquad \square$$

The elements of the quotient of X determined by an equivalence E on X are called the (*equivalence*) *classes* of the equivalence E. The equivalence classes of the equivalence on X determined by a map $f: X \rightarrow Y$ are just the non-null fibres of f.

Products on a set

Let G and H be sets. A map of the form $G \times G \rightarrow H$ is called a *binary operation* or *product* on the set G with *values* in H, a map $G \times G \rightarrow G$ being called, simply, a *product* on G.

For example, the map $X^X \times X^X \rightarrow X^X$; $(f,g) \rightsquigarrow fg$ is a product on the set X^X of maps $X \rightarrow X$, while the map $X! \times X! \rightarrow X!$; $(f,g) \rightsquigarrow fg$ is a product on the set $X!$ of invertible maps $X \rightarrow X$, since the composite of any two invertible maps is invertible.

Prop. 1.25. Let $G \times G \rightarrow G$; $(a,b) \rightsquigarrow ab$ be a product on a set G, and suppose that there is an element e of G such that each of the maps $G \rightarrow G$; $a \rightsquigarrow ea$ and $a \rightsquigarrow ae$ is the identity map 1_G. Then e is unique.

Proof Let e and e' be two such elements. Then $e' = e'e = e$. $\square$

Such a product is said to be a *product with unity* e.

A product $G \times G \rightarrow G$; $(a,b) \rightsquigarrow ab$ is said to be *associative* if, for all $a, b, c \in G$, $(ab)c = a(bc)$.

Prop. 1.26. Let $G \times G \rightarrow G$; $(a,b) \rightsquigarrow ab$ be an associative product on G with unity e. Then, for any $g \in G$ there is at most one element $h \in G$ such that $gh = e = hg$.

Proof Let h and h' be two such elements. Then $h' = h'e = h'(gh) = (h'g)h = eh = h$. $\square$

The element h, if it exists, is said to be the *inverse* of g with respect to the product. If each element of G has an inverse, the product is said to *admit inverses.*

A product $G \times G \to G$; $(a,b) \rightsquigarrow ab$ is said to be *commutative* if, for all $a, b \in G$, $ba = ab$.

Suppose, for example, that G is a subset of X^X, for some set X, and that the product is composition. Then the product is commutative if, and only if, for each $a, b \in G$, the diagram

$$\begin{array}{ccc} X & \xrightarrow{a} & X \\ \big\downarrow{\scriptstyle b} & & \big\downarrow{\scriptstyle b} \\ X & \xrightarrow{a} & X \end{array}$$

is commutative. This example connects the two uses of the word 'commutative'.

For examples of products on the set of natural numbers see page 20.

The detailed study of products on a set is deferred until Chapter 2. *Bilinear* products are introduced in Chapter 3.

Union and intersection

Let $\mathscr{S}$ be any set of sets. We may then form its *union* $\bigcup \mathscr{S}$. An element $x \in \bigcup \mathscr{S}$ if, and only if, $x \in X$ for some $X \in \mathscr{S}$. If $\mathscr{S} \neq \emptyset$ we may also form its *intersection* $\bigcap \mathscr{S}$. An element $x \in \bigcup \mathscr{S}$ if, and only if, $x \in X$ for all $X \in \mathscr{S}$. This latter set is clearly a subset of any member of $\mathscr{S}$. If $\mathscr{S} = \{X,Y\}$ we write

$$\bigcup \mathscr{S} = X \cup Y, \quad \text{the } \textit{union} \text{ of } X \text{ and } Y$$

and

$$\bigcap \mathscr{S} = X \cap Y, \quad \text{the } \textit{intersection} \text{ of } X \text{ and } Y.$$

If $X \cap Y \neq \emptyset$, X intersects Y, while, if $X \cap Y = \emptyset$, X and Y are mutually disjoint. The *difference* $X \setminus Y$ of sets X and Y is defined to be the set $\{x \in X : x \notin Y\}$. It is a subset of X. (Y need not be a subset of X.) When Y is a subset of X, the difference $X \setminus Y$ is also called the *complement* of Y in X.

Prop. 1.27. Let $f: X \to Y$ be a map, let A, $B \in \operatorname{Sub} X$ and let $\mathscr{S} \subset \operatorname{Sub} X$. Then $f_{\vdash}(\emptyset) = \emptyset$,

$$A \subset B \Rightarrow f_{\vdash}(A) \subset f_{\vdash}(B), \quad f_{\vdash}(\bigcup \mathscr{S}) = \bigcup (f_{\vdash})_{\vdash} \mathscr{S}$$

and $f_{\vdash}(A \cap B) \subset f_{\vdash}(A) \cap f_{\vdash}(B)$, with $f_{\vdash}(A \cap B) = f_{\vdash}A \cap f_{\vdash}B$ for all A, $B \in \operatorname{Sub} X$ if, and only if, f is injective.

(All but the last part is easy to prove. The proof of the last part may be modelled on the proof of Prop. 1.14.) □

It is easier to go backwards than to go forwards.

Prop. 1.28. Let $f: X \to Y$ be a map, let $A, B \in \operatorname{Sub} Y$ and let $\mathscr{S} \subset \operatorname{Sub} Y$.

Then $f^{-1}(\emptyset) = \emptyset, \quad f^{-1}(Y) = X,$

$$A \subset B \Rightarrow f^{-1}(A) \subset f^{-1}(B),$$
$$f^{-1}(\bigcup \mathscr{S}) = \bigcup (f^{-1})_{\vdash}(\mathscr{S})$$

and, if $\mathscr{S} \neq \emptyset$, $\quad f^{-1}(\bigcap \mathscr{S}) = \bigcap (f^{-1})_{\vdash}(\mathscr{S})$. □

Prop. 1.29. Let $f: X \to Y$ be a map, let $\mathscr{T} \subset \operatorname{Sub} X$ be such that for all $\mathscr{S} \subset \mathscr{T}$, $\bigcup \mathscr{S} \in \mathscr{T}$ and let $\mathscr{V} = (f^{-1})^{-1}(\mathscr{T})$. Then, for all $\mathscr{U} \subset \mathscr{V}$, $\bigcup \mathscr{U} \in \mathscr{V}$.

Proof

$$\begin{aligned} \mathscr{U} \subset \mathscr{V} = (f^{-1})^{-1}(\mathscr{T}) &\Rightarrow (f^{-1})_{\vdash}(\mathscr{U}) \subset \mathscr{T}, \quad \text{by Prop. 1.12,} \\ &\Rightarrow f^{-1}(\bigcup \mathscr{U}) = \bigcup((f^{-1})_{\vdash}(\mathscr{U})) \in \mathscr{T} \\ &\Rightarrow \bigcup \mathscr{U} \in (f^{-1})^{-1}(\mathscr{T}) = \mathscr{V}. \end{aligned}$$ □

To conclude this section there is a long list of standard results, all easy to prove.

Prop. 1.30. Let X, Y and Z be sets. Then

$$\begin{aligned} X \cup Y &= Y \cup X \\ X \cup (Y \cup Z) &= (X \cup Y) \cup Z \\ X \cup X &= X \\ X \cup \emptyset &= X \\ X &\subset X \cup Y \\ X \subset Y &\Leftrightarrow X \cup Y = Y \\ X \subset Z \text{ and } Y \subset Z &\Leftrightarrow X \cup Y \subset Z, \\ X \cap Y &= Y \cap X \\ X \cap (Y \cap Z) &= (X \cap Y) \cap Z \\ X \cap X &= X \\ X \cap \emptyset &= \emptyset \\ X \cap Y &\subset X \\ X \subset Y &\Leftrightarrow X \cap Y = X \\ (X \cup Y) \cap Z &= (X \cap Z) \cup (Y \cap Z) \\ (X \cap Y) \cup Z &= (X \cup Z) \cap (Y \cup Z), \\ X \setminus X &= \emptyset \\ X \setminus (X \setminus Y) &= X \cap Y \\ X \setminus (X \cap Y) &= X \setminus Y \\ X \setminus (Y \setminus Z) &= (X \setminus Y) \cup (X \cap Z) \\ (X \setminus Y) \setminus Z &= X \setminus (Y \cup Z) = (X \setminus Y) \cap (X \setminus Z) \\ X \setminus (Y \cap Z) &= (X \setminus Y) \cup (X \setminus Z) \\ Y \subset Z &\Rightarrow (X \setminus Z) \subset (X \setminus Y), \end{aligned}$$

$$X \times (Y \cup Z) = (X \times Y) \cup (X \times Z)$$
$$X \times (Y \cap Z) = (X \times Y) \cap (X \times Z)$$
$$X \times (Y \setminus Z) = (X \times Y) \setminus (X \times Z)$$
$$(X \times Y) \subset (X \times Z) \Leftrightarrow X = \emptyset \quad \text{or} \quad Y \subset Z. \qquad \square$$

Prop. 1.31. Let X be a set and let $\mathscr{S}$ be a non-null subset of Sub X. Then

$$X \setminus (\bigcup \mathscr{S}) = \bigcap \{X \setminus A;\, A \in \mathscr{S}\}$$
$$X \setminus (\bigcap \mathscr{S}) = \bigcup \{X \setminus A;\, A \in \mathscr{S}\}. \qquad \square$$

Exercise 1.32. Let X, Y, Z, X' and Y' be sets. Is it always true or is it in general not true

(i) that $Y \times X = X \times Y$,
(ii) that $X \times (Y \times Z) = (X \times Y) \times Z$,
(iii) that $(X \times Y) \cup (X' \times Y') = (X \cup X') \times (Y \cup Y')$,
(iv) that $(X \times Y) \cap (X' \times Y') = (X \cap X') \times (Y \cap Y')$? $\square$

Natural numbers

We have already encountered the sets, or numbers, $0 = \emptyset$, $1 = \{0\} = 0 \cup \{0\}$ and $2 = \{0,1\} = 1 \cup \{1\}$. It is our final assumption in this chapter that this construction can be indefinitely extended. To be precise, we assert that there exists a unique set ω such that

(i) : $0 \in \omega$
(ii) : $n \in \omega \Rightarrow n \cup \{n\} \in \omega$
(iii) : the only subset ω' of ω such that $0 \in \omega'$ and such that $n \in \omega' \Rightarrow n \cup \{n\} \in \omega'$ is ω itself.

The elements of ω are called the *natural* or *finite numbers*. For any $n \in \omega$ the number $n \cup \{n\}$, also written $n + 1$, is called *the successor* of n, and n is said to be the *predecessor* of $n \cup \{n\}$—we prove in Prop. 1.33 below that every natural number, with the exception of 0, has a predecessor, this predecessor then being proved in Cor. 1.36 to be unique. The number 0 clearly has no predecessor for, for any $n \in \omega$, $n \cup \{n\}$ is non-null.

Prop. 1.33. Let $n \in \omega$. Then $n = 0$ or n has a predecessor.

Proof Let $\omega' = \{n \in \omega\colon n = 0$, or n has a predecessor$\}$. Our aim is to prove that $\omega' = \omega$ and by (iii) it is enough to prove (i) and (ii) for the set ω'.

(i) : We are told explicitly that $0 \in \omega'$.
(ii) : Let $n \in \omega'$. Then $n \cup \{n\}$ has n as a predecessor. So $n \cup \{n\} \in \omega'$. $\square$

Axiom (iii) is known as the *principle* of *mathematical induction*, and any proof relying on it, such as the one we have just given, is said to be an *inductive proof*, or a *proof by induction*. The truth of (i) for the set ω' is called the *basis* of the induction, and the truth of (ii) for ω' is called the *inductive step*.

The set $\omega \setminus \{0\}$ will be denoted by ω^+, and an element of ω^+ will be said to be *positive*.

The next proposition develops a remark made at the foot of page 8.

Prop. 1.34. Let $m, n \in \omega$. Then $m \in n \Rightarrow m \subset n$.

Proof Let $\omega' = \{n \in \omega : m \in n \Rightarrow m \subset n\}$. Since 0 has no members, $0 \in \omega'$. This is the basis of the induction.

Let $n \in \omega'$ and let $m \in n \cup \{n\}$. Then either $m \in n$ or $m = n$. In either case $m \subset n$, and therefore $m \subset n \cup \{n\}$. That is $n \cup \{n\} \in \omega'$. This is the inductive step.

So $\omega' = \omega$, which is what we had to prove. □

Cor. 1.35. Let $n \in \omega$. Then $\bigcup (n \cup \{n\}) = n$. □

Cor. 1.36. The predecessor of a natural number, when it exists, is unique. □

Cor. 1.37. The map $\omega \to \omega^+$; $n \rightsquigarrow n \cup \{n\}$ is bijective. □

Prop. 1.38. Let m and $n \in \omega$ and let there be a bijection $f : n \to m$. Then $m = n$.

Proof Let $\omega' = \{n \in \omega : m \in \omega \text{ and } f : n \to m \text{ a bijection} \Rightarrow m = n\}$. Now, since the image by any map of the null set is the null set, a map $f : 0 \to m$ is bijective only if $m = 0$. So $0 \in \omega'$.

Let $n \in \omega'$ and let $f : n \cup \{n\} \to m$ be a bijection. Since $m \neq 0$, $m = k \cup \{k\}$, for some $k \in \omega$. Define $f' : n \to k$ by $f'(i) = f(i)$ if $f(i) \neq k$, and $f'(i) = f(n)$ if $f(i) = k$. Then f' is bijective, with $f'^{-1}(j) = f^{-1}(j)$ if $j \neq f(n)$ and $f'^{-1}(j) = f^{-1}(k)$ if $j = f(n)$. Since $n \in \omega'$ it follows that $k = n$. Therefore $n \cup \{n\} = k \cup \{k\} = m$. That is, $n \cup \{n\} \in \omega'$.

So $\omega' = \omega$. □

A *sequence* on a set X is a map $\omega \to X$. A sequence on X may be denoted by a single letter, a, say, or by such notations as $n \rightsquigarrow a_n$, or $(a_n : n \in \omega)$ or $(a_n)_{n \in \omega}$, or simply, but confusingly, a_n. The symbol a_n strictly denotes the nth *term* of the sequence, that is, its value $a(n)$ at n. In some contexts one avoids the single-letter notation as a name for the sequence, the letter being reserved for some other use, as for example to denote the limit of the sequence when it is convergent. For convergence see Chapter 2, page 42.

A sequence may be defined *recursively*, that is by a formula defining the $(n+1)$th term, for each $n \in \omega$, in terms either of the nth term or of all the terms of the sequence up to and including the nth. The 0th and possibly several more terms of the sequence have, of course, to be stated explicitly.

A map $n \longrightarrow X$, or *n-tuple* of X, where $n \in \omega$, is often called a *finite sequence* on X, the word finite frequently being omitted when it can be inferred from the context.

A set X is said to be *finite* if there is, for some natural number n, a bijection $n \longrightarrow X$. By Prop. 1.38 the number n, if it exists, is uniquely determined by the set X. It is denoted by $\#X$ and called the *number* of elements or the *cardinality* of the set X. If X is not finite, we say X is *infinite*.

Prop. 1.39. The set ω is infinite. □

A set X is said to be *countably infinite* if there is a bijection $\omega \longrightarrow X$. A set is said to be *countable* if it is either finite or countably infinite.

Prop. 1.40. A set X is *countable* if, and only if, there is a surjection $\omega \longrightarrow X$ or equivalently, by the axiom of choice, an injection $X \longrightarrow \omega$. □

Cor. 1.41. If a set is not countable, then the elements of ω or of any subset of ω cannot be used to label the elements of the set. □

Prop. 1.42. The set 2^ω is not countable.

Proof Suppose that there is an injection $i: 2^\omega \longrightarrow \omega$. Then each $n \in \omega$ is the image of at most one sequence $b_{(n)}: \omega \longrightarrow 2$. Now let a sequence

$$a: \omega \longrightarrow 2; \quad n \rightsquigarrow a_n$$

be constructed as follows. If $n \notin \operatorname{im} i$, let $a_n = 0$, while, if $n = i(b_{(n)})$, let $a_n = 0$ if $(b_{(n)})_n = 1$ and let $a_n = 1$ if $(b_{(n)})_n = 0$. By this construction, some term of a differs from some term of b, for every $b \in 2^\omega$. But $a \in 2^\omega$, so that there is a contradiction. There is, therefore, no such injection i. So 2^ω is not countable. □

Products on ω

There are three standard products on ω, *addition*, *multiplication* and *exponentiation*. In this section we summarize their main properties. For a full treatment the reader is referred, for example, to [21].

Addition: $\omega \times \omega \longrightarrow \omega$; $(m,n) \rightsquigarrow m+n$ is defined, for all $(m,n) \in \omega \times \omega$, by the formula

$$m + n = \#((m \times \{0\}) \cup (n \times \{1\}))$$

or recursively, for each $m \in \omega$, by the formula

$$m + 0 = m, \quad m + 1 = \#(m \cup \{m\})$$

and, for all $k \in \omega$, $m + (k + 1) = (m + k) + 1$, $m + n$ being said to be the *sum* of m and n. It can be proved that these alternative definitions are in agreement, that addition is associative and commutative, and that, for any $m, n, p \in \omega$,

$$m + p = n + p \;\Rightarrow\; m = n,$$

this implication being commonly referred to as the *cancellation* of p. The number 0 is unity for addition and is the only natural number with a natural number, namely itself, as additive inverse.

If m, n and p are natural numbers such that $n = m + p$, then p, uniquely determined by m and n, is called the *difference* of m and n and denoted by $n - m$. The difference $n - m$ exists if, and only if, $m \in n$ or $m = n$.

Multiplication: $\omega \times \omega \longrightarrow \omega$; $(m,n) \rightsquigarrow mn$ is defined, for all $(m,n) \in \omega \times \omega$, by the formula

$$mn = \#(m \times n),$$

mn being called the *product* of m and n. The number mn is denoted also sometimes by $m \times n$. This is the original use of the symbol $\times$. Its use to denote cartesian product is much more recent. Multiplication may also be defined, recursively, for each $n \in \omega$ by the formula

$$0n = 0, \quad 1n = n \quad \text{and, for all } k \in \omega, \quad (k + 1)n = kn + n.$$

It can be proved that these two definitions are in agreement, that multiplication is associative and commutative, and also that it is *distributive* over addition, that is, for all $m, n, p \in \omega$,

$$(m + n)p = mp + np,$$

and that, for any $m, n \in \omega$ and any $p \in \omega^+$,

$$mp = np \;\Rightarrow\; m = n,$$

this implication being referred to as the *cancellation* of the non-zero number p. The number 1 is unity for multiplication and is the only natural number with a natural number, namely itself, as multiplicative inverse.

For any $n, p \in \omega$, n is said to *divide* p or to be a *divisor* or *factor* of p if $p = mn$, for some $m \in \omega$.

Exponentiation: $\omega \times \omega \longrightarrow \omega$; $(m,n) \rightsquigarrow m^n$ is defined, for all $(m,n) \in \omega \times \omega$, by the formula,

$$m^n = \#(m^n).$$

The notation m^n is here used in two senses, to denote both the set of maps $n \longrightarrow m$ and the cardinality of this set. The latter usage is the original one. The use of the notation Y^X to denote the set of maps $X \longrightarrow Y$ is much more recent and was suggested by the above formula, by analogy with the use of $\times$ to denote cartesian product. The number m^n is called the nth *power* of m. It may be defined recursively, for each $m \in \omega$, by the formula

$$m^0 = 1, \quad m^1 = m$$

and, for all $k \in \omega$, $m^{k+1} = (m^k)m$. For all $m, n, p \in \omega$,

$$(mn)^p = m^p n^p, \quad m^{n+p} = m^n m^p \quad \text{and} \quad m^{np} = (m^n)^p.$$

Exponentiation is neither associative nor commutative, nor is there a unity element.

Σ and Π

Let $n \longrightarrow \omega$; $i \rightsquigarrow a_i$ be a finite sequence on ω. Then $\sum_{i \in n} a_i$, or $\sum_{i=0}^{n-1} a_i$, the *sum* of the sequence, and $\prod_{i \in n} a_i$, or $\prod_{i=0}^{n-1} a_i$, the *product* of the sequence, are defined recursively by

$$\sum_{i \in 0} a_i = 0, \quad \prod_{i \in 0} a_i = 1.$$

and for all $j \in n - 1$,

$$\sum_{i \in j+1} a_i = (\sum_{i \in j} a_i) + a_j \quad \text{and} \quad \prod_{i \in j+1} a_i = (\prod_{i \in j} a_i)a_j.$$

Prop. 1.43. Let $n \longrightarrow \omega$; $i \rightsquigarrow a_i$ be a finite sequence on ω and let $k \in n + 1$. Then

$$\sum_{i \in n} a_i = (\sum_{i \in k} a_i) + (\sum_{j \in n-k} a_{k+j})$$

and

$$\prod_{i \in n} a_i = (\prod_{i \in k} a_i) \; (\prod_{j \in n-k} a_{k+j}). \qquad \square$$

Prop. 1.44. Let $n \longrightarrow \omega$; $i \rightsquigarrow a_i$ be a finite sequence on ω and let $\pi : n \longrightarrow n$; $i \rightsquigarrow \pi i$ be any permutation of n. Then

$$\sum_{i \in n} a_{\pi i} = \sum_{i \in n} a_i \quad \text{and} \quad \prod_{i \in n} a_{\pi i} = \prod_{i \in n} a_i. \qquad \square$$

Prop. 1.45. Let $n \in \omega$. Then

$$2(\sum_{k \in n} k) = n^2 - n$$

and
$$2(\sum_{k\in n+1} k) = 2(\sum_{k\in n}(k+1)) = n^2 + n. \qquad \square$$

Prop. 1.46. Let $n \in \omega$. Then
$$\prod_{k\in n}(k+1) = \#(n!). \qquad \square$$

The number $\prod_{k\in n}(k+1)$ is denoted also by $n!$ and is called *n factorial.* This is the original use of the symbol ! in mathematics, the earlier notation $\underline{|n}$ for $n!$ having fallen into disuse through being awkward to print. As we remarked on page 11, our use of $X!$ to denote the set of permutations of X is non-standard.

Similar notational conventions to those introduced here for $\sum$ and $\prod$ apply in various analogous cases. For example, let $i \rightsquigarrow A_i$ be a finite sequence of sets with domain n, say. Then $\bigcup_{i\in n} A_i = \bigcup\{A_i : i \in n\}$ and $\bigcap_{i\in n} A_i = \bigcap\{A_i : i \in n\}$, while $\bigtimes_{i\in n} A_i = \{(a_i : i \in n) : a_i \in A_i\}$.

Order properties of ω

If m, n and p are natural numbers, the statement $m \in n$ is also written $m < n$ or, equivalently, $n > m$, and m is said to be *less than* n or, equivalently, n is said to be *greater than* m. Both notations will be used throughout this book, the notation $m \in n$ being reserved from now on for use in those situations in which n is thought of explicitly as the standard set with n elements. The symbol $\leqslant$ means *less than or equal to* and the symbol $\geqslant$ means *greater than or equal to.*

Prop. 1.47. For any $m, n \in \omega$,
$$m < n, \quad m = n, \quad \text{or} \quad m > n$$
(or, equivalently, $m \in n$, $m = n$, or $n \in m$), these possibilities being mutually exclusive. $\square$

This proposition is referred to as the *trichotomy* of $<$ (or, equivalently, $\in$) on ω. The word is derived from the Greek words 'tricha' meaning 'in three' and 'tomé' meaning 'a cut'.

Cor. 1.48. For any $m, n \in \omega$,
$$m \leqslant n \Leftrightarrow m \not> n \quad \text{and} \quad m \geqslant n \Leftrightarrow m \not< n. \qquad \square$$

Note that it is *true* that $0 \leqslant 1$, this being equivalent to the statement that $0 \not> 1$.

Prop. 1.49. For any $m, n \in \omega \cup \{\omega\}$,
$$m \in n, \quad m = n \quad \text{or} \quad n \in m,$$
these possibilities being mutually exclusive. $\square$

Prop. 1.50. For any $m, n, p \in \omega$,

$$m < n \text{ and } n < p \Rightarrow m < p$$

(or, equivalently,

$$m \in n \text{ and } n \in p \Rightarrow m \in p). \qquad \square$$

This is referred to as the *transitivity* of $<$ (or, equivalently, $\in$) on ω.

Prop. 1.51. For any $m, n, p \in \omega \cup \{\omega\}$,

$$m \in n \text{ and } n \in p \Rightarrow m \in p. \qquad \square$$

Let A be a subset of ω. Then there is at most one element a of A such that, for all $x \in A$, $a \leqslant x$. This element a, if it exists, is called the *least* element of A. The *greatest* element of A is similarly defined.

Prop. 1.52. Every non-null subset of ω has a least element. $\square$

Proposition 1.52 is called the *well-ordered* property of ω.

Prop. 1.53. A subset A of ω has a greatest element if, and only if, it is finite and non-null. $\square$

The remaining propositions relate $<$ to addition and multiplication, ω^+ denoting, as before, the set $\omega \setminus \{0\}$ of positive natural numbers.

Prop. 1.54. For any $m, n, p \in \omega$,

$$m < n \Rightarrow m + p < n + p. \qquad \square$$

Prop. 1.55. For any $m, n \in \omega$ and any $p \in \omega^+$,

$$m < n \Rightarrow mp < np. \qquad \square$$

Prop. 1.56. For any $m, n \in \omega^+$,

$$m + n \in \omega^+ \text{ and } mn \in \omega^+. \qquad \square$$

Prop. 1.57. For each $a \in \omega$ and each $b \in \omega^+$ there exists $n \in \omega$ such that $nb > a$. $\square$

Proposition 1.57 is called the *archimedean* property of ω.

Prop. 1.58. For each $a \in \omega$ and each $b \in \omega^+$ there exist unique numbers $h \in \omega$ and $k \in b$ such that $a = hb + k$. $\square$

The number k in Prop. 1.58 is called the *remainder* of a *modulo* b. ('Modulo' is the ablative of 'modulus', meaning 'divisor'.)

An infinite sequence on ω, $a : \omega \longrightarrow \omega$, can be summed if all but a finite number of the terms are zero or, equivalently, if $a_n = 0$ for all sufficiently large n, say for all $n \geqslant m$. One sets

$$\sum_{n \in \omega} a_n = \sum_{n \in m} a_n.$$

In a similar way one can form the product of an infinite sequence on ω, provided that all but a finite number of the terms are equal to 1.

Prop. 1.59. Let $b \in \omega^+$. Then any finite number can be uniquely expressed in the form $\sum_{n \in \omega} a_n b^n$, where $a_n \in b$, for all n, and $a_n = 0$ for all sufficiently large n. □

It is common practice to set b equal to ten.

Further Exercises

1.60. A set is said to be *normal* if it does not contain itself as a member. Otherwise it is said to be *abnormal*. Is the set of all normal sets normal or abnormal? (Cf. [52], page 76.) □

1.61. Show that any map $f: X \rightarrow Y$ may be expressed as the composite gh of an injection $g: W \rightarrow Y$ following a surjection $h: X \rightarrow W$. (First construct a suitable set W.) □

1.62. Maps $f: W \rightarrow X$, $g: X \rightarrow Y$ and $h: Y \rightarrow Z$ are such that gf and hg are bijective. Prove that g is bijective. □

1.63. Let $f: X \rightarrow Y$ be a surjection such that, for every $y \in Y$, $f^{\dashv}\{y\} = y$. Prove that f is a partition of X. □

1.64. Let $f: X \rightarrow Y$ be a map. Prove that, for all $A \subset X$ and all $B \subset Y$, $f_{\vdash}(A \cap f^{\dashv}(B)) = f_{\vdash}(A) \cap B$. □

1.65. Give examples of maps $f: \omega \rightarrow \omega$ which are (i) injective but not surjective, (ii) surjective but not injective. □

1.66. Show by an example that it is possible to have a bijection from a set X to a proper subset Y of X. □

1.67. Let X and Y be countable sets. Prove that $X \times Y$ is countable. □

1.68. Let X and Y be sets and let $f: X \rightarrow Y$ and $g: Y \rightarrow X$ be injections. Then, for any $x \in X$, call x the *zeroth* ancestor of x, $g_{\text{sur}}^{-1}(x)$, if $x \in \operatorname{im} g$, the *first* ancestor of x, $f_{\text{sur}}^{-1} g_{\text{sur}}^{-1}(x)$, if $x \in \operatorname{im} gf$, the *second* ancestor of x and so on. Also let X_0, X_1 and X_ω, respectively, denote the set of points in X whose ultimate ancestor lies in X, lies in Y, or does not exist. Similarly for Y: denote by Y_0, Y_1 and Y_ω the set of points whose ultimate ancestor lies in Y, lies in X, or does not exist.

Show that $f_{\vdash}(X_0) = Y_1$, $g_{\vdash}(Y_0) = X_1$ and $f_{\vdash}(X_\omega) = Y_\omega$, and construct a bijection from Y to X. (Schröder–Bernstein.) □

1.69. Let X be a set for which there is a surjection $f: X \longrightarrow X^2$. First construct a surjection $X \longrightarrow X^3$ and then construct a surjection $X \longrightarrow X^n$, for any finite n. (Here X^n denotes the cartesian product of n copies of X.) □

1.70. Let S be a non-null subset of ω such that, for any $a, b \in S$, with $a \leqslant b$, it is true that $a + b$ and $b - a \in S$. Prove that there exists $d \in \omega$ such that $S = \{nd : n \in \omega\}$. □

CHAPTER 2

REAL AND COMPLEX NUMBERS

In this chapter we discuss briefly the definitions of *group*, *ring* and *field* and the additive, multiplicative and order properties of the *ring* of *integers* **Z** and the *rational*, *real* and *complex fields* **Q**, **R** and **C**. Since there are many detailed accounts of these topics available, most of the proofs are omitted. For more details of the construction of **R** see, for example, [12].

Groups

Products on a set were defined in Chapter 1, and we have already encountered several specimens, for example, composition either on X^X or on $X!$, where X is any set. Further examples are union and intersection on Sub X, and addition, multiplication and exponentiation on the set of natural numbers ω.

A *group* consists of a set, G, say, and a product with unity on G that is associative and admits inverses, such a product being said to be a *group structure* for the set G.

For example, composition is a group structure for $X!$. This group is called the *group of permutations of* X. On the other hand composition is not a group structure for X^X, unless $\# X = 0$ or 1, for in all other cases there exist non-invertible transformations of X.

A set may have many group structures. When we are only interested in one such structure it is common practice to use the same letter, G, say, to denote both the group and the underlying set, and to use the notational conventions of ordinary multiplication or of composition. For example, the product of a pair (a,b) of elements of G is usually denoted by ab, unity is denoted by $1_{(G)}$, or simply by 1 when there is no risk of confusion, and the multiplicative inverse of any element $a \in G$ is denoted by a^{-1}. Exponential notations are in common use. For any $a \in G$ and any $n \in \omega$, a^n is defined recursively by setting

$$a^0 = 1_{(G)}, \quad a^1 = a \quad \text{and, for all } k \in \omega, \quad a^{k+1} = (a^k)a,$$

the inverse of a^n being denoted by a^{-n}. The element a^n is called the nth *power* of a, a^2 also being called the *square* of a. The properties of the exponential notation are summarized in Prop. 2.29, later in the chapter.

A group G is said to be *abelian* if its product is *commutative*, that is if, for any a and $b \in G$, $ba = ab$, the word 'abelian' being derived from the name of N. Abel, one of the founders of modern group theory.

An *additive* group G is just an abelian group with the product of any two elements a and b of the group denoted by $a + b$ and called the *sum* of a and b. The unity element is then denoted by $0_{(G)}$ (or simply by 0 when there is no risk of confusion) and is called *zero*. The *additive inverse* or *negative* of an element a is denoted by $-a$ and the additive analogue of a^n, for any $n \in \omega$, is denoted by na, with $0a = 0_{(G)}$ and $1a = a$, the negative of na being denoted simply by $-na$. For any $a, b \in G$, $a + (-b)$ may also be denoted by $a - b$.

A *group map* $t : G \rightarrow H$ is a map between groups G and H that respects the products on G and H, that is, is such that, for all $a, b \in G$,

$$t(ab) = t(a)\,t(b),$$

or, equivalently, is such that, for all $a, b, c \in G$,

$$ab = c \;\Rightarrow\; t(a)\,t(b) = t(c).$$

Prop. 2.1. Let $t : G \rightarrow H$ be a group map. Then $t(1_{(G)}) = 1_{(H)}$ and, for any $g \in G$, $(t(g))^{-1} = t(g^{-1})$. □

A *group reversing map* $t : G \rightarrow H$ is a map between groups G and H that *reverses* multiplication, that is, is such that, for all $a, b \in G$,

$$t(ab) = t(b)\,t(a).$$

Prop. 2.2. Let G be any group. Then the map $G \rightarrow G;\, g \rightsquigarrow g^{-1}$ is a group reversing map. □

There is an important criterion for a group map to be injective.

Prop. 2.3. A group map $t : G \rightarrow H$ is injective if, and only if, for all $g \in G$,

$$t(g) = 1_{(H)} \;\Rightarrow\; g = 1_{(G)},$$

that is if, and only if, $t^{\dashv}\{1_{(H)}\} = \{1_{(G)}\}$.

Proof $\Rightarrow$: Suppose that t is injective. Then, by Prop. 2.1,

$$t(g) = 1_{(H)} \;\Rightarrow\; t(g) = t(1_{(G)}) \;\Rightarrow\; g = 1_{(G)}.$$

$\Leftarrow$: Suppose that t is not injective. Then there are distinct elements a and b of G such that $t(a) = t(b)$. Since t is a group map, $t(a\,b^{-1}) = t(a)\,t(b^{-1}) = t(a)\,t(b)^{-1} = 1_{(H)}$. But $a\,b^{-1} \neq 1_{(G)}$. So $t(g) = 1_{(H)} \not\Rightarrow g = 1_{(G)}$. □

Prop. 2.4. (Cayley) Let G be a group. Then, for any $a \in G$, the map $a_{\mathrm{L}} : G \rightarrow G;\, g \rightsquigarrow ag$ is bijective and the map $G \rightarrow G!;\, a \rightsquigarrow a_{\mathrm{L}}$ is an injective group map.

Proof Let $a, b \in G$. Then, for all $g \in G$, $a(bg) = (ab)g$. That is, $a_L b_L = (ab)_L$. In particular, for any $a \in G$, $a_L^{-1} a_L = (a^{-1}a)_L = 1_G$ and, similarly, $a_L a_L^{-1} = 1_G$. That is, for each $a \in G$ the map a_L is bijective and the map $a \rightsquigarrow a_L$ is a group map. To prove injectivity it is enough, by Prop. 2.3, to remark that, if $ag = g$ for all, indeed for any, $g \in G$, then $a = 1_{(G)}$. □

A group map $t : G \longrightarrow H$ is said to be a (*group*) *isomorphism* if it is bijective and if the inverse map $t^{-1} : H \longrightarrow G$ also is a group map. The second condition is redundant and is inserted here only for emphasis.

Prop. 2.5. A group map $t : G \longrightarrow H$ is an isomorphism if, and only if, it is bijective. □

The word *isomorphism* is derived from two Greek words, 'isos', meaning 'equal', and 'morphé', meaning 'form'. Two groups G and H are said to be (mutually) *isomorphic* if they have the same form, that is, if there exists an isomorphism $t : G \longrightarrow H$.

The word *homomorphism* or simply *morphism* is frequently used to denote a map preserving structure, in the present context to denote a group map. The prefix is derived from a Greek word 'homos' meaning 'same'. The word morphism is also used with other prefixes. For example, a group isomorphism $t : G \longrightarrow G$ is said to be an *automorphism* of G. A more complete list is given later, on page 59.

A subset F of a group G is said to be a *subgroup* of G if there is a group structure for F such that the inclusion $F \longrightarrow G$ is a group map. Such a structure is necessarily unique, the product in F of a pair of elements (a,b) in F coinciding with the product ab in G.

For example, for any $n \in \omega$, the subset of the permutations of the set $n + 1$ that leave the element n fixed is a subgroup of the group $(n + 1)!$ isomorphic to $n!$.

The following proposition enables one in practice to decide readily whether a given subset of a group is a subgroup.

Prop. 2.6. Let G be a group and F a subset of G. Then F is a subgroup of G if, and only if,

(i) $1 \in F$,
(ii) for all $a, b \in F$, $ab \in F$,
(iii) for all $a \in F$, $a^{-1} \in F$.

Proof The three conditions are satisfied if F is a subgroup of G. It remains to prove the converse.

Suppose, therefore, that they are satisfied. Then by (ii) the map

$$F^2 \to F;\ (a,b) \rightsquigarrow ab$$

is well-defined. This product is associative on F as on G, $1 \in F$ by (i) and is unity for the product on F, while, for all $a \in F$, $a^{-1} \in F$ by (iii) and is the inverse of a. □

(The case $F = \emptyset$ shows that (i) is not deducible from (ii) and (iii).)

Prop. 2.7. Let $t: G \to H$ be a group map. Then $t^{-1}\{1\}$ is a subgroup of G and im t is a subgroup of H.

Proof In either case conditions (i), (ii) and (iii) of Prop. 2.6 follow directly from the remark that, for any $a, b \in G$, $t(1) = 1$, $t(ab) = t(a)\,t(b)$ and $t(a^{-1}) = (t(a))^{-1}$. □

The group $t^{-1}\{1\}$ is called the *kernel* of t and denoted by ker t.

The *product* $G \times H$ of groups G and H is defined to be the group consisting of the set product $G \times H$ with multiplication defined, for any $(g,h), (g',h') \in G \times H$ by the formula $(g,h)(g',h') = (gg',hh')$. It is readily verified that this is a group and that $1_{(G\times H)} = (1_{(G)},1_{(H)})$.

In particular, a group structure on a set G induces a group structure on G^2.

Rings

Frequently one is concerned at the same time with two or more products on a set and with the interrelationships of the various products with each other.

A *ring* consists of an additive group, X, say, and a product $X^2 \to X$; $(a,b) \rightsquigarrow ab$ that is *distributive* over addition, that is, is such that, for all $a, b, c \in X$, $(a + b)c = ac + bc$ and $ab + ac = a(b + c)$. The product is also required to be associative unless there is explicit mention to the contrary, in which case the ring is said to be *non-associative*. The ring is said to be *commutative* (the word *abelian* not being used in this context) if its product is commutative, and to *have unity* if its product has unity. Frequently, when there is no risk of confusion, one uses the same letter to denote both the ring and the underlying abelian group or the underlying set.

Example 2.8. Let $n \in \omega$, and let addition and multiplication be defined on the set n, by Prop. 1.58, by defining the *sum* of any $a, b \in n$ to be the remainder $(a + b)_{(n)}$ of their sum $a + b$ in ω modulo n and by defining their *product* to be the remainder $(ab)_{(n)}$ of their product ab in ω modulo n. For $n \geqslant 1$, the set n with this addition is an additive group with 0 as zero, and with either $n - a$ or 0 as the additive inverse of a, according as $a \neq 0$ or $a = 0$, while, for $n \geqslant 2$, this group with the stated multiplication is a commutative (and associative) ring with unity, namely 1, distinct from 0. □

The ring constructed in Example 2.8 is called the *ring n*, or the *ring of remainders modulo n*, or, most commonly, the *ring* $\mathbf{Z}_n$.

In working with the ring $\mathbf{Z}_n$ it is usual to use the ordinary notational conventions to denote addition and multiplication, adding the phrase 'modulo n' or 'mod n' in parentheses wherever this is necessary to prevent confusion.

A ring is said to be *without divisors of zero* if the product of any two non-zero elements of the ring is non-zero.

For example, the rings $\mathbf{Z}_2$ and $\mathbf{Z}_3$ are without divisors of zero, as is easily verified, but the ring $\mathbf{Z}_4$ is not, since $2 \times 2 = 0$ (modulo 4).

Prop. 2.9. Let X be a commutative ring without divisors of zero. Then, for any $a, b \in X$,

$$a^2 = b^2 \Rightarrow a = b \text{ or } -b. \qquad \square$$

The *product* $X \times Y$ of rings X and Y is defined to be the ring consisting of the set product $X \times Y$, with addition and multiplication defined, for any $(x,y), (x',y') \in X \times Y$ by the formulae $(x,y) + (x',y') = (x + x', y + y')$ and $(x,y)(x',y') = (xx', yy')$. It is readily verified that this is a ring and that if X and Y have unity elements $1_{(X)}$ and $1_{(Y)}$ respectively, then the ring $X \times Y$ has $(1_{(X)}, 1_{(Y)})$ as unity.

In particular, a ring structure on a set X induces a ring structure on the set X^2.

The set Y^X of maps from a set X to a ring Y becomes a ring when the *sum* $f + g$ and the *product* $f \cdot g$ of any pair (f,g) of elements of Y^X are defined by the formulae

$$(f + g)(x) = f(x) + g(x), \quad \text{for all } x \in X,$$

and

$$(f \cdot g)(x) = f(x)\, g(x), \quad \text{for all } x \in X.$$

(We have to write $f \cdot g$ here rather than fg, since fg denotes the composite of f and g wherever f and g are composable, as would be the case if $X = Y$. An alternative convention is to denote the product of f and g by fg and the composite by $f \circ g$. We prefer the former convention.)

The ring structure just defined is the *standard* ring structure on Y^X.

Let X and Y be rings. Then a map $t : X \to Y$ is said to be a *ring map* if addition and multiplication are each respected by t, that is if, for any $a, b \in X$,

$$t(a + b) = t(a) + t(b) \quad \text{and} \quad t(ab) = t(a)\, t(b),$$

and to be a *ring-reversing map* if, for any $a, b \in X$,

$$t(a + b) = t(a) + t(b) \quad \text{and} \quad t(ab) = t(b)\,t(a).$$

A ring map need not respect unity.

Prop. 2.10. Let X and Y be rings with unity and let $t : X \to Y$ be a ring map. Then

$$t(1_{(X)}) = 1_{(Y)} \quad \Leftrightarrow \quad 1_{(Y)} \in \operatorname{im} t. \qquad \square$$

An example of a ring map that does not respect unity is the zero map $X \to X$; $x \rightsquigarrow 0$, where X is any ring with unity different from zero.

A ring map $t : X \to Y$ is said to be a (*ring*) *isomorphism* if it is bijective, for this implies, as was the case with a group bijection, that the inverse map $t^{-1} : Y \to X$ also is a ring map.

Subrings of a ring are defined in the obvious way. (For *ideals* see page 89.)

The use of the symbols $\sum$ and $\prod$ and the conventions governing their use carry over from sequences on ω to sequences on any ring X, the definition in either case being the obvious recursive one.

Polynomials

Let X be a ring with unity. A polynomial over X is a sequence $a : \omega \to X$, all but a finite number of whose terms are zero. The greatest number m for which a_m is non-zero is called the *degree*, $\deg a$, of the polynomial, a_m being called the *leading term* or *leading coefficient* of the polynomial. The degree $\deg a$ exists, provided that $a \neq 0$.

The *ring* of *polynomials* over X, which will be denoted by Pol X, consists of the set of polynomials over X with addition and multiplication defined, for any two polynomials a and b, by the formulae

$$(a + b)_k = a_k + b_k, \qquad \text{for all } k \in \omega,$$

and

$$(ab)_k = \sum_{i \in k+1} a_i b_{k-i}, \qquad \text{for all } k \in \omega.$$

It is readily verified that this is a ring.

The polynomial, all of whose terms, including the 0th, are zero, except for the 1st, which is 1, will usually be denoted, without further comment, by the letter x. This is often indicated in practice by writing $X[x]$ for Pol X.

Prop. 2.11. Let X be a ring with unity. Then, for any polynomial a over X, $a = \sum_{k \in \omega} a_k x^k$. (The sum is well-defined, since all but a finite number of the terms of the sequence to be summed are zero.) $\square$

Prop. 2.12. Let a and b be non-zero polynomials over a ring with unity. Then $\deg(ab) = \deg a \deg b$, the leading term of the polynomial

ab being the product of the leading terms of the polynomials a and b. □

Cor. 2.13. Let X be a ring with unity and without divisors of zero. Then Pol X is a ring with unity and without divisors of zero. □

Prop. 2.14. Let X be a ring with unity. Then the map

$$X \longrightarrow \text{Pol } X;\ a \rightsquigarrow a\ (= ax^0)$$

is a ring injection. □

As has already been anticipated by the notations, this map is normally regarded as an inclusion.

Prop. 2.15. Let X be a ring with unity. Then the map

$$\text{Pol } X \longrightarrow X;\ \sum_{k \in \omega} a_k x^k \rightsquigarrow a_0$$

is a ring map. □

Any polynomial $a = \sum_{k \in \omega} a_k x^k$ over a ring with unity X induces a *polynomial map*, or a *polynomial function*, $X \longrightarrow X$; $x \rightsquigarrow \sum_{k \in \omega} a_k x^k$, the sum being well-defined, for each $x \in X$, since all but a finite number of the terms of the sequence to be summed are zero. Note that we have just used the letter x in two quite distinct ways, to denote a particular, and rather special, polynomial over X and to denote an element of X. The double use of the letter x in this context is traditional.

Prop. 2.16. Let X be a ring with unity. Then the map Pol $X \longrightarrow X^X$ associating to any polynomial over X the induced polynomial map is a ring map. □

The ring map defined in Prop 2.16 need not be injective. For example, if $X = \mathbf{Z}_3$ the polynomial $x^3 - x$ is not the zero polynomial, but the map $\mathbf{Z}_3 \longrightarrow \mathbf{Z}_3$; $x \rightsquigarrow x^3 - x$ is the zero map, since, for all $x \in \mathbf{Z}_3$,

$$x^3 - x = x(x-1)(x+1) = x(x-1)(x-2)\ (\text{mod } 3).$$

However, if X is commutative and if the set X is infinite, then the map is injective, by Prop. 2.3 and by Prop. 2.18 below. In this case the formal distinction between polynomials and polynomial maps may be ignored.

We need a preliminary lemma.

Prop. 2.17. Let X be a commutative ring with unity, let $c = \sum_{k \in \omega} c_k x^k$ be a polynomial over X of degree $n + 1$, and let $a \in X$. Then $\sum_{k \in \omega} c_k a^k = 0$ if, and only if, there exists a polynomial d of degree n such that $c =$

$(x - a)d$. (If $\sum_{k\in\omega} c_k a^k = 0$, then $\sum_{k\in\omega} c_k x^k = \sum_{k\in\omega} c_k(x^k - a^k)$. Also, for any positive k, $x^k - a^k = (x - a) \sum_{i\in k} x^i a^{k-1-i}$.) □

Prop. 2.18. Let X be a commutative ring with unity and let c be a non-zero polynomial over X of degree m. Then there are at most m elements of X at which the induced polynomial map $x \rightsquigarrow \sum_{k\in\omega} c_k x^k$ is zero.

The proof is by induction, using Prop. 2.17. □

Ordered rings

An *ordered ring* (X,X^+) consists of a ring X and a subset X^+ of X such that, for all $a, b \in X^+$, $a + b$ and $ab \in X^+$ and, for all $a \in X$, $a \in X^+$ or $a = 0$ or $-a \in X^+$, these three possibilities being mutually exclusive.

The statement $a - b \in X^+$ is also written $a > b$ or $b < a$, while the statement $a - b \notin X^+$ is also written $a \leqslant b$ or $b \geqslant a$. An element a of X is said to be *positive* if $a \in X^+$ and *negative* if $-a \in X^+$.

Prop. 2.19. Let (X,X^+) be an ordered ring. Then, for all $a, b, c \in X$,

(i) $a > b$ and $b > c \Rightarrow a > c$,
(ii) $a > b \Rightarrow a + c > b + c$,
(iii) $a > b$ and $c > 0 \Rightarrow ac > bc$ and $ca > cb$,
and (iv) $a > b$, $b > a$ or $a = b$, these three possibilities being mutually exclusive. □

When there is no risk of confusion one speaks simply of the *ordered ring X*.

Prop. 2.20. The square a^2 of any non-zero element a of an ordered ring X is positive. □

Cor. 2.21. Let X be an ordered ring with unity, 1, distinct from zero. Then $1 > 0$. □

Cor. 2.22. Let $n \to X;\ i \rightsquigarrow a_i$ be a finite sequence on X, an ordered field, such that $\sum_{i\in n} a_i^2 = 0$. Then, for all $i \in n$, $a_i = 0$. □

Prop. 2.23. An ordered ring is without divisors of zero. □

Absolute value

Let X be an ordered ring. Then a map $X \to X;\ x \rightsquigarrow |x|$ is defined by setting $|0| = 0$, $|x| = x$, if $x > 0$, and $|x| = -x$ if $-x > 0$. The element $|x|$ is said to be the *absolute value* of x and, for any $a, b \in X$, $|b - a|$ is said to be the *absolute difference* of a and b.

Prop. 2.24. Let X be an ordered ring. Then, for all $a, b \in X$,

$$|a| \geqslant 0,$$
$$|a| = 0 \Leftrightarrow a = 0,$$
$$|a|^2 = a^2,$$
$$|a| \leqslant |b| \Leftrightarrow a^2 \leqslant b^2,$$
$$|a + b| \leqslant |a| + |b|$$
$$|a - b| \geqslant ||a| - |b||,$$

and

$$|ab| = |a||b|. \qquad \square$$

An element b of an ordered ring X is said to lie *between* elements a and c of X if $a < b < c$ or if $c < b < a$.

Prop. 2.25. An element b of an ordered ring X lies between elements a and c of X if, and only if, a, b and c are mutually distinct and

$$|a - c| = |a - b| + |b - c|. \qquad \square$$

Prop. 2.26. Let a and b be elements of an ordered ring X and let $\varepsilon \in X^+$. Then $|a - b| \leqslant \varepsilon$ if, and only if, b lies between $a - \varepsilon$ and $a + \varepsilon$. $\square$

Let X and Y be ordered rings. Then a ring map $f: X \to Y$ is said to be an *ordered ring map* if, for any $a \in X^+$, $f(a) \in Y^+$, and to be an *ordered ring isomorphism* if it is also bijective, the inverse map $f^{-1}: Y \to X$ being then also an ordered ring map.

Prop. 2.27. Let $f: X \to Y$ be a ring map, X and Y being ordered rings. Then f is an ordered ring map if, and only if, for all $a, b \in X$,

$$a < b \Rightarrow f(a) < f(b). \qquad \square$$

An *ordered subring*, or more correctly, a *sub-ordered-ring*, of an ordered ring is defined in the obvious way.

The ring of integers

It has been rather difficult to avoid explicit mention of the ring of integers before now. The *ring of integers* $\mathbf{Z}$ is an ordered ring with unity which contains the set of natural numbers ω in such a way that addition and multiplication on ω agrees with addition and multiplication on $\mathbf{Z}$ and which has the property that the map

$$\theta : \omega \times \omega \to \mathbf{Z};\ (m,n) \rightsquigarrow m - n$$

is surjective. Each element of **Z** is called an *integer*. (The letter Z is the initial letter of the German word 'Zahl', meaning 'number'.)

Prop. 2.28. $\mathbf{Z}^+ = \omega^+$, 0 being zero in **Z** and 1 unity in **Z**. □

The ordered ring **Z** exists and is unique up to isomorphism. The usual method of proof is to construct from ω an ordered ring Z isomorphic as an ordered ring to **Z**. We omit the details, observing only that it is usual to define Z to be the coimage of θ and that the various stages of the construction are based on the following statements about any m, n, m' and n' belonging to ω, namely

$$m - n = m' - n' \Leftrightarrow m + n' = m' + n,$$

which enables us to define the fibres of θ as the classes of an equivalence on $\omega \times \omega$, defined in terms of ω alone,

$$(m - n) + (m' - n') = (m + m') - (n + n'),$$
$$m - m = 0 \quad \text{and} \quad -(m - n) = n - m,$$

leading to the additive structure,

$$(m - n)(m' - n') = (mm' + nn') - (mn' + m'n),$$

leading to the multiplicative structure,

$$m - n > 0 \Leftrightarrow m - n \in \omega^+,$$

leading to the order structure, and, finally,

$$m - 0 = m,$$

leading to an injection $\omega \rightarrow Z$ that preserves addition, multiplication and order.

In most applications the precise method of constructing the ring of integers **Z** is unimportant. It is its structure as an ordered ring which is vital.

Much of the terminology introduced for ω extends in an obvious way for **Z**. For example, a map $\mathbf{Z} \rightarrow X$; $n \rightsquigarrow a_n$ is often called a *doubly-infinite sequence* on the set X.

The exponential notations introduced for groups make more sense if the indices are interpreted as elements of **Z**.

Prop. 2.29. Let G be a group and let the map $\mathbf{Z} \times G \rightarrow G$; $(n,a) \rightsquigarrow a^n$ be defined by defining a^n to be the nth power of a, for all non-negative n and by defining a^n to be the $(-n)$th power of a^{-1} for all negative n. Then, for all $a \in G$ and all $m, n \in \mathbf{Z}$,

$$a^{m+n} = a^m a^n \quad \text{and} \quad a^{mn} = (a^m)^n,$$

while if G is also abelian, then, for all $a, b \in G$ and all $n \in \mathbf{Z}$,

$$(ab)^n = a^n b^n. \qquad \square$$

It is useful to see the form this proposition takes for an additive (abelian) group.

Prop. 2.30. Let X be an additive group. Then the map $\mathbf{Z} \times X \rightarrow X$; $(n,a) \rightsquigarrow na$, defined by defining na to be the nth additive power of a, for all non-negative n and by defining na to be $(-n)(-a)$ for all negative n, is also such that, for all $a, b \in X$ and all $m, n \in \mathbf{Z}$,

$$n(a + b) = na + nb, \quad (m + n)a = ma + na \quad \text{and} \quad (mn)a = m(na).$$

If X also has a ring structure, then also, for all $a, b \in X$ and all $m, n \in \mathbf{Z}$,

$$(mn)ab = (ma)(nb). \quad \square$$

It is a corollary of the last part of Prop. 2.30 that if X is without divisors of zero and if, for some $n \in \omega$ and $a \in X \backslash \{0\}$, $na = 0$, then, for every $b \in X$, $nb = 0$. The least *positive* number n such that, for all $a \in X$, $na = 0$ is said to be the *characteristic* of the ring X. If such a positive number does not exist, then X is said illogically to have *characteristic zero.*

Prop. 2.31. Let X be a ring with unity. Then the map

$$\mathbf{Z} \rightarrow X;\ n \rightsquigarrow n\,1_{(X)}$$

is a ring map. □

Prop. 2.32. Let X be a ring with unity and without divisors of zero. Then X has characteristic zero if, and only if, the ring map $\mathbf{Z} \rightarrow X$; $n \rightsquigarrow n\,1_{(X)}$ is injective. □

Prop. 2.33. Any ordered ring with unity has characteristic zero. □

Fields

A *field* $\mathbf{K}$ is a ring such that multiplication is a group structure for the subset $\mathbf{K}^* = \mathbf{K} \backslash \{0\}$; that is, a ring with unity distinct from zero such that each non-zero element of $\mathbf{K}$ has a multiplicative inverse. A field is also required to be commutative unless there is express mention to the contrary, in which case the field is said to be *non-commutative.*

An *ordered field* is a field $\mathbf{K}$ which is ordered as a ring.

Prop. 2.34. Let $\mathbf{K}$ be an ordered field and let a be a positive element of $\mathbf{K}$. Then $a^{-1} > 0$.

Proof Suppose that $a^{-1} \leqslant 0$. Then $1 \leqslant 0$, a contradiction. □

Prop. 2.35. Let $\mathbf{K}$ be an ordered field. Then 2^{-1} exists and $0 < 2^{-1} < 1$. □

Prop. 2.36. Let a, an element of an ordered field **K**, be such that, for all positive ε in **K**, $a \leqslant \varepsilon$. Then $a \leqslant 0$.

Proof Suppose $a > 0$. Then $a > (2^{-1})a > 0$, contrary to the hypothesis that a is not greater than any positive element of **K**. So $a \leqslant 0$. □

Exercise 2.37. Let a and b be elements of an ordered field **K**. Determine the truth or the falsity of each of the statements:

$$\text{(i) } a \leqslant b + \varepsilon \quad \text{for all } \varepsilon > 0 \;\Rightarrow\; a \leqslant b,$$
$$\text{(ii) } a < b + \varepsilon \quad \text{for all } \varepsilon > 0 \;\Rightarrow\; a < b.$$

(The most effective method of disproof is a well-chosen counter-example.) □

Proposition 2.36 will be used repeatedly later, as one of the standard methods of proving that two elements a and b of an ordered field are equal is to prove that their absolute difference is not greater than each positive ε. For examples, see the proofs of Prop. 2.56, Prop. 15.38 and Lemma 18.11.

Finally, two remarks about fields in general.

Prop. 2.38. Let **K** and **L** be fields. Then a ring map $t : \mathbf{K} \to \mathbf{L}$ either is the zero map or is injective.

Proof In any field the only elements x of the field satisfying the equation $x^2 = x$ are 0 and 1. For if $x \neq 0$ we may multiply either side of the equation by x^{-1}. So, if t is not the zero map, $t(1) = 1$. But it follows in this case that no non-zero element of **K** can be sent by t to 0. For suppose $t(a) = 0$, where $a \neq 0$. Then $1 = t(1) = t(a\,a^{-1}) = t(a)\,t(a^{-1}) = 0$, a contradiction. The injectivity then follows from the additive form of Prop. 2.3. □

Prop. 2.39. The ring product of two fields is *not* a field. □

The rational field

It has been almost as difficult to avoid mention of the rational field as it has been to avoid mention of the ring of integers.

The *rational field* **Q** is an ordered field which contains **Z** as an ordered subring and which has the property that the map

$$\mathbf{Z} \times \mathbf{Z}^{+} \to \mathbf{Q};\ (m,n) \rightsquigarrow m\,n^{-1}$$

is surjective. Each element of **Q** is called a *rational number*, the number $m\,n^{-1}$ also being denoted by m/n or by $\dfrac{m}{n}$ and called the *quotient* of m by n.

(The letter Q is the initial letter of the word 'quotient', **R** being reserved as a notation for the *real* field.)

The ordered field **Q** is unique up to isomorphism. The usual method of proof is to use **Z** to construct an ordered field $\underset{\sim}{Q}$ isomorphic as an ordered field to **Q**. The details are again omitted. We observe only that the various stages are based on the following statements about any $m, m' \in \mathbf{Z}$ and any $n, n' \in \mathbf{Z}^+$, namely that

$$\frac{m}{n} = \frac{m'}{n'} \Leftrightarrow mn' = m'n,$$

$$\frac{m}{n} + \frac{m'}{n'} = \frac{mn' + m'n}{nn'}, \quad \frac{0}{n} = 0 \quad \text{and} \quad -\left(\frac{m}{n}\right) = \frac{-m}{n},$$

$$\left(\frac{m}{n}\right)\left(\frac{m'}{n'}\right) = \frac{mm'}{nn'}, \quad \frac{n}{n} = 1 \quad \text{and} \quad \left(\frac{n}{n'}\right)^{-1} = \frac{n'}{n},$$

$$\frac{m}{n} > 0 \Leftrightarrow m > 0,$$

and, finally, $\dfrac{m}{1} = m.$

Prop. 2.40. The rational field is archimedean; that is, for any $a, b \in \mathbf{Q}^+$ there exists $n \in \omega$ such that $na > b$.

Proof Let $a = a'/a''$ and $b = b'/b''$ where a', a'', b' and $b'' \in \omega^+$. Then $na > b \Leftrightarrow na'b'' > b'a''$.

Since ω is archimedean, the proposition follows. □

Cor. 2.41. Let $a \in \mathbf{Q}^+$. Then there exists $n \in \omega$ such that $\dfrac{1}{n} < a$. □

Prop. 2.42. The rational field is not well-ordered; that is, a subset of $\mathbf{Q}^+$ need not have a least member.

Proof $\mathbf{Q}^+$ itself does not have a least member. □

Bounded subsets

A subset A of an ordered field X is said to be *bounded above* if there exists $b \in X$ such that, for all $x \in A$, $x \leqslant b$. The element b is then said to be an *upper bound* for A. An element $c \in X$ is said to be *the supremum* of A, denoted $\sup A$, if every $b > c$ is an upper bound for A and no $b < c$ is an upper bound for A. It may readily be proved that the supremum of a subset A, if it exists, is unique (justifying the use of the definite article) and is itself an upper bound for A, the *least upper bound* for A.

A bounded subset of **Q** may or may not have a supremum in **Q**. For example, the subset $\{x \in \mathbf{Q} : x \leqslant 0\}$ has supremum 0, as has the subset $\{x \in \mathbf{Q} : x < 0\}$, the latter example showing that the supremum of a subset need not itself belong to the subset.

Prop. 2.43. The subset $\{a/b \in \mathbf{Q} : a,b \in Z \text{ and } a^2 \leqslant 2b^2\}$ has no supremum in **Q**.

Proof This falls into two parts. First it can be shown that if s were a rational supremum for this set, then $s^2 = 2$. For if $s^2 < 2$ then, by Cor. 2.41, for suitably large $n \in \omega$, $s^2 < s^2\left(1 + \frac{1}{n}\right)^2 < 2$, while if $s^2 > 2$ then, for suitably large $n \in \omega$, $2 < s^2\left(1 - \frac{1}{n}\right)^2 < s^2$, this leading in either case to a contradiction.

Secondly, by an argument attributed to Euclid, it can be shown that there is no rational number a/b such that $a^2/b^2 = 2$. For suppose such a number exists. Then it may be supposed that the integers a and b are positive and have no common integral divisor greater than 1. On the other hand, since $a^2 = 2b^2$, a is of the form $2k$, where $k \in \omega$, implying that $b^2 = 2k^2$, that is that b is of the form $2k$, where $k \in \omega$. So both a and b are divisible by 2, a contradiction. □

The terms *bounded below*, *lower bound* and *infimum* (abbreviated to inf) or *greatest lower bound* are defined analogously in the obvious way. A subset A of an ordered field X is said to be *bounded* if it is both bounded above and bounded below.

Prop. 2.44. The supremum, when it exists, of a subset A of an ordered field X is the infimum of the set of upper bounds of A in X. □

The $\rightarrowtail$ notation

Often in the sequel we shall be concerned with maps whose domain is only a subset of some set already known to us and already named. To avoid having continually to introduce new notations, it will be convenient to let $f: X \rightarrowtail Y$ denote a map with target Y whose domain is a subset of X. The subset X will be referred to as the *source* of the map. We are then free to speak, for example, of the map $\mathbf{Q} \rightarrowtail \mathbf{Q};\ x \rightsquigarrow x^{-1}$, the convention being that in such a case, unless there is an explicit statement to the contrary, the domain shall be the largest subset of the source for which the definition is meaningful. In the above instance, therefore, the domain is $\mathbf{Q}^*$.

It is often convenient also to extend the concept of the composite of two maps to include the case of maps $f: X \rightarrowtail Y$ and $g: W \rightarrowtail X$. Their composite fg is then the map $W \rightarrowtail Y$; $w \rightsquigarrow f(g(w))$, with domain $g^{-1}(\operatorname{dom} f)$. For any map $f: X \rightarrowtail Y$ and any subset A of X, the set $f_\vdash(A)$ is, by definition, the same as the set $f_\vdash(A \cap \operatorname{dom} f)$.

The first place where the $\rightarrowtail$ notation is convenient is in the statement of Prop. 2.47. It is of considerable use in some of the later chapters.

The real field

We are at last in a position to describe the field of real numbers.

The *real field* **R** is an ordered field containing **Q** as an ordered subfield and such that each subset A of **R** that is non-null and bounded above has a supremum (this last requirement is known as the *upper bound axiom*). Each element of **R** is called a *real* number.

The power of the upper bound axiom is illustrated by the proof of the following proposition.

Prop. 2.45. The ordered field **R** is archimedean; that is, for any $a, b \in \mathbf{R}^+$ there exists $n \in \omega$ such that $na > b$.

Proof Consider the subset $\{ka \in \mathbf{R} : k \in \omega, ka \leqslant b\}$ of **R**. Since $0 = 0a$ belongs to it, the subset is non-null. It is also bounded above. So it has a supremum which, in this case, must be a member of the subset, that is, of the form ma, where $m \in \omega$. Then $(m+1)a > b$. □

Cor. 2.46. Between any two distinct elements a and b of **R** there lies a rational number.

Proof By Corollary 2.41 there is a natural number n such that $1/n < |b - a|$. □

The real field exists and is unique up to isomorphism. Yet once again we omit proof. Some clues may be afforded by the following proposition, which indicates one of the several ways in which a copy of **R** may be built from the rational field **Q**.

Prop. 2.47. Let f be the map $\operatorname{Sub} \mathbf{Q} \rightarrowtail \mathbf{R}$; $A \rightsquigarrow \sup A$, with domain the set of non-null subsets of **Q** with upper bounds, **Q** being regarded as a subset of **R**. Then the map $\mathbf{R} \to \operatorname{Sub} \mathbf{Q}$; $x \rightsquigarrow \bigcup f^{-1}(\{x\})$ is a section of f.

(Corollary 2.46 may be of help in proving that f is surjective.) □

Other methods of constructing a copy of **R** from **Q** are hinted at in the section on *convergence* below.

The geometrical intuition for **R** is an unbounded straight line, each point of the line corresponding to a real number and vice versa, with the order relation on **R** corresponding to the intuitive order in which the representative points lie on the line. The correspondence is uniquely determined once the positions of 0 and 1 are fixed. The absolute difference $|a - b|$ of a pair of real numbers (a,b) is defined to be the *distance* between a and b, the distance between any two successive integers being 1 and the absolute value $|a|$ of any real number a being its distance from 0. This corresponds to the ordinary concept of distance on a line when the distance between the points chosen to represent 0 and 1 is taken as the *unit distance*. The upper bound axiom for **R** corresponds to the intuition that the line has no gaps. It is, in fact, the prototype of a *connected space* (cf. Chapter 16).

The correspondence between the field of real numbers and the intuitive line is central to the building of intuitive *geometrical* models of mathematical concepts and, conversely, to the applicability of the real numbers and systems formed from them in physics and engineering. ('Geometry' is, etymologically, the science of earth measurement.) The other central idea is, of course, the Cartesian correspondence between $\mathbf{R}^2$ and the plane and between $\mathbf{R}^3$ and three-dimensional space.

Standard figures in any text on geometry are the line itself, and figures based on two lines in the plane. For example, the figure

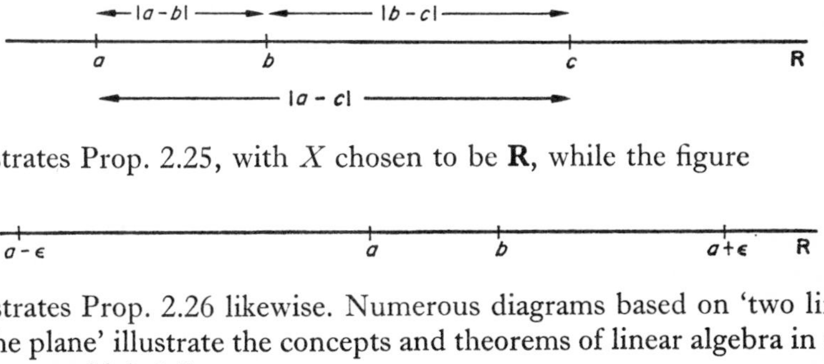

illustrates Prop. 2.25, with X chosen to be **R**, while the figure

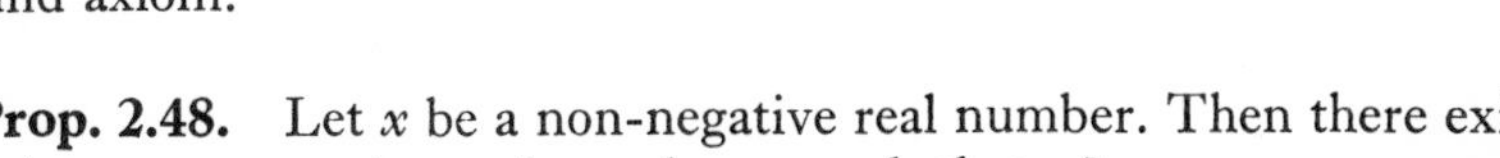

illustrates Prop. 2.26 likewise. Numerous diagrams based on 'two lines in the plane' illustrate the concepts and theorems of linear algebra in the chapters which follow.

The following proposition illustrates the application of the upper bound axiom.

Prop. 2.48. Let x be a non-negative real number. Then there exists a unique non-negative real number w such that $w^2 = x$.

(The subset A of **R** consisting of all the non-negative reals whose square is not greater than x is non-null, since it contains 0 and is bounded above by $1 + \frac{1}{2}x$. Let $w = \sup A$, and prove that $w^2 = x$. The uniqueness is a corollary of Prop. 2.9 applied to **R**.) □

Cor. 2.49. Any field isomorphic to the field **R** can be made into an ordered field in only one way. □

The unique non-negative number w such that $w^2 = x$ is called the *square root* of x and is denoted by $\sqrt{x}$. No negative real number is a square, by Prop. 2.20. Therefore the map $\mathbf{R} \rightarrowtail \mathbf{R};\ x \rightsquigarrow \sqrt{x}$ has as its domain the set of non-negative reals.

A subset A of **R** is said to be an *interval* of **R** if, for any $a, b \in A$, and any x lying between a and b, x also is an element of A. The following proposition lists the various possible types of interval.

Prop. 2.50. For any $a, b \in \mathbf{R}$, the subsets

$$\begin{aligned} [a\,, b] &= \{x \in \mathbf{R}\colon a \leqslant x \leqslant b\}, \\ ]a\,, b[&= \{x \in \mathbf{R}\colon a < x < b\}, \\ [a\,, b[&= \{x \in \mathbf{R}\colon a \leqslant x < b\}, \\ ]a\,, b] &= \{x \in \mathbf{R}\colon a < x \leqslant b\}, \\ [a\,, +\infty[&= \{x \in \mathbf{R}\colon a \leqslant x\}, \\ ]a\,, +\infty[&= \{x \in \mathbf{R}\colon a < x\}, \\ ]-\infty\,, b] &= \{x \in \mathbf{R}\colon x \leqslant b\}, \end{aligned}$$

and

$$]-\infty\,, b[= \{x \in \mathbf{R}\colon x < b\}$$

are intervals of **R**, and any interval of **R** other than $\emptyset$, or **R** itself, is of one of these eight types. □

The symbol ∞, called *infinity*, is used here purely as a convenient notation. There is no number $+\infty$, nor any number $-\infty$.

Convergence

A sequence $n \rightsquigarrow x_n$ on an ordered field X is said to be *convergent* with *limit* x if $x \in X$ and, for each positive element ε of X, there is a natural number m such that, for any natural number n,

$$n \geqslant m \Rightarrow x - x_n \leqslant \varepsilon,$$

that is, such that the absolute difference $|\, x - x_n \,|$ can be made as small as we please by choosing n to be sufficiently large.

(Recall that $|\, x - x_n \,| \leqslant \varepsilon$ if, and only if, $x - \varepsilon \leqslant x_n \leqslant x + \varepsilon$.

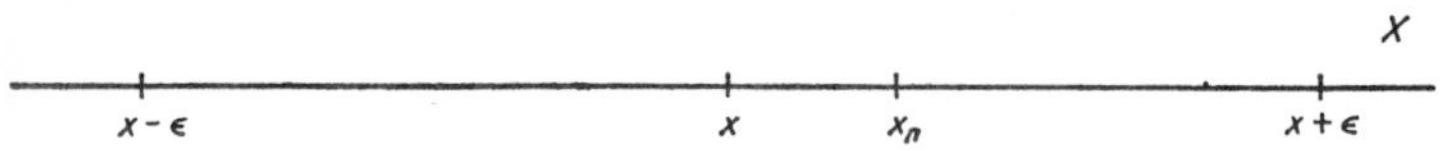

).

Prop. 2.51. Let $n \rightsquigarrow x_n$ be a convergent sequence on **R** with limit x. Then

$$x = \sup\,\{a \in \mathbf{R} : x_n < a \text{ for only a finite set of numbers } n\}.$$

Proof Let $A = \{a \in \mathbf{R} : x_n < a$ for only a finite set of numbers $n\}$. Then A is non-null, since $x - \varepsilon \in A$ for all $\varepsilon > 0$ and A is bounded above by $x + \varepsilon$ for all $\varepsilon > 0$. So sup A exists, necessarily equal to x. □

Cor. 2.52. The limit of a convergent sequence on $\mathbf{R}$ is unique. □

Prop. 2.53. For any real number r between 0 and 1 the sequence $n \rightsquigarrow r^n$ is convergent, with limit 0.

Proof For any real $s > 0$ and any $n \geqslant 0$,

$$(1 + s)^n \geqslant 1 + n\,s > n\,s \quad \text{(by induction)}.$$

So $r^n < \dfrac{1}{n\,s}$ for any $n > 0$, provided that $r = \dfrac{1}{1+s}$, that is, if $s = \dfrac{1-r}{r}$. The proposition follows. □

Cor. 2.54. For any real number r such that $|\,r\,| < 1$, the sequence $n \rightsquigarrow r^n$ is convergent, with limit 0. □

Prop. 2.55. For any real number r such that $|\,r\,| < 1$ the sequence $n \rightsquigarrow \sum_{k \in n} r^k$ is convergent with limit $(1 - r)^{-1}$.

Proof For any $n \in \omega$,

$$\begin{aligned}(1 - r)^{-1} - \sum_{k \in n} r^k &= (1 - r)^{-1}(1 - (1 - r)(\sum_{k \in n} r^k)) \\ &= (1 - r)^{-1} r^n.\end{aligned}$$

The proposition follows, by Cor. 2.54. □

Prop. 2.56. Let $n \rightsquigarrow x_n$ be a convergent sequence on an ordered field X, with limit x, let $\varepsilon > 0$ and let $m \in \omega$ be such that, for all $p, q > m$, $|\,x_p - x_q\,| \leqslant \varepsilon$. Then, for all $q \geqslant m$, $|\,x - x_q\,| \leqslant \varepsilon$.

Proof For any $\eta > 0$ there exists $p \geqslant m$ such that $|\,x - x_p\,| \leqslant \eta$. This implies that, for all $q \geqslant m$ and for all $\eta > 0$,

$$|\,x - x_q\,| \leqslant |\,x - x_p\,| + |\,x_p - x_q\,| \leqslant \eta + \varepsilon.$$

Hence the proposition, by Prop. 2.36 or, rather, by the true part of Exercise 2.37. □

A sequence $n \rightsquigarrow x_n$ on an ordered field X such that for each $\varepsilon > 0$ there is a natural number m such that, for all $p, q \in \omega$,

$$p, q \geqslant m \quad \Rightarrow \quad |\,x_p - x_q\,| \leqslant \varepsilon$$

is said to be a *Cauchy sequence* on X.

Prop. 2.57. Any convergent sequence on an ordered field X is Cauchy.

Proof Let $n \rightsquigarrow x_n$ be a convergent sequence on X, with limit x. Then, for any $\varepsilon > 0$, there exists a number m such that

$$p \geqslant m \Rightarrow |x_p - x| \leqslant \tfrac{1}{2}\varepsilon,$$

and therefore such that

$$p, q \geqslant m \Rightarrow |x_p - x_q| \leqslant |x_p - x| + |x_q - x| \leqslant \varepsilon. \qquad \square$$

Prop. 2.58. Let $k \rightsquigarrow a_k$ be a sequence on an ordered field X such that the sequence $n \rightsquigarrow \sum_{k \in n} |a_k|$ is Cauchy. Then the sequence $n \rightsquigarrow \sum_{k \in n} a_k$ is Cauchy. □

Prop. 2.59. Every real number is the limit of a Cauchy sequence on **Q**.

(Show first that any positive real number is the limit of a Cauchy sequence on **Q** of the form $n \rightsquigarrow \sum_{k \in n} u_k$, where $u_k = a_k b^{-k}$, with b a fixed natural number greater than 1 (so that $b^{-1} < 1$), with $a_0 \in \omega$ and with $a_k \in b$, for all positive k.) □

This leads to the following proposition, to be compared with the contrasting situation for the field of complex numbers described on page 48 below.

Prop. 2.60. Let $f: \mathbf{R} \to \mathbf{R}$ be a field isomorphism. Then $f = 1_{\mathbf{R}}$.

Proof Necessarily f sends 0 to 0 and 1 to 1, from which it follows by an easy argument that f sends each element of **Q** to itself. Also, by Props. 2.20 and 2.48, the order of the elements of **R** is determined by the field structure. So f also respects order and, in particular, limits of sequences. Since each real number is, by Prop. 2.59, the limit of a convergent sequence of rational numbers and since, by Cor. 2.52, the limit of a convergent sequence on **R** is unique, it follows that f sends each element of **R** to itself. □

When $b = 2$, a sequence of the type constructed in Prop. 2.59 is said to be a *binary expansion* for the real number, and when $b = 10$ the sequence is said to be a *decimal expansion* for the real number.

Exercise 2.61. Discuss to what extent the binary and the decimal expansions for a given real number are unique. □

Exercise 2.62. Prove that the set of real numbers is uncountable. (A proof that the interval [0,1] is uncountable may be based on Prop. 1.42, taking the conclusions of Exercise 2.61 into account.) □

Various constructions of the field of real numbers may be based on Prop. 2.59.

The following proposition, together with the archimedean property (Prop. 2.45), is equivalent to the upper bound axiom for **R**, and in some treatments of the field of real numbers it is preferred as an intuitive starting point. (Cf. [12], p. 95.)

Prop. 2.63. (The *general principle of convergence.*)
Every Cauchy sequence on **R** is convergent. (In the language of Chapter 15, **R** is *complete* with respect to the absolute value norm.)

Proof The method of proof is suggested by Prop. 2.51.
Let $n \rightsquigarrow x_n$ be a Cauchy sequence on **R**, and let $A = \{a \in \mathbf{R};\, x_n < a$ for only a finite set of numbers $n\}$. Since the sequence is Cauchy, there exists a number n such that, for all $p \geqslant n$, $|\, x_p - x_n \,| \leqslant 1$, that is $x_n - 1 \leqslant x_p \leqslant x_n + 1$. So $x_n - 1 \in A$, implying that A is non-null, while $x_n + 1$ is an upper bound for A.
Let $x = \sup A$, existing by the upper bound axiom. It then remains to be proved that x is the limit of the sequence. Let $\varepsilon > 0$. Then, for some number m and any p, q,

$$p, q \geqslant m \;\Rightarrow\; |\, x_p - x_q \,| \leqslant \tfrac{1}{2}\varepsilon,$$

while, for some particular $r \geqslant m$, $|\, x - x_r \,| \leqslant \frac{1}{2}\varepsilon$. So, for all $p \geqslant m$,

$$|\, x - x_p \,| \leqslant |\, x - x_r \,| + |\, x_p - x_r \,| \leqslant \varepsilon.$$

That is, x is the limit of the sequence. □

Cor. 2.64. Let $k \rightsquigarrow a_k$ be a sequence on **R** such that the sequence $n \rightsquigarrow \sum_{k \in n} |\, a_k \,|$ is convergent. Then the sequence $n \rightsquigarrow \sum_{k \in n} a_k$ is convergent.

Proof This follows at once from Prop. 2.57, Prop. 2.58 and Prop. 2.63. □

The complex field

There is more than one useful ring structure on the set $\mathbf{R}^2$.
First there is the ring product of **R** with itself. Addition, defined, for all (a,b), $(c,d) \in \mathbf{R}^2$, by the formula $(a,b) + (c,d) = (a + c\,,\, b + d)$ is an abelian group structure with zero $(0,0)$; multiplication, defined by the formula $(a,b)(c,d) = (ac,bd)$, is both commutative and associative, with unity $(1,1)$; and there is an injective ring map $\mathbf{R} \to \mathbf{R}^2$; $\lambda \rightsquigarrow (\lambda,\lambda)$ inducing a multiplication

$$\mathbf{R} \times \mathbf{R}^2 \to \mathbf{R}^2 \times \mathbf{R}^2 \to \mathbf{R}^2;\ (\lambda,(a,b)) \rightsquigarrow ((\lambda,\lambda),(a,b)) \rightsquigarrow (\lambda a,\lambda b).$$

This ring will be denoted by ${}^2\mathbf{R}$. Though **R** is a field, the ring ${}^2\mathbf{R}$ is not. For example, the element $(1,0)$ does not have an inverse.

Secondly, and more importantly, there is a field structure on $\mathbf{R}^2$ that provides the solution to the problem of finding a field containing $\mathbf{R}$ as a subfield, but such that every element of the field is a square.

Experience, as fossilized in the familiar formula for the roots of a quadratic equation, suggests that it may be sufficient to adjoin to $\mathbf{R}$ an element whose square is -1 in such a way that the field axioms still hold. Then at least every *real* number will be a square.

Suppose i is such an element. Then the new field must contain all elements of the form $a + ib$, where a and b are real, and if $a + ib$ and $c + id$ are two such elements, then we must have

$$(a + ib) + (c + id) = (a + c) + i(b + d)$$

and
$$(a + ib)(c + id) = (ac - bd) + i(ad + bc)$$

with, in particular, $(a + ib)(a - ib) = a^2 + b^2$.

In fact, it may be readily verified that $\mathbf{R}^2$ is assigned a field structure by the decree that, for all (a,b), $(c,d) \in \mathbf{R}^2$, $(a,b)(c,d) = (a + c, b + d)$ and $(a,b)(c,d) = (ac - bd, ad + bc)$. Unity is $(1,0)$ and the inverse of any non-zero element (a,b) is $((a^2 + b^2)^{-1}a, -(a^2 + b^2)^{-1}b)$. As with the previous ring structure for $\mathbf{R}^2$, there is an injective ring map $\mathbf{R} \to \mathbf{R}^2$; $\lambda \rightsquigarrow (\lambda,0)$, inducing a multiplication

$$\mathbf{R} \times \mathbf{R}^2 \to \mathbf{R}^2 \times \mathbf{R}^2 \to \mathbf{R}^2;\ (\lambda,(a,b)) \rightsquigarrow ((\lambda,0),(a,b)) \rightsquigarrow (\lambda a,\lambda b),$$

the same one as before, though composed differently. Moreover, if $\mathbf{C}$ is any ring consisting of a set of the form $\{a + ib : a,b \in \mathbf{R}\}$ with addition and multiplication as stated above, then the map

$$\mathbf{R}^2 \to \mathbf{C};\ (a,b) \rightsquigarrow a + ib$$

is a bijective ring map. That the map is a surjective ring map is clear. That it is also injective follows from the fact that, if $a + ib = 0$, then, since $(a + ib)(a - ib) = a^2 + b^2$, $a^2 + b^2 = 0$ and therefore, since a and $b \in \mathbf{R}$, $(a,b) = 0$ by Cor. 2.22. The map is, therefore, a field isomorphism.

To conclude, such a field $\mathbf{C}$ exists and is unique up to isomorphism. It may be constructed by first constructing the above ring structure on $\mathbf{R}^2$ and then by identifying $(a,0)$ with a, for each $a \in \mathbf{R}$. The field $\mathbf{C}$ is called the field of *complex numbers* and any element of $\mathbf{C}$ is called a complex number.

The map $\mathbf{C} \to \mathbf{C}$; $a + ib \rightsquigarrow a - ib$, where $(a,b) \in \mathbf{R}^2$, is called *conjugation*, $a - ib$ being called the *conjugate* of $a + ib$. The conjugate of a complex number z is denoted by $\bar{z}$.

Prop. 2.65. Conjugation is an automorphism of the field $\mathbf{C}$. That is, for any $z, z' \in \mathbf{C}$, $\overline{z + z'} = \bar{z} + \bar{z}'$, $\overline{zz'} = \bar{z}\,\bar{z}'$, $\bar{1} = 1$ and, for any $z \neq 0$, $\overline{z^{-1}} = (\bar{z})^{-1}$. Also, for any $z \in \mathbf{C}$, $\overline{(\bar{z})} = z$; $\bar{z} + z$ and $i(\bar{z} - z)$ are real numbers and $\bar{z}\,z$ is a non-negative real number, $\bar{z}$ being equal to z if, and only if, z is real. □

For any $z = x + iy \in \mathbf{C}$, with $x, y \in \mathbf{R}$, $x = \frac{1}{2}(\bar{z} + z)$ is said to be the *real part* of z and $y = \frac{1}{2}i(\bar{z} - z)$ is said to be the *pure imaginary part* of z. The real part of z will be denoted by re z and the pure imaginary part of z will be denoted by pu z (the letters im being reserved as an abbreviation for 'image'). The square root of the non-negative real number $\bar{z}\,z$ is said to be the *absolute value* (or *modulus* or *norm*) of z, and denoted by $|z|$.

Prop. 2.66. For any $z, z' \in \mathbf{C}$,

$$|z| \geqslant 0, \text{ with } |z| = 0 \Leftrightarrow z = 0,$$
$$\bar{z}\,z = |z|^2,$$
$$|\bar{z}| = z$$
$$|z + z'| \leqslant |z| + |z'|$$
$$||z| - |z'|| \leqslant |z - z'|,$$
$$|z\,z'| = |z|\,|z'|,$$

and, if $z \neq 0$,

$$|z^{-1}| = |z|^{-1}.$$

Also, for any $z \in \mathbf{R}$, $|z| = \sqrt{(z^2)}$, in agreement with the earlier definition of $|\ \ |$ on an ordered field. (Note that $z^2 \neq |z|^2$, unless $z \in \mathbf{R}$.) □

It will be proved in Chapter 19 that any non-zero polynomial map $\mathbf{C} \to \mathbf{C}$ is surjective (the *fundamental theorem of algebra*). An *ad hoc* proof that every complex number is a square runs as follows.

Let z be a non-zero complex number, and let $w = z/|z|$. Then $z = |z|\,w$, with $|z|$ a non-negative real number and w a complex number of absolute value 1. Since $|z|$ is a square, it is enough to prove that w is a square. However,

$$\begin{aligned}(1 + w)^2 &= 1 + 2w + w^2\\ &= \bar{w}\,w + (1 + \bar{w}\,w)w + w^2\\ &= (1 + \bar{w})(1 + w)w\\ &= |1 + w|^2\,w,\end{aligned}$$

from which it follows that if $w \neq -1$, then $w = ((1 + w)/|1 + w|)^2$. Finally, $-1 = i^2$.

Convergence for a sequence of complex numbers is defined in the same way as for a sequence of real numbers, the absolute value map on $\mathbf{R}$ being replaced in the definition by the absolute value map on $\mathbf{C}$. The definition of a Cauchy sequence also generalizes at once to sequences of complex numbers.

Prop. 2.67. A sequence $\omega \to \mathbf{C}$; $n \rightsquigarrow c_n = a_n + ib_n$ is convergent or Cauchy if, and only if, each of the sequences $\omega \to \mathbf{R}$; $n \rightsquigarrow a_n$ and $n \rightsquigarrow b_n$ is, respectively, convergent or Cauchy. □

However, unlike the real field, the field **C** cannot be made into an ordered field. For Prop. 2.20 implies that, if **C** were ordered, then both $1 = 1^2$ and $-1 = i^2$ would be positive, contradicting trichotomy.

Unlike **R** also, the field **C** has many automorphisms. (Cf. [53], and also [15], page 122.) Conjugation has already been noted as an example. This map sends each real number to itself and is the only field isomorphism $\mathbf{C} \to \mathbf{C}$ with this property other than the identity, since i must be sent either to i or to $-$i. However, a field isomorphism $\mathbf{C} \to \mathbf{C}$ need not send each real number to itself. It is true, by part of the proof of Prop. 2.60, that each *rational* number must be sent to itself, but the remainder of the proof of that proposition is no longer applicable. Indeed, one of the non-standard automorphisms of **C** sends $\sqrt{2}$ to $-\sqrt{2}$. What is implied by these remarks is that the real subfield of **C** is not uniquely determined by the field structure of **C** alone. The field injection or inclusion $\mathbf{R} \to \mathbf{C}$ is an additional piece of structure. In practice the additional structure is usually taken for granted; that is, **C** is more usually thought of as a *real algebra*—see page 67 for the definition—rather than as a *field*. It is unusual to say so explicitly!

The relationship between the complex field and the group of rotations of $\mathbf{R}^2$ is best deferred until we have discussed rotations, which we do in Chapter 9. The matter is dealt with briefly at the beginning of Chapter 10.

The exponential maps

Occasional reference will be made (mainly in examples) to the real and complex exponential maps, and it is convenient to define them here and to state some of their properties without proof. A full discussion will be found in most books on the analysis of real-valued functions of one real variable.

Prop. 2.68. Let $z \in \mathbf{C}$. Then the sequence $\omega \to \mathbf{C}$; $n \rightsquigarrow \sum_{k \in n} \frac{z^n}{n!}$ is convergent. □

The limit of the sequence $n \rightsquigarrow \sum_{k \in n} \frac{z^n}{n!}$ is denoted, for each $z \in \mathbf{C}$, by e^z, the map $\mathbf{R} \to \mathbf{R}$; $x \rightsquigarrow e^x$ being called the *real exponential map* (or *real exponential function*) and the map $\mathbf{C} \to \mathbf{C}$; $z \rightsquigarrow e^z$ the *complex exponential map*.

The following theorem states several important algebraic properties of these maps, $\mathbf{R}^+$ being a notation for the multiplicative group of positive real numbers and $\mathbf{C}^*$ a notation for the multiplicative group of non-zero complex numbers.

Theorem 2.69. For any $x \in \mathbf{R}$, $e^x \in \mathbf{R}^+$ and, for any $z \in \mathbf{C}$, $e^z \in \mathbf{C}^*$. The map $\mathbf{R} \to \mathbf{R}^+;\ x \rightsquigarrow e^x$ is a group isomorphism of the additive group $\mathbf{R}$ to the multiplicative group $\mathbf{R}^+$, while the map $\mathbf{C} \to \mathbf{C}^*;\ z \rightsquigarrow e^z$ is a group surjection of the additive group $\mathbf{C}$ to the multiplicative group $\mathbf{C}^*$, with kernel the image of the additive group injection $\mathbf{Z} \to \mathbf{C};\ n \rightsquigarrow 2\pi i n$, π being a positive real number uniquely determined by this property. For any $z \in \mathbf{C}$, $\overline{e^z} = e^{\bar{z}}$. The image by the complex exponential map of the additive group of all complex numbers with zero real part is the multiplicative subgroup of $\mathbf{C}^*$ consisting of all complex numbers with absolute value 1. □

The map $\mathbf{R} \to \mathbf{R}^+;\ x \rightsquigarrow e^x$ is not the only such group isomorphism. For example, the map $\mathbf{R} \to \mathbf{R}^+;\ x \rightsquigarrow e^{kx}$ also is a group isomorphism, for any $k \in \mathbf{R}^* = \mathbf{R} \setminus \{0\}$. The reason for singling out the former one only becomes clear when one looks at the topological and differential properties of the map (cf. Chapters 15, 16 and 18). What one can show is that the only continuous isomorphisms of $\mathbf{R}$ to $\mathbf{R}^+$ are those of the form $x \rightsquigarrow e^{kx}$, where $k \neq 0$. Each of these is differentiable, the differential coefficient of the map $x \rightsquigarrow e^{kx}$ at any $x \in \mathbf{R}$ being ke^{kx}. The exponential map $x \rightsquigarrow e^x$ is therefore distinguished by the property that its differential coefficient at 0 is 1, or, in the language of Chapter 18, that its differential at 0 is the identity map $\mathbf{R} \to \mathbf{R} : x \rightsquigarrow x$. It is an order-preserving map.

The exponential maps may be used to define several others which have standard names. For example, for any $x \in \mathbf{R}$, one defines

$$\cos x = \tfrac{1}{2}(e^{ix} - e^{-ix}), \text{ the real part of } e^{ix},$$

$$\sin x = \frac{1}{2i}(e^{ix} - e^{-ix}), \text{ the pure imaginary part of } e^{ix},$$

$$\cosh x = \tfrac{1}{2}(e^x + e^{-x}) \text{ and } \sinh x = \tfrac{1}{2}(e^x - e^{-x}).$$

The maps $\mathbf{R} \to \mathbf{R};\ x \rightsquigarrow \cos x$ and $x \rightsquigarrow \sin x$ are *periodic*, with minimum *period* 2π; that is, for each $n \in \omega$,

$$\cos(x + 2n\pi) = \cos x \quad \text{and} \quad \sin(x + 2n\pi) = \sin x, \quad \text{for all } x \in \mathbf{R},$$

no smaller positive number than 2π having this property. Moreover, since $\cos^2 x + \sin^2 x = 1$, $|\cos x| \leqslant 1$ and $|\sin x| \leqslant 1$, for all $x \in \mathbf{R}$. The maps $[0,\pi] \to [-1,1];\ x \rightsquigarrow \cos x$ and $[-\frac{1}{2}\pi,\frac{1}{2}\pi] \to [-1,1];\ x \rightsquigarrow \sin x$ are bijective, the former being order-reversing and the latter order-preserving. By contrast, $\cosh^2 x - \sinh^2 x = 1$, with $\cosh x \geqslant 1$, for all $x \in \mathbf{R}$. The map $[0, +\infty[\longrightarrow [1, +\infty[;\ x \rightsquigarrow \cosh x$ is bijective and order-preserving. The map $\mathbf{R} \longrightarrow \mathbf{R};\ x \rightsquigarrow \sinh x$ also is bijective and order-preserving, with $\sinh 0 = 0$.

Further Exercises

2.70. Let G be a group, let $a, b, c \in G$ and let $x = ba^{-1}$, $y = ab^{-1}c$ and $z = c^{-1}b$. Express a, b and c in terms of x, y and z. Hence express $ba^{-1}cac^{-1}b$ in terms of x, y and z. □

2.71. Let a, b, c, x, y be elements of a group G and let $ax = by = c$. Express $a^{-1}cb^{-1}ac^{-1}b$ in terms of x and y. □

2.72. Let G be a group such that, for all $x \in G$, $x^2 = 1$. Prove that G is abelian. □

2.73. Let G be a finite group. Prove that, for all $a \in G$, the set $\{a^n : n \in \omega\}$, with $a^0 = 1$, is a subgroup of G. □

2.74. The cardinality of the underlying set of a group G is said to be its *order*, denoted by $\#G$.

Let F be a subgroup of a group G and, for any $a \in G$, let $aF = \{af : f \in F\}$. Prove that, for any $a \in G$, the map $F \longrightarrow aF;\ f \rightsquigarrow af$ is bijective and that, for any $a, b \in G$, either $aF = bF$ or $aF \cap bF = \emptyset$. Hence prove that, if $\#G$ is finite, $\#F$ divides $\#G$. □

2.75. A natural number p, not 0 or 1, is said to be ***prime*** if its only divisors are 1 and p. Prove that any finite group of prime order p is isomorphic to the additive group $\mathbf{Z}_p$. □

2.76. Prove that any group of order 4 is isomorphic either to $\mathbf{Z}_4$ or to $\mathbf{Z}_2 \times \mathbf{Z}_2$. □

2.77. Prove that any group of order 6 is isomorphic either to $\mathbf{Z}_6$ or to $3!$ □

2.78. For any $n \in \omega$ let $a \in n!$ be the map sending each even $k \in n$ to its successor and each odd k to its predecessor, except that, when n is odd, $a(n-1) = n-1$, and let $b \in n!$ be the map sending each odd $k \in n$ to its successor and each even k to its predecessor, except that $b(0) = 0$ and, when n is even, $b(n-1) = n-1$. Prove that $a^2 = b^2 = (ab)^n = 1$ and that the subset $\{(ab)^k : k \in n\} \cup \{(ab)^k a : k \in n\}$ is a subgroup of $n!$ of order $2n$. □

2.79. Let S be a non-null subset of $\mathbf{Z}$ such that, for any $a, b \in S$,

$a + b$ and $b - a \in S$. Prove that there exists $d \in \omega$ such that $S = \{nd : n \in \mathbf{Z}\}$. (Cf. Exercise 1.69.) □

2.80. Let p be a prime number and let n be a positive number that is not a multiple of p. Prove that there exist $h, k \in \mathbf{Z}$ such that $hn + kp = 1$. (Apply Exercise 2.79 to the set $\{hn + kp : h,k \in \mathbf{Z}\}$.) □

2.81. Prove that $\mathbf{Z}_p$ is a field if, and only if, p is a prime. □

2.82. Let p be a prime number. Prove that, for all $a, b \in \omega$, if p divides ab, then p divides either a or b. □

2.83. Prove that any field not of characteristic zero has prime characteristic. □

2.84. Give examples of injections $]-1,1[\to [-1,1]$ and $[-1,1] \to\ ]-1,1[$. Hence construct a bijection $]-1,1[\to [-1,1]$. (Cf. Exercise 1.68.) □

2.85. Prove that the maps $]-1,1[\to \mathbf{R};\ x \rightsquigarrow \dfrac{x}{1 - |x|}$ and $x \rightsquigarrow \dfrac{x}{1 - x^2}$ are bijective. □

2.86. Let A, B and C be intervals of $\mathbf{R}$ with at least one common point. Prove that one of the intervals is a subset of the union of the other two. □

2.87. Prove that, for all $a, b \in \mathbf{R}$, $a < b \iff a^2 < b^2$ if, and only if, $a + b > 0$ (or $a + b = 0$, with $a \geqslant 0$). □

2.88. Let a and b be real numbers such that, for all real numbers x and y, $ax + by = 0$. Prove that $a = b = 0$. (Choose $(x,y) = (1,0)$ and $(0,1)$.) □

2.89. Let a and b be real numbers such that, for all real numbers x and y, $ax + by \leqslant \sqrt{(x^2 + y^2)}$. Prove that $|a| \leqslant 1$ and $|b| \leqslant 1$. Show by an example that the converse is false. □

2.90. Let a be a real number. Prove that

$$ab + 1 \geqslant 0 \quad \text{for all real } b < 1$$

if, and only if, $-1 \leqslant a \leqslant 0$. □

2.91. Express the sets $\left\{x \in \mathbf{R} : \dfrac{1}{(2 - x)^2} > x^2\right\}$ and $\left\{x \in \mathbf{R} : \dfrac{4}{3 - x} \leqslant x^2\right\}$ as unions of intervals. □

2.92. Let $n \rightsquigarrow x_n$ be a convergent sequence on $\mathbf{R}$. Prove that the sequence $n \rightsquigarrow |x_n|$ also is convergent. □

2.93. Which of the following are subfields of $\mathbf{R}$:

$$\{a + b\sqrt{2} : a,b \in \mathbf{Z}\},$$
$$\{a + b\sqrt{2} : a,b \in \mathbf{Q}\},$$
$$\{a + b\sqrt{2} + c\sqrt{3} : a,b,c \in \mathbf{Q}\},$$
$$\{a + b\sqrt{2} + c\sqrt{3} + d\sqrt{6} : a,b,c,d \in \mathbf{Q}\}\ ?\qquad \square$$

2.94. Find the real and imaginary parts and the modulus of each of the following complex numbers:

$$(5 + \mathrm{i})(4 - 3\mathrm{i}),\quad \frac{1 + \mathrm{i}}{5 - 12\mathrm{i}},\quad \frac{2 + 3\mathrm{i}}{3 + 4\mathrm{i}},\quad \frac{1 + \mathrm{e}^{\mathrm{i}\alpha}}{1 + \mathrm{e}^{\mathrm{i}\beta}},$$

where α, β are real. $\square$

2.95. Let a and b be complex numbers. Show that if $a + b$ and ab are both real, then either a and b are real or $a = \bar{b}$. $\square$

2.96. Prove that multiplication is a group structure for the set of complex numbers of modulus 1. (For reasons discussed later, at the beginning of Chapter 10, this group is called the *circle group* and denoted by S^1.) $\square$

2.97. A product on $\mathbf{C}^2$ is defined by the rule

$$(z_0,w_0)(z_1,w_1) = (z_0z_1 - w_0\bar{w}_1\,,\, z_0w_1 + w_0\bar{z}_1)$$

where z_0, z_1, w_0, $w_1 \in \mathbf{C}$. Prove that this product is associative and has unity, but is not commutative. Show also, by consideration of $(z,w)(\bar{z},-w)$, that if z and w are not both zero, then (z,w) has a unique inverse. $\square$

For further exercises on complex numbers, see page 195.

CHAPTER 3

LINEAR SPACES

At the end of Chapter 2 two ring structures for $\mathbf{R}^2$ were introduced, there being in each case also a natural ring injection $\mathbf{R} \to \mathbf{R}^2$. What both these structures have in common, apart from the additive structure, is the map

$$\mathbf{R} \times \mathbf{R}^2 \to \mathbf{R}^2;\ (\lambda,(a,b)) \rightsquigarrow (\lambda a,\lambda b).$$

This map is known as the *scalar multiplication* on $\mathbf{R}^2$, and addition and scalar multiplication together form what is known as the *standard linear structure* for $\mathbf{R}^2$. Addition is naturally an abelian group structure for $\mathbf{R}^2$, while, for any λ, $\mu \in \mathbf{R}^2$ and any (a,b), $(c,d) \in \mathbf{R}^2$,

$$\lambda((a,b) + (c,d)) = \lambda(a,b) + \lambda(c,d)$$
$$(\lambda + \mu)(a,b) = \lambda(a,b) + \mu(a,b)$$

and

$$(\lambda\mu)(a,b) = \lambda(\mu(a,b)).$$

Note also that the restriction of the scalar multiplication to $\mathbf{Z} \times \mathbf{R}^2$ coincides with the multiplication $\mathbf{Z} \times \mathbf{R}^2 \to \mathbf{R}^2$ induced by the additive

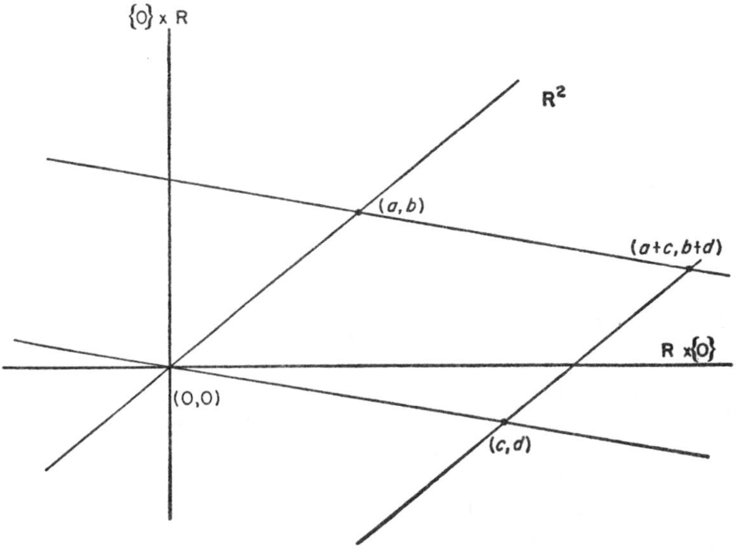

group structure on $\mathbf{R}^2$, according to Prop. 2.30. In particular for any $(a,b) \in \mathbf{R}^2$, $1(a,b) = (a,b)$, $0(a,b) = (0,0)$ and $(-1)(a,b) = (-a, -b)$.

The word *linear* derives from the standard intuitive picture of $\mathbf{R}^2$, a plane of unbounded extent, with the cartesian product structure induced by two *lines*, each a copy of $\mathbf{R}$, intersecting at their respective zeros and each lying in the plane.

In this picture the set $\{\lambda(a,b) : \lambda \in \mathbf{R}\}$ is, for any non-zero $(a,b) \in \mathbf{R}^2$, a line through the origin $(0,0)$, and any line through $(0,0)$ is so describable. The sum $(a+c, b+d)$ of any two elements (a,b) and (c,d) of $\mathbf{R}^2$ not lying on the same line through $(0,0)$ also has a geometrical interpretation, this point being the vertex opposite $(0,0)$ of the parallelogram whose other three vertices are (a,b), (c,d) and $(0,0)$.

The practical benefits of this intuition are fourfold. First, we have a method for bringing algebra to bear on geometry. Secondly, we have available a large geometrical vocabulary for use in any situation where we are concerned with a structure analogous to the linear structure for $\mathbf{R}^2$, that is, any situation where we have objects which can be added together in some reasonable way and be multiplied by elements of some given field. Thirdly, many general theorems on linear structures may be illustrated vividly by considering in detail the particular case in which the linear structure is taken to be the standard linear structure on $\mathbf{R}^2$. Finally, and more particularly, we have a serviceable intuitive picture of the field of complex numbers.

This chapter is concerned with those properties of linear spaces and linear maps that follow most directly from the definition of a linear structure. The discussion of dimension is deferred until Chapter 6, for logical reasons, but most of that chapter could usefully be read concurrently with this one, since many of the most vivid examples of linear spaces and maps involve finite-dimensional spaces. Further examples on the material of this chapter and the two which follow are scattered throughout the book, the interest of linear spaces lying not so much in themselves as in the more complicated geometrical or algebraic structures that can be built with them, as in Chapters 8, 9 and 12, or in their applicability, as in the theory of linear, or more strictly affine, approximation in Chapters 18 and 19. Various features are highlighted also in generalizations of the material, such as the theory of modules touched on briefly at the end of this chapter and, in particular, the theory of quaternionic linear spaces outlined in Chapter 10.

It should be noted that concepts such as *distance* or *angle* are *not* inherent in the concept of linear structure. For example, if X is a linear space isomorphic to $\mathbf{R}^2$ (that is, a two-dimensional linear space), then it is not meaningful to say that two lines of X are at right angles to (or orthogonal to) one another. For this, extra structure is required (see Chapter 9). It requires a little practice to use $\mathbf{R}^2$ as a typical linear space for illustrative purposes and at the same time to leave out of consideration all its metric features.

Linear spaces

Let X be an (additive) abelian group and let $\mathbf{K}$ be a commutative field. A $\mathbf{K}$-*linear structure* for X consists of a map

$$\mathbf{K} \times X \to X;\ (\lambda, x) \rightsquigarrow \lambda x,$$

called *scalar multiplication*, such that for all $\lambda, \lambda' \in X$, and all $\lambda, \lambda' \in K$,

$$\left.\begin{array}{ll}\text{(i)} & \lambda(x + x') = \lambda x + \lambda x' \\ \text{(ii)} & (\lambda + \lambda')x = \lambda x + \lambda' x\end{array}\right\},\quad \text{distributivity,}$$

$$\text{(iii)}\quad \lambda'(\lambda x) = (\lambda'\lambda)x,\qquad \text{associativity,}$$

$$\text{(iv)}\quad 1x = x,\qquad \text{unity.}$$

A $\mathbf{K}$-linear structure for a *set* X consists of an abelian group structure (*addition*) for X and a $\mathbf{K}$-linear structure for the abelian group.

Examples 3.1.

1. The null set has no linear structure.

2. For any finite n, a $\mathbf{K}$-linear structure is defined on $\mathbf{K}^n$ by the formulae

$$(x + x')_i = x_i + x'_i \quad \text{and} \quad (\lambda x)_i = \lambda x_i,$$

where $x, x' \in \mathbf{K}^n$, $\lambda \in \mathbf{K}$, and $i \in n$.

3. Let A be a set and X a linear space, and let X^A denote the set of maps of A to X. A linear structure is defined on X^A by the formulae

$$(f + g)(a) = f(a) + g(a) \quad \text{and} \quad (\lambda f)(a) = \lambda f(a)$$

where $f, g \in X^A$, $\lambda \in \mathbf{K}$ and $a \in A$. □

The linear structures defined in Examples 2 and 3 are referred to as the *standard* or *canonical* linear structures on $\mathbf{K}^n$ and X^A respectively, the linear structure on $\mathbf{R}^2$ described in the introduction to this chapter being its standard linear structure. Note that Example 2 is a special case of Example 3.

An abelian group with a prescribed $\mathbf{K}$-linear structure is said to be a $\mathbf{K}$-*linear space* or a *linear space over* $\mathbf{K}$. In applications the field $\mathbf{K}$ will usually be either the field of real numbers $\mathbf{R}$ or the field of complex numbers $\mathbf{C}$. While much of what we do will hold for any commutative field, except possibly fields of characteristic 2, we shall for simplicity restrict attention to fields of characteristic zero, and this will be assumed tacitly in all that follows.

Since an abelian group is rarely assigned more than one linear structure, it is common practice to denote a linear space, the underlying abelian group and, indeed, the underlying set all by the same letter; but care is required in working with complex linear spaces, for such a space can also be regarded as a real linear space by disregarding part of the structure (as in Prop. 7.32). When several linear spaces are under discussion at the same time it is assumed, unless there is explicit mention to the contrary, that they are all defined over the same field. The elements of the field are frequently referred to as *scalars*. (Hence the term 'scalar multiplication'.) The elements of a linear space may be referred to as *points* or as *vectors*. A linear space is often called a *vector space*, but we prefer to use this term only in the context of affine spaces, as discussed in Chapter 4.

The (additive) neutral element of a linear space X is called the *origin* or *zero* of X and is denoted by $0_{(X)}$, or simply by 0. From the context one can usually distinguish this use of the symbol 0 from its use as the scalar zero, as for example in the statement and proof of the next proposition.

Prop. 3.2. Let X be a $\mathbf{K}$-linear space, let $x \in X$ and let $\lambda \in \mathbf{K}$. Then $\lambda x = 0 \Leftrightarrow \lambda = 0$ or $x = 0$.

Proof $\Leftarrow$: $0x + 0x = (0 + 0)x = 0x = 0x + 0$.
Cancelling $0x$ from each side, $0x = 0$.

$$\lambda 0 + \lambda 0 = \lambda(0 + 0) = \lambda 0 = \lambda 0 + 0.$$

Cancelling $\lambda 0$ from each side, $\lambda 0 = 0$.
$\Rightarrow$: Let $\lambda x = 0$. Then either $\lambda = 0$ or $x = \lambda^{-1}(\lambda x) = \lambda^{-1}0 = 0$, as we have just shown. □

Note that in proving '$\Rightarrow$' we have made use of the existence of a multiplicative inverse of a non-zero scalar.

The additive inverse of an element x is denoted by $-x$.

Prop. 3.3. Let X be a linear space and let $x \in X$. Then $(-1)x = -x$.

Proof $x + (-1)x = 1x + (-1)x = (1 + (-1))x = 0x = 0$. Therefore $(-1)x = -x$. □

By definition $x' - x = x' + (-x)$ $(= x' + (-1)x)$, for all $x, x' \in X$. The map $X \times X \to X$; $(x',x) \rightsquigarrow x' - x$ is called *subtraction*.

Let x be any element of the linear space X and n any natural number. Then, for $n \geqslant 1$, nx may be defined either by means of addition or by means of scalar multiplication, by regarding n as a scalar. The unity axiom guarantees that the two definitions agree, as one sees from the next proposition.

Prop. 3.4. Let X be a linear space, let $x \in X$ and let $n \in \omega$. Then

$$(n + 1)x = nx + x.$$

Proof $(n + 1)x = nx + 1x = nx + x$. □

This is the necessary inductive step. The basis of the induction may be taken to be $0x = 0$ or $1x = x$.

Linear maps

A group map was defined in Chapter 2 to be a map from one group to another that respected the group structure. A linear map is defined analogously to be a map from one linear space to another that respects the linear structure. To be precise, a map $t : X \longrightarrow Y$ between linear spaces X and Y is said to be *linear* if, and only if, for any $a, b \in X$ and any scalar λ,

$$t(a + b) = t(a) + t(b) \quad \text{and} \quad t(\lambda a) = \lambda t(a).$$

The following proposition provides some elementary examples.

Prop. 3.5. Let X be a $\mathbf{K}$-linear space. Then, for any $a \in X$ and any $\mu \in \mathbf{K}$, the maps $a_{\mathbf{K}} : \mathbf{K} \longrightarrow X;\ \lambda \rightsquigarrow \lambda a$ and $\mu_X : X \longrightarrow X;\ x \rightsquigarrow \mu x$ are linear. In particular, the identity map 1_X is linear. □

When $\mathbf{K} = \mathbf{R}$, the map $a_{\mathbf{R}} : \mathbf{R} \longrightarrow X$ can be thought of intuitively as laying $\mathbf{R}$ along the line in X through 0 and a, with 0 laid on 0 and 1 laid on a.

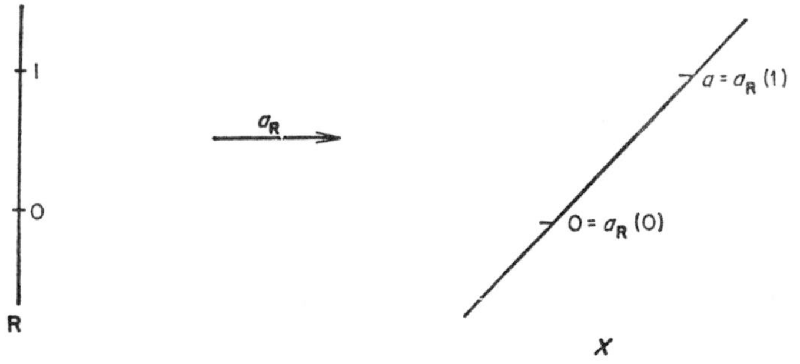

A linear map from $\mathbf{K}^n$ to $\mathbf{K}^m$ is defined by a set of m linear equations in n variables. Consider, for example, the particular case where $X = \mathbf{R}^3$, $Y = \mathbf{R}^2$, and let $t : \mathbf{R}^3 \longrightarrow \mathbf{R}^2$ be linear. Then for all $x = (x_0, x_1, x_2) \in \mathbf{R}^3$ we have

$$\begin{aligned} t(x) &= t(x_0, x_1, x_2) = t(x_0(1,0,0) + x_1(0,1,0) + x_2(0,0,1)) \\ &= x_0 t(1,0,0) + x_1 t(0,1,0) + x_2 t(0,0,1), \end{aligned}$$

by the linearity of t.

Let $t(1,0,0) = (t_{00},t_{10})$, $t(0,1,0) = (t_{01},t_{11})$ and $t(0,0,1) = (t_{02},t_{12})$. Then, writing $t(x) = (y_0,y_1)$, we have

$$(y_0,y_1) = x_0(t_{00},t_{10}) + x_1(t_{01},t_{11}) + x_2(t_{02},t_{12}),$$

that is, $$y_0 = t_{00}x_0 + t_{01}x_1 + t_{02}x_2$$

and $$y_1 = t_{10}x_0 + t_{11}x_1 + t_{12}x_2.$$

It is easy to see, conversely, that any pair of equations of this form, with $t_{ij} \in \mathbf{R}$ for all $(i,j) \in 2 \times 3$, determines a unique linear map $t : \mathbf{R}^3 \longrightarrow \mathbf{R}^2$.

The next few propositions are concerned with the role of 0 and with the composition of linear maps.

Prop. 3.6. Let $t : X \longrightarrow Y$ be a linear map. Then $t(0) = 0$.

Proof $t(0) = 0t(0) = 0$. □

Cor. 3.7. A linear map $t : X \longrightarrow Y$ is constant if, and only if, its sole value is 0. □

Prop. 3.8. Let $t : X \longrightarrow Y$ and $u : W \longrightarrow X$ be linear maps such that the composite $tu = 0$. Then if u is surjective, $t = 0$, and if t is injective, $u = 0$. □

Prop. 3.9. Let $t : X \longrightarrow Y$ and $u : W \longrightarrow X$ be linear maps. Then the composite $tu : W \longrightarrow Y$ is linear.

Proof For any $a, b \in W$ and any scalar λ,

$$tu(a + b) = t(u(a) + u(b)) = tu(a) + tu(b)$$

and $$tu(\lambda a) = t(\lambda u(a)) = \lambda tu(a).$$ □

Prop. 3.10. Let W, X and Y be linear spaces and let $t : X \longrightarrow Y$ and $u : W \longrightarrow X$ be maps whose composite $tu : W \longrightarrow Y$ is linear. Then

(i) if t is a linear injection, u is linear;
and (ii) if u is a linear surjection, t is linear.

Proof (i) Let t be a linear injection. Then for any $a, b \in W$, and any scalar λ,

$$\begin{aligned} tu(a + b) &= tu(a) + tu(b), \quad \text{since } tu \text{ is linear,} \\ &= t(u(a) + u(b)), \quad \text{since } t \text{ is linear,} \end{aligned}$$

and $tu(\lambda a) = \lambda tu(a) = t(\lambda u(a))$, for the same reasons. Since t is injective, it follows that $u(a + b) = u(a) + u(b)$ and $u(\lambda a) = \lambda u(a)$. That is, u is linear.

(ii) Exercise. □

Cor. 3.11. The inverse t^{-1} of a linear bijection $t : X \longrightarrow Y$ is linear. □

Such a map is called an *invertible linear map* or *linear isomorphism*. Two linear spaces X and Y are said to be (*mutually*) *isomorphic*, and either is said to be a (*linear*) *model* or *copy* of the other if there exists a linear isomorphism $t : X \to Y$. This is denoted by $X \cong Y$. One readily proves that the relation $\cong$ is an equivalence on any set of linear spaces.

The terms *morphism* or *homomorphism* for linear map, *monomorphism* for injective linear map, *epimorphism* for surjective linear map, *endomorphism* for a linear transformation of a linear space and *automorphism* for an invertible linear transformation of a linear space, or isomorphism of the space to itself, are all in common use, the adjective *linear* being implied in each case by the context.

For any linear space X the maps 1_X and -1_X are automorphisms of X. Indeed, λ_X is an automorphism of X for any non-zero scalar λ.

An automorphism $t : X \to X$ of a linear space X such that $t^2 = 1_X$ is also called a (*linear*) *involution* of X. The maps 1_X and -1_X are involutions of X.

Linear sections

Let $t : X \to Y$ and $u : Y \to X$ be linear maps such that $tu = 1_Y$. Then u is said to be a *linear section* of t, t in such a case being necessarily surjective and u injective by Cor. 1.4.

For example, the linear injection $i : \mathbf{R} \to \mathbf{R}^2$; $x \rightsquigarrow (x,0)$ is a linear section of the linear surjection $p : \mathbf{R}^2 \to \mathbf{R}$; $(x,y) \rightsquigarrow x$.

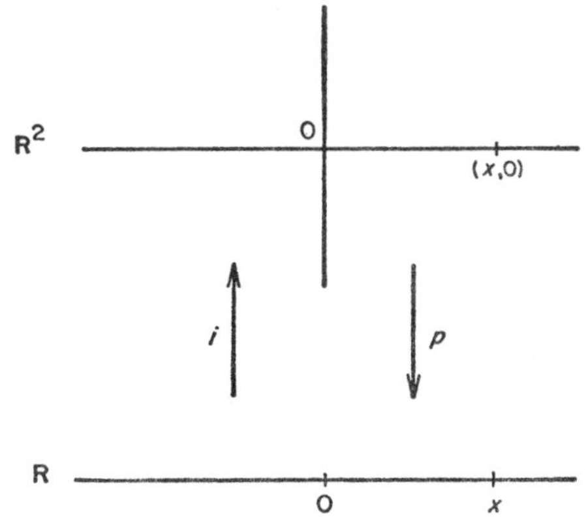

Prop. 3.12. Let $m \in \mathbf{R}$. Then the map $\mathbf{R} \to \mathbf{R}^2$; $x \rightsquigarrow (x,mx)$ is a linear section of the linear surjection $\mathbf{R}^2 \to \mathbf{R}$; $(x,y) \rightsquigarrow x$. □

This is a special case of Prop. 3.26 below.

Linear subspaces

Let X be a $\mathbf{K}$-linear space and W a subset of X. The set W is said to be a *linear subspace* of X if there exists a linear structure for W such that the inclusion $W \to X$; $w \leadsto w$ is linear.

The linearity of the inclusion is equivalent to the statement that, for any $w, w' \in W$ and any $\lambda \in \mathbf{K}$, $w + w'$ and λw are the same, whether with respect to the linear structure for W or with respect to the given linear structure for X. The linear structure for a linear subspace W of X is therefore unique; it is called the *natural* linear structure for W. A linear subspace of a linear space is tacitly assigned its natural linear structure.

The next proposition provides a practical test as to whether or not a given subset of a linear space is a linear subspace.

Prop. 3.13. A subset W of a linear space X is a linear subspace of X if, and only if,

(i) $0 \in W$,
(ii) for all $a, b \in W$, $a + b \in W$,
(iii) for any $a \in W$ and any scalar λ, $\lambda a \in W$.

Proof The three conditions are satisfied if W is a linear subspace of X. It remains to prove the converse.

Suppose therefore that they are satisfied. Then, by (ii) and by (iii), the maps

$$W^2 \to W;\ (a,b) \leadsto a + b$$

and

$$\mathbf{K} \times W \to W;\ (\lambda,a) \leadsto \lambda a$$

are well-defined, $\mathbf{K}$ being the field of scalars. This addition for W is associative and commutative as on X, $0 \in W$ by (i), while, for all $a \in W$, $-a = (-1)a \in W$ by (iii). Also, all the scalar multiplication axioms hold on W as on X. So W has a linear structure such that the inclusion $W \to X$ is linear. □

Note that we may not dispense with (i), for $\emptyset$, which has no linear structure, satisfies (ii) and (iii). Note also that (i) and (ii) by themselves are not a guarantee that addition is an abelian group structure for W.

Examples 3.14.

1. For any $(a,b) \in \mathbf{R}^2$, the set of scalar multiples of (a,b) is a linear subspace of $\mathbf{R}^2$. In particular, $\{(0,0)\}$ is a linear subspace of $\mathbf{R}^2$.
2. For any $(a,b) \in \mathbf{R}^2$, the set $\{(x,y) \in \mathbf{R}^2 : ax + by = 0\}$ is a linear subspace of $\mathbf{R}^2$.
3. The interval $[-1,1]$ is *not* a linear subspace of $\mathbf{R}$.

4. The set complement $X \setminus W$ of a linear subspace W of a linear space X is *not* a linear subspace of X. Neither is the subset $(X \setminus W) \cup \{0\}$.

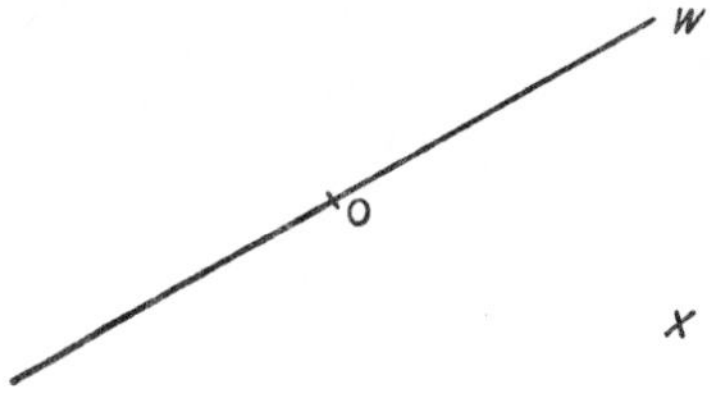

□

The *linear union* or *sum* $V + W$ of linear subspaces V and W of a linear space X is the subset $\{v + w \in X : v \in V, w \in W\}$ of X.

Prop. 3.15. Let V and W be linear subspaces of a linear space X. Then $V + W$ also is a linear subspace of X. By contrast, $V \cup W$, the set union of V and W, is a linear subspace of X if, and only if, one of the subspaces V or W is a subspace of the other. □

Intersections of linear spaces behave more nicely.

Prop. 3.16. Let $\mathscr{W}$ be a non-null set of linear subspaces of a linear space X. Then $\bigcap \mathscr{W}$ is a linear subspace of X. (Note that there is no assumption that the set $\mathscr{W}$ of linear subspaces is finite, nor even countable.) □

Linear injections and surjections

Prop. 3.17. Let $t : X \longrightarrow Y$ be a linear map. Then $t^{-1}\{0\}$ is a linear subspace of X and $\operatorname{im} t$ is a linear subspace of Y.

Proof In either case conditions (i), (ii) and (iii) of Prop. 3.13 follow directly from the remark that, for any $a, b \in X$ and any scalar λ, $t(0) = 0$, $t(a + b) = t(a) + t(b)$ and $t(\lambda a) = \lambda t(a)$. □

The linear subspace $t^{-1}\{0\}$ is called the *kernel* of t and is denoted by $\ker t$.

Prop. 3.18. A linear map $t : X \longrightarrow Y$ is injective if, and only if, $\ker t = \{0\}$. □

This is in fact just a special case of the criterion for injectivity which was noted for group maps (Prop. 2.3).

Prop. 3.19. Let $t : X \longrightarrow Y$ be a linear map. Then $t_{sur} : X \longrightarrow \operatorname{im} t$ is linear.

Proof This immediately follows from the fact that, for all $x \in X$, $t_{sur}(x) = t(x)$. Alternatively, since $t = t_{inc}t_{sur}$ and since t_{inc} is a linear injection, the linearity of t_{sur} follows from Prop. 3.10. □

Prop. 3.20. Let $u: X \to X$ be a linear map such that $u^2 = u$ and let $s = u_{sur}$, $i = u_{inc}$. Then

$$si = 1_{\text{im } u} = sui.$$

Proof By two applications of Prop. 3.8, $si = 1$, since $i(si - 1)s = u^2 - u = 0$.

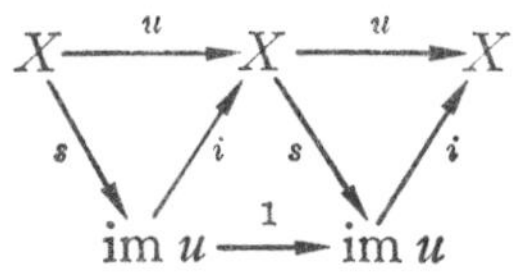

Then $sui = sisi = 1$.

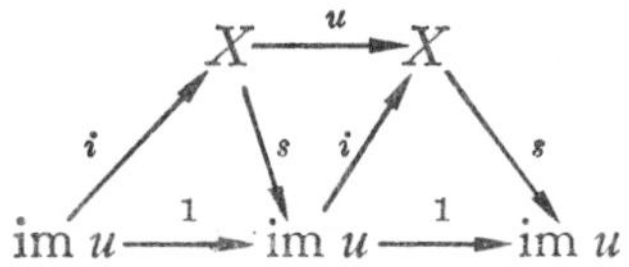

□

This is of use in the proof of Theorem 11.32.

Prop. 3.21. Let $t: X \to Y$ be a linear map. Then there exists a unique linear structure for coim t, the set of non-null fibres of t, such that the maps $t_{par}: X \to \text{coim } t$ and $t_{bij}: \text{coim } t \to \text{im } t$ are linear.

Proof The requirement that the bijection t_{bij} be linear determines a unique linear structure for coim t. Then, since $t_{par} = (t_{bij})^{-1}t_{sur}$, t_{par} also is linear. □

The further study of linear injections and surjections, and in particular the study of linear partitions, is deferred until Chapter 5.

Linear products

The product $X \times Y$ of two linear spaces X and Y has a natural linear structure defined by the formulae

$$(x,y) + (x',y') = (x + x', y + y')$$
$$\lambda(x,y) = (\lambda x, \lambda y)$$

for any (x,y), $(x',y') \in X \times Y$ and any scalar λ, the origin of $X \times Y$ being the element $(0,0)$. The set $X \times Y$ with this linear structure is said to be the (*linear*) *product* of the linear spaces X and Y. The product of any positive number of linear spaces is defined analogously.

The product construction assigns the standard linear structure to $\mathbf{K}^n$, for any positive number n.

The notation used in the next proposition to denote a map to a product is that first used in Chapter 1.

Prop. 3.22. Let W, X and Y be linear spaces. Then a map $(u,v) : W \to X \times Y$ is linear if, and only if, $u : W \to X$ and $v : W \to Y$ are linear.

Proof For any $a, b \in W$ and for any scalar λ,

$$\begin{aligned} (u,v)(a+b) &= (u,v)(a) + (u,v)(b) \\ \Leftrightarrow (u(a+b), v(a+b)) &= (u(a),v(a)) + (u(b),v(b)) \\ &= (u(a)+u(b), v(a)+v(b)) \end{aligned}$$

and

$$\begin{aligned} (u,v)(\lambda a) &= \lambda(u,v)(a) \\ \Leftrightarrow (u(\lambda a),v(\lambda a)) &= \lambda(u(a),v(a)) \\ &= (\lambda u(a),\lambda v(a)). \quad \square \end{aligned}$$

Cor. 3.23. For any linear spaces X and Y the maps

$$\begin{aligned} i : X &\to X \times \{0\}; & x &\rightsquigarrow (x,0), \\ j : Y &\to \{0\} \times Y; & y &\rightsquigarrow (0,y), \\ p : X \times Y &\to X; & (x,y) &\rightsquigarrow x \end{aligned}$$

and

$$q : X \times Y \to Y; \quad (x,y) \rightsquigarrow y \quad \text{are linear.}$$

Proof $i = (1_X, 0)$, $j = (0, 1_Y)$ and $(p,q) = 1_{X \times Y}$. $\square$

Prop. 3.24. Let X, Y and Z be linear spaces. Then each linear map from $X \times Y$ to Z is uniquely expressible as a map $(x,y) \rightsquigarrow a(x) + b(y)$, where a is a linear map from X to Z and b is a linear map from Y to Z. $\square$

This linear map will be denoted by $(a \;\; b)$.

These last two propositions generalize in the obvious way to n-fold products of linear spaces, for any positive number n.

Prop. 3.25. Let X and Y be linear spaces. Then the natural linear structure for $X \times Y$ is the only linear structure for $X \times Y$ such that the projection maps $p : X \times Y \to X$ and $q : X \times Y \to Y$ are linear. $\square$

Prop. 3.26. Let $t : X \to Y$ be a linear map. Then the map $X \to X \times Y$; $x \rightsquigarrow (x, t(x))$ is a linear section of the projection $p : X \times Y \to X$. $\square$

Prop. 3.27. A map $t : X \to Y$ between linear spaces X and Y is linear if, and only if, graph t is a linear subspace of $X \times Y$. $\square$

The study of linear products is continued in Chapter 8.

Linear spaces of linear maps

Let X and Y be linear spaces. Then the set of linear maps of X to Y will be denoted by $\mathscr{L}(X,Y)$ and the set of invertible linear maps or linear isomorphisms of X to Y will be denoted by $\mathscr{GL}(X,Y)$. An alternative notation for $\mathscr{L}(X,Y)$ is Hom (X,Y), 'Hom' being an abbreviation for '(linear) homomorphism'.

Prop. 3.28. Let X and Y be linear spaces. Then $\mathscr{L}(X,Y)$ is a linear subspace of Y^X. □

Prop. 3.29. Let X and Y be linear spaces. Then the natural linear structure for $\mathscr{L}(X,Y)$ is the only linear structure for the set such that for each $x \in X$ the map

$$\mathscr{L}(X,Y) \longrightarrow Y; \quad y \rightsquigarrow t(x)$$

is linear. □

Prop. 3.30. Let X be a $\mathbf{K}$-linear space. Then the maps

$$X \longrightarrow \mathscr{L}(\mathbf{K},X); \ a \rightsquigarrow a_{\mathbf{K}} \text{ and } \mathscr{L}(\mathbf{K},X) \longrightarrow X; \ u \rightsquigarrow u(1)$$

are linear isomorphisms, each being the inverse of the other. □

In practice one frequently identifies $\mathscr{L}(\mathbf{K},X)$ with X by these isomorphisms. In particular $\mathscr{L}(\mathbf{K},\mathbf{K})$ is frequently identified with $\mathbf{K}$, and $\mathscr{GL}(\mathbf{K},\mathbf{K})$ with $\mathbf{K}^*$.

The linear space $\mathscr{L}(X,\mathbf{K})$ is of more interest. This space is called the (*linear*) *dual* of the $\mathbf{K}$-linear space X for reasons that will be given later (cf. page 100). The dual space $\mathscr{L}(X,\mathbf{K})$ will be denoted also by $X^{\mathscr{L}}$. (More usual notations are X^*, $\hat{X}$ or $\check{X}$. The reason for adopting the notation $X^{\mathscr{L}}$ here will become clear only in Chapter 15, where we draw a distinction between $\mathscr{L}(X,Y)$, the set of linear maps from the real linear space X to the real linear space Y, and the linear subspace $L(X,Y)$ of $\mathscr{L}(X,Y)$, consisting of the *continuous* linear maps from X to Y. The notation X^L is then available to denote $L(X,\mathbf{R})$.)

Any linear map $t: X \longrightarrow Y$ induces a map $t^{\mathscr{L}}: Y^{\mathscr{L}} \longrightarrow X^{\mathscr{L}}$, defined by $t^{\mathscr{L}}(\beta) = \beta t$ for all $\beta \in Y^{\mathscr{L}}$. This definition is more vividly displayed by the diagram

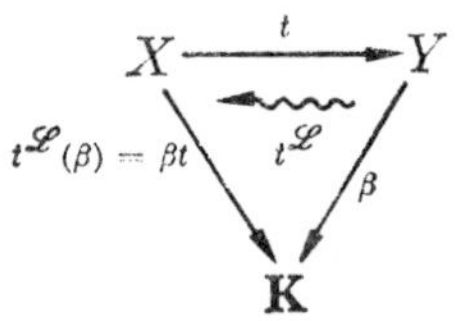

where the arrow labelled β represents an element of $Y^{\mathscr{L}}$ and the arrow labelled $t^{\mathscr{L}}(\beta)$ represents the image of β in $X^{\mathscr{L}}$ by the map $t^{\mathscr{L}}$. The map $t^{\mathscr{L}}$ is called the (*linear*) *dual* of the map t.

Prop. 3.31. The linear dual $t^{\mathscr{L}}: Y^{\mathscr{L}} \to X^{\mathscr{L}}$ of a linear map $t: X \to Y$ is linear.

Proof For any $\beta, \beta' \in Y^{\mathscr{L}}$, any scalar λ and any $x \in X$,

$$t^{\mathscr{L}}(\beta + \beta')(x) = (\beta + \beta')t(x) = \beta t(x) + \beta' t(x) = t^{\mathscr{L}}\beta(x) + t^{\mathscr{L}}\beta'(x)$$

and $t^{\mathscr{L}}(\lambda\beta)(x) = \lambda\beta t(x) = \lambda t^{\mathscr{L}}\beta(x)$.

Therefore $t^{\mathscr{L}}(\beta + \beta') = t^{\mathscr{L}}\beta + t^{\mathscr{L}}\beta'$ and $t^{\mathscr{L}}(\lambda\beta) = \lambda t^{\mathscr{L}}\beta$. □

Prop. 3.32. Let X and Y be linear spaces. Then the map $\mathscr{L}(X,Y) \to \mathscr{L}(Y^{\mathscr{L}},X^{\mathscr{L}});\ t \rightsquigarrow t^{\mathscr{L}}$ is linear. □

Prop. 3.33. Let $t: X \to Y$ and $u: W \to X$ be linear maps. Then $(tu)^{\mathscr{L}} = u^{\mathscr{L}}t^{\mathscr{L}}$. □

The ordered pair notation for a map to a product is further justified in the case of linear maps by the following proposition.

Prop. 3.34. Let X, Y_0, Y_1 be linear spaces. Then the map

$$\mathscr{L}(X,Y_0) \times \mathscr{L}(X,Y_1) \to \mathscr{L}(X,Y_0 \times Y_1); \quad (t_0,t_1) \rightsquigarrow (t_0,t_1)$$

is a linear isomorphism. □

There is a companion proposition to this one, involving linear maps *from* a product.

Prop. 3.35. Let X_0, X_1 and Y be linear spaces. Then the map

$$\mathscr{L}(X_0,Y) \times \mathscr{L}(X_1,Y) \to \mathscr{L}(X_0 \times X_1,Y); \quad (a_0,a_1) \rightsquigarrow (a_0 \quad a_1)$$

is an isomorphism. □

In the particular case that $Y = \mathbf{K}$, this becomes an isomorphism between $X_0^{\mathscr{L}} \times X_1^{\mathscr{L}}$ and $(X_0 \times X_1)^{\mathscr{L}}$.

Bilinear maps

A map $\beta: X \times Y \to Z$ is said to be *bilinear* or 2-linear if for each $a \in X$ and each $b \in Y$ the maps

$$X \to Z; \quad x \rightsquigarrow \beta(x,b) \quad \text{and} \quad Y \to Z; \quad y \rightsquigarrow \beta(a,y)$$

are linear, X, Y and Z being linear spaces.

For example, scalar multiplication on a linear space X is bilinear. For further examples see Exercise 3.55. It should be noted that a bilinear map is, in general, not linear, the only linear bilinear maps being the zero maps.

The definition of an *n-linear* map, for any $n \in \omega$, is the obvious generalization of this one, a *multilinear* map being a map that is n-linear, for some n.

Prop. 3.36. Let W, X and Y be linear spaces. Then composition

$$\mathscr{L}(X,Y) \times \mathscr{L}(W,X) \to \mathscr{L}(W,Y); \quad (t,u) \rightsquigarrow tu$$

is bilinear.

Proof For all $t, t' \in \mathscr{L}(X,Y)$, for all $u, u' \in \mathscr{L}(W,X)$, for any $w \in W$ and any scalar λ,

$$(t + t')u(w) = tu(w) + t'u(w) = (tu + t'u)(w)$$

and

$$(\lambda t)u(w) = \lambda(tu)(w) = (\lambda(tu))(w),$$

that is,

$$(t + t')u = tu + t'u \quad \text{and} \quad (\lambda t)u = \lambda(tu),$$

while, since t is linear,

$$t(u + u')(w) = t(u(w) + u'(w)) = tu(w) + tu'(w) = (tu + tu')(w)$$

and

$$t(\lambda u)(w) = t(\lambda u(w)) = \lambda(tu(w)) = (\lambda(tu))(w),$$

that is,

$$t(u + u') = tu + tu' \quad \text{and} \quad t(\lambda u) = \lambda(tu). \qquad \square$$

Prop. 3.37. Let X be a $\mathbf{K}$-linear space. Then the composition map $\mathscr{L}(X,X)^2 \to \mathscr{L}(X,X)$; $(t,u) \rightsquigarrow tu$ is associative, with unit 1_X, and distributive over addition both on the left and on the right. Also, the map $\mathbf{K} \to \mathscr{L}(X,X)$; $\lambda \rightsquigarrow \lambda_X$ is a linear injection such that, for all $\lambda, \mu \in \mathbf{K}$ and all $t \in \mathscr{L}(X,X)$, $\lambda_X \mu_X = (\lambda\mu)_X$ and $\lambda_X t = \lambda t$.

Proof Routine checking. $\square$

This shows that $\mathscr{L}(X,X)$ not only is a linear space with respect to addition and scalar multiplication but is also a ring with unity with respect to addition and composition, the ring injection $\mathbf{K} \to \mathscr{L}(X,X)$ sending 1 to 1_X and transforming scalar multiplication into ring multiplication.

The linear space $\mathscr{L}(X,X)$ with this additional structure is denoted by $\mathscr{L}(X)$ and called the *algebra of endomorphisms* of X. The notation End X is also used.

It is a corollary of Prop. 3.36 that composition is a group structure for the set $\mathscr{GL}(X)$ of invertible linear endomorphisms of the linear space X, that is, the set of automorphisms of X. The *group of automorphisms* $\mathscr{GL}(X)$ is also denoted by Aut X. See also page 106.

Algebras

The word algebra has just been used in its technical sense. The precise definition of a *linear algebra* runs as follows.

Let **K** be a commutative field. Then a *linear algebra* over **K** or **K**-linear algebra is, by definition, a linear space A over **K** together with a bilinear map $A^2 \to A$, the *algebra product* or the *algebra multiplication*.

An algebra A may, or may not, have unity, and the product need be neither commutative nor associative, though it is usual, as in the case of rings, to mention explicitly any failure of associativity. By the bilinearity of the product, multiplication is distributive over addition; that is, multiplication is a ring structure for the additive group A. Unity, if it exists, will be denoted by $1_{(A)}$, the map $\mathbf{K} \to A;\ \lambda \rightsquigarrow \lambda 1_{(A)}$ being injective. It frequently simplifies notations to identify $1 \in \mathbf{K}$ with $1_{(A)} \in A$, and, more generally, to identify any $\lambda \in \mathbf{K}$ with $\lambda 1_{(A)} \in A$.

Examples of associative algebras over **R** include ${}^2\mathbf{R}$ and **C**, as well as the algebra of linear endomorphisms End X of any real linear space X.

Examples of non-associative algebras include the Cayley algebra, discussed in Chapter 14, and Lie algebras, discussed in Chapter 20.

The *product* $A \times B$ of two **K**-linear algebras A and B is a **K**-linear algebra in the obvious way. Other concepts defined in the obvious way include *subalgebras*, *algebra maps* and *algebra-reversing maps*, *algebra isomorphisms* and *algebra anti-isomorphisms*, an *algebra anti-isomorphism*, for example, being a linear isomorphism of one algebra to another that reverses multiplication. The sth power of a **K**-algebra A will be denoted by ${}^sA$, not by A^s, which will be reserved as a notation for the underlying **K**-linear space.

There will be interest later, for example in Chapters 10, 11 and 13, in certain automorphisms and anti-automorphisms of certain linear algebras. An *automorphism* of an algebra A is a linear automorphism of A that respects multiplication, and an *anti-automorphism* of the algebra A is a linear automorphism of A that reverses multiplication. An automorphism or anti-automorphism t of A such that $t^2 = 1_A$ is said to be, respectively, an *involution* or an *anti-involution* of A.

The *centre* of an algebra A is the subset of A consisting of all those elements of A that commute with each element of A.

Prop. 3.38. The only algebra automorphisms of the real algebra **C** are the identity and conjugation. Both are involutions. □

Prop. 3.39. The centre of an algebra A is a subalgebra of A. □

A subset of an algebra A that is a group with respect to the algebra multiplication will be called a *subgroup* of A.

Matrices

Let X_0, X_1, Y_0 and Y_1 be **K**-linear spaces and let $t: X_0 \times X_1 \rightarrow Y_0 \times Y_1$ be a linear map. Then, by Prop. 3.24, there exist unique linear maps $t_{ij} \in \mathscr{L}(X_j, Y_i)$, for all i, $j \in 2 \times 2$, such that, for all $(x_0,x_1) \in X_0 \times X_1$,

$$t_0(x_0,x_1) = t_{00}x_0 + t_{01}x_1$$

and

$$t_1(x_0,x_1) = t_{10}x_0 + t_{11}x_1.$$

The array $\begin{pmatrix} t_{00} & t_{01} \\ t_{10} & t_{11} \end{pmatrix}$ of linear maps is said to be the *matrix* of t. Strictly speaking, this is a map with domain the set 2×2, associating to each $(i,j) \in 2 \times 2$ an element t_{ij} of $\mathscr{L}(X_j, Y_i)$. Conversely, any such matrix represents a unique linear map of $X_0 \times X_1$ to $Y_0 \times Y_1$.

The matrix notation just introduced is analogous to the notation $(u_0 \;\; u_1)$ for a linear map of the form $u: X_0 \times X_1 \rightarrow Y$ and it may be further generalized in an obvious way to linear maps of an n-fold product to an m-fold product of **K**-linear spaces, for any finite m and n, such a map being represented by a matrix with m *rows* and n *columns*. In particular, a linear map $(t_0,t_1): X \rightarrow Y_0 \times Y_1$ is represented by a *column matrix* $\begin{pmatrix} t_0 \\ t_1 \end{pmatrix}$. Moreover, if the linear spaces X_0, X_1 and $X_0 \times X_1$ are identified with the spaces $\mathscr{L}(\mathbf{K},X_0)$, $\mathscr{L}(\mathbf{K},X_1)$ and $\mathscr{L}(\mathbf{K},X_0 \times X_1)$, respectively, in accordance with the remark following Prop. 3.30, then any point $(x_0,x_1) \in X_0 \times X_1$ also may be represented as a column matrix, namely the matrix $\begin{pmatrix} x_0 \\ x_1 \end{pmatrix}$, there being an analogous matrix representation for a point of an n-fold product of linear spaces, for any finite n.

The matrix representation of a linear map t is most commonly used when the source and target of t are both powers of the field **K**. In this case each term of the matrix is a linear map of **K** to **K**, identifiable with an element of **K** itself. The matrix of a linear map $t: \mathbf{K}^n \rightarrow \mathbf{K}^m$ may therefore be defined to be the map

$$m \times n \rightarrow \mathbf{K}; \quad (i,j) \rightsquigarrow t_{ij}$$

such that, for all $x \in \mathbf{K}^n$, and for all $i \in m$,

$$t_i(x) = \sum_{j \in n} t_{ij}x_j,$$

or, equivalently, such that, for all $j \in n$ and all $i \in m$,

$$t_i(e_j) = t_{ij}$$

where $e_j \in \mathbf{K}^n$ is defined by $(e_j)_j = 1$ and $(e_j)_k = 0$, for all $k \in n \setminus \{j\}$.

In particular, the identity map $1_{\mathbf{K}_n} : \mathbf{K}^n \to \mathbf{K}^n$ is represented by the matrix, denoted by ${}^n1$, or sometimes simply by 1, all of whose *diagonal terms* $(1_{\mathbf{K}_n})_{ii}$, $i \in n$, are equal to 1, the *off-diagonal terms* $(1_{\mathbf{K}_n})_{ij}$, $i, j \in n$, $i \neq j$, all being zero. For example, ${}^31 = \begin{pmatrix} 1 & 0 & 0 \\ 0 & 1 & 0 \\ 0 & 0 & 1 \end{pmatrix}$.

Prop. 3.40. Let $\mathbf{C}$ be identified with $\mathbf{R}^2$ in the standard way. Then, for any $c = a + ib \in \mathbf{C}$ with $(a,b) \in \mathbf{R}^2$, the real linear map $\mathbf{C} \to \mathbf{C}$; $z \rightsquigarrow cz$ has matrix $\begin{pmatrix} a & -b \\ b & a \end{pmatrix}$. □

The operations of addition, scalar multiplication and composition for linear maps induce similar operations for the matrices representing the linear maps, the matrix representing the composite of two linear maps normally being referred to as the *product* of the matrices of the linear maps in the appropriate order.

Prop. 3.41. Let $t, t' \in \mathscr{L}(\mathbf{K}^n,\mathbf{K}^m)$, $u \in \mathscr{L}(\mathbf{K}^p,\mathbf{K}^n)$ and $\lambda \in \mathbf{K}$. Then, for all $(i,j) \in m \times n$,

$$(t + t')_{ij} = t_{ij} + t'_{ij}$$

and

$$(\lambda t)_{ij} = \lambda t_{ij},$$

while, for all $(i,k) \in m \times p$,

$$(tu)_{ik} = \sum_{j \in n} t_{ij} u_{jk}. \quad \square$$

As an example of this last formula,

$$\begin{pmatrix} a & d \\ b & e \\ c & f \end{pmatrix}\begin{pmatrix} u & x \\ v & y \end{pmatrix} = \begin{pmatrix} au + dv & ax + dy \\ bu + ev & bx + ey \\ cu + fv & cx + fy \end{pmatrix},$$

for any $a, b, c, d, e, f, u, v, x, y \in \mathbf{K}$.

It often simplifies notations to identify the elements of $\mathbf{K}^n$ with the $n \times 1$ column matrices in accordance with the remarks on column matrices above. It is usual at the same time to identify the elements of the dual space $(\mathbf{K}^n)^{\mathscr{L}}$ with the $1 \times n$ *row matrices* representing them. Then, for any $x \in \mathbf{K}^n$ and any $\alpha \in (\mathbf{K}^n)^{\mathscr{L}}$, $\alpha(x)$ becomes, simply, αx, the product being matrix multiplication.

To the endomorphism algebra End $\mathbf{K}^n$ there corresponds the *matrix algebra* $\mathbf{K}^{n \times n}$. Like the algebra End $\mathbf{K}^n$ to which it is isomorphic, this is an associative algebra with unity. Each of these algebras will also be denoted ambiguously by $\mathbf{K}(n)$, and ${}^n1$ will denote the unity element of either.

Prop. 3.42. For any finite n, the centre of the algebra $\mathbf{K}(n)$ consists of the scalar multiples of the identity, the subalgebra

$$\{\lambda({}^n1) : \lambda \in \mathbf{K}\}. \qquad \square$$

One map that is conveniently described in terms of matrices is *transposition*. Let $t \in \mathscr{L}(\mathbf{K}^n,\mathbf{K}^m)$. Then the *transpose* t^τ of t is the linear map of $\mathbf{K}^m$ to $\mathbf{K}^n$, with matrix the *transpose* of the matrix for t, that is, the $n \times m$ matrix $(t_{ij} : (j,i) \in n \times m)$.

For example, for any $a, b, c, d \in K$, $\begin{pmatrix} a & c \\ b & d \end{pmatrix}^\tau = \begin{pmatrix} a & b \\ c & d \end{pmatrix}$.

Prop. 3.43. For any finite m, n, the map

$$\mathscr{L}(\mathbf{K}^n,\mathbf{K}^m) \longrightarrow \mathscr{L}(\mathbf{K}^m,\mathbf{K}^n); \quad t \rightsquigarrow t^\tau$$

is a linear isomorphism, with

$$(t^\tau)^\tau = t, \quad \text{for all } t \in \mathscr{L}(\mathbf{K}^n,\mathbf{K}^m).$$

In particular, for any finite n the map

$$\mathbf{K}^n \cong \mathscr{L}(\mathbf{K},\mathbf{K}^n) \longrightarrow (\mathbf{K}^n)^{\mathscr{L}} = \mathscr{L}(\mathbf{K}^n,\mathbf{K}); \quad x \rightsquigarrow x^\tau$$

is a linear isomorphism. $\square$

Prop. 3.44. Let $t \in \mathscr{L}(\mathbf{K}^n,\mathbf{K}^m)$ and let $u \in \mathscr{L}(\mathbf{K}^p,\mathbf{K}^n)$, m, n and p being finite. Then $(ut)^\tau = t^\tau u^\tau$. $\square$

Cor. 3.45. For any finite n the map

$$\mathscr{L}(\mathbf{K}^n,\mathbf{K}^n) \longrightarrow \mathscr{L}(\mathbf{K}^n,\mathbf{K}^n); \quad t \rightsquigarrow t^\tau$$

is an anti-involution of the algebra $\mathscr{L}(\mathbf{K}^n,\mathbf{K}^n)$. $\square$

The algebras ${}^s\mathbf{K}$

For any commutative field $\mathbf{K}$ and any $s \in \omega$, the sth power of $\mathbf{K}$, ${}^s\mathbf{K}$, is a commutative $\mathbf{K}$-linear algebra.

Prop. 3.46. For any $s \in \omega$, the map $\alpha : {}^s\mathbf{K} \longrightarrow \mathbf{K}(s)$ defined for all $\lambda \in {}^s\mathbf{K}$, by the formula

$$(\alpha(\lambda))_{ii} = \lambda_i, \quad \text{for all } i \in s,$$
$$(\alpha(\lambda))_{ij} = 0, \quad \text{for all } i, j \in s, i \neq j,$$

is an algebra injection. $\square$

One-sided ideals

Let A be an associative $\mathbf{K}$-linear algebra with unity. A *left ideal* $\mathscr{J}$ of A is a linear subspace $\mathscr{J}$ of A such that, for all $x \in \mathscr{J}$ and all $a \in A$, $ax \in \mathscr{J}$. *Right ideals* are similarly defined.

Example 3.47. Let X be a **K**-linear space and let $t \in \operatorname{End} X$. Then the subset

$$\mathscr{I}(t) = \{at \in \operatorname{End} X : a \in \operatorname{End} X\}$$

is a left ideal of End X. □

A left ideal of A is said to be *minimal* if the only proper subset of A which is a left ideal of A is $\{0\}$. The minimal left ideals of the endomorphism algebra of a finite-dimensional **K**-linear space are described at the end of Chapter 6.

Two-sided ideals are defined on page 89.

Modules

In later chapters we shall be concerned with various generalizations of the material of this chapter. It is convenient here to indicate briefly one such generalization, involving the replacement of the field **K** either by a commutative ring with unity or, more particularly, by a commutative algebra with unity.

Let X be an additive group and let Λ be either a commutative (and associative) ring with unity or a commutative (and associative) algebra with unity. Then a map

$$\Lambda \times X \to X; \quad (\lambda, x) \rightsquigarrow \lambda x$$

is said to be a Λ-*module* structure for X if the same four axioms hold as in the definition of a **K**-linear space on page 55, the field **K** being replaced simply by the ring or algebra Λ. The reader should work through the chapter carefully to see how much of it does in fact generalize to commutative ring or algebra modules. In Chapter 8 we consider in some detail the particular case where Λ is the commutative algebra ${}^2\mathbf{K}$ over the field **K**.

Modules may also be defined over non-commutative rings or algebras. We shall have something to say about this in Chapter 10.

Further Exercises

3.48. Let $t : \mathbf{R}^2 \to \mathbf{R}$ be a linear map. Prove that there are unique real numbers a and b such that, for all $(x,y) \in \mathbf{R}^2$, $t(x,y) = ax + by$. Describe the fibres of t, supposing that a and b are not both zero. Prove that if $x \in \mathbf{R}^2$ is such that, for all $t \in L(\mathbf{R}^2,\mathbf{R})$, $t(x) = 0$, then $x = 0$. □

3.49. Let $t : X \to Y$ be a linear map such that, for any linear space W and any linear map $u : W \to X$, $tu = 0 \Rightarrow u = 0$. Prove that t is injective. □

3.50. Maps $t: \mathbf{R}^2 \longrightarrow \mathbf{R}^3$ and $u: \mathbf{R}^3 \longrightarrow \mathbf{R}^2$ are defined by

$$t(x) = (3x_0 - 2x_1, 2x_0 - x_1, -x_0 + x_1)$$

and $$u(y) = (-2y_0 + 3y_1 - y_2, 3y_0 - 2y_1 + 5y_2),$$

for all $x \in \mathbf{R}^2$, $y \in \mathbf{R}^3$. Verify that t and u are linear, that $ut = {}^21$ but that $tu \neq {}^31$. □

3.51. Show, by an example, that it is possible for the composite ut of a composable pair of linear maps t, u to be 0 even if neither t nor u is 0. □

3.52. Let U and V be linear subspaces of a linear space X, with $X = U + V$, and suppose that $t: X \longrightarrow Y$ is a linear surjection with kernel U. Prove that $t \mid V$ is surjective. □

3.53. Let U, V and W be linear subspaces of a linear space X. Prove that $U \cap V + U \cap W$ is a linear subspace of $U \cap (V + W)$. Show, by an example, that $U \cap V + U \cap W$ and $U \cap (V + W)$ need not be equal. □

3.54. Let X be a $\mathbf{K}$-linear space and, for each $x \in X$, let $\varepsilon_x \in X^{\mathscr{L}}$ be defined by the formula

$$\varepsilon_x(t) = t(x), \quad \text{for all } t \in X^{\mathscr{L}}.$$

Prove that the map

$$\varepsilon_X : X \longrightarrow X^{\mathscr{L}\mathscr{L}}; \quad x \rightsquigarrow \varepsilon_x$$

is a $\mathbf{K}$-linear map.

Let $u: X \longrightarrow Y$ be a $\mathbf{K}$-linear map. Prove that the diagram

$$\begin{array}{ccc} X & \xrightarrow{\varepsilon_X} & X^{\mathscr{L}\mathscr{L}} \\ \downarrow{\scriptstyle u} & & \downarrow{\scriptstyle u^{\mathscr{L}\mathscr{L}}} \\ Y & \xrightarrow{\varepsilon_Y} & Y^{\mathscr{L}\mathscr{L}} \end{array}$$

is commutative. (Cf. Exercise 6.44.) □

3.55. Prove that the maps

$$\mathbf{R}^3 \times \mathbf{R}^3 \longrightarrow \mathbf{R};$$

$$((x,y,z), (x',y',z')) \rightsquigarrow xx' + yy' + zz'$$

and $$\mathbf{R}^3 \times \mathbf{R}^3 \longrightarrow \mathbf{R}^3;$$

$$((x,y,z), (x',y',z')) \rightsquigarrow (yz' - y'z, zx' - z'x, xy' - x'y)$$

are bilinear. □

3.56. Let X, Y, Z be linear spaces. Prove that a bilinear map $X \times Y \longrightarrow Z$ is linear if, and only if, $X = \{0\}$ or $Y = \{0\}$. □

3.57. Let $\beta : X_0 \times X_1 \to Y$ be a bilinear map and $t : Y \to Z$ a linear map. Prove that the map $t\beta$ is bilinear. □

3.58. Let $\mathscr{BL}(X \times Y, Z)$ denote the set of bilinear maps of $X \times Y$ to Z, X, Y and Z being linear spaces. Prove that $\mathscr{BL}(X \times Y, Z)$ is a linear subspace of the linear space $Z^{X \times Y}$. □

3.59. (For practice!) Write down an example of a 2×3 matrix a, two 3×3 matrices b and c and a 3×2 matrix d. Compute all possible products of two possibly equal matrices out of the set $\{a,b,c,d\}$. Check also, by evaluating both sides, that $(ab)d + a(cd) = a((b + c)d)$. □

3.60. Give an example of a 3×3 matrix a such that $a^2 \neq 0$, but $a^3 = 0$. □

3.61. Prove that the set of all matrices of the form $\begin{pmatrix} a & a \\ a & a \end{pmatrix}$, where $a \in \mathbf{R}^*$, forms a group with respect to matrix multiplication. □

3.62. Find the inverse of each of the matrices a, b, ab and ba, where

$$a = \begin{pmatrix} 1 & 2 & -1 \\ 2 & -1 & 1 \\ -1 & 1 & 2 \end{pmatrix}, \quad b = \begin{pmatrix} 1 & -3 & 2 \\ -2 & 4 & -3 \\ 3 & -7 & 4 \end{pmatrix}.$$

(Solve the equations $y = ax$ and $y = bx$.) □

3.63. Let $g : \mathbf{R}^3 \to \mathbf{R}^3$ be the linear map with matrix

$$\begin{pmatrix} -1 & 3 & 2 \\ 1 & 3 & 1 \\ 2 & 4 & 1 \end{pmatrix}.$$

Find $t^{-1}\{(0,0,0)\}$, $t^{-1}\{(0,1,2)\}$ and $t^{-1}\{(1,2,3)\}$. □

3.64. A *principal circle* in $\mathbf{C} = \mathbf{R}^2$ is defined (for the purposes of this exercise) to be a subset of $\mathbf{C}$ of the form $\{z \in \mathbf{C} : |z| = r\}$, where r is a non-negative real number (we include the case $r = 0$). Prove that, if $t : \mathbf{C} \to \mathbf{C}$ is linear over $\mathbf{C}$, then the image by $t_\vdash$ of any principal circle is a principal circle.

If, conversely, t is linear over $\mathbf{R}$ and $t_\vdash$ sends principal circles to principal circles, does it follow that t is also linear over $\mathbf{C}$? □

CHAPTER 4

AFFINE SPACES

Roughly speaking, an affine space is a linear space with its origin 'rubbed out'. This is easy to visualize, for our picture of a line or plane or three-dimensional space does not require any particular point to be named as origin—the choice of origin is free. It is not quite so simple to give a precise definition of an affine space, and there are various approaches to the problem. The one we have chosen suits the applications which we make later, for example in Chapters 18 and 19. For further comment see page 81.

In the earlier parts of the chapter the field **K** may be any commutative field. Later, in the section on *lines*, it has to be supposed that the characteristic is not equal to 2, while in the section on *convexity*, which involves line-segments, the field must also be *ordered*. For illustrative purposes **K** is taken to be **R** throughout the chapter. This is, in any case, the most important case in the applications.

Affine spaces

Let X be a non-null set and X_* a linear space. An *affine structure* for X with *vector space* X_* is a map

$$\theta : X \times X \to X_*; \quad (x,a) \rightsquigarrow x \mathbin{\dot{-}} a$$

such that

(i) for all $a \in X$, the map $\theta_a : X \to X_*; x \rightsquigarrow x \mathbin{\dot{-}} a$ is bijective,

(ii) for all $a \in X$, $a \mathbin{\dot{-}} a = 0$,

(iii) for all $a, b, x \in X$, $(x \mathbin{\dot{-}} b) + (b \mathbin{\dot{-}} a) = x \mathbin{\dot{-}} a$.

Axiom (iii) is called the *triangle axiom*. It follows from (ii) and (iii) with $x = a$ that, for all $a, b \in X$, $b \mathbin{\dot{-}} a = -(a \mathbin{\dot{-}} b)$.

The map θ is called *subtraction* and the elements of X_* are called *vectors* or *increments* on X. The vector $x \mathbin{\dot{-}} a$ is called the *difference (vector)* of the pair (x,a) of points of X.

An *affine space* (X,θ) is a non-null set X with a prescribed affine structure θ, (X,θ) being frequently abbreviated to X in practice.

Frequently capital letters are used in elementary texts to denote points of an affine space and bold-face lower-case letters to denote vectors. The difference vector $B \dot{-} A$ of two points A and B of the space is then usually denoted by $\overrightarrow{AB}$, this symbol also being used to denote the corresponding tangent vector at A (see below).

Normally, in subsequent chapters, the ordinary subtraction symbol $-$ will be used in place of $\dot{-}$. However, to emphasize logical distinctions, the symbol $\dot{-}$ will continue to be the one used throughout the present chapter.

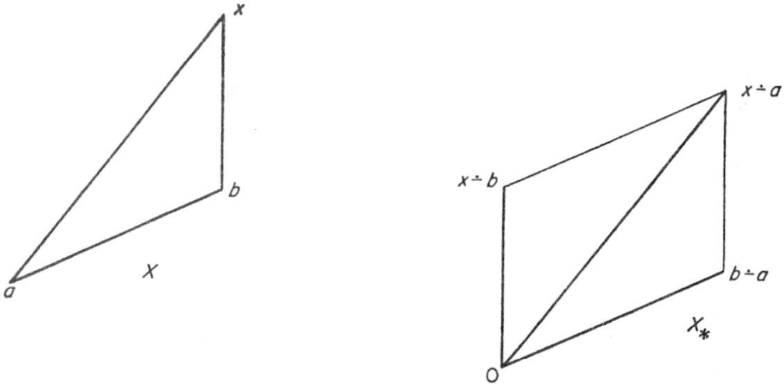

There are various linear spaces associated with an affine space X, each isomorphic to its vector space X_*. For example, axioms (i) and (ii) provide a linear structure for X itself with any chosen point $a \in X$ as origin, the linear structure being that induced by the bijection $\theta_a : X \to X_*;\ x \rightsquigarrow x \dot{-} a$. The choice is indicated in practice by saying that the point a has been *chosen* as *origin* or *set equal* to 0. With this structure chosen it follows, by axiom (iii), that, for any $x, b \in X$,

$$\theta_a(x - b) = \theta_a(x) - \theta_a(b) = (x \dot{-} a) - (b \dot{-} a) = x \dot{-} b;$$

that is, $x - b$ may be identified with $x \dot{-} b$.

The map $X \times \{a\} \to X_*;\ (x,a) \rightsquigarrow x \dot{-} a$ also is bijective, for any chosen $a \in X$, and induces a linear structure on $X \times \{a\}$, with (a,a) as origin. The induced linear space is called the *tangent space* to X at a and denoted by TX_a. Its elements are called *tangent* (or *bound*) *vectors* to X at a. The reason for the term 'tangent' will become apparent in Chapters 19 and 20, when tangent spaces to manifolds are discussed.

Lastly, the map $\theta : X \times X \to X_*$ is surjective, by axiom (i), and so the map $\theta_{\text{inj}} : \operatorname{coim} \theta \to X_*$ is bijective, inducing a linear structure on $\operatorname{coim} \theta$, the set of fibres of the affine structure θ. The induced linear space is called the *free vector space* on X, and its elements are called *free vectors* on X.

Tangent vectors $(x',a') \in X \times \{a'\}$ and $(x,a) \in X \times \{a\}$ are said to be *equivalent* (or *parallel*) if they are representatives of the same free vector. This is written $(x',a') \,||\, (x,a)$. Clearly,

$$(x',a') \,||\, (x,a) \quad \Leftrightarrow \quad x' - a' = x - a.$$

Prop. 4.1. Let $(x',a') \,||\, (x,a)$, where x', a', x and a belong to the affine space X. Then

(i) $(x',x) \,||\, (a',a)$; (ii) $(x',a) = (x,a) + (a',a)$. □

Prop. 4.2. A free vector of an affine space X has a unique representative in every tangent space to X. □

Translations

Let X be an affine space, let $a \in X$ and let $h \in X_*$. By axiom (i) there is a unique $x \in X$ such that $x \dot{-} a = h$. The point x is denoted by $h \dot{+} a$ (or $a \dot{+} h$), and the map $\tau^h : X \to X; a \rightsquigarrow h \dot{+} a$ is called the *translation* of X by h. By Prop. 4.3 below, each translation of X is induced by a unique vector and the set of translations of X is an abelian subgroup of the group $X!$ of permutations of X.

Prop. 4.3. Let X be an affine space, let X_* be the vector space of X regarded simply as an additive group and let $h \in X_*$. Then τ^h is bijective, the map $\tau : X_* \to X!; h \rightsquigarrow \tau^h$ is an injective group map and, for all $a, b \in X$, $\tau^h(a) \dot{-} \tau^h(b) = a \dot{-} b$.

Proof For any $h, k \in X_*$, $\tau^{k+h} = \tau^k\tau^h$; for, if $a \in X$, if $b = h \dot{+} a$ and if $x = k \dot{+} b$, then, by (iii), $x \dot{-} a = k + h$; that is, $(k + h) \dot{+} a = k \dot{+} (h \dot{+} a)$. Also, $\tau^0 = 1_X$, by (ii). So, for any $h \in X_*$, $\tau^{-h} = (\tau^h)^{-1}$; that is, τ^h is bijective.

From the equation $\tau^{k+h} = \tau^k\tau^h$ it also follows that τ is a group map. This map is injective, for, if $h \in X_*$ and if $a \in X$, then $h \dot{+} a = a \Rightarrow h = a \dot{-} a = 0$; that is, $\tau^h = 1_X \Rightarrow h = 0$, from which the injectivity of τ follows by Prop. 3.18.

Finally, for all $a, b \in X$ and $h \in X_*$,

$$h \dot{+} a = h \dot{+} ((a \dot{-} b) \dot{+} b) = (a \dot{-} b) + (h \dot{+} b).$$

that is,

$$\tau^h(a) \dot{-} \tau^h(b) = (h \dot{+} a) - (h \dot{+} b) = a \dot{-} b. \qquad \square$$

The next proposition relates translations to parallel tangent vectors.

Prop. 4.4. Let x, a, x' and a' belong to an affine space X. Then $(x,a) \,||\, (x',a')$ if, and only if, for some $h \in X_*$, $x' = \tau^h(x)$ and $a' = \tau^h(a)$.

Proof $\Leftarrow$: The equations $x' = \tau^h(x)$ and $a' = \tau^h(a)$ imply that

$$x' \dot{-} a' = \tau^h(x) \dot{-} \tau^h(a) = x \dot{-} a.$$

$\Rightarrow$: Suppose that $(x,a) \,||\, (x',a')$, and let $h \in X_*$ be such that $x' = h \dot{+} x$. Then $a' \dot{-} a = x' \dot{-} x = h$. That is, $a' = h \dot{+} a$. □

Finally, by Prop. 4.5, a translation of X may be regarded as a change of origin.

Prop. 4.5. Let a and b belong to an affine space X and let θ_a and θ_b be the maps $X \to X_*$; $x \rightsquigarrow x \dot{-} a$ and $x \rightsquigarrow x \dot{-} b$, corresponding to the choice of a or b, respectively, as 0 in X. Then $\theta_a^{-1}\theta_b = \tau^{a \dot{-} b}$. □

The image $(\tau^h)_\vdash(A)$ of a subset A of an affine space X is said to be a *translate* of A.

Affine maps

A map $t : X \to Y$ between affine spaces X and Y is said to be *affine* if there exists a linear map $t_* : X_* \to Y_*$ such that, for all $x, a \in X$,

$$t(x) \dot{-} t(a) = t_*(x \dot{-} a).$$

The map t_*, if it exists, is unique—for all $h \in X_*$ and any $a \in X$, $t_*(h) = t(a \dot{+} h) - t(a)$—and is called the *linear part* of the affine map t.

For example, if $t : X \to Y$ is constant, $t_* = 0$ and if $t : X \to X$ is a translation of X, $t_* = 1_X$. Indeed, these conditions are necessary as well as sufficient.

Prop. 4.6. Let $t : X \to Y$ and $u : Y \to Z$ be affine maps. Then ut is affine and $(ut)_* = u_* t_*$.

Proof For all $x, a \in X$,

$$ut(x) \dot{-} ut(a) = u_*(t(x) \dot{-} t(a)) = u_* t_*(x \dot{-} a),$$

and $u_* t_*$ is linear, by Prop. 3.9. □

Prop. 4.7. Let $t : X \to Y$ be an affine bijection. Then t^{-1} is affine, and $(t^{-1})_* = (t_*)^{-1}$.

Proof For all $y, b \in Y$,

$$t_*(t^{-1}(y) \dot{-} t^{-1}(b)) = t\,t^{-1}(y) \dot{-} t\,t^{-1}(b) = y \dot{-} b$$
$$= t_*(t_*)^{-1}(y \dot{-} b).$$

Since t_* is injective, $t^{-1}(y) \dot{-} t^{-1}(b) = (t_*)^{-1}(y \dot{-} b)$. □

An affine bijection is said to be an *affine isomorphism*.

Two affine structures on a set X are said to be *equivalent* if the set identity map from either of the affine spaces to the other is an affine isomorphism. In particular, if (X,θ) is an affine space with vector space X_* and if $\gamma : X_* \to Y$ is a linear isomorphism, then the affine space $(X,\gamma\theta)$ with vector space Y is equivalent to the affine space (X,θ). It is often convenient to replace an affine space by an equivalent one in the course of an argument. For example, when a point a of an affine space X is chosen to be 0 it is common practice to replace X_* tacitly by the tangent space to X at a, with which X also has been identified, γ, in this case, being the inverse of θ_a.

Prop. 4.8. An affine map $t : X \to Y$ between linear spaces X and Y is linear if, and only if, $t(0) = 0$.

Proof $\Rightarrow$: Prop. 3.6.
$\Leftarrow$: Suppose $t(0) = 0$. Then for all $x \in X$

$$t(x) = t(x) - t(0) = t_*(x - 0) = t_*(x). \qquad \square$$

Cor. 4.9. Let $t : X \to Y$ be a map between affine spaces X and Y. Then t is affine if, and only if, for some choice of 0 in X and with $t(0) = 0$ in Y, t is linear. $\square$

Since, for all $x \in X$, $t(x) = t(0) + t_*(x)$, any affine map $t : X \to Y$ between linear spaces X and Y may be regarded either as the sum of a constant map and a linear map or as the composite of a linear map and a translation. Conversely, by Prop. 4.6, any such map is affine.

For example, any map of the form

$$\mathbf{R}^2 \to \mathbf{R}^2 : (x,y) \rightsquigarrow (ax + by + c\,,\, a'x + b'y + c')$$

is affine, a, b, c, a', b' and c' being real numbers. This map may be regarded either as the sum of the linear map $\mathbf{R}^2 \to \mathbf{R}^2$; $(x,y) \rightsquigarrow (ax + by\,,\, a'x + b'y)$ and the constant map $\mathbf{R}^2 \to \mathbf{R}^2$; $(x,y) \rightsquigarrow (c,c')$, or as the composite of the same linear map

$$\mathbf{R}^2 \to \mathbf{R}^2 : (x,y) \rightsquigarrow (x',y') = (ax + by\,,\, a'x + b'y)$$

and the translation

$$\mathbf{R}^2 \to \mathbf{R}^2 : (x',y') \rightsquigarrow (x' + c\,,\, y' + c') = (x',y') + (c,c').$$

Affine subspaces

A subset W of an affine space X is said to be an *affine subspace* of X if there is a linear subspace W_* of the vector space X_* and an affine structure for W with vector space W_* such that the inclusion $W \to X$ is affine. Such an affine structure, if it exists, is unique—for any w, $c \in W$, the vector $w \dot{-} c$ is the same, whether with respect to the affine structure for W or with respect to the affine structure for X—and the subspace W is tacitly assigned this structure. (Cf. Exercise 4.22.)

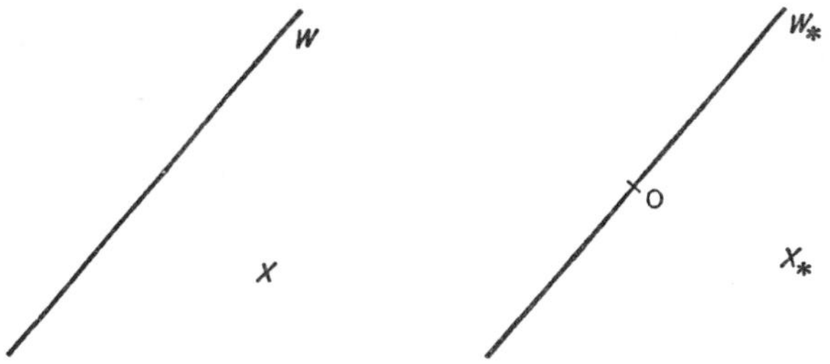

Prop. 4.10. A subset W of an affine space X is an affine subspace of X if, and only if, for some choice of a point of W as 0 in X, W is a linear subspace of X. □

Prop. 4.11. Let $t: X \rightarrow Y$ be an affine map between affine spaces X and Y. Then im t is an affine subspace of Y and each non-null fibre of t is an affine subspace of X. □

Intuitively the affine subspaces of $\mathbf{R}^2$ consist of all the points of $\mathbf{R}^2$, all the lines of $\mathbf{R}^2$ and the plane $\mathbf{R}^2$ itself, while, similarly, the affine subspaces of $\mathbf{R}^3$ consist of the points, the lines and the planes of $\mathbf{R}^3$ and the space $\mathbf{R}^3$ itself.

Affine subspaces of a linear space

The following proposition characterizes the affine subspaces of a linear space.

Prop. 4.12. A subset W of a linear space X is an affine subspace of X if, and only if, it is a translate of a linear subspace of X. □

The set of translates of a linear subspace W of a linear space X is denoted by X/W, and called the *linear quotient* of X by W.

For example, let $X = \mathbf{R}^2$ and let $W = \mathbf{R} \times \{0\}$, the 'vertical axis' in the figure below.

The elements of the linear quotient X/W are then the lines parallel to W in X. Note that each point of X lies on exactly one of the elements of the quotient. That this is true in the general case is proved in the next proposition.

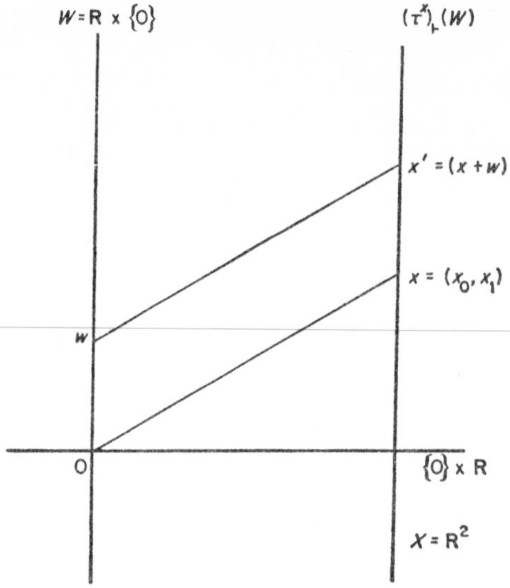

Prop. 4.13. Let W be a linear subspace of the linear space X. Then the map

$$\pi : X \to X/W;\quad x \rightsquigarrow (\tau^x)_\vdash(W)$$

is a partition of X.

Proof Each element of X/W is a non-null subset of X. Also, for any $x, x' \in X$,

$$\begin{aligned} x' \in (\tau^x)_\vdash(W) &\Leftrightarrow \text{for some } w \in W,\ x' = x + w \\ &\Leftrightarrow (\tau^{x'})_\vdash(W) = (\tau^x)_\vdash(W). \end{aligned}$$

That is, π is a partition of X, by Prop. 1.9. □

Prop. 4.14. Let $t : X \to Y$ be a linear map, X and Y being linear spaces. Then each translate of $\ker t$ in X is a fibre of t, and conversely each non-null fibre of t is a translate of $\ker t$ in X. □

In practical terms this states that if $x = a$ is any particular solution of the linear equation $t(x) = b$, where $b \in Y$, then every solution of the equation is of the form $x = a + w$, where $x = w$ is a solution of the equation $t(x) = 0$. Conversely, for any $w \in \ker t$, $x = a + w$ is a solution of the equation.

Further discussion of linear quotients, and in particular the possibility of assigning a linear structure to a linear quotient, is deferred until the next chapter.

Lines in an affine space

The next proposition is concerned with the description of the line through two distinct points a and b of an affine space in terms of the affine structure. This leads on to an alternative definition of affine maps and affine subspaces (valid for any field **K** of characteristic other than 2).

Prop. 4.15. Let X be an affine space. Then, for any $a, b \in X$ and any $\lambda, \mu \in \mathbf{K}$ such that $\lambda + \mu = 1$,

$$a \dotplus \lambda(b \dot{-} a) = b \dotplus \mu(a \dot{-} b).$$

Proof With the point a set equal to 0 the equation reduces to $\lambda b = (1 - \mu)b$. □

When $\lambda + \mu = 1$ we *define* $\mu a \dotplus \lambda b$ to be either side of the above equation, the subset $\{\mu a \dotplus \lambda b : \lambda + \mu = 1\}$ being called the *line* through a and b, whenever a and b are distinct.

Prop. 4.16. A map $t : X \rightarrow Y$ between affine spaces X and Y is affine if, and only if, for all $a, b \in X$ and all $\lambda, \mu \in \mathbf{K}$ such that $\lambda + \mu = 1$,

$$t(\mu a \dotplus \lambda b) = \mu t(a) \dotplus \lambda t(b).$$

Proof $\Rightarrow$: Choose 0 in X and set $t(0) = 0$ in Y. The map t is then linear and the equation follows.

$\Leftarrow$: Again choose 0 in X and set $t(0) = 0$ in Y. Then, with $a = 0$, $t(\lambda b) = \lambda t(b)$, for all $b \in X$ and all $\lambda \in \mathbf{K}$. Also with $\lambda = \mu = \frac{1}{2}$ we find that $t(\frac{1}{2}a + \frac{1}{2}b) = \frac{1}{2}t(a) + \frac{1}{2}t(b)$. Together with what we have just proved, this implies that $t(a + b) = t(a) + t(b)$, for all $a, b \in X$. So t is linear and therefore affine. □

Prop. 4.17. A subset W of an affine space X is an affine subspace of X if, and only if, for all $a, b \in W$ and all $\lambda, \mu \in \mathbf{K}$ such that $\lambda + \mu = 1$, $\mu a \dotplus \lambda b \in W$. □

Convexity

Until now the only restriction on the field **K** that we have had to make has been that it be of characteristic not equal to 2. In this section it is supposed that **K** also is *ordered*—so **K** may be **R**, but not **C**.

Let X be an affine space over such a field and let $a, b \in X$. The subset

$$[a,b] = \{(1 - \lambda)a \dotplus \lambda b : \lambda \in [0,1]\}$$

of the line through a and b is called the *line-segment* joining a to b. When $a = b$, $[a,b] = \{a\}$. A subset A of X is said to be *convex* if, for all a, $b \in A$, $[a,b]$ is a subset of A. For example, any affine subspace of X is convex. On the other hand, in the figure which follows, the subset A of the affine space X is not convex.

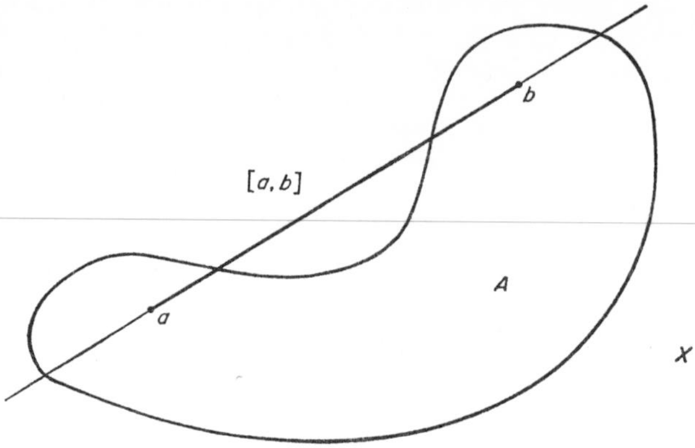

Prop. 4.18. Let $t : X \longrightarrow Y$ be an affine map and let A be a convex subset of X. Then $t_{\vdash}(A)$ is a convex subset of Y.

Proof It has to be proved that, for any a, $b \in A$ and $\lambda \in [0,1]$, $(1 - \lambda)t(a) \dotplus \lambda t(b) \in t_{\vdash}(A)$. For any such a, b set $a = 0$ in X, $t(a) = 0$ in Y. What then has to be shown is that, for any $\lambda \in [0,1]$, $\lambda t(b) \in t_{\vdash}(A)$; but $\lambda t(b) = t(\lambda b)$, since t is now linear, and $\lambda b \in A$ for any such λ, by hypothesis, and this proves the statement. □

Prop. 4.19. Let $t : X \longrightarrow Y$ be an affine map and let B be a convex subset of Y. Then $t^{\dashv}(B)$ is a convex subset of X. □

Further examples of convex sets will be given later, for example in Chapter 15, where it is remarked that any ball in a normed affine space is convex.

Affine products

The product $X \times^{\cdot} Y$ of two affine spaces X and Y has a natural affine structure, with vector space $X_* \times Y_*$, defined by the formula

$$(x',y') \mathbin{\dot{-}} (x,y) = (x' \mathbin{\dot{-}} x\,,\, y' \mathbin{\dot{-}} y)$$

for all (x,y), $(x',y') \in X \times Y$. The various axioms are verified at once. If X, Y and W are affine spaces, a map $(t,u) : W \longrightarrow X \times Y$ is affine if, and only if, t and u are affine. In particular, the projection maps p and q, where $(p,q) = 1_{X \times Y}$, are affine.

Comment

There are many points of view on affine spaces in the literature. To some the term affine space is just another name for a linear space, used when considering properties of the linear space invariant under the group of translations. For example, any translate of a convex set is convex, and therefore in discussions involving convexity such a translation may be made at any time without affecting the argument essentially. We feel that it is simpler to have the origin out of the way at the start of such a discussion. This has the advantage that at the appropriate stage the origin can be *chosen* (not *shifted*) in such a way as to make the subsequent algebra as simple as possible. This is the procedure we adopt in Chapter 18.

An alternative satisfactory definition of affine space axiomatizes the linear space of translations. This has the disadvantage of putting overmuch emphasis on the translations. A vector on an affine space is of course frequently thought of as a translation, but not always.

FURTHER EXERCISES

4.20. Justify the use of the word 'equivalent' in the term 'equivalent tangent vectors'. □

4.21. Express the definition of an affine map by means of a commutative diagram. □

4.22. Let X be an affine space with vector space X_*, let W be a subset of X, let W_* be a linear subspace of X_* and let there be an affine structure for W with vector space W_* such that the inclusion $j: W \to X$ is affine. Prove that j_* is the inclusion of W_* in X_*. □

4.23. Let X be a set, let V be a linear space, and for each $h \in V$ let there be a map $\tau^h : X \to X$ such that, for all $h, k \in V$,

$$\tau^0 = 1_X, \quad \tau^{k+h} = \tau^k \tau^h \quad \text{and} \quad \tau^h = \tau^k \Leftrightarrow h = k.$$

Show that there exists a unique affine structure for X with vector space V and translations $\{\tau^h : h \in V\}$. □

4.24. Let X be an affine space over a field $\mathbf{K}$ with vector space X_*, $\mathbf{K}^*$ denoting the group of non-zero elements of $\mathbf{K}$. Show that the set $(\mathbf{K}^* \times X) \cup X_*$ has a linear structure with respect to which X_* is a linear subspace and $\{1\} \times X$ is an affine subspace parallel to X_*. □

4.25. Let A and B be subsets of an affine space X with vector space X_* and let

$$A \dot{-} B = \{a \dot{-} b \in X_* : a \in A,\ b \in B\}.$$

Prove that, if A and B are affine subspaces of X, $A \dot{-} B$ is an affine subspace of X_*, and is a linear subspace of X_* if, and only if, A and B intersect. □

4.26. Let A be a subset of the real affine space X. The intersection of all the convex subsets of X containing A as a subset is called the *convex hull* of A in X. Prove that the convex hull of A is convex. □

4.27. A line in $\mathbf{R}^2$ is said to pass *between* two subsets A and B of $\mathbf{R}^2$ if it intersects neither set but intersects the line segment $[a,b]$ joining any $a \in A$ to any $b \in B$. A *triangle* in $\mathbf{R}^2$ is the convex hull of a set of three distinct non-collinear points of $\mathbf{R}^2$.

Prove that a line can be drawn between any two disjoint triangles in $\mathbf{R}^2$. □

CHAPTER 5

QUOTIENT STRUCTURES

Topics discussed in this chapter include linear quotients, quotient groups, quotient rings and exact sequences. Group actions and orbits are defined at the end of the chapter.

Professor S. MacLane [39] traces exact sequences back to a paper by W. Hurewicz in 1940. The arrow notation for a map developed about the same time.

Linear quotients

Linear quotients were briefly introduced in Chapter 4. In practice they often occur in the following way.

Suppose that $t : X \longrightarrow Y$ is a linear map constant on a linear subspace W of X. Then t must be constant on each of the affine subspaces of X parallel to W; for since $t(0) = 0$ and since $0 \in W$, $t(w) = 0$ for each $w \in W$ and therefore, for any $x \in X$ and any $w \in W$, $t(x + w) = t(x)$; that is, t has the value $t(x)$ at every point of the parallel to W through x.

It follows from this that t has the decomposition $X \xrightarrow{\pi} X/W \xrightarrow{t'} Y$, where X/W is the set of translates of W in X and π is the partition defined in Prop. 4.13, the map t' being uniquely defined by the requirement that $t'\pi(x) = t(x)$ for all $x \in X$. If there is a linear structure for X/W such that the surjection π is linear, then by Prop. 3.10 the map t' also will be linear. What we shall now prove is that such a linear structure does exist and that it is unique.

To define a linear structure, the operations of addition and scalar multiplication have first to be defined and then the axioms for a linear structure have to be checked. This checking is usually straightforward—the main interest lies in the definition.

Theorem 5.1. Let W be a linear subspace of a linear space X. Then there is a unique linear structure for X/W such that the partition $\pi : X \longrightarrow X/W;\ x \rightsquigarrow (\tau^x)_{\vdash}(W)$ is linear, with $\ker \pi = W$.

Proof Suppose first that such a structure exists. Then it must be unique, for the linearity of π provides formulae both for addition and for scalar multiplication, namely, for any $\pi(x)$ and $\pi(x') \in X/W$ and for any $\lambda \in \mathbf{K}$,

$$\pi(x) + \pi(x') = \pi(x + x') \quad \text{and} \quad \lambda\pi(x) = \pi(\lambda x).$$

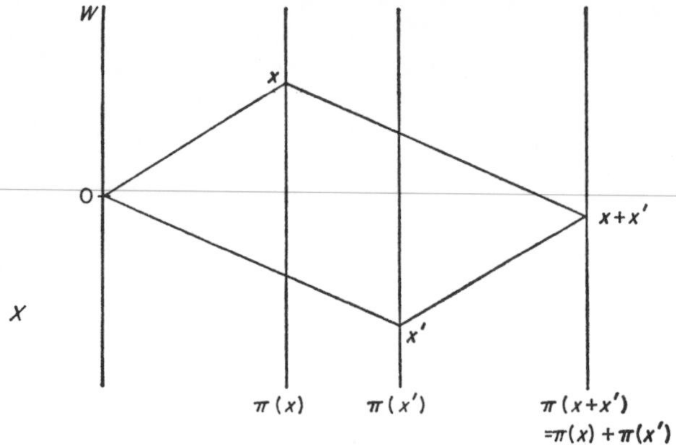

To prove the existence of such a structure, it has first to be checked that these formulae define $\pi(x) + \pi(x')$ and $\lambda\pi(x)$ independently of the choice of x to represent $\pi(x)$ and x' to represent $\pi(x')$. However, by the commutativity of addition,

$$(x + w) + (x' + w') = (x + x') + (w + w'),$$

for any $w, w' \in W$.

So, for any $x + w \in \pi(x)$ and any $x' + w' \in \pi(x')$,

$$\pi((x + w) + (x' + w')) = \pi(x + x')$$

and
$$\pi(\lambda(x + w)) = \pi(\lambda x + \lambda w) = \pi(\lambda x).$$

It remains to verify that the addition and scalar multiplication so defined satisfy all the axioms. As we have already remarked, this is a routine check. The origin in X/W is $\pi(0) = W$ since, for any $\pi(x) \in X/W$,

$$\pi(x) + \pi(0) = \pi(x + 0) = \pi(x),$$

while the additive inverse of any $\pi(x) \in X/W$ is $\pi(-x)$, since

$$\pi(x) + \pi(-x) = \pi(x - x) = \pi(0).$$

The verification of the remaining axioms is left to the reader. □

The linear quotient X/W of a linear space X by a linear subspace W is tacitly assigned this linear structure.

Cor. 5.2. Every linear subspace W of a linear space X is the kernel of some linear surjection $t: X \to Y$. □

By Prop. 4.14 the coimage of any linear map $t: X \to Y$ is a linear quotient of X, namely $X/\ker t$. A second linear quotient, associated to the map t, to which it is also convenient to give a name, is its *cokernel*, $\operatorname{coker} t = Y/\operatorname{im} t$.

Quotient groups

The preceding discussion can be regarded as an analysis of the structure of a linear surjection. Such a surjection $t: X \to Y$ has a canonical decomposition $X \xrightarrow{t_{\text{par}}} X/\ker t \xrightarrow{t_{\text{bij}}} Y$, with $X/\ker t = \operatorname{coim} t$, t_{par} being a linear partition of X.

Surjective group maps can be similarly analysed. Although only one operation, the group product, is involved rather than two, the failure of commutativity highlights some of the details of the argument in the linear case.

We begin by defining the analogues, for a group G, of the translates in a linear space X of a linear subspace W of X.

Let G be a group and F a subgroup of G. Then, for any $g \in G$ the sets

$$gF = \{gf : f \in F\}$$

and

$$Fg = \{fg : f \in F\}$$

are called, respectively, the *left* and *right cosets* of F in G, the sets of left and right cosets of F in G being denoted respectively by $(G/F)_L$ and $(G/F)_R$.

For example, let G be the group of permutations of the set $\{0,1,2\}$ and let a denote the transposition of 0 and 1 and b the transposition of 1 and 2. Then the elements of the group are $1_{(G)}$, a, b, ab, ba and $aba = bab$. The left cosets of the subgroup $\{1_{(G)},a\}$ are $\{1_{(G)},a\}$, $\{b,ba\}$ and $\{ab,aba\}$, while the right cosets are $\{1_{(G)},a\}$, $\{b,ab\}$ and $\{ba,aba\}$. It follows from this example that a left coset is not necessarily a right coset, and vice versa.

Prop. 5.3. Let G be a group and F a subgroup of G. Then the maps

$$G \to (G/F)_L \quad g \rightsquigarrow gF$$

and

$$G \to (G/F)_R \quad g \rightsquigarrow Fg$$

are partitions of G. □

Prop. 5.4. Let $t: G \to H$ be a surjective group map, G and H being groups. Then each fibre of t is both a left and a right coset of $\ker t$ in G. Conversely, each left or right coset of $\ker t$ in G is a fibre of t.

Proof For all $g, g' \in G$

$$\begin{aligned} t(g') = t(g) &\Rightarrow (t(g))^{-1}t(g') = 1 = t(g')(t(g))^{-1} \\ &\Rightarrow t(g^{-1}g') = 1 = t(g'g^{-1}) \\ &\Rightarrow g^{-1}g', g'g^{-1} \in \ker t \\ &\Rightarrow g' \in gF \quad \text{and} \quad g' \in Fg \end{aligned}$$

where $F = \ker t$, and conversely, if $g' \in gF$ or Fg, then $t(g') = t(g)$. So, for all $g \in G$,

$$t^{-1}\{t(g)\} = gF = Fg. \qquad \square$$

A subgroup F of G such that each left coset of F is also a right coset of F in G is said to be a *normal subgroup* of G. The set of cosets in G of a normal subgroup F is denoted simply by G/F.

The analogue of Theorem 5.1 is now the following.

Theorem 5.5. Let F be a normal subgroup of a group G. Then there is a unique group structure for G/F such that the partition $\pi : G \to G/F;\ g \rightsquigarrow gF$ is a group map, with $\ker \pi = F$.

Proof The first part of the proof is as before. If such a structure exists it must be unique, for the requirement that π be a group map provides a formula for the group multiplication, namely, for any $\pi(g)$ and $\pi(g') \in G/F$,

$$\pi(g)\,\pi(g') = \pi(gg').$$

The next part is slightly trickier, because of the absence of commutativity. To prove existence it has to be checked that the formula defines $\pi(g)\,\pi(g')$ independently of the choice of g to represent $\pi(g)$ and g' to represent $\pi(g')$. However, for any $f, f' \in F$,

$$(gf)(g'f') = g(fg')f'$$

and since $Fg' = g'F$ there exists an element $f'' \in F$ such that $fg' = g'f''$, so that

$$(gf)(g'f') = (gg')(f''f').$$

That is, for any $gf \in \pi(g)$ and any $g'f' \in \pi(g')$,

$$\pi(gf)(g'f') = \pi(gg').$$

Finally, there is the routine check that the axioms for a group structure are satisfied, that π is a group map, and in particular that $\ker \pi = F$ is the neutral element for the group structure. $\square$

Cor. 5.6. A subgroup F of a group G is the kernel of some group surjection $t : G \to H$ if, and only if, F is a normal subgroup of G. $\square$

The group quotient G/F, where F is normal in G, is tacitly assigned the group structure defined in Theorem 5.5 and is then called the *quotient group* of G with kernel F.

One point to watch in the case of normal subgroups is that the concept is not transitive. One can have groups F, G and H such that F is a normal subgroup of G and G is a normal subgroup of H, but F is not a normal subgroup of H. (Cf. Exercise 5.35.)

Non-normal subgroups are not unimportant in practice. As we shall see later, for example in Chapter 12 and in Chapter 17, many spaces of interest are representable as the set of left cosets $(G/F)_L$ of a not necessarily normal subgroup F of some larger group G. In these later applications we shall abbreviate notations, writing simply G/F in place of $(G/F)_L$.

Ideals

Surjective ring maps can be subjected to a similar analysis.

Let $t: A \twoheadrightarrow B$ be a surjective ring map and let $C = \ker t$. Then C is a subring of A. Moreover, for any $a \in A$ and any $c \in C$,

$$t(ca) = t(c)\, t(a) = 0$$

and

$$t(ac) = t(a)\, t(c) = 0,$$

since $t(c) = 0$.

Therefore $CA \subset C$ and $AC \subset C$. A subring C of A with this property is said to be a *two-sided ideal* of A.

Prop. 5.7. Let A be a ring and C a two-sided ideal of A. Then the (additive) abelian group A/C has a unique ring structure such that the natural projection $\pi: A \twoheadrightarrow A/C$ is a ring map.

Proof For any $a, a' \in A$ we must have

$$(\{a\} + C)(\{a'\} + C) = (\{aa'\} + C),$$

and this is legitimate, since for any $c, c' \in C$,

$$(a + c)(a' + c') = aa' + c'',$$

where $c'' = ca' + ac' + cc' \in C$.

The remaining details are readily checked. □

Two-sided ideals of an algebra are similarly defined, but with the additional condition that the ideal be a linear subspace of the algebra. One-sided ideals of an algebra have been introduced already, at the end of Chapter 3.

Exact sequences

Let s and t be linear maps such that the target of s is also the source of t. Such a pair of maps is said to be *exact* if $\operatorname{im} s = \ker t$. Note that this is stronger than the assertion that $ts = 0$, which is equivalent to the condition $\operatorname{im} s \subset \ker t$ only. A possibly doubly infinite sequence of linear maps such that the target of each map coincides with the source of its successor is said to be *exact* if each pair of adjacent maps is exact.

Suppose, for example, that s is the map $\mathbf{R} \to \mathbf{R}^2$; $y \rightsquigarrow (0,y)$ and that t is the map $\mathbf{R}^2 \to \mathbf{R}$; $(x,y) \rightsquigarrow x$.

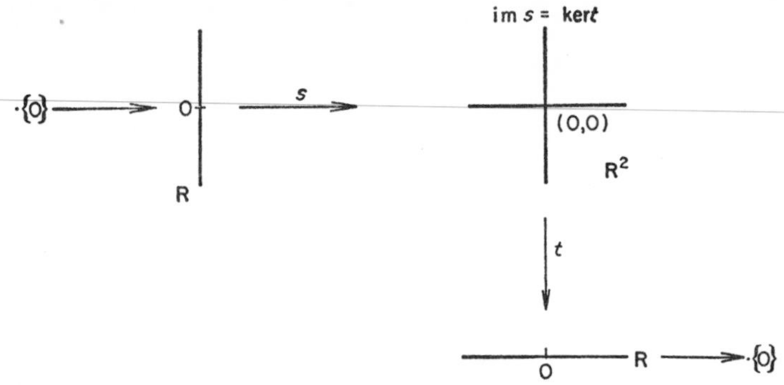

Then the sequence of linear maps

$$\{0\} \to \mathbf{R} \xrightarrow{s} \mathbf{R}^2 \to \mathbf{R} \to \{0\}$$

is exact. Here, as in the following propositions, $\{0\}$ denotes the linear space whose sole element is 0. For any linear space X, the linear maps $\{0\} \to X$ and $X \to \{0\}$ are uniquely defined and it is not usually necessary to name them.

Prop. 5.8. The sequence of linear maps

$$\{0\} \to W \xrightarrow{s} X$$

is exact if, and only if, s is injective.

Proof The sequence is exact if, and only if, $\ker s = \{0\}$; but, by Prop. 3.18, $\ker s = \{0\}$ if, and only if, s is injective. □

Prop. 5.9. The sequence of linear maps

$$X \to Y \to \{0\}$$

is exact if, and only if, t is surjective.

Proof The sequence is exact if, and only if, $\operatorname{im} t = Y$; but this is just the assertion that t is surjective. □

Cor. 5.10. The sequence of linear maps

$$\{0\} \to X \xrightarrow{t} Y \to \{0\}$$

is exact if, and only if, t is an isomorphism. □

Prop. 5.11. Let X and Y be linear spaces, let

$$i = (1_X, 0): X \to X \times Y, \quad j = (0, 1_Y): Y \to X \times Y$$

and let $(p,q) = 1_{X \times Y}$. Then the sequences

$$\{0\} \to X \xrightarrow{i} X \times Y \xrightarrow{q} Y \to \{0\}$$

and
$$\{0\} \to Y \xrightarrow{j} X \times Y \xrightarrow{p} X \to \{0\}$$

are exact, with $pi = 1_X$ and $qj = 1_Y$. □

Prop. 5.12. Let X and Y be linear subspaces of a linear space V. Then the sequence

$$\{0\} \to X \cap Y \xrightarrow{s} X \times Y \xrightarrow{t} X + Y \to \{0\},$$

where, for all $w \in X \cap Y$, $s(w) = (w, -w)$ and, for all $(x,y) \in X \times Y$, $t(x,y) = x + y$, is exact. □

Prop. 5.13. Let W be a linear subspace of a linear space X. Then the sequence of linear maps

$$\{0\} \to W \xrightarrow{i} X \xrightarrow{\pi} X/W \to \{0\},$$

where i is the inclusion and π the partition, is exact.

Proof The sequence is exact at W, since i is injective, exact at X, since $\operatorname{im} i = W = \ker \pi$, and exact at X/W, since π is surjective. □

An exact sequence of linear maps of the form

$$\{0\} \to W \xrightarrow{s} X \xrightarrow{t} Y \to \{0\}$$

is said to be a *short exact sequence.*

Prop. 5.14. Let $\{0\} \to W \xrightarrow{s} X \xrightarrow{t} Y \to \{0\}$ be a short exact sequence. The diagram of maps

$$\begin{array}{ccccccccc}
\{0\} & \to & W & \xrightarrow{s} & X & \xrightarrow{t} & Y & \to & \{0\} \\
 & & \downarrow{\scriptstyle s_{\text{sur}}} & & \downarrow{\scriptstyle 1_X} & & \uparrow{\scriptstyle t_{\text{inj}}} & & \\
\{0\} & \to & \operatorname{im} s & \xrightarrow{s_{\text{inc}}} & X & \xrightarrow{t_{\text{par}}} & \operatorname{coim} t & \to & \{0\}
\end{array}$$

is commutative (cf. page 10). The vertical maps are isomorphisms, and the lower sequence is an exact sequence of the type discussed in Prop. 5.13, with $\operatorname{coim} t = X/\operatorname{im} s$. □

In practice one often takes advantage of this proposition and regards any short exact sequence as being essentially one involving a subspace and a quotient space. Given the short exact sequence

$$\{0\} \to W \to X \to Y \to \{0\},$$

one thinks of W as a subspace of X and of Y as the quotient space X/W.

Diagram-chasing

The following proposition is a slight generalization of the remarks with which we opened this chapter, and it may, in fact, be proved as a corollary to Prop. 5.14. Instead, we give a direct proof, as the argument is typical of many arguments involving exact sequences. The proposition will be useful in Chapter 19.

Prop. 5.15. Let $t : X \to Y$ be a **K**-linear surjection, let $W = \ker t$, and let $\beta : X \to \mathbf{K}$ be a linear map whose restriction to W is zero. Then there exists a unique linear map $\gamma : Y \to \mathbf{K}$ such that $\beta = \gamma t$.

Proof During the proof we 'chase around' the diagram of linear maps

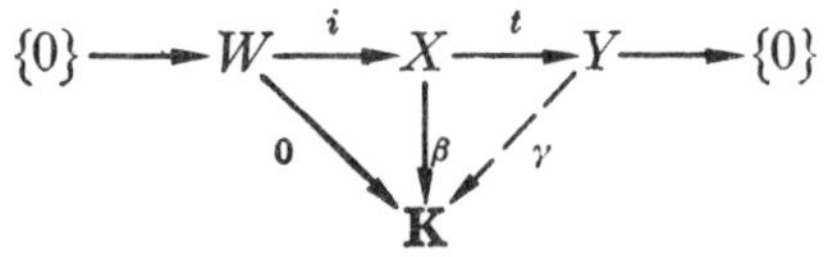

where i is the inclusion map, the row is exact and $\beta i = 0$.

Since t is surjective, any element of Y is of the form $t(x)$ where $x \in X$. Also, for any $x_1 \in X$, $t(x_1) = t(x)$ if, and only if, $t(x_1 - x) = 0$, that is if, and only if, $x_1 - x \in W$. From the first of these remarks it follows that if there is a map γ such that $\beta = \gamma t$, then, for all $t(x) \in Y$, $\gamma t(x) = \beta(x)$, that is, γ is *unique*. The *existence* of such a map then follows from the second remark, since $x_1 - x \in W \Rightarrow \beta(x_1 - x) = 0 \Rightarrow \beta(x_1) = \beta(x)$, implying that if $t(x_1) = t(x)$, then $\beta(x_1) = \beta(x)$.

Finally, by Prop. 3.10, γ is linear, since t is surjective and since t and β are linear. □

The dual of an exact sequence

Prop. 5.16. Let $\{0\} \to W \to X \to Y \to \{0\}$ be an exact sequence of linear maps. Then the dual sequence

$$\{0\} \to Y^{\mathscr{L}} \xrightarrow{t^{\mathscr{L}}} X^{\mathscr{L}} \xrightarrow{s^{\mathscr{L}}} W^{\mathscr{L}} \to \{0\}$$

is exact at $Y^{\mathscr{L}}$ and at $X^{\mathscr{L}}$. In particular, the dual of a linear surjection is a linear injection.

(We shall prove in Chapter 6 that when the linear spaces involved are finite-dimensional, the dual of a linear injection is a linear surjection, implying that the dual sequence is exact also at $W^{\mathscr{L}}$.)

Proof Exactness at $Y^{\mathscr{L}}$:

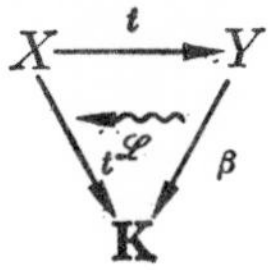

What has to be proved is that $t^{\mathscr{L}}$ is injective, or equivalently that if, for any $\gamma \in Y^{\mathscr{L}}$, $\gamma t = t^{\mathscr{L}}(\gamma) = 0$, then $\gamma = 0$, the map t being surjective. This is just Prop. 3.8.

Exactness at $X^{\mathscr{L}}$: there are two things to be proved. First, $\operatorname{im} t^{\mathscr{L}} \subset \ker s^{\mathscr{L}}$, for $s^{\mathscr{L}} t^{\mathscr{L}} = (ts)^{\mathscr{L}} = 0^{\mathscr{L}} = 0$. Secondly, $\ker s^{\mathscr{L}} \subset \operatorname{im} t^{\mathscr{L}}$, by Prop. 5.15. So $\ker s^{\mathscr{L}} = \operatorname{im} t^{\mathscr{L}}$. □

More diagram-chasing

Proposition 5.15 is a special case of the following proposition, also proved by diagram-chasing.

Prop. 5.17. Let

$$\begin{array}{ccccccccc} \{0\} & \longrightarrow & W & \xrightarrow{s} & X & \longrightarrow & Y & \longrightarrow & \{0\} \\ & & \downarrow{\scriptstyle\alpha} & & \downarrow{\scriptstyle\beta} & & \downarrow{\scriptstyle\gamma} & & \\ \{0\} & \longrightarrow & W' & \xrightarrow{s'} & X' & \xrightarrow{t'} & Y' & \longrightarrow & \{0\} \end{array}$$

be a diagram of linear maps such that the rows are exact and $\beta s = s'\alpha$, that is, the square formed by these maps is commutative. Then there exists a unique linear map $\gamma : Y \to Y'$ such that $\gamma t = t'\beta$, and if α and β are isomorphisms, then γ also is an isomorphism.

Proof—Uniqueness of γ Suppose γ exists. By hypothesis $\gamma t(x) = t'\beta(x)$, for all $x \in X$. Since for each $y \in Y$ there exists x in $t^{-1}\{y\}$, $\gamma(y) = t'\beta(x)$ for any such x; that is, γ is uniquely determined.

Existence of γ Let $y \in Y$ and let $x, x_1 \in t^{-1}\{y\}$. Then $x_1 - x \in \ker t = \operatorname{im} s$ and so $x_1 = x + s(w)$, for some $w \in W$. Then

$$\begin{aligned} t'\beta(x') &= t'\beta(x + s(w)) = t'\beta(x) + t'\beta s(w) \\ &= t'\beta(x) + t's'\alpha(w) \\ &= t'\beta(x), \quad \text{since } t's' = 0. \end{aligned}$$

The prescription $\gamma(y) = t'\beta(x)$, for any x in $t^{-1}\{y\}$, does therefore determine a map $\gamma : Y \to Y'$ such that $\gamma t = t'\beta$. Since $t'\beta$ is linear and t is a linear surjection, γ is linear, by Prop. 3.10.

Now suppose α and β are isomorphisms and let $\eta : Y' \to Y$ be the unique linear map such that $\eta t' = t\beta^{-1}$. Then applying the uniqueness part of the proposition to the diagram

$$\begin{array}{ccccccccc} \{0\} & \to & W & \xrightarrow{s} & X & \xrightarrow{t} & Y & \to & \{0\} \\ & & \downarrow{\scriptstyle 1_W} & & \downarrow{\scriptstyle 1_X} & & \downarrow{\scriptstyle \eta\gamma} & & \\ \{0\} & \to & W & \xrightarrow{s} & X & \xrightarrow{t} & Y & \to & \{0\}, \end{array}$$

yields $\eta\gamma = 1_Y$. Similarly $\gamma\eta = 1_{Y'}$. That is, $\eta = \gamma^{-1}$ and so γ is an isomorphism. □

This proposition is in its turn a special case of the following proposition.

Prop. 5.18. Let $\begin{array}{ccc} X & \xrightarrow{t} & Y \\ \downarrow{\scriptstyle \alpha} & & \downarrow{\scriptstyle \beta} \\ X' & \xrightarrow{t'} & Y' \end{array}$ be a commutative square of linear maps.

Then there exists a diagram of linear maps:

$$\begin{array}{ccccccccccc} & & \{0\} & & \{0\} & & \{0\} & & \{0\} & & \\ & & \downarrow & & \downarrow & & \downarrow & & \downarrow & & \\ \{0\} & \to & \ker \omega & \to & \ker \alpha & \xrightarrow{s} & \ker \beta & \to & \ker \gamma & \to & \{0\} \\ & & \downarrow & & \downarrow & & \downarrow & & \downarrow & & \\ \{0\} & \to & \ker t & \to & X & \xrightarrow{t} & Y & \to & \operatorname{coker} t & \to & \{0\} \\ & & \downarrow{\scriptstyle \omega} & & \downarrow{\scriptstyle \alpha} & & \downarrow{\scriptstyle \beta} & & \downarrow{\scriptstyle \gamma} & & \\ \{0\} & \to & \ker t' & \to & X' & \xrightarrow{t'} & Y' & \to & \operatorname{coker} t' & \to & \{0\} \\ & & \downarrow & & \downarrow & & \downarrow & & \downarrow & & \\ \{0\} & \to & \operatorname{coker} \omega & \to & \operatorname{coker} \alpha & \xrightarrow{u} & \operatorname{coker} \beta & \to & \operatorname{coker} \gamma & \to & \{0\} \\ & & \downarrow & & \downarrow & & \downarrow & & \downarrow & & \\ & & \{0\} & & \{0\} & & \{0\} & & \{0\} & & \end{array}$$

with exact rows and columns and commutative squares. The precise definition of the various maps in this diagram is left to the reader, as is the proof! □

There are many useful special cases of Prop. 5.18 other than those already mentioned. Suppose, for example, that $W^\perp$ is a linear subspace of a linear space X and that W is a linear subspace of $W^\perp$. Then the three inclusions induce the following commutative diagram of exact sequences:

$$
\begin{array}{ccccccccc}
 & & \{0\} & & \{0\} & & & & \\
 & & \downarrow & & \downarrow & & & & \\
\{0\} & \longrightarrow & W & \xrightarrow{1} & W & \longrightarrow & \{0\} & & \\
 & & \downarrow{\scriptstyle\text{inc}} & & \downarrow{\scriptstyle\text{inc}} & & \downarrow & & \\
\{0\} & \longrightarrow & W^\perp & \xrightarrow{\text{inc}} & X & \xrightarrow{\text{par}} & X/W^\perp & \longrightarrow & \{0\}\\
 & & \downarrow{\scriptstyle\text{par}} & & \downarrow{\scriptstyle\text{par}} & & \downarrow{\scriptstyle 1} & & \\
\{0\} & \longrightarrow & W^\perp/W & \longrightarrow & X/W & \longrightarrow & X/W^\perp & \longrightarrow & \{0\}\\
 & & \downarrow & & \downarrow & & \downarrow & & \\
 & & \{0\} & & \{0\} & & \{0\} & &
\end{array}
$$

This will be applied in Chapter 9 (page 163).

Sections of a linear surjection

Prop. 5.19. Let $t' : Y \to X$ be a not necessarily linear section of a surjective linear map $t : X \to Y$ and let $s : W \to X$ be an injective map with image $\ker t$. Then the map

$$(s \quad t') : W \times Y \to X; \quad (w,y) \rightsquigarrow s(w) + t'(y)$$

is bijective.

Proof Let $x \in X$. Then $x = (x - t't(x)) + t't(x)$, and $t(x - t't(x)) = 0$, since $tt' = 1_Y$. That is, $x = s(w) + t'(y)$ where $w = x - t't(x)$ and $y = t(x)$. So $(s \quad t')$ is surjective.

Suppose next that, for any $w, w' \in W$, $y, y' \in Y$,

$$s(w) + t'(y) = s(w') + t'(y').$$

Then, since t is linear, since $ts = 0$, and since $tt' = 1_Y$, $y = y'$. So $s(w) = s(w')$ and, since s is injective, $w = w'$. That is, $(s \quad t')$ is injective.

So $(s \quad t')$ is bijective. □

Cor. 5.20. If, in addition, s and t' are linear, then $(s \quad t')$ is a linear isomorphism and there is a unique linear surjection $s' : X \to W$ such that $(s' \quad t) = (s \quad t')^{-1}$. Moreover, $\operatorname{im} t' = \ker s'$ and $s's = 1_W$. □

A short exact sequence of linear maps

$$\{0\} \to W \xrightarrow{s} X \xrightarrow{t} Y \to \{0\}$$

with a prescribed linear section $t' : Y \to X$ of t is said to be a *split exact sequence*, the map t' being the *splitting* of the sequence.

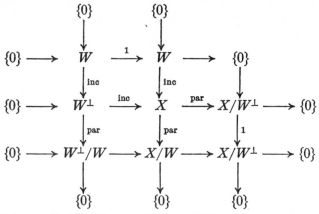

Prop. 5.21. Let u, u' be linear sections of a surjective linear map $t: X \longrightarrow Y$. Then $\operatorname{im}(u' - u) \subset \ker t$.

Proof For all $y \in Y$, $t(u' - u)(y) = tu'(y) - tu(y) = 0$. □

We shall denote by $u' \dot{-} u$ the map $u' - u$ with target restricted to $\ker t$.

Prop. 5.22. Let U denote the set of linear sections of a surjective linear map $t: X \longrightarrow Y$ with kernel W. Then, provided that U is non-null, the map

$$\theta : U \times U \longrightarrow \mathscr{L}(Y,W); \quad (u',u) \rightsquigarrow u' \dot{-} u$$

is an affine structure for U, with linear part $\mathscr{L}(Y,W)$.

Proof

(i) For all u, $u' \in U$ and all $v \in \mathscr{L}(Y,W)$, $v = u' \dot{-} u \Leftrightarrow jv = u' - u \Leftrightarrow u' = u + jv$, where $j: W \longrightarrow X$ is the inclusion. That is, the map $U \longrightarrow \mathscr{L}(Y,W)$; $u' \rightsquigarrow u' \dot{-} u$ is bijective.

(ii) For all $u \in U$, $u \dot{-} u = 0$, since $u - u = 0$.

(iii) For all u, u', $u'' \in U$, $(u'' \dot{-} u') + (u' - u) = (u'' \dot{-} u)$, since $(u'' - u') + (u' - u) = u'' - u$. □

Cor. 5.23. The set of linear sections of a linear partition $\pi : X \longrightarrow X/W$ has a natural affine structure with linear part $\mathscr{L}(X/W, W)$, provided that the set of linear sections is non-null. □

The study of linear sections is continued in Chapter 8.

Analogues for group maps

The *definition* of an exact sequence goes over without change to sequences of group maps as do several, but by no means all, of the propositions listed for linear exact sequences. The reader should check through each carefully to find out which remain valid.

In work with multiplicative groups the symbol $\{1\}$ is usually used in place of $\{0\}$ to denote the one-element group.

Certain results on surjective group maps even extend to left (or right) coset partitions. In particular the following extension of the concept of a short exact sequence is convenient.

Let F and G be groups. Let H be a set and let

$$F \xrightarrow{s} G \xrightarrow{t} H$$

be a pair of maps such that s is a group injection and t is a surjection whose fibres are the left cosets of the image of F in G. The pair will then be said to be *left-coset exact*, and the bijection $t_{\text{inj}} : G/F \longrightarrow H$ will be said to be a (*left*) *coset space representation* of the set H. Numerous examples are given in Chapter 11 (Theorem 11.55) and in Chapter 12.

The following are analogues for left-coset exact pairs of maps of part of Prop. 5.18 and of Prop. 5.19 and Cor. 5.20 respectively.

Prop. 5.24. Let F, G, F' and G' be groups, let H, H', M and N be sets and let

$$\begin{array}{ccccc} F & \xrightarrow{s} & G & \longrightarrow & H \\ \downarrow{\scriptstyle \alpha} & & \downarrow{\scriptstyle \beta} & & \\ F' & \xrightarrow{s'} & G' & \xrightarrow{t'} & H' \\ \downarrow{\scriptstyle \mu} & & \downarrow{\scriptstyle \nu} & & \\ M & & N & & \end{array}$$

be a commutative diagram of maps whose rows and columns are left-coset exact. Then if there is a (necessarily unique) bijection $u : M \longrightarrow N$ such that $u\mu = \nu s'$, there is a unique bijection $\gamma : H \longrightarrow H'$ such that $\gamma t = t'\beta$. If, moreover, H and H' are groups and if t and t' are group maps, then γ is a group isomorphism. □

Prop. 5.25. Let F and G be groups, let H be a set, let $F \xrightarrow{s} G \xrightarrow{t} H$ be a left-coset exact pair of maps and let $t' : H \longrightarrow G$ be a section of t. Then the map

$$F \times H \longrightarrow G; \quad (f,h) \rightsquigarrow t'(h)\, s(f)$$

is bijective.

Moreover, if H is a group, if t and t' are group maps and if each element of im t' *commutes* with every element of im s, then the bijection $F \times H \longrightarrow G$ is a group isomorphism. □

A short exact sequence of group maps

$$\{1\} \longrightarrow F \xrightarrow{s} G \xrightarrow{t} H \longrightarrow \{1\}$$

with a prescribed group section $t' : H \longrightarrow G$ of t satisfying the condition of the last paragraph of Prop. 5.25 is said to be a *split* exact sequence, the map t' being the *splitting* of the sequence.

For examples of Prop. 5.24 and Prop. 5.25, see Exercises 9.38, 9.39, 11.63 and 11.65.

Orbits

An extension of the quotient notation is the following.

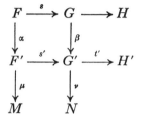

Suppose that X is a set and H a group and that there is a *group action*

of H on X *on the right*, that is, a map

$$X \times H \longrightarrow X; \quad (x,h) \rightsquigarrow xh$$

such that, for all $x \in X$, $x1_{(H)} = x$ and, for all $h, h' \in H$, $(xh)h' = x(hh')$.

Then the relation $\sim$ on X defined by

$$x \sim x' \Leftrightarrow \text{for some } h \in H, x' = xh$$

is an equivalence. Each equivalence class is said to be an *orbit* of H in X and the set of orbits is denoted by X/H.

This is in accord with the notation G/H when H is a subgroup of G and the action $G \times H \longrightarrow G$ is the restriction of the group product on G to $G \times H$.

Similar remarks and notations apply to group actions *on the left*.

Further Exercises

5.26. Let X and Y be linear spaces and let $t: X \longrightarrow Y$ be a linear map such that, for each linear space Z and each linear map $u: Y \longrightarrow Z$, $ut = 0 \Rightarrow u = 0$. Prove that t is surjective. (Let $Z = Y/\operatorname{im} t$, and let u be the associated linear partition.) □

(This complements Exercise 3.49.)

5.27. (The Four Lemma.) Let

$$\begin{array}{ccccccc} W & \longrightarrow & X & \xrightarrow{t} & Y & \longrightarrow & Z \\ \downarrow{\scriptstyle a} & & \downarrow{\scriptstyle b} & & \downarrow{\scriptstyle c} & & \downarrow{\scriptstyle d} \\ W' & \longrightarrow & X' & \xrightarrow{u} & Y' & \longrightarrow & Z' \end{array}$$

be a commutative diagram of linear maps with exact rows, a being surjective and d injective. Prove that $\ker c = t_{\vdash}(\ker b)$ and that $\operatorname{im} b = u^{\dashv}(\operatorname{im} c)$. □

5.28. (The Five Lemma.) Let

$$\begin{array}{ccccccccc} V & \longrightarrow & W & \longrightarrow & X & \longrightarrow & Y & \longrightarrow & Z \\ \downarrow{\scriptstyle a} & & \downarrow{\scriptstyle b} & & \downarrow{\scriptstyle c} & & \downarrow{\scriptstyle d} & & \downarrow{\scriptstyle e} \\ V' & \longrightarrow & W' & \longrightarrow & X' & \longrightarrow & Y' & \longrightarrow & Z' \end{array}$$

be a commutative diagram of linear maps with exact rows, a, b, d and e being isomorphisms. Prove that c is an isomorphism. □

5.29. Let $W \xrightarrow{s} X \xrightarrow{t} Y \xrightarrow{u} Z$ be an exact sequence of linear maps. Find an isomorphism $\operatorname{coker} s \longrightarrow \ker u$. □

5.30. Let $t: G \longrightarrow H$ be an injective group map. Prove that, for any group F and group maps $s, s': F \longrightarrow G$, $ts = ts'$ if, and only if, $s = s'$.

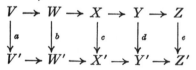

Prove, conversely, that if $t: G \to H$ is not injective, then there are a group F and distinct group maps $s, s': F \to G$ such that $ts = ts'$. □

5.31. Let $t: G \to H$ be a group map, with image H', let $i: H' \to H$ be the inclusion, and let $\pi: H \to H''$ be the partition of H to the set of left cosets $H'' = H/H'$ of H' in H. Let π' be a section of π. Prove that there exists a unique map $i': H \to H'$ such that, for all $h \in H$, $ii'(h)\pi'\pi(h) = h$ and verify that, for all $h \in H$, $h' \in H'$, $i'(h'h) = h'i'(h)$.

Suppose that α is a permutation of H'' whose only fixed point is H'. Prove that the map $\beta: h \rightsquigarrow ii'(h)\pi'\alpha\pi(h)$ is a permutation of H, and verify that, for all $h \in H$, $h_L\beta = \beta h_L$ if, and only if, $h \in H'$. (Cf. Prop. 2.4.) Show that the maps $H \to H!$; $h \rightsquigarrow h_L$ and $h \rightsquigarrow \beta^{-1}h_L\beta$ are group maps that agree for, and only for, $h \in H'$. □

5.32. Prove that a group map $t: G \to H$ is surjective if, and only if, for all groups K and group maps $u, u': H \to K$, $ut = u't$ if, and only if, $u = u'$.

(There are two cases, according as $\operatorname{im} t$ is or is not a normal subgroup of H. In the former case note that the partition $\pi: H \to H/\operatorname{im} t$ is a group map. In the latter case apply Exercise 5.31, remembering to establish the existence of α in this case.) □

5.33. Let F be a subgroup of a finite group G such that $\#G = 2(\#F)$. Prove that F is a normal subgroup of G. □

5.34. Let A be a non-null subset of a finite set X, let h be any permutation of X and let $B = h_\vdash(A)$. Also let $A!$ and $B!$ be identified with subgroups of $X!$ in the obvious way. Prove that $h(A!) = (B!)h$. Hence show that, if $\#X > 2$ and if A is a proper subset of X, $A!$ is not a normal subgroup of $X!$. □

5.35. Let A be a subset of a finite set X and suppose that $\#A = 3$ and $\#X = 4$. Find a subgroup G of $X!$, with $\#G = 12$, such that $A!$ is a subgroup of G. Hence, show that normality for subgroups is not a transitive concept. (Cf. Theorem 7.9 and Theorem 7.11.) □

CHAPTER 6

FINITE-DIMENSIONAL SPACES

A *finite-dimensional* linear space over a commutative field $\mathbf{K}$ is a linear space over $\mathbf{K}$ that is isomorphic to $\mathbf{K}^n$, for some finite number n. For example, any linear space isomorphic to $\mathbf{R}^2$ is finite-dimensional. Among the many corollaries of the main theorem of the chapter, Theorem 6.12, is Corollary 6.13 which states that, for any finite m, n, the $\mathbf{K}$-linear spaces $\mathbf{K}^m$ and $\mathbf{K}^n$ are isomorphic if, and only if, $m = n$. It is therefore meaningful to say that a $\mathbf{K}$-linear space isomorphic to $\mathbf{K}^n$ has *dimension* n or is *n-dimensional,* and, in particular, to say that a real linear space isomorphic to $\mathbf{R}^2$ is two-dimensional.

The theory of finite-dimensional linear spaces is in some ways simpler than the theory of linear spaces in general. For example, if X and Y are linear spaces each isomorphic to $\mathbf{K}^n$, then a linear map $t: X \to Y$ is injective if, and only if, it is surjective (Cor. 6.33). Therefore, to prove that t is an isomorphism it is necessary only to verify that X and Y have the same dimension and that t is injective. Any finite-dimensional linear space has the same dimension as its dual space and may be identified with the dual of its dual (Exercise 6.44), this being the origin of the term 'dual' space.

We begin with a discussion of linear dependence and independence.

Linear dependence

Let A be any subset of a $\mathbf{K}$-linear space X, there being no assumption that A is finite, nor even countable. A *coefficient system* for A is defined to be a map $\lambda : A \to \mathbf{K}$; $a \rightsquigarrow \lambda_a$ such that the set $\lambda^{-1}(\mathbf{K} \setminus \{0\})$ is finite. A point $x \in X$ is then said to *depend* (*linearly*) *on* A if, and only if, there is a coefficient system λ for A such that

$$x = \sum_{a \in A} \lambda_a a.$$

(Strictly speaking, the summation is over the set $\lambda^{-1}(\mathbf{K} \setminus \{0\})$.) For example, the origin 0 of X depends on each subset of X, including, by convention, the null set. The point $x \in X$ is said to depend *uniquely* on A is the coefficient system λ is uniquely determined by x.

The subset of X consisting of all those points of X dependent on a particular subset A will be called the *linear image* of A and be denoted by im A, the subset A being said to *span* im A (linearly). It is readily proved that im A is a linear subspace of X. If im $A = X$, that is, if A spans X, then A is said to be a *spanning subset* of X.

For example, the set $\{(1,0,0), (1,1,0), (1,1,1)\}$ spans $\mathbf{R}^3$, since, for any $(x,y,z) \in \mathbf{R}^3$,

$$(x,y,z) = (x - y)(1,0,0) + (y - z)(1,1,0) + z(1,1,1).$$

On the other hand, the set $\{(1,1,3), (1,2,0), (1,1,1)\}$ does not span $\mathbf{R}^3$, since $(0,0,1)$ does not depend on it.

A point $x \in X$ which is not linearly dependent on a subset A of X is said to be *linearly free* or *linearly independent of* A. A subset A of a linear space X is said to be *linearly free* or *linearly independent* in X, if, for each $a \in A$, a is free of $A \setminus \{a\}$.

For example, the set $\{(1,0,0), (0,1,0)\}$ is free in $\mathbf{R}^3$, since neither element is a real multiple of the other. On the other hand, the set $\{(1,0,0), (0,1,0), (1,2,0)\}$ is not free in $\mathbf{R}^3$, since $(1,2,0) = (1,0,0) + 2(0,1,0)$.

There are various details to be noted if error is to be avoided. For example, there can be points a, b in a linear space X such that a is free of $\{b\}$, yet b depends on $\{a\}$. For example, $(1,0,0)$ is free of $\{(0,0,0)\}$ in $\mathbf{R}^3$, but $(0,0,0)$ depends on $\{(1,0,0)\}$.

Another common error, involving three points a, b, c of a linear space X, is to suppose that if c is free of $\{a\}$, and if c is also free of $\{b\}$, then c is free of $\{a,b\}$. That this is false is seen by setting $a = (1,0)$, $b = (0,1)$ and $c = (1,1)$, all three points being elements of the linear space $\mathbf{R}^2$.

The null set is a free subset of every linear space.

Prop. 6.1. Let X be a linear space, let B be a subset of X, and let A be a subset of B. Then

(i) if B is free in X, A is free in X

(ii) if A spans X, B spans X. □

The following propositions are useful in determining whether or not a subset A of a linear space X is free in X.

Prop. 6.2. A subset A of a linear space X is free in X if, and only if, for each coefficient system λ for A,

$$\sum_{a \in A} \lambda_a a = 0 \Rightarrow \lambda = 0.$$

Proof $\Rightarrow$: Suppose that A is free in X and that λ is not zero, say $\lambda_b \neq 0$, for some $b \in A$. Then $\sum_{a \in A} \lambda_a a \neq 0$; for otherwise

$$b = -\lambda_b^{-1}\Big(\sum_{a \in A \setminus \{b\}} \lambda_a a\Big),$$

that is, b is dependent on $A \setminus \{b\}$, and A is not free in X. The implication follows.

$\Leftarrow$: Suppose that A is not free in X. Then there exists an element $b \in A$ and a coefficient system λ for A such that $b = \sum_{a \in A \setminus \{b\}} \lambda_a a$ and $\lambda_b = -1$, that is, such that $\sum_{a \in A} \lambda_a a = 0$, but $\lambda \neq 0$. The implication follows. □

In other words, a subset A of X is free in X if, and only if, 0 depends uniquely on A.

Prop. 6.3. A subset A of a linear space X is free in X if, and only if, each element of im A depends uniquely on A.

Proof Let A be free in X and let $\lambda, \mu : A \to \mathbf{K}$ be coefficient systems for A such that $\sum_{a \in A} \lambda_a a = \sum_{a \in A} \mu_a a$. Then $(\lambda - \mu) : A \to \mathbf{K}$ is a system of coefficients for A and $\sum_{a \in A} (\lambda - \mu)_a a = \sum_{a \in A} (\lambda_a - \mu_a) a = 0$. Therefore, by Prop. 6.2, $\lambda - \mu = 0$. That is, $\lambda = \mu$.

Conversely, as we have just remarked, if 0 depends uniquely on A, then A is free in X. □

Exercise 6.4. Suppose that a, b and c are three distinct elements of a linear space X such that $\{a,b\}$, $\{b,c\}$ and $\{c,a\}$ are free subsets of X. Is $\{a,b,c\}$ necessarily a free subset of X?

Prop. 6.5. Let X be a $\mathbf{K}$-linear space and let A be free in X. Then, for any $x \in X$ free of A, $A \cup \{x\}$ is free in X. (To prevent any possible confusion we remind the reader that $\{x\}$ denotes the set whose sole element is x.)

Proof Suppose that $A \cup \{x\}$ is not free of X. Then there exists a non-zero coefficient system $\lambda : A \cup \{x\} \to \mathbf{K}$ such that $\sum_{a \in A} \lambda_a a + \lambda_x x = 0$. Now $\lambda_x \neq 0$, for otherwise $\sum_{a \in A} \lambda_a a = 0$ with $\lambda \mid A \neq 0$, contrary to the hypothesis that A is free in X. So $x = -(\lambda_x)^{-1}(\sum_{a \in A} \lambda_a a)$, and x depends on A.

The truth of the assertion follows. □

A free subset A of a linear space X is said to be *maximal* if there is no $x \in X \setminus A$ such that $A \cup \{x\}$ is free.

A spanning subset A of X is said to be *minimal* if there is no $a \in A$ such that $A \setminus \{a\}$ spans X.

A subset A of X is said to be a *basis* for X if A is free and spans X, that is, if each $x \in X$ depends uniquely on A.

For example, the set $\{(1,0), (0,1)\}$ is a basis for $\mathbf{R}^2$ and the set $\{(1,0,0), (0,1,0), (0,0,1)\}$ is a basis for $\mathbf{R}^3$, these bases being the most obvious bases to choose for $\mathbf{R}^2$ and $\mathbf{R}^3$. Similarly, for any field $\mathbf{K}$ and any positive number n, the set $E = \{e_j : j \in n\}$ is a basis for $\mathbf{K}^n$, where for each $j \in n$, $e_j = (e_{ij} : i \in n)$, with $e_{ij} = 0$ for $i \neq j$, and $e_{ii} = 1$, for all $i \in n$. This basis is defined to be the *standard basis* for $\mathbf{K}^n$.

In the sequel we shall mostly be concerned with linear spaces with a finite basis.

Exercise 6.6. Let a, b, c and $c' \in \mathbf{R}^3$ be such that neither of the sets $\{a,b,c\}$, $\{a,b,c'\}$ is a basis for $\mathbf{R}^3$. Prove that $\{a,b,(c + c')\}$ is not a basis for $\mathbf{R}^3$. □

The following proposition shows how a basis for a linear space X may be used to construct linear maps with the linear space X as source.

Prop. 6.7. Let X and Y be linear spaces, let A be a basis for X and let $s : A \rightarrow Y$ be any map. Then there is a unique linear map t from X to Y such that $t \mid A = s$, namely the map

$$t : X \rightarrow Y; \quad \sum_{a \in A} \lambda_a a \rightsquigarrow \sum_{a \in A} \lambda_a s(a),$$

where λ denotes a coefficient system for A. □

This is very clear in the case that $X = \mathbf{K}^n$ and $Y = \mathbf{K}^m$, with A the standard basis for $\mathbf{K}^n$. For in this case any linear map $t : X \rightarrow Y$ is uniquely determined by its matrix and, for each $j \in n$, the jth column of this matrix is the column matrix representing the image in Y of the jth basis vector e_j of the standard basis A. This is worth emphasizing: *the columns of the matrix of a linear map $t : \mathbf{K}^n \rightarrow \mathbf{K}^m$ are the images in $\mathbf{K}^m$ of the vectors of the standard basis for $\mathbf{K}^n$.* If the columns are determined, then so is the matrix, and so is the map.

An application of Prop. 6.7 is to the construction of sections of a linear map.

Prop. 6.8. Let $t : X \rightarrow Y$ be a linear surjection and let there be a basis B for Y. Then there exists a linear map $t' : Y \rightarrow X$ such that $tt' = 1_Y$.

Proof Let $s : B \rightarrow X$ be a section of t over B, that is, a map such that $(t \mid B)s = 1_B$. Then define $t' : Y \rightarrow X$ to be the unique linear map such that $t' \mid B = s$. This is a section of t. For let $\sum_{b \in B} \mu_b b$ be any element of Y; then $tt'(\sum_{b \in B} \mu_b b) = \sum_{b \in B} \mu_b\, ts(b) = \sum_{b \in B} \mu_b b$. □

Prop. 6.9. Let $t : X \to Y$ be a linear map, let A be a free subset of X and B a spanning subset of X. Then, if t is injective, $t_\vdash(A)$ is free in Y and, if t is surjective, $t_\vdash(B)$ spans Y. □

Prop. 6.10. Let X and Y be linear spaces and let A be a basis for X and B be a basis for Y. Then $(A \times \{0\}) \cup (\{0\} \times B)$ is a basis for $X \times Y$. □

Prop. 6.11. The following conditions on a subset A of a linear space X are equivalent:

(a) A is a basis for X;
(b) A is a maximal free subset of X;
(c) A is a minimal spanning subset of X.

Proof (a) $\Leftrightarrow$ (b): Let A be a basis for X. Then, for any $x \in X \setminus A$, x depends on A and $A \cup \{x\}$ is not free. That is, A is a maximal free subset of X. Conversely, let A be a maximal free subset of X. Then A spans X, for otherwise, by Prop. 6.5, $A \cup \{x\}$ is free, for any x free of A. That is, A is a basis for X.

(a) $\Leftrightarrow$ (c): Let A be a basis for X and let $a \in A$. Since A is free in X, a is free of $A \setminus \{a\}$ and so $A \setminus \{a\}$ does not span X. That is, A is a minimal spanning subset of X. Conversely, let A be a minimal spanning subset of X. Then, for all $a \in A$, a is free of $A \setminus \{a\}$. That is, A is a free subset of X and so a basis for X. □

The basis theorem

The following theorem is the central theorem of the chapter.

Theorem 6.12. Let X be a linear space, let A be a free subset of X and suppose that B is a *finite* subset of X spanning X. Then A is finite and $\#A \leqslant \#B$. When $\#A = \#B$, both A and B are bases for X.

Proof For the first part it is sufficient to prove that $\#A \geqslant \#B$ implies $\#A = \#B$. Suppose, therefore, that $\#A \geqslant \#B$. The idea is to replace the elements of B one by one by elements of A, ensuring at each stage that one has a subset of $A \cup B$ spanning X. One eventually obtains a subset of A spanning X. This must be A itself, and so $\#A = \#B$.

The details of the induction are as follows: Let P_k be the proposition that there exists a subset B_k of $A \cup B$ spanning X such that

$$\text{(i) } \#B_k = \#B, \quad \text{(ii) } \#(B_k \cap A) \geqslant k.$$

P_0 is certainly true—take $B_0 = B$. It remains to prove that, for all $k \in \#B$, $P_k \Rightarrow P_{k+1}$.

Suppose therefore P_k, let a_i, $i \in k$, be k distinct elements of $B_k \cap A$ and let a_k be a further element of A. If $a_k \in B_k$, define $B_{k+1} = B_k$. If not, let $B'_k = B_k \cup \{a_k\}$. Since B_k spans X, B'_k spans X, but since it is not a minimal spanning subset it is not free in X. Therefore, if the elements of B'_k are ordered, beginning with those in A, there exists some element b_k, say, linearly dependent on those preceding it, for otherwise B'_k is free, by Prop. 6.5. Since any subset of A is free, $b_k \notin A$. Now define $B_{k+1} = B'_k \setminus \{b_k\}$. In either case, B_{k+1} spans X, $\#B_{k+1} = \#B$ and $\#(B_{k+1} \cap A) \geqslant k + 1$. That is, P_{k+1}.

So P_n, where $n = \#B$. That is, there is a subset B_n of $A \cup B$ spanning X, such that $\#B_n = n$ and $\#(B_n \cap A) \geqslant n$. It follows that $B_n \subset A$. Since A is free, no subset of A spans X other than A itself. So $B_n = A$, A is a basis for X and $\#A = \#B$.

Since, by what has just been proved, no spanning subset of X has fewer than n elements, B is a minimal spanning subset of X and so is a basis for X also.

This conclude the proof of the theorem. □

Cor. 6.13. Let X be a $\mathbf{K}$-linear space isomorphic to $\mathbf{K}^n$ for some finite number n. Then any basis for X has exactly n members.

Proof Let $t : \mathbf{K}^n \to X$ be an isomorphism, let E be the standard basis for $\mathbf{K}^n$, let $A = t_\vdash(E)$ and let B be any basis for X. Since t is injective, $\#A = n$. By Prop. 6.9, A is a basis for X and therefore spans X, while B is free in X. Therefore, by Theorem 6.12, $\#B \leqslant \#A$, implying that B also is finite. It then follows, by the same argument, that $\#A \leqslant \#B$. That is, $\#B = \#A = n$. □

In particular, $\mathbf{K}^m$ is isomorphic to $\mathbf{K}^n$ if, and only if, $m = n$.

A finite-dimensional $\mathbf{K}$-linear space X that is isomorphic to $\mathbf{K}^n$ is said to be of *dimension* n over $\mathbf{K}$, $\dim_{\mathbf{K}} X = n$, the subscript $\mathbf{K}$ being frequently omitted.

Prop. 6.14. Let X be a finite-dimensional complex linear space. Then

$$\dim_{\mathbf{R}} X = 2 \dim_{\mathbf{C}} X,$$

$\dim_{\mathbf{R}} X$ being the dimension of X regarded as a real linear space. □

A one-dimensional linear subspace of a linear space X is called a *line through* 0 or a *linear line* in X and a two-dimensional linear subspace of X is called a *plane through* 0 or a *linear plane* in X.

An affine space X is said to have finite dimension n if its vector space X_* is of finite dimension n. A one-dimensional affine space is called an *affine line* and a two-dimensional affine space an *affine plane*.

An affine or linear subspace of dimension k of an affine or linear space X of dimension n is said to have *codimension* $n - k$ in X, an affine or linear subspace of codimension 1 being called, respectively, an *affine* or *linear hyperplane* of X.

The following proposition gives the dimensions of some linear spaces constructed from linear spaces of known dimension.

Prop. 6.15. Let X be an n-dimensional linear space and let Y be a p-dimensional linear space, n and p being any finite numbers. Then

$$\dim (X \times Y) = n + p \quad \text{and} \quad \dim \mathscr{L}(X,Y) = np.$$

In particular, $\dim X^{\mathscr{L}} = \dim X$.

These results follow from Props. 6.10, 6.7, and Cor. 6.13. □

When we are dealing only with finite-dimensional linear spaces, we frequently write $L(X,Y)$ in place of $\mathscr{L}(X,Y)$ for the linear space of linear maps of the linear space X to the linear space Y, and we frequently write X^L in place of $X^{\mathscr{L}}$ for the linear dual of the linear space X. The reason for this is that in Chapter 15 a distinction has to be made between linear maps $X \to Y$ that are continuous and those that are not, in the case that X and Y are not finite-dimensional. The notation $L(X,Y)$ will then be used to denote the linear subspace of $\mathscr{L}(X,Y)$ of continuous linear maps of X to Y. It is a theorem of that chapter (Prop. 15.27) that any linear map between finite-dimensional linear spaces is continuous.

It is convenient also in the finite-dimensional case to extend the use of the GL notation and to denote by $GL(X,Y)$ the set of injective or surjective linear maps of the finite-dimensional linear space Y. In particular, $GL(X,X)$, often abbreviated to $GL(X)$, denotes the group of automorphisms of X. For any finite n, $GL(\mathbf{K}^n)$ is also denoted by $GL(n;\mathbf{K})$ and referred to as the *general linear group* of *degree n*.

Just as $L(\mathbf{K},X)$ is often identified with X, so also $GL(\mathbf{K},X)$ is often identified with $X \setminus \{0\}$, to simplify notations.

Prop. 6.16. Any free subset of a finite-dimensional linear space X is a subset of some basis for X.

Proof Let A be free in X. By Theorem 6.12, $\#A \leqslant n$, where $n = \dim X$. Let $\#A = k$. Now adjoin, successively, $n - k$ members of X to A, each being free of the union of A and the set of those already adjoined. This is possible, by Prop. 6.5, since by Theorem 6.12 a free subset of X is maximal if, and only if, it has n members. The set thus formed is therefore a maximal free subset and so a basis for X. This basis contains A as a subset. □

Another way of expressing this is to say that any free subset of X can be *extended* to a basis for X.

Cor. 6.17. Let W be a linear subspace of a finite-dimensional linear space X, with $\dim W = \dim X$. Then $W = X$. □

Cor. 6.18. Any linear subspace W of a finite-dimensional linear space X is finite-dimensional, with $\dim W \leqslant \dim X$. □

Cor. 6.19. Let $s : W \to X$ be an injective linear map, X being a finite-dimensional linear space. Then W is finite-dimensional and $\dim W \leqslant \dim X$. □

Prop. 6.20. Let $s : W \to X$ be an injective linear map, X being a finite-dimensional linear space, and let $\alpha : W \to Z$ also be linear. Then there exists a linear map $\beta : X \to Z$ such that $\alpha = \beta s$.

Proof Let A be a basis for W. Then this can be extended to a basis $s_{\vdash}(A) \cup B$ for X, with $s_{\vdash}(A) \cap B = \emptyset$. Now send each element a of A to $\alpha(a) \in Z$ and each element of B to 0 and let β be the linear extension of this map. Then β is a map of the required type. □

Note that β is not in general unique. It depends on the choice of the set B.

Cor. 6.21. Let $s : W \to X$ be an injective linear map, X being finite-dimensional. Then the dual map $s^L : X^L \to W^L$ is surjective.

Proof This is just the particular case of Prop. 6.20 obtained by taking $Z = \mathbf{K}$. □

Cor. 6.22. Let $\{0\} \to W \xrightarrow{s} X \xrightarrow{t} Y \to \{0\}$ be an exact sequence of linear maps, X being finite-dimensional. Then the sequence

$$\{0\} \to Y^L \xrightarrow{t^L} X^L \xrightarrow{s^L} W^L \to \{0\}$$

is exact.

Proof This follows from the preceding corollary, together with Prop. 5.16. □

Prop. 6.23. Let X be a finite-dimensional linear space and let B be a subset spanning X. Then some subset of B is a basis for X.

Proof Let A be a free subset of B, maximal in B. Since the null set is free in X and since by Theorem 6.12 no free subset of X contains more than n members where $\dim X = n$, such a set A can be constructed in at most n steps, one new member of B being adjoined at each step. Then $B \subset \operatorname{im} A$ and so $\operatorname{im} A = X$. That is, A is a basis for X. □

Cor. 6.24. Any quotient space Y of a finite-dimensional linear space X is finite-dimensional, and $\dim Y \leqslant \dim X$.

Proof Let $\pi : X \rightarrow Y$ be the linear partition and let A be a basis for X. Then, by Prop. 6.9, $\pi_\vdash(A)$ spans Y and $\#(\pi_\vdash(A)) \leqslant \#A = \dim X$. Hence Y is finite-dimensional and $\dim Y \leqslant \#(\pi_\vdash(A)) \leqslant \dim X$. □

Cor. 6.25. Let $t : X \rightarrow Y$ be a surjective linear map, X being a finite-dimensional linear space. Then Y is finite-dimensional and $\dim Y \leqslant \dim X$. □

Prop. 6.26. Let

$$\{0\} \rightarrow W \xrightarrow{s} X \rightarrow Y \rightarrow \{0\}$$

be an exact sequence of linear maps, W, X and Y being finite-dimensional linear spaces. Then

$$\dim X = \dim W + \dim Y.$$

Proof The linear space Y has a (finite) basis and this can be used to construct a linear section $t' : Y \rightarrow X$ of t, as in Prop. 6.8. Then, by Cor. 5.20, the map

$$(s \quad t') : W \times Y \rightarrow X; \quad (w,y) \rightsquigarrow s(w) + t'(y)$$

is a linear isomorphism. So, by Prop. 6.15,

$$\dim X = \dim W + \dim Y.$$

An alternative proof consists in taking a basis A for W, extending the free subset $s_\vdash(A)$ of X to a basis $s_\vdash(A) \cup B$ of X, where $s_\vdash(A) \cap B = \emptyset$ and then proving that $t_\vdash(B)$ is a basis for Y, with $\#(t_\vdash(B)) = \#B$. The injectivity of s implies that $\#s_\vdash(A) = \#A$, and the result then follows at once. □

Cor. 6.27. Let W be a linear subspace of a finite-dimensional linear space X. Then

$$\dim X/W = \dim X - \dim W. \quad \square$$

The *dual annihilator* $W^@$ of a linear subspace W of a linear space X is, by definition, the kernel of the map $i^{\mathscr{L}}$ dual to the inclusion $i : W \rightarrow X$. That is, $W^@ = \{\beta \in X^{\mathscr{L}} : \text{for all } w \in W, \beta(w) = 0\}$.

Prop. 6.28. Let W be a linear subspace of a finite-dimensional linear space X. Then

$$\dim W^@ = \dim X - \dim W.$$

Proof By Prop. 6.26 and Prop. 6.15,

$$\dim W^@ = \dim X^L - \dim W^L = \dim X - \dim W. \quad \square$$

The annihilator of a linear subspace has a role to play in Chapters 9 and 11.

Prop. 6.29. Let X and Y be linear subspaces of a finite-dimensional linear space. Then

$$\dim (X \cap Y) + \dim (X + Y) = \dim X + \dim Y.$$

(Apply Prop. 6.26 to the exact sequence of Prop. 5.12, or, alternatively, select a basis A for $X \cap Y$, extend it to bases $A \cup B$ and $A \cup C$ for X and Y respectively, where $A \cap B = A \cap C = \emptyset$, and then show that $B \cap C = \emptyset$ and that $A \cup B \cup C$ is a basis for $X + Y$). □

For example, the linear subspace spanned by two linear planes X and Y in $\mathbf{R}^4$ is the whole of $\mathbf{R}^4$ if, and only if, the intersection of X and Y is $\{0\}$, is three-dimensional if, and only if, the intersection of X and Y is a line, and is two-dimensional if, and only if, $X = Y$.

Exercise 6.30. Let X and Y be linear planes in a four-dimensional linear space V. Prove that there exists a linear plane W in V such that

$$V = W + X = W + Y. \quad \square$$

For affine subspaces of an affine space, the situation is more complicated. The analogue of the linear sum $X + Y$ of the linear subspaces X and Y of the linear space V is the *affine join* $\operatorname{jn}(X, Y)$ of the affine subspaces X and Y of the affine space V, this being, by definition, the smallest affine subspace of V containing both X and Y as subspaces, the intersection of the set of all the affine subspaces of V containing both X and Y as subspaces. The dimension of $\operatorname{jn}(X, Y)$ is determined precisely by the dimensions of X, Y and $X \cap Y$ only when $X \cap Y$ is non-null.

Prop. 6.31. Let X and Y be affine subspaces of a finite-dimensional affine space. Then if X and Y intersect,

$$\dim (X \cap Y) + \dim \operatorname{jn}(X, Y) = \dim X + \dim Y,$$

while if X and Y do not intersect, then

$$\sup\{\dim X, \dim Y\} \leqslant -1 + \dim \operatorname{jn}(X, Y) \leqslant \dim X + \dim Y,$$

either bound being attained for suitable X and Y. □

Rank

A linear map $t: X \longrightarrow Y$ is said to be of *finite rank* if $\operatorname{im} t$ is finite-dimensional, the number $\dim \operatorname{im} t$ being called the *rank* of t and denoted by $\operatorname{rk} t$. The map t is said to be of *finite kernel rank* if $\ker t$ is finite-dimensional, the number $\dim \ker t$ being called the ***kernel rank***, or ***nullity***, of t and denoted by $\operatorname{kr} t$.

Prop. 6.32. Let $t: X \to Y$ be a linear map, X being finite-dimensional. Then t has finite rank and kernel rank and

$$\dim X = \operatorname{rk} t + \operatorname{kr} t.$$

Proof The formula follows, by Prop. 6.26, from the exactness of the sequence of linear maps

$$\{0\} \to \ker t \to X \to \operatorname{im} t \to \{0\},$$

$\ker t$ and $\operatorname{im} t$ being finite-dimensional, by Cor. 6.18 and Cor. 6.25, respectively. □

Proposition 6.32 has the following very useful corollary.

Cor. 6.33. Let $t: X \to Y$ be a linear map, the linear spaces X and Y being finite-dimensional with $\dim X = \dim Y$. Then t is injective if, and only if, t is surjective.

Proof Suppose t is injective. Then $\operatorname{kr} t = 0$. So $\operatorname{rk} t = \dim X = \dim Y$, by hypothesis, from which it follows, by Cor. 6.17, that t is surjective. Conversely, if t is surjective, $\operatorname{rk} t = \dim Y = \dim X$. So $\operatorname{kr} t = 0$ and t is injective. □

This corollary can be reformulated as follows.

Cor. 6.34. Let $t \in L(X,Y)$ and $u \in L(Y,X)$ be such that $ut = 1_X$, X and Y being finite-dimensional linear spaces, with $\dim X = \dim Y$. Then $tu = 1_Y$. In particular, when $X = Y$, $ut = 1_X$ if, and only if, t is invertible and $t^{-1} = u$. □

In particular, when X is finite-dimensional, any linear injection $X \to X^L$ is a linear isomorphism. Such isomorphisms will be studied in detail in Chapters 9 and 10. An example, for $X = \mathbf{K}^n$, is the transposition map $\mathbf{K}^n \to (\mathbf{K}^n)^L$; $x \rightsquigarrow x^\tau$, introduced in Chapter 3.

In the following proposition the rank and kernel rank of the composite of two linear maps are related to the rank and kernel rank of the components.

Prop. 6.35. Let $t: X \to Y$ and $u: W \to X$ be linear maps, W, X and Y being finite-dimensional. Then

$$\operatorname{rk} tu + \operatorname{kr}(t \mid \operatorname{im} u) = \operatorname{rk} u,$$

$$\operatorname{rk} tu \leqslant \inf\{\operatorname{rk} t, \operatorname{rk} u\}$$

and

$$\operatorname{kr} tu \leqslant \operatorname{kr} t + \operatorname{kr} u.$$

If also $\dim W = \dim X$, then

$$\text{kr } tu \geqslant \sup\{\text{kr } t, \text{kr } u\}. \qquad \square$$

Prop. 6.36. Let $u : X \rightarrow W$ be a linear surjection and let $v : W \rightarrow Y$ be a linear injection, where W is finite-dimensional. Then

$$\text{rk}\,(vu) = \dim W.$$

Proof Since u is surjective,

$$\text{rk}\,(vu) = \dim\,(\text{im}\,(vu)) = \dim\,(\text{im}\,v)$$

and since v is injective,

$$\dim\,(\text{im}\,v) = \dim W. \qquad \square$$

Prop. 6.37. Let $t : X \rightarrow Y$ be a linear map, where X and Y are finite-dimensional. Then $\text{rk}\, t^L = \text{rk}\, t$.

Proof The map t is equal to the linear surjection $t_{\text{sur}} : X \rightarrow \text{im}\, t$ followed by the linear injection $t_{\text{inc}} : \text{im}\, t \rightarrow Y$. Hence the dual map t^L is the composite of the linear surjection $t_{\text{inc}}^L : Y^L \rightarrow (\text{im}\, t)^L$ and the linear injection $t_{\text{sur}}{}^L : (\text{im}\, t)^L \rightarrow X^L$. Therefore, by Prop. 6.36, $\text{rk}\, t^L = \dim\,(\text{im}\, t)^L$. Since, by Prop. 6.15, $\dim\,(\text{im}\, t)^L = \dim\,(\text{im}\, t) = \text{rk}\, t$, the result follows. $\square$

Matrices

Matrix notations are of great use in discussing particular examples of linear maps between finite-dimensional linear spaces X and Y. For, for each choice of isomorphisms $\alpha : \mathbf{K}^n \rightarrow X$ and $\beta : \mathbf{K}^m \rightarrow Y$, any linear map $t : X \rightarrow Y$ may be identified with the map $\beta^{-1}t\alpha : \mathbf{K}^n \rightarrow \mathbf{K}^m$, the map

$$L(X, Y) \rightarrow L(\mathbf{K}^n, \mathbf{K}^m); \quad t \rightsquigarrow \beta^{-1}t\alpha$$

being a linear isomorphism. It is however important to notice that the map $\beta^{-1}t\alpha$, and, clearly, the $m \times n$ matrix representing it, both depend on the choice of the isomorphisms α and β. Nevertheless, for different choices α, α' and β, β', the maps $\beta^{-1}t\alpha$ and $\beta'^{-1}t\alpha'$ will share many properties. They will, for example, have the same rank. These remarks prompt the following definitions.

First, let t and u be linear maps of $\mathbf{K}^n$ to $\mathbf{K}^m$ for some finite n and m. Then t and u are said to be *equivalent* if, for some $\alpha \in GL(\mathbf{K}^n)$ and for some $\beta \in GL(\mathbf{K}^m)$, $u = \beta^{-1}t\alpha$, that is, if the diagram of linear maps

$$\begin{array}{ccc} \mathbf{K}^n & \xrightarrow{t} & \mathbf{K}^m \\ \uparrow{\scriptstyle \alpha} & & \uparrow{\scriptstyle \beta} \\ \mathbf{K}^n & \xrightarrow{u} & \mathbf{K}^m \end{array}$$

is commutative.

Secondly, let t and u be linear maps of $\mathbf{K}^n$ to $\mathbf{K}^n$ for some finite n. Then t and u are said to be *similar* if, for some $\alpha \in GL(\mathbf{K}^n)$, $u = \alpha^{-1}t\alpha$, that is, if the diagram of linear maps

$$\begin{array}{ccc} \mathbf{K}^n & \longrightarrow & \mathbf{K}^n \\ \uparrow \alpha & & \uparrow \alpha \\ \mathbf{K}^n & \xrightarrow{u} & \mathbf{K}^n \end{array}$$

is commutative. Note that 'similarity' is a stronger equivalence relation on the set $L(\mathbf{K}^n,\mathbf{K}^n)$ than 'equivalence'. Two elements of $L(\mathbf{K}^n,\mathbf{K}^n)$ may well be equivalent and yet not be similar.

Finite-dimensional algebras

A $\mathbf{K}$-linear algebra A is said to be of finite dimension n if the linear space A is of finite dimension n over $\mathbf{K}$. For example, the dimension over $\mathbf{R}$ of the real matrix algebra $\mathbf{R}(2)$ is 4, each of the sets

$$\left\{\begin{pmatrix}1 & 0\\0 & 0\end{pmatrix}, \begin{pmatrix}0 & 0\\1 & 0\end{pmatrix}, \begin{pmatrix}0 & 1\\0 & 0\end{pmatrix}, \begin{pmatrix}0 & 0\\0 & 1\end{pmatrix}\right\}$$

and
$$\left\{\begin{pmatrix}1 & 0\\0 & 1\end{pmatrix}, \begin{pmatrix}1 & 0\\0 & -1\end{pmatrix}, \begin{pmatrix}0 & 1\\1 & 0\end{pmatrix}, \begin{pmatrix}0 & -1\\1 & 0\end{pmatrix}\right\}$$

being a basis.

In practice, for example in Chapter 13, one often wishes to construct an algebra map of one algebra, A, to another, B, and such a map, in so far as it must be linear, will be determined by its restriction to any basis for A, by Prop. 6.7. However, the converse is no longer true—we are not free to assign arbitrarily the values in B of a map of the basis for A to B and then to extend this to an algebra map of the whole of A to B. In general such an extension will not be possible.

There is, in fact, no easy answer here. What one normally starts with is a subset S of A that generates A either as a ring or as a $\mathbf{K}$-algebra, the subset S being said to *generate A as a ring* if each element of A is expressible, possibly in more than one way, as the sum of a finite sequence of elements of A each of which is the product of a finite sequence of elements of S, and *as an algebra* if the word 'sum' in the above definition is replaced by the words 'linear combination'. For example, the set of matrices $\left\{\begin{pmatrix}1 & 0\\0 & -1\end{pmatrix}, \begin{pmatrix}0 & 1\\1 & 0\end{pmatrix}\right\}$ generates $\mathbf{R}(2)$ as an algebra. The following is then true.

Prop. 6.38. Let A and B be algebras over a field $\mathbf{K}$ and let S be a subset of A that generates A as an algebra. Then any algebra or algebra-reversing map $t : A \to B$ is uniquely determined by its restriction $t \mid S$. □

Since a subset S of an algebra A generating A as an algebra can normally be chosen to be a proper subset of a basis, the chances of a map defined on S being extendible to an algebra map with domain A are thereby increased.

Minimal left ideals

Minimal left ideals of an associative algebra with unity were defined at the end of Chapter 3.

Theorem 6.39. Let X be a finite-dimensional $\mathbf{K}$-linear space. Then the minimal left ideals of the $\mathbf{K}$-algebra End X are the left ideals of End X of the form

$$\mathscr{I}(t) = \{at : a \in \operatorname{End} X\},$$

where $t \in \operatorname{End} X$ and $\operatorname{rk} t = 1$.

Proof Suppose first that $\mathscr{I}$ is a minimal left ideal of End X. Then, for any $t \in \mathscr{I}$, $\mathscr{I}(t)$ is a left ideal of End X and a subset of $\mathscr{I}$. Since $\mathscr{I}$ is minimal, it follows that $\mathscr{I} = \mathscr{I}(t)$ for any non-zero $t \in \mathscr{I}$.

Now, suppose $\operatorname{rk} t > 1$. Then, for any $s \in \operatorname{End} X$ with $\operatorname{rk}(st) = 1$, $\mathscr{I}(st)$ is a proper subset of $\mathscr{I}(t)$. Since there is such an s, it follows that $\mathscr{I}(t)$ is not minimal. So $\mathscr{I}(t)$ is minimal if, and only if, $\operatorname{rk} t = 1$. □

The minimal left ideals remain the same even if End X is regarded as an algebra over any subfield of the field $\mathbf{K}$.

Similar remarks may be made about minimal right ideals.

For an application see the proof of Theorem 11.32.

Further Exercises

6.40. Let $t : X \to Y$ and $u : W \to Y$ be linear maps, W being finite-dimensional, and t being surjective. Prove that there exists a linear map $s : W \to X$ such that $u = ts$. □

6.41. Let X and X' be two-dimensional linear subspaces of a four-dimensional linear space V. Prove that there exists a two-dimensional linear subspace Y of X such that $V = X + Y = X' + Y$. □

6.42. Let X and Y be finite-dimensional linear spaces, and, for each $x \in X$, let s_x denote the map

$$L(X,Y) \to Y \times L(X,Y); \quad t \rightsquigarrow (t(x),t).$$

Prove that, for each non-zero $a \in X$,

$$Y \times L(X,Y) = \operatorname{im} s_0 + \operatorname{im} s_a. \qquad \square$$

6.43. Let A and B be affine subspaces of a finite-dimensional affine space X. Is there any relationship between $\dim (A \dotdiv B)$, $\dim A$, $\dim B$ and $\dim X$? (Cf. Exercise 4.22.) $\square$

6.44. Let X be a finite-dimensional real linear space, and for each $x \in X$, let ε_x be the map $X^L \to \mathbf{R}$; $t \leadsto t(x)$. Prove that the map

$$\varepsilon_X : X \to X^{LL}; \quad x \leadsto \varepsilon_x$$

is an injective linear map, and deduce that ε_X is a linear isomorphism. (See also Exercise 3.54.) $\square$

6.45. Let X, Y and Z be finite-dimensional $\mathbf{K}$-linear spaces. Verify that the map

$$\pi : X^L \times Y \to L(X, Y); \quad (\alpha, y) \leadsto y_{\mathbf{K}}\alpha$$

is bilinear, and prove that the map

$$L(L(X,Y),Z) \to BL(X^L \times Y, Z); \quad s \leadsto s\pi$$

is a linear isomorphism, where $BL(X^L \times Y, Z)$ denotes the linear space of bilinear maps $X^L \times Y \to Z$. $\square$

6.46. Let t be a linear endomorphism of $\mathbf{K}^n$ of rank 1, n being any positive number. Find $u \in L(\mathbf{K},\mathbf{K}^n)$ and $v \in L(\mathbf{K}^n,\mathbf{K})$ such that $tuv = t$.

6.47. Find the centre of the subalgebra of $\mathbf{R}(4)$ generated by the matrices of the form

$$\begin{pmatrix} 1 & 0 & 0 & 0 \\ a & 1 & 0 & 0 \\ b & 0 & 1 & 0 \\ c & -b & a & 1 \end{pmatrix},$$

where a, b and c are real numbers. $\square$

6.48. Prove that the algebra $\mathbf{R}(n)$ has no two-sided ideals other than itself and $\{0\}$, and therefore has no quotient algebras other than itself and $\{0\}$, n being any finite number.

(Prove, for example, that the two-sided ideal generated by any non-zero element of $\mathbf{R}(n)$ contains, for each $(i,j) \in n \times n$, a matrix all of whose entries are zero other than the (i,j)th entry which is non-zero. The result follows, since these matrices span $\mathbf{R}(n)$.) $\square$

6.49. Let U, V and W be linear subspaces of a linear space X, and let $u \in U$, $v \in V$ and $w \in W$ be such that $u + v + w = 0$ and such that

$u \notin U \cap V + U \cap W$. (Cf. Exercise 3.53.) Prove that $v \notin V \cap W + V \cap U$. □

6.50. Let U, V and W be linear subspaces of a finite-dimensional linear space X. Prove that

$$\dim (U \cap (V + W))/(U \cap V + U \cap W) = \dim (V \cap (W + U))/(V \cap W + V \cap U).$$

(Choose a basis for $U \cap V \cap W$ and extend this to bases for $U \cap V$, $U \cap W$ and eventually to a basis for $U \cap (V + W)$. Then use an appropriate generalization of Exercise 6.49 to deduce an inequality one way.) □

CHAPTER 7

DETERMINANTS

This chapter is concerned with the problem of determining whether or not a given linear map $\mathbf{K}^n \longrightarrow \mathbf{K}^n$ is invertible, and related problems. Throughout the chapter the field $\mathbf{K}$ will be supposed to be commutative.

Frames

In the study of a $\mathbf{K}$-linear space X, $\mathbf{K}$ being a commutative field, it is sometimes convenient to represent a linear map $a : \mathbf{K}^k \longrightarrow X$ by its k-tuple of *columns*

$$\text{col } a = (a_j : j \in k) = (a(e_j) : j \in k),$$

k being finite. The use of the term 'column' is suggested by the case where $X = \mathbf{K}^n$, n also being finite, in which case a_j is the jth column of the $n \times k$ matrix for a. By Prop. 6.7, the map

$$\text{col} : L(\mathbf{K}^k, X) \longrightarrow X^k; \quad a \rightsquigarrow \text{col } a.$$

is a linear isomorphism.

Prop. 7.1. Let $a : \mathbf{K}^k \longrightarrow \mathbf{K}^k$ and $b : \mathbf{K}^k \longrightarrow X$ be linear maps. Then, for each $j \in n$,

$$(ba)_j = \sum_{i \in n} a_{ij} b_i.$$

Proof For each $j \in n$,

$$(ba)_j = ba(e_j) = b(\sum_{i \in n} a_{ij} e_i) = \sum_{i \in n} a_{ij} b(e_i) = \sum_{i \in n} a_{ij} b_i. \qquad \square$$

An injective linear map $a : \mathbf{K}^k \longrightarrow X$ will be called a *k-framing* on X and col a will then be called a *k-frame* on X. When a is an isomorphism both a and col a will be said to be *basic.*

A k-frame on X is just an ordered linearly independent subset of X of cardinality k, as Cor. 7.3 shows.

Prop. 7.2. A linear map $a : \mathbf{K}^k \longrightarrow X$ is a framing on X if, and only if, for each $\lambda \in \mathbf{K}^k$,

$$a(\lambda) = \sum_{j \in k} \lambda_j a_j = 0 \;\Rightarrow\; \lambda = 0.$$

This is just a particular case of Prop. 3.18. $\square$

Cor. 7.3. A k-tuple $(a_j : j \in k)$ of elements of X is a frame on X if, and only if, the set $\{a_j : j \in k\}$ has k elements, and is free in X.

Proof When $\{a_j : j \in k\}$ has k elements, which will, in particular, be the case when a is injective, any $\lambda \in \mathbf{K}^k$ may be regarded as a system of coefficients for $\{aj : j \in k\}$, and we may apply Prop. 6.2. □

The free subset $\{a_j : j \in k\}$ of X is said to be *represented* by the framing a, and any finite free subset of X may be so represented.

A basic framing on $\mathbf{K}^n$, for any finite n, is just an automorphism of $\mathbf{K}^n$, an element of the general linear group $GL(n;\mathbf{K})$.

Elementary basic framings

The standard basic framing on $\mathbf{K}^n$ is the identity $e = {}^n1 : \mathbf{K}^n \to \mathbf{K}^n$, with col $e = (e_i : i \in n)$.

For any $\lambda \in \mathbf{K}$ and any $i \in n$ let ${}^\lambda e_i : \mathbf{K}^n \to \mathbf{K}^n$ be the map defined in terms of its columns by the formula

$$({}^\lambda e_i)_k = \begin{cases} \lambda e_i, & \text{when } k = i \\ e_k, & \text{when } k \neq i. \end{cases}$$

For example, if $n = 2$, ${}^\lambda e_0$ is the map with matrix $\begin{pmatrix} \lambda & 0 \\ 0 & 1 \end{pmatrix}$.

Prop. 7.4. If $\lambda \neq 0$, ${}^\lambda e_i$ is a basic framing on $\mathbf{K}^n$.

Proof When $\lambda \neq 0$, the map ${}^\lambda e_i$ has inverse ${}^{\lambda^{-1}}e_i$. □

The map ${}^\lambda e_i$, when $\lambda \neq 0$, will be said to be an *elementary framing* on $\mathbf{K}^n$ of the *first kind*.

For any $\mu \in \mathbf{K}$ and any $i, j \in n$, with $i \neq j$, let ${}^\mu e_{ij} : \mathbf{K}^n \to \mathbf{K}^n$ be the map defined in terms of its columns by the formula

$$({}^\mu e_{ij})_k = \begin{cases} \mu e_i + e_j, & \text{when } k = j \\ e_k, & \text{when } k \neq j. \end{cases}$$

For example, if $n = 2$, ${}^\mu e_{01}$ is the map with matrix $\begin{pmatrix} 1 & \mu \\ 0 & 1 \end{pmatrix}$.

Prop. 7.5. The map ${}^\mu e_{ij}$ is a basic framing on $\mathbf{K}^n$.

Proof The map ${}^\mu e_{ij}$ has inverse ${}^{-\mu}e_{ij}$. □

The map ${}^\mu e_{ij}$ will be said to be an *elementary framing* on $\mathbf{K}^n$ of the *second kind*. Any element of $\mathbf{K}(n)$ that is the composite of a finite number of elementary framings of the second kind is said to be *unimodular*. When $n = 1$ no elementary framings of the second kind exist. In this case the identity is defined to be unimodular.

Prop. 7.6. For any $\mu \in \mathbf{K}$ and $i, j \in n$ with $i \neq j$,

$$({}^{\mu}e_i)({}^{1}e_{ij})({}^{\mu^{-1}}e_i) = {}^{\mu}e_{ij}.$$

Proof The ith and jth columns transform as follows:

$$(e_i, e_j) \rightsquigarrow (\mu e_i, e_j) \rightsquigarrow (\mu e_i, \mu e_i + e_j) \rightsquigarrow (e_i, \mu e_i + e_j). \qquad \square$$

Prop. 7.7. For any $a \in \mathbf{K}(n)$, λ, $\mu \in \mathbf{K}$ and $i, j \in n$, with $i \neq j$,

$$a({}^{\lambda}e_i)_k = \begin{cases} \lambda a_i, & \text{when } k = i \\ a_k, & \text{when } k \neq i, \end{cases}$$

$$a({}^{\mu}e_{ij})_k = \begin{cases} \mu a_i + a_j, & \text{when } k = j \\ a_k, & \text{when } k \neq j, \end{cases}$$

and

$$a({}^{1}e_{ij})({}^{-1}e_i)({}^{1}e_{ji})({}^{-1}e_i)({}^{1}e_{ij})({}^{-1}e_i) = \begin{cases} a_j, & \text{when } k = i \\ a_i, & \text{when } k = j \\ a_k, & \text{when } k \neq i \text{ or } j. \end{cases} \qquad \square$$

Theorem 7.8. Any basic framing a on $\mathbf{K}^n$ is of the form bu, when b is an elementary framing on $\mathbf{K}^n$ of the first kind, and u is unimodular.

Proof The proof is by induction, the theorem being obvious when $n = 1$.

Suppose the theorem true for $\mathbf{K}^m$, where $m \geqslant 1$, and let a be a basic framing on $\mathbf{K}^{m+1}$. Now by composing a with elementary framings of the second kind on the right, it is possible to alter the framing a step by step, some multiple of any column of a being added at each step to some other column. At each stage the set of columns is free. We claim that by a finite succession of such steps we can make $a_{mm} = 1$ and $a_{im} = a_{mi} = 0$ for all $i < m$.

This may be done as follows. First make $a_{m0} \neq 0$. Then, by adding $(1 - a_{mm})a_{m0}^{-1}$ times the 0th column to the mth column, replace the original a_{mm} by 1. Next, by adding a suitable multiple of the last column to each of the others, make the last row consist entirely of 0s, with the exception of a_{mm}, which remains as 1.

Finally, let $b_j = (a_{ij} : i \in m)$, for all $j \in m + 1$. Since $(a_j : j \in m)$ is a frame on $\mathbf{K}^{m+1}$, and since $a_{mj} = 0$ for all $j \in m$, $(b_j : j \in m)$ is a frame on $\mathbf{K}^m$. So b_{m+1} is a linear combination of the set $\{b_j : j \in m\}$ and may therefore be killed by a further series of elementary framings on $\mathbf{K}^{m+1}$ of the second kind that leave the mth row untouched.

That is, there exists a unimodular map $v : \mathbf{K}^m \longrightarrow \mathbf{K}^m$ such that av is of the form $\begin{pmatrix} a' & 0 \\ 0 & 1 \end{pmatrix}$, where a' is a basic framing of $\mathbf{K}^m$, $\mathbf{K}^{m+1}$ here being identified with $\mathbf{K}^m \times \mathbf{K}$. By the inductive hypothesis, $a' = b'u'$ where b' is an elementary framing on $\mathbf{K}^m$ of the first kind and u' is unimodular.

So

$$av = \begin{pmatrix} b' & 0 \\ 0 & 1 \end{pmatrix}\begin{pmatrix} u' & 0 \\ 0 & 1 \end{pmatrix}.$$

That is, $a = bu$, where $b = \begin{pmatrix} b' & 0 \\ 0 & 1 \end{pmatrix}$ and $u = \begin{pmatrix} u' & 0 \\ 0 & 1 \end{pmatrix} v^{-1}$. □

Permutations of *n*

Let n be any finite number, and let $\pi : n \longrightarrow n$ be a permutation of n, the group of permutations of n being denoted by $n!$, as in Chapter 2. For each $i, j \in n$ with $i \neq j$, let $\zeta_\pi(i,j) = 1$ if $\pi(i) < \pi(j)$ and -1 if $\pi(i) > \pi(j)$ and define the *sign* of π, $\operatorname{sgn} \pi$, by the formula

$$\operatorname{sgn} \pi = \prod_{i \in j \in n} \zeta_\pi(i,j),$$

π being said to be *even* if $\operatorname{sgn} \pi = 1$ and *odd* if $\operatorname{sgn} \pi = -1$.

This definition may be made more vivid by an example. Consider the permutation of 5:

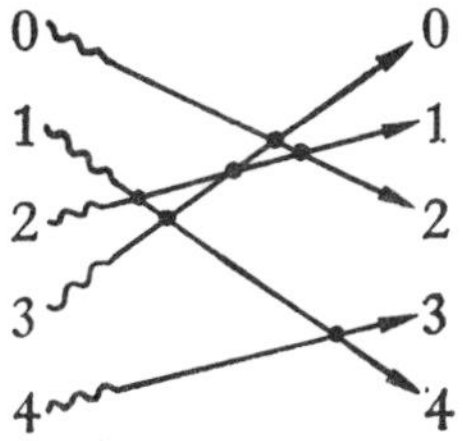

The sign of the permutation is the parity of the number of intersections of arrows in the diagram. In the example there are six intersections, so the permutation is even.

A practical method for computing the *parity* of a permutation is outlined in Exercise 7.38. The method relies on Theorem 7.9 or, rather, its generalization, Theorem 7.11.

Theorem 7.9. For any finite number n the map

$$n! \longrightarrow \{1, -1\}; \quad \pi \rightsquigarrow \operatorname{sgn} \pi$$

is a group map.

Proof Let $\pi, \pi' \in n!$. Then for any $i, j \in n$, with $i \neq j$,

$$\zeta_{\pi'\pi}(i,j) = \zeta_{\pi'}(\pi i, \pi j)\, \zeta_\pi(i,j).$$

Therefore

$$\operatorname{sgn} \pi'\pi = \operatorname{sgn} \pi' \operatorname{sgn} \pi. \qquad \square$$

The kernel of the group map in Theorem 7.9 is called the *alternating group* of degree n.

Cor. 7.10. For any $\pi \in n!$, $\operatorname{sgn} \pi^{-1} = \operatorname{sgn} \pi$. □

Theorem 7.11 extends the definition of the sign of a permutation to permutations of an arbitrary finite set.

Theorem 7.11. Let X be a set of finite cardinality n and let $\alpha : n \to X$ be a bijection. Then the map

$$X! \to \{-1,1\}; \quad \pi \rightsquigarrow \operatorname{sgn}(\alpha^{-1}\pi\alpha)$$

is a group map, and is independent of the choice of α.

Proof For any $\pi, \pi' \in X!$,

$$\begin{aligned} \operatorname{sgn}(\alpha^{-1}\pi'\pi\alpha) &= \operatorname{sgn}(\alpha^{-1}\pi'\alpha)(\alpha^{-1}\pi\alpha) \\ &= \operatorname{sgn}(\alpha^{-1}\pi'\alpha)\operatorname{sgn}(\alpha^{-1}\pi\alpha). \end{aligned}$$

So the given map is a group map.

Now suppose $\beta : n \to X$ also is a bijection, inducing a map $\pi \rightsquigarrow \operatorname{sgn}(\beta^{-1}\pi\beta)$.

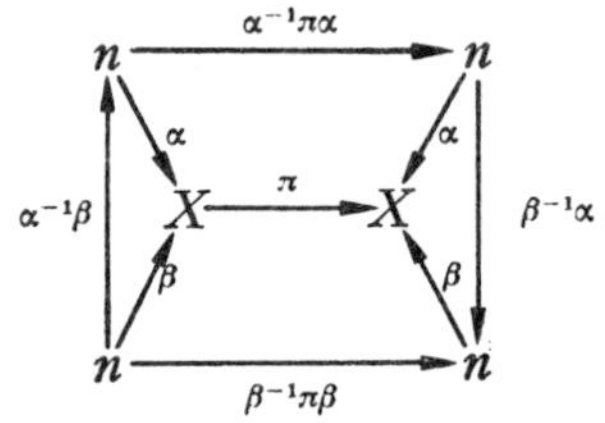

Since, for any $\pi \in X!$, $\beta^{-1}\pi\beta = (\alpha^{-1}\beta)^{-1}(\alpha^{-1}\pi\alpha)(\alpha^{-1}\beta)$, and since $\{1,-1\}$ is an abelian group and sgn a group map, it follows that $\operatorname{sgn}(\beta^{-1}\pi\beta) = \operatorname{sgn}(\alpha^{-1}\pi\alpha)$. So the maps coincide. □

The map defined in Theorem 7.11 is also denoted by sgn, $\operatorname{sgn} \pi$ $(= \operatorname{sgn}(\alpha^{-1}\pi\alpha))$ being called the *sign* of π.

The determinant

The practical problem of *determining* whether or not a given linear map $a : \mathbf{K}^n \to \mathbf{K}^n$ is invertible, or whether or not the corresponding n-tuple col a of $\mathbf{K}^n$ is a basic frame for $\mathbf{K}^n$, is solved by the following theorem in which $\mathbf{K}(n)$ denotes, as before, the algebra End $\mathbf{K}^n$ $= L(\mathbf{K}^n, \mathbf{K}^n)$, with unity ${}^n1$, whose elements may be represented, if one so wishes it, by $n \times n$ matrices over $\mathbf{K}$.

Theorem 7.12. For any finite number n there is a unique map

$$\det : \mathbf{K}(n) \to \mathbf{K}; \quad a \rightsquigarrow \det a,$$

such that

(i) for any $\lambda \in \mathbf{K}$ and any $i \in n$, $\det a({}^{\lambda}e_i) = \lambda \det a$,
(ii) for any distinct $i, j \in n$, $\det a({}^{1}e_{ij}) = \det a$,
(iii) $\det 1_{(n)} = 1$.

The map is defined, for all $a \in \mathbf{K}(n)$, by the formula

$$\det a = \sum_{\pi \in n!} \operatorname{sgn} \pi \prod_{j \in n} a_{\pi(j), j},$$

and has the following further properties:

(iv) for any $a, b \in \mathbf{K}(n)$, $\det ba = \det b \det a$,
(v) for any invertible $a \in \mathbf{K}(n)$, $\det a^{-1} = (\det a)^{-1}$,
(vi) for any $a \in \mathbf{K}(n)$, a is invertible if, and only if, $\det a$ is invertible, that is, if, and only if, $\det a \neq 0$.

The map det is called the *determinant* on $\mathbf{K}(n)$.

Plan of the proof The proof occupies pages 121–124. From (i), (ii) and (iii) it is easy to deduce several further properties which det must possess and so to construct the formula stated in the theorem. This establishes the uniqueness of det. To prove existence it only remains to verify the three conditions for the unique candidate. The various additional properties listed are proved by the way.

The proof is presented as a series of lemmas. Throughout these lemmas it is assumed that det is a map from $\mathbf{K}(n)$ to $\mathbf{K}$ satisfying conditions (i) and (ii). Condition (iii) is first introduced in the crucial Cor. 7.20.

Some of the proofs may appear formidable on a first reading, because of the proliferation of indices and summation signs. The way to master any of them is to work through, in detail, the special case when $n = 3$. For example, Lemma 7.16 reduces, in that case, to Exercise 6.6.

Lemma 7.13. Let a and b be elements of $\mathbf{K}(n)$ differing only in that, for some particular $j \in n$, $b_j = \mu a_i + a_j$, where $i \neq j$ and $\mu \in \mathbf{K}$. Then $\det b = \det a$.

Proof If $\mu = 0$, there is nothing to be proved. If $\mu \neq 0$, apply Prop. 7.6 and axioms (i) and (ii). □

Lemma 7.14. Let $a \in \mathbf{K}(n)$ be such that, for some particular $i, j \in n$, $a_j = a_i$, with $j \neq i$. Then $\det a = 0$.

Proof Set $\mu = -1$ in Lemma 7.13 and apply (ii) with $\lambda = 0$. □

Lemma 7.15. An element a of $\mathbf{K}(n)$ is invertible if $\det a \neq 0$. (That is, if a is not invertible, then $\det a = 0$.)

Proof Suppose that a is not invertible. Then col a is not a frame on $\mathbf{K}^n$. So, for some non-zero $\lambda \in \mathbf{K}^n$, $\sum_{k \in n} \lambda_k a_k = 0$. Suppose $\lambda_j \neq 0$. Then, since

$$\left(\sum_{i \neq j} \lambda_i a_i\right) + \lambda_j a_j = 0,$$

it follows, by (ii) and by Lemma 7.13, that $\lambda_j(\det a) = 0$. Since $\lambda_j \neq 0$, $\det a = 0$. □

Lemma 7.16. Let $a, b, c \in \mathbf{K}(n)$ differ only in that, for one particular $j \in n$, their jth columns are not necessarily equal but are instead related by the equation $c_j = a_j + b_j$. Then, if col a and col b are not basic frames for $\mathbf{K}^n$, neither is col c.

Proof Let C be the set of columns of c and let $D = C \setminus \{c_j\}$. Then either a_j and $b_j \in \operatorname{im} D$, in which case $c_j \in \operatorname{im} D$, or $\dim(\operatorname{im} D) < n - 1$. In either case it follows that $\operatorname{rk} c = \dim(\operatorname{im} C) < n$ and that c is not a basic framing for $\mathbf{K}^n$. □

Lemma 7.17. With a, b and c as in Lemma 7.16, $\det c = \det a + \det b$.

Proof If a and b are not basic framings for $\mathbf{K}^n$, then neither is c, by Lemma 7.16, and so $\det c$ and $\det a + \det b$ are each zero, by Lemma 7.15. Suppose, on the other hand, that a is a basic framing for $\mathbf{K}^n$. Then, for some $\lambda \in \mathbf{K}^n$, $c_j = \sum_{k \neq j} \lambda_k a_k$, from which it at once follows that both $\det c$ and $\det a + \det b$ are equal to $(1 + \lambda_j)\det a$, and therefore to each other. □

Lemma 7.18. Let a and b be elements of $\mathbf{K}(n)$ differing only in that, for two distinct $i, j \in n$, $b_i = a_j$ and $b_j = a_i$. Then $\det b = -\det a$.

Proof Apply Prop. 7.7 and (i) and (ii). □

Lemma 7.17, with (i), implies that, if det exists, then $\det \operatorname{col}^{-1}$ is n-linear, while Lemma 7.18 implies that $\det \operatorname{col}^{-1}$ is *alternating*, that is, transposing any two components of its source changes its sign.

We are now in a position to establish the formula for det, and hence its uniqueness.

Lemma 7.19. For any $a, b \in \mathbf{K}(n)$

$$\det ba = (\det b)\left(\sum_{\pi \in n!} \operatorname{sgn} \pi \prod_{j \in n} a_{\pi j, j}\right).$$

Proof Let $a, b \in \mathbf{K}(n)$. Then, by Prop. 7.1, for any $j \in n$,

$$(ba)_j = \sum_{i \in n} a_{ij} b_i,$$

or, writing πj for i,

$$(ba)_j = \sum_{\pi j \in n} a_{\pi j, j} b_{\pi j}.$$

Since det col^{-1} is n-linear, it follows that

$$\begin{aligned}\det (ba) &= \det \mathrm{col}^{-1}(\mathrm{col}\, ba)\\ &= \sum_{\pi \in n^n} (\prod_{j \in n} a_{\pi j, j}) \det \mathrm{col}^{-1}(b_{\pi j} : j \in n),\end{aligned}$$

where n^n denotes the set of maps $n \longrightarrow n$.

For example, when $n = 2$,

$$(ba)_0 = a_{00}b_0 + a_{10}b_1 \quad \text{and} \quad (ba)_1 = a_{01}b_0 + a_{11}b_1,$$

and

$$\begin{aligned}&\det \mathrm{col}^{-1}(a_{00}b_0 + a_{10}b_1\,, a_{01}b_0 + a_{11}b_1)\\ &\quad = a_{00}a_{01} \det \mathrm{col}^{-1}(b_0,b_0) + a_{00}a_{11} \det \mathrm{col}^{-1}(b_0,b_1)\\ &\qquad + a_{10}a_{01} \det \mathrm{col}^{-1}(b_1,b_0) + a_{10}a_{11} \det \mathrm{col}^{-1}(b_1,b_1).\end{aligned}$$

If π is not a permutation of n, then $\pi i = \pi j$ for some $i \neq j$ and det $\mathrm{col}^{-1}(b_{\pi j} : j \in n) = 0$, by Lemma 7.14. If π is a permutation of n, then, by Lemma 7.18, and by Theorem 7.9,

$$\begin{aligned}\det \mathrm{col}^{-1}(b_{\pi j} : j \in n) &= \operatorname{sgn} \pi \det \mathrm{col}^{-1}(b_j : j \in n)\\ &= \operatorname{sgn} \pi \det b.\end{aligned}$$

In conclusion, therefore,

$$\det (ba) = (\det b)(\sum_{\pi \in n!} \operatorname{sgn} \pi \prod_{j \in n} a_{\pi j, j}).$$

For example, when $n = 2$,

$$\det (ba) = (\det b)(a_{00}a_{11} - a_{10}a_{01}). \quad \square$$

The above argument should also be written out in detail for the case $n = 3$.

Cor. 7.20. The map det, if it exists, is unique, with

$$\det a = \sum_{\pi \in n!} \operatorname{sgn} \pi \prod_{j \in n} a_{\pi j, j},$$

for any $a \in \mathbf{K}(n)$.

Proof Set $b = {}^n1$ in Lemma 7.19, and use (iii). □

Cor. 7.21. For any $a, b \in \mathbf{K}(n)$,

$$\det ba = \det b \det a.$$

Proof Combine Lemma 7.19 with Cor. 7.20. □

Cor. 7.22. For any invertible $a \in \mathbf{K}(n)$,

$$\det a^{-1} = (\det a)^{-1}.$$

Proof Set $b = a^{-1}$ in Cor. 7.21. □

Cor. 7.23. For any invertible $a \in \mathbf{K}(n)$, $\det a \neq 0$. (This is the converse of Lemma 7.15.)

Proof Apply Cor. 7.22. □

Cor. 7.24. An element a of $\mathbf{K}(n)$ has determinant 1 if, and only if, it is unimodular.

Proof Apply Theorem 7.8, (i), (ii) and Cor. 7.21. □

The group of unimodular maps in $\mathbf{K}(n)$ is called the ***unimodular group*** of degree n and denoted by $SL(n;\mathbf{K})$.

To prove the existence of det, one has only to verify that the map defined by the formula satisfies axioms (i), (ii) and (iii). To prove (i) and (iii) is easy.

Proof of (ii) What has to be proved is that $\det b - \det a = 0$, where $a \in \mathbf{K}(n)$ and $b = a({}^1e_{ij})$, i and j being distinct elements of n.

Let $c \in \mathbf{K}(n)$ be formed from a by replacing a_j by a_i. Both b and c then differ from a in one column only, the jth, with $c_j = b_j - a_j$, and two columns of c, namely c_i and c_j, are equal. Then

$$\det b - \det a = \sum_{\pi \in n!} \operatorname{sgn} \pi \prod_{k \in n} b_{\pi k,k} - \sum_{\pi \in n!} \operatorname{sgn} \pi \prod_{k \in n} a_{\pi k,k}$$

$$= \sum_{\pi \in n!} \operatorname{sgn} \pi \prod_{k \in n} c_{\pi k,k}$$

$$= \sum_{\text{even } \pi} \prod_{k \in n} c_{\pi k,k} - \sum_{\text{odd } \pi} \prod_{k \in n} c_{\pi k,k}.$$

Now

$$\prod_{k \in n} c_{\pi k,k} = \prod_{k \in n} c_{\pi\nu k,\nu k} = \prod_{k \in n} c_{\pi\nu k,k},$$

where $\nu : n \to n$ is the permutation of n interchanging i and j and leaving every other number fixed. Since ν is odd, $\pi\nu$ is even when π is odd. So

$$\sum_{\text{odd } \pi} \prod_{k \in n} c_{\pi k,k} = \sum_{\text{even } \pi} \prod_{k \in n} c_{\pi k,k}.$$

Therefore $\det b - \det a = 0$.

This also should be followed through in detail for $n = 2$ and $n = 3$.

This completes the proof of Theorem 7.12. □

Transposition

Prop. 7.25. Let $a \in \mathbf{K}(n)$ and let b be the transpose of a. Then $\det b = \det a$.

Proof Since $\operatorname{sgn} \pi^{-1} = \operatorname{sgn} \pi$, for each $\pi \in n!$, by Cor. 7.10,

$$\det b = \sum_{\pi \in n!} \operatorname{sgn} \pi \prod_{j \in n} b_{\pi j,j}$$

$$= \sum_{\pi \in n!} \operatorname{sgn} \pi \prod_{j \in n} a_{j,\pi j}$$
$$= \sum_{\pi^{-1} \in n!} \operatorname{sgn} \pi^{-1} \prod_{j \in n} a_{\pi^{-1} j, j} = \det a. \quad \square$$

Determinants of endomorphisms

Any basic framing $a : \mathbf{K}^n \rightarrow X$ on an n-dimensional linear space X induces a map

$$\text{End } X = L(X,X) \rightarrow \mathbf{K}; \quad t \rightsquigarrow \det(a^{-1}ta)$$

called the *determinant* on End X and also denoted by det. This map is independent of the choice of basic framing on X, as the following proposition shows.

Prop. 7.26. Let a and $b : \mathbf{K}^n \rightarrow X$ be basic framings on the n-dimensional linear space X and let $t \in \text{End } X$. Then

$$\det(b^{-1}tb) = \det(a^{-1}ta).$$

Proof

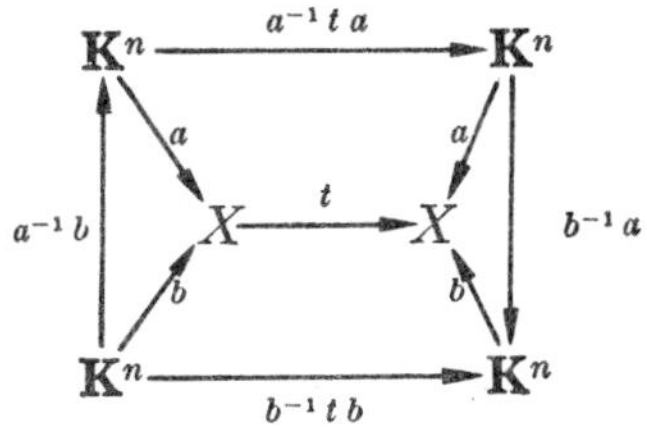

Since
$$b^{-1}tb = (b^{-1}a)(a^{-1}ta)(b^{-1}a)^{-1},$$
$$\det(b^{-1}tb) = \det(b^{-1}a)\det(a^{-1}ta)(\det(b^{-1}a))^{-1}$$
$$= \det(a^{-1}ta). \quad \square$$

Prop. 7.27. Let $t \in \text{End } X$, where $X \cong \mathbf{K}^n$. Then $\det t \neq 0$ if, and only if, t is invertible, and the map

$$\text{Aut } X \rightarrow \text{Aut } \mathbf{K} \ (= \mathbf{K}^*); \quad t \rightsquigarrow \det t$$

is a group map. $\square$

The absolute determinant

In some applications it is the absolute determinant that is important, and not the determinant. For simplicity we suppose that $\mathbf{K} = \mathbf{R}$ or $\mathbf{C}$.

Theorem 7.28. Let n be any finite number. Then there exists a unique map

$$\Delta : \mathbf{K}(n) \rightarrow \mathbf{R}; \quad a \rightsquigarrow \Delta(a)$$

such that

(i) for each $a \in \mathbf{K}(n)$, each $i \in n$ and each $\lambda \in \mathbf{K}$,
$\Delta(a({}^{\lambda}e_i)) = \Delta(a) \mid \lambda \mid$

(ii) for each $a \in \mathbf{K}(n)$, and each distinct $i, j \in n$,
$\Delta(a({}^{1}e_{ij})) = \Delta(a)$

(iii) $\Delta({}^{n}1) = 1$.

Proof The existence of such a map is a corollary of Theorem 7.12, for the map

$$\mathbf{K}(n) \longrightarrow \mathbf{R}; \quad t \rightsquigarrow \mid \det t \mid$$

satisfies (i), (ii) and (iii).

The uniqueness argument we gave before for det depended on first showing that det, if it existed, was alternating multilinear, from which a formula could be deduced. However, Prop. 7.17 fails when Δ replaces det, since, in general, for $\lambda \in \mathbf{K}$, $1 + \mid \lambda \mid \neq \mid 1 + \lambda \mid$.

There is, fortunately, an alternative argument, which is also valid applied to det, but which we found it convenient to suppress earlier on! It is based on Theorem 7.8, and the reader is invited to find it for himself. □

The map Δ will be called the *absolute determinant* on $\mathbf{K}(n)$.

Applications

The two basic operations $\mathbf{K}(n) \longrightarrow \mathbf{K}(n)$; $a \rightsquigarrow a({}^{\lambda}e_i)$ and $a \rightsquigarrow a({}^{1}e_{ij})$ which we have used to characterize the determinant and the absolute determinant on $\mathbf{K}(n)$, occur, thinly disguised, in many situations.

For example, the set of solutions of a set of m linear equations over $\mathbf{K}$

$$\sum_{j \in n} b_{ij} x_j = y_i, \quad i \in m,$$

where, for all $i \in m$ and all $(i,j) \in m \times n$, y_i and $b_{ij} \in \mathbf{K}$, is unaltered if one of the equations is multiplied by a non-zero element of $\mathbf{K}$, or if one of the equations is added to another.

Again, to take a particular case of a more general situation which we shall shortly discuss in detail, if W is a two-dimensional *real* linear subspace of a *real* three-dimensional linear space X and if c is a basic framing for X such that $W = \operatorname{im}\{c_0, c_1\}$, then, for any $w \in W$ and any $\nu > 0$, $c_2 + w$ and νc_2 both lie on the same side of W in X as c_2, while, for any $\nu < 0$, νc_2 lies on the opposite side.

Finally, to take again a particular case, and without being precise about the definition of area, for to be precise would lead us too far afield, let $A(a)$ denote the area of the convex parallelogram with vertices 0,

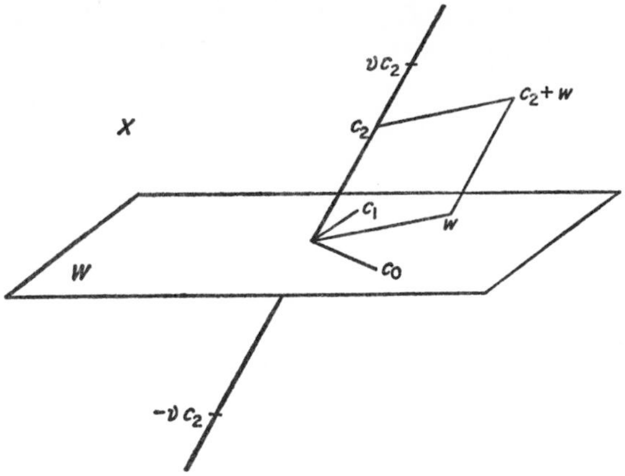

a_0, a_1 and $a_0 + a_1$, where a is any basic framing of $\mathbf{R}^2$. Then, for any non-zero real λ,

$$A(a({}^{\lambda}e_i)) = |\lambda| A(a), \quad \text{for } i \in 2,$$

while
$$A(a({}^{1}e_{01})) = A(a({}^{1}e_{10})) = A(a)$$

and
$$A({}^{2}1) = 1.$$

The following diagram illustrates the assertion that $A(a({}^{1}e_{01})) = A(a)$.

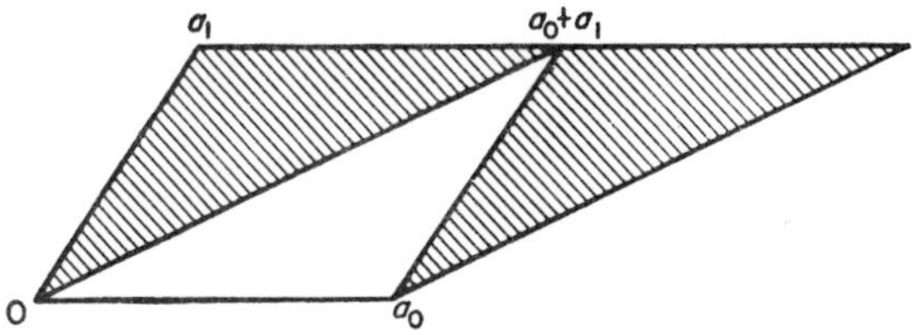

These examples indicate that the determinant may be expected to play an important role in the solution of sets of linear equations, in the classification of basic framings on a finite-dimensional real linear space and in the theory of area or measure on finite-dimensional real linear spaces and therefore in the theory of integration.

As we have just hinted, the third of these applications is outside our scope, while the first is an obvious application of Theorem 7.12(vi). The second of the applications requires further study here.

The sides of a hyperplane

A linear hyperplane in a finite-dimensional *real* linear space X has two sides. This is clear when $\dim X = 1$. In this case the only linear hyperplane of X is $\{0\}$ and the *sides* of $\{0\}$ in X are the two parts into which $X \setminus \{0\}$ is divided by 0. Two points a and b of X lie *on the same side* of $\{0\}$ if, and only if, $0 \notin [a,b]$.

Now let dim $X > 1$ and suppose that W is a linear hyperplane of X. Then points a and b of $X \setminus W$ are said to lie *on the same side* of W if, and only if, $[a,b] \cap W = \emptyset$, or equivalently, if, and only if, $\pi(a)$ and $\pi(b)$ lie on the same side of $\pi(0)$ in X/W, $\pi : X \to X/W$ being the linear partition of X with kernel W. Otherwise they are said to lie *on opposite sides* of W.

Prop. 7.29. Let W be a linear hyperplane of a finite-dimensional real linear space X and let $a \in X \setminus W$. Then, for all positive real λ and all $w \in W$, both λa and $a + w$ lie on the same side of W as a, while $-\lambda a$ lies on the opposite side of W. □

The two sides of an affine hyperplane in a finite-dimensional real affine space are defined in the obvious way.

Exercise 7.30. Is there any sense in which a two-dimensional linear subspace W in a four-dimensional real linear subspace X can be said to be two-sided? □

Orientation

Let $X \cong \mathbf{R}^n$, for any finite n. Then the following proposition shows that the set of basic framings of X divides naturally into two disjoint subsets.

Prop. 7.31. There is a unique map

$$\zeta : \mathbf{R}(n) \to \{-1,0,1\},$$

namely the map defined by the formula

$$\zeta(a) = \begin{cases} -1 \text{ when } \det a < 0 \\ 0 \text{ when } \det a = 0 \\ 1 \text{ when } \det a > 0, \end{cases}$$

such that

(i) $\zeta(a) \neq \emptyset \Leftrightarrow a$ is a basic framing on $\mathbf{R}^n$

(ii) $\zeta({}^n1) = 1$

(iii) if a and b are basic framings of $\mathbf{R}^n$ such that each column of a is equal to the corresponding column of b, with the exception of one, say the jth, then $\zeta(b) = \zeta(a)$ if, and only if, a_j and b_j lie on the same side of the linear hyperplane $\operatorname{im}\{a_k : k \in n \setminus \{j\}\}$ in $\mathbf{R}^n$.

Moreover, the map

$$GL(n;\mathbf{R}) \to \{1,-1\}; \quad a \rightsquigarrow \zeta(a)$$

is a group map.

Proof That the map defined by the formula has properties (i), (ii) and (iii) is clear from Theorem 7.12, and the final statement is also true for this map.

The uniqueness of ζ is a corollary of Theorem 7.8 and Prop. 7.6, which together imply that any basic framing of $\mathbf{R}^n$ is the composite of a finite number of elementary framings of $\mathbf{R}^n$ either of the form ${}^{\lambda}e_j$, where $\lambda \neq 0$ and $j \in n$, or of the form ${}^{1}e_{ij}$, where $i, j \in n$, with $i \neq j$. Now, for any basic framing a of $\mathbf{R}^n$, if $\lambda > 0$, then both λa_j and $a_i + a_j$ lie on the same side of $\operatorname{im}\{a_k : k \in n \setminus \{j\}\}$ as a_j, while $-\lambda a_j$ lies on the opposite side, and therefore, by (iii),

$$\zeta(a({}^{\lambda}e_j)) = \zeta(a({}^{1}e_{ij})) = \zeta(a) \quad \text{and} \quad \zeta(a({}^{-\lambda}e_j)) = -\zeta(a).$$

It follows from this that ζ is uniquely determined. □

The sets $\zeta^{\dashv}\{1\}$ and $\zeta^{\dashv}\{-1\}$ are called, respectively, the *positive* and *negative* orientations for $\mathbf{R}^n$, two basic framings a and b on $\mathbf{R}^n$ being said to be *like-oriented* if $\zeta(a) = \zeta(b)$, that is, if $\det(b^{-1}a) > 0$, and *oppositely oriented* if $\zeta(a) = -\zeta(b)$, that is, if $\det(b^{-1}a) < 0$.

The same holds for an arbitrary n-dimensional real linear space X. Two basic framings a and b on X are said to be *like-oriented* if $\zeta(b^{-1}a) = 1$, that is, if $\det(b^{-1}a) > 0$, and *oppositely oriented* if $\zeta(b^{-1}a) = -1$, that is, if $\det(b^{-1}a) < 0$, and the two classes of basic framings on X so induced are called the *orientations* of X. Only this time, unlike the case where $X = \mathbf{R}^n$, there is no natural preference for either against the other. An automorphism $t: X \to X$ of the linear space X is said to *preserve orientations* if for one, and therefore for every, basic framing a on X the basic framings a and ta are like-oriented.

To round off this string of definitions, a finite-dimensional real linear space with a chosen orientation is said to be an *oriented* linear space, while, if X and Y are oriented linear spaces of the same dimension, a linear isomorphism $t: X \to Y$ is said to *preserve orientations* if, for one, and therefore for every, basic framing a of the chosen orientation for X the framing ta belongs to the chosen orientation for Y.

The orientations for a line are often referred to as the *right* and the *left* orientations for the line, the orientations for a plane are said to be *positive* and *negative*, while the orientations for three-dimensional space are said to be *right-handed* and *left-handed*. In every case one has in mind a basis for the space, with the elements of the basis taken in a particular order.

Prop. 7.32. Let X be a finite-dimensional complex linear space and let $X_{\mathbf{R}}$ be the underlying real linear space. Then if $t: X \to X$ is a complex linear map,

$$\det_{\mathbf{R}} t = |\det_{\mathbf{C}} t|^2,$$

where $\det_{\mathbf{C}} t$ is the determinant of t regarded as a complex linear map, and $\det_{\mathbf{R}} t$ the determinant of t regarded as a real linear map.

Proof By Cor. 7.21, Theorem 7.8 and Prop. 2.66 it is enough first to assume that $X = \mathbf{C}^n$ and then to check the formula for the elementary complex framings on $\mathbf{C}^n$. This is easily done, by Prop. 3.40. □

Cor. 7.33. Let X be as in Prop. 7.32, and let $t : X \longrightarrow X$ be a complex linear map. Then $\det_{\mathbf{R}} t \geqslant 0$. □

Cor. 7.34. Let X and $X_{\mathbf{R}}$ be as in Prop. 7.32, and let $t : X \longrightarrow X$ be a complex linear isomorphism. Then t preserves the orientations of $X_{\mathbf{R}}$. □

Further Exercises

7.35. Let $n \geqslant 3$, let $i, j, k \in n$, no two of i, j, k being equal, and let $\lambda, \mu \in \mathbf{K}$. Prove that

$$^{\lambda}e_{ij}\,^{\mu}e_{jk}\,^{-\lambda}e_{ij}\,^{-\mu}e_{jk} = {}^{\lambda\mu}e_{ik}.$$

Hence, by setting $\lambda = 1$, prove that, if $\Delta : \mathbf{K}(n) \longrightarrow \mathbf{K}$ is a non-zero map such that, for all $a, b \in \mathbf{K}(n)$, $\Delta(ab) = \Delta(a)\,\Delta(b)$, n being not less than 3, then, for any unimodular $a \in \mathbf{K}(n)$, $\Delta(a) = 1$. □

7.36. For any $m, n \in \omega$ and any $a \in \mathbf{K}(m)$, $b \in \mathbf{K}(n)$ let $a \times b$ denote the element of $\mathbf{K}(m + n)$ with matrix $\begin{pmatrix} a & 0 \\ 0 & b \end{pmatrix}$, $\mathbf{K}^{m+n}$ having been identified with $\mathbf{K}^m \times \mathbf{K}^n$, to simplify notations. Let

$$\Delta : \bigcup_{n \in \omega} \mathbf{K}(n) \longrightarrow \mathbf{K}$$

be a map such that

(i) for any $n \in \omega$ and any $a, b \in \mathbf{K}(n)$, $\Delta(ab) = \Delta(a)\,\Delta(b)$

(ii) for any $m, n \in \omega$ and any $a \in \mathbf{K}(m)$, $b \in \mathbf{K}(n)$,

$$\Delta(a \times b) = \Delta(a)\,\Delta(b)$$

(iii) for any $\lambda \in \mathbf{K}(1) = \mathbf{K}$, $\Delta(\lambda) = \lambda$.

Prove that Δ is unique. □

7.37. Let X be a finite set and let $n = \#X$. A permutation π of X is said to be *cyclic* if there is a bijection $s : n \longrightarrow X$ such that, for all $k \in n - 1$, $\pi(s(k)) = s(k + 1)$ and $\pi(s(n - 1)) = s(0)$. Prove that a cyclic permutation of X is even or odd according as n is odd or even. □

7.38. A permutation π of a finite set X is said to be a *cycle* if $(\pi \mid W)_{\text{sur}} : W \longrightarrow W$ is cyclic, W being the complement in X of the elements of X left fixed by π. We might call W the *wheel* of the cycle.

Show that any permutation of X may be expressed as the composite of a set of cycles whose wheels are mutually disjoint.

(The construction of such a decomposition is, in practice, the most efficient way of computing the parity of a permutation. Given the decomposition, one applies Exercise 7.37 and Theorem 7.11.) □

7.39. (Pivotal condensation.) Prove that, for any finite n and any $a \in \mathbf{K}(n+1)$ with a_{nn} (the pivot) non-zero,

$$(a_{nn})^{n-2} \det a = \det b,$$

where $b \in \mathbf{K}(n)$ is defined, for all $i, j \in n$, by

$$b_{ij} = a_{ij}a_{nn} - a_{in}a_{nj}.$$

Formulate an analogue of this, with the (i,j)th entry as pivot, for any $i, j \in n+1$. □

7.40. Write down a 4×4 matrix and compute its determinant by pivotal condensation. □

7.41. Let $\{0\} \rightarrow W \xrightarrow{s} X \xrightarrow{t} Y \rightarrow \{0\}$ be an exact sequence of linear maps, W, X and Y being finite-dimensional, and let $\alpha \in \operatorname{End} W$, $\beta \in \operatorname{End} X$ and $\gamma \in \operatorname{End} Y$ be such that $\beta s = s\alpha$ and $\gamma t = t\beta$. Prove that $\det \beta = \det \alpha \det \gamma$. □

CHAPTER 8

DIRECT SUM

In this chapter the field **K** remains commutative and may be taken, for simplicity, to be either **R** or **C**. The algebra $^s\mathbf{K}$, for any $s \in \omega$, is the product of s copies of **K** according to the definition at the end of Chapter 3. The algebra $^2\mathbf{K}$ will be called a *double field*.

The main part of the chapter is concerned with the elementary properties of direct sum decompositions of linear spaces and of the set of linear complements of a linear subspace of a linear space. This leads on in the later part of the chapter to the definition of Grassmannians and, in particular, projective spaces. These will be the subject of further study in later chapters, for example in Chapters 12, 17 and 20.

Direct sum

A linear space V is the sum $X + Y$ of two of its linear subspaces X and Y if the linear map

$$\alpha : X \times Y \rightarrow V; \quad (x,y) \rightsquigarrow x + y$$

is surjective. If α is also injective, that is, if α is a linear isomorphism, V is said to be the *direct sum* $X \oplus Y$ of its subspaces X and Y. That is, V is the direct sum of X and Y if, and only if, each element $v \in V$ is uniquely expressible in the form $x + y$, where $x \in X$, $y \in Y$.

A pair (X, Y) of linear subspaces X and Y of V such that $V = X \oplus Y$ is said to be a *direct sum decomposition* of V. Abuses of language, as in 'Let $X \oplus Y$ be a direct sum decomposition of the linear space V . . .' are common and should not lead to confusion.

If $V = X \oplus Y$ is finite-dimensional, then $\dim V = \dim X + \dim Y$, by Prop. 6.15, since $\alpha : X \times Y \rightarrow X \oplus Y$ is an isomorphism.

Direct sum decompositions of a linear space V, with a number of components greater than 2, are defined analogously. For direct sum decompositions with only two components one has:

Prop. 8.1. Let X and Y be linear subspaces of a linear space V. Then $V = X \oplus Y$ if, and only if, (i) $V = X + Y$ and (ii) $X \cap Y = \{0\}$.

Proof By Prop. 3.18 the map α is injective if, and only if, $X \cap Y = \{0\}$. □

When $V = X \oplus Y$ there is a natural tendency to regard the isomorphism $\alpha : X \times Y \rightarrow X \oplus Y$ as an identification. In the reverse direction, also, there is a tendency, with a linear product $X \times Y$, to identify X with $X \times \{0\}$ and Y with $\{0\} \times Y$, and to write $X \oplus Y$ in place of $X \times Y$, ignoring the distinction between them. Strictly speaking, of course,

$$X \times Y = (X \times \{0\}) \oplus (\{0\} \times Y).$$

Most of us have been conditioned to make these identifications ever since we were first introduced to 'graphs' at school. There is a benefit from both sides, for $X \oplus Y$ is easier to picture than $X \times Y$, while notationally (x,y) is less confusing than $x + y$ (or $x \oplus y$). So we draw the diagram

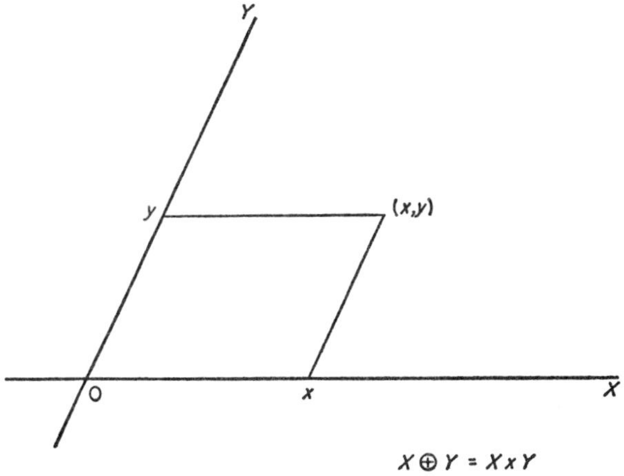

and shift from the one aspect to the other as and when it suits us. For example if $t \in L(X, Y)$ it is often convenient to think of graph t as a subset of $X \oplus Y$ rather than as a subset of $X \times Y$.

This ambivalence is, however, only possible when the direct sum decomposition is fixed throughout the argument. Later in this chapter we shall be involved in a comparison of different direct sum decompositions of the same linear space. In such a context we have sometimes to forgo cartesian product habits.

^{2}K-modules and maps

Modules over a commutative ring with unity were defined on page 71. A direct sum decomposition $X_0 \oplus X_1$ of a **K**-linear space X may be regarded as a ^{2}**K**-*module structure* for X by setting, for all $x \in X$ and all $(\lambda,\mu) \in {}^2\mathbf{K}$,

$$(\lambda,\mu)x = \lambda x_0 + \mu x_1.$$

The various axioms are readily verified. Conversely, any ^{2}**K**-module structure for X determines a direct sum decomposition $X_0 \oplus X_1$ of X in which $X_0 = (1,0)X$ $(=\{(1,0)x : x \in X\})$ and $X_1 = (0,1)X$. For $X = X_0 + X_1$, since $(1,1) = (1,0) + (0,1)$, while $X_0 \cap X_1 = \{0\}$, since $(1,0)(0,1) = (0,0)$.

Prop. 8.2. Let $t : X \to X$ be a linear involution of the **K**-linear space X. Then a ^{2}**K**-module structure, and therefore a direct sum decomposition, is defined for X by setting, for any $x \in X$,

$$(1,0)x = \tfrac{1}{2}(x + t(x)) \quad \text{and} \quad (0,1)x = \tfrac{1}{2}(x - t(x)).$$

Proof The various axioms have to be checked. In particular, for any $x \in X$ and any (λ,μ), $(\lambda',\mu') \in {}^2\mathbf{K}$,

$$\begin{aligned}(\lambda',\mu')((\lambda,\mu)x) &= \tfrac{1}{2}(\lambda' + \mu')(\tfrac{1}{2}(\lambda + \mu)x + \tfrac{1}{2}(\lambda - \mu)\, t(x)) \\ &\quad + \tfrac{1}{2}(\lambda' - \mu')(\tfrac{1}{2}(\lambda - \mu)x + \tfrac{1}{2}(\lambda + \mu)\, t(x)) \\ &\qquad\qquad \text{since } t^2 = 1_X \\ &= \tfrac{1}{2}(\lambda'\lambda + \mu'\mu)x + \tfrac{1}{2}(\lambda'\lambda - \mu'\mu)\, t(x) \\ &= (\lambda'\lambda, \mu'\mu)x\end{aligned}$$

while $$(1,1)x = \tfrac{1}{2}(x + t(x)) + \tfrac{1}{2}(x - t(x)) = x. \qquad \square$$

^{2}**K**-*module maps* and ^{2}**K**-*submodules* are defined in the obvious ways. The set of ^{2}**K**-module maps of the form $t : X \to Y$, where X and Y are ^{2}**K**-modules, will be denoted by $\mathscr{L}_{^2\mathbf{K}}(X,Y)$. This set is assigned the obvious ^{2}**K**-module structure. For any ^{2}**K**-module X, the ^{2}**K**-module $\mathscr{L}_{\mathbf{K}^2}(X,{}^2\mathbf{K})$ is called the ^{2}**K**-dual of X and is also denoted by $X^{\mathscr{L}}_{^2\mathbf{K}}$, or simply by $X^{\mathscr{L}}$ when there is no danger of confusion (see Prop. 8.4 below!).

In working with a ^{2}**K**-module map $t : X \to Y$ it is often convenient to represent X and Y each as the product of its components and then to use notations associated with maps between products, as, for example, in the next two propositions.

Prop. 8.3. Let $t : X \to Y$ be a ^{2}**K**-module map. Then t is of the form $\begin{pmatrix} a_0 & 0 \\ 0 & a_1 \end{pmatrix}$ where $a_0 \in \mathscr{L}(X_0, Y_0)$ and $a_1 \in \mathscr{L}(X_1, Y_1)$. Conversely, any map of this form is a ^{2}**K**-module map. $\square$

Prop. 8.4. Let X be a ^{2}**K**-module, with ^{2}**K**-dual $X^{\mathscr{L}}_{^2\mathbf{K}}$, and let $X^{\mathscr{L}}_{\mathbf{K}}$ be the **K**-linear dual of X formed by regarding X as a **K**-linear space. Then the map

$$X^{\mathscr{L}}_{^2\mathbf{K}} \to X^{\mathscr{L}}_{\mathbf{K}};\quad \begin{pmatrix} a_0 & 0 \\ 0 & a_1 \end{pmatrix} \rightsquigarrow (a_0 \;\; a_1)$$

is a **K**-linear isomorphism. □

Chapter 6 does not generalize directly to ^{2}**K**-modules. To begin with, it is necessary to make a distinction between 'linearly free' and 'linearly independent', 'linearly free' being the stronger notion. The definitions of linear dependence and independence run as before, but we say that an element x of a ^{2}**K**-module X is *linearly free* of a subset A of X if, and only if, $(1,0)x$ is free of $(1,0)A$ in the **K**-linear space $(1,0)X$ and $(0,1)x$ is free of $(0,1)A$ in the **K**-linear space $(0,1)X$. For example, in ^{2}**K** itself, $(1,1)$ is linearly independent of the set $\{(1,0)\}$, but is not free of $\{(1,0)\}$.

With this definition of freedom, the ^{2}**K**-analogues of Prop. 6.3 and Prop. 6.7 hold. On the other hand, the implications (b) ⇒ (a) and (c) ⇒ (a) of Prop. 6.11 fail. For example, let X be the ^{2}**R**-module $\mathbf{R}^2 \times \mathbf{R}$ with scalar multiplication defined by

$$(\lambda,\mu)(a,b) = (\lambda a,\mu b), \quad \text{for all } (\lambda,\mu) \in {}^2\mathbf{R},\ a \in \mathbf{R}^2,\ b \in \mathbf{R}.$$

Then $\{((1,0)\,,1)\}$ is a maximal free subset of X and $\{((1,0)\,,1), ((0,1)\,,0)\}$ is a minimal spanning subset of X. Yet neither is a basis for X.

The following is the *basis theorem* for ^{2}**K**-modules.

Theorem 8.5. Let X be a ^{2}**K**-module with a basis, A. Then $(1,0)X$ and $(0,1)X$ are isomorphic as **K**-linear spaces, the set $(1,0)A$ being a basis for the **K**-linear space $(1,0)X$ and the set $(0,1)A$ being a basis for the **K**-linear space $(0,1)X$.

Moreover, any two finite bases for X have the same number of elements.

Any ^{2}**K**-module with a finite basis is isomorphic to the ^{2}**K**-module $^2\mathbf{K}^n = (^2\mathbf{K})^n$, n being the number of elements in the basis. □

A ^{2}**K**-module X such that the **K**-linear spaces $(1,0)X$ and $(0,1)X$ are isomorphic will be called a ^{2}**K**-*linear space.* A ^{2}**K**-module map $X \to Y$ between ^{2}**K**-linear spaces X and Y will be called a ^{2}**K**-*linear map.*

It should be noted that not every point of a ^{2}**K**-linear space X spans a ^{2}**K**-line. For this to happen, each component of the point must be non-zero. A point that spans a line will be called a *regular* point of X. Similar considerations show that if $t: X \to Y$ is a ^{2}**K**-linear map, with X and Y each a ^{2}**K**-linear space, then im t and ker t, though necessarily ${}^2\mathbf{K}$-submodules of Y and X, respectively, are not necessarily ${}^2\mathbf{K}$-linear subspaces of Y and X.

All that has been said about ${}^2\mathbf{K}$-modules and maps and ${}^2\mathbf{K}$-linear spaces extends in the obvious way, for any positive s, to ${}^s\mathbf{K}$-modules and maps and ${}^s\mathbf{K}$-linear spaces.

Linear complements

When $V = X \oplus Y$ we say that X is *a linear complement* of Y in V. Distinguish between a linear complement of Y and the set complement $V \setminus Y$, which is not even a linear subspace of V. Confusion should not arise, provided that one's intuition of direct sum is firmly based on the figure on p. 133.

Prop. 8.6. Every linear subspace X of a finite-dimensional linear space V has a linear complement in V.

(Extend a basis for X to a basis for V.) □

Prop. 8.7. Let X and X' be linear subspaces of a finite-dimensional space V, with $\dim X = \dim X'$. Then X and X' have a common linear complement in V.

Proof Extend a basis A for the space $X \cap X'$ to bases $A \cup B$, $A \cup B'$, $A \cup B \cup B'$ and $A \cup B \cup B' \cup C$ for X, X', $X + X'$ and V, respectively, the sets A, B, B' and C being mutually disjoint. Since $\dim X = \dim X'$, $\#B = \#B'$. Choose some bijection $\beta;\ B \rightarrow B'$ and let $B'' = \{b + \beta(b) : b \in B\}$. Then $B'' \cup C$ spans a linear complement both of X and of X' in V. □

Complements and quotients

A linear complement X of a linear subspace Y of a linear space V may be regarded as a model of the quotient space V/Y, as the next proposition shows.

Prop. 8.8. Let X be a linear complement of the linear subspace Y of the linear space V. Then the map $\pi \mid X : X \rightarrow V/Y$ is a linear isomorphism, $\pi : V \rightarrow V/Y$ being the linear partition with kernel Y.

Proof

Since π is linear, $\pi | X$ is linear. Secondly, $V = X + Y$. So for any $v \in V$ there exist $x \in X$ and $y \in Y$ such that $v = x + y$, and in particular such that $\pi(v) = \pi(x + y) = \pi(x)$. So $\pi | X$ is surjective. Finally,

$X \cap Y = \{0\}$. So, for any $x \in X$,

$$\pi(x) = 0 \Leftrightarrow x \in Y \Leftrightarrow x = 0.$$

That is, $\pi | X$ is injective. So $\pi | X$ is a linear isomorphism.

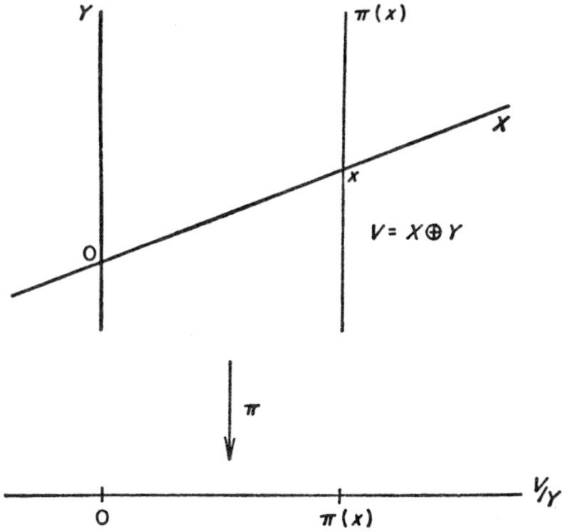

□

Another way of putting this is that any linear complement X of Y in V is the image of a linear section of the linear partition π, namely $(\pi | X)^{-1}$. The following proposition is converse to this.

Prop. 8.9. Let Y be a linear subspace of the linear space V and let $s : V/Y \longrightarrow V$ be a linear section of the linear partition $\pi : V \longrightarrow V/Y$. Then $V = \operatorname{im} s \oplus Y$.

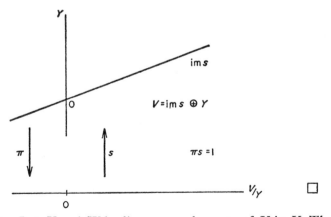

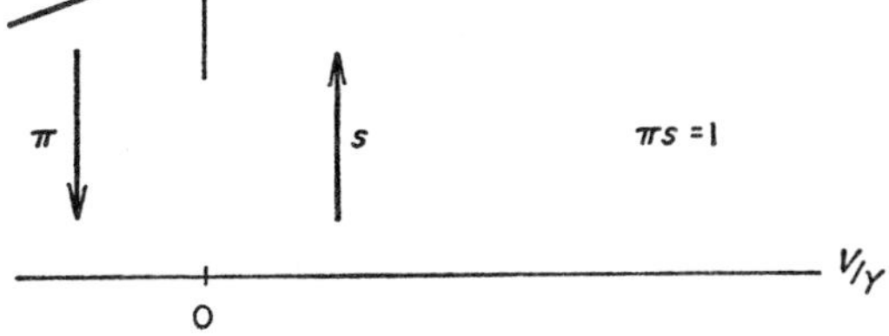

□

Prop. 8.10. Let X and W be linear complements of Y in V. Then there exists a unique linear map $t : X \longrightarrow Y$ such that $W = \operatorname{graph} t$.

Proof

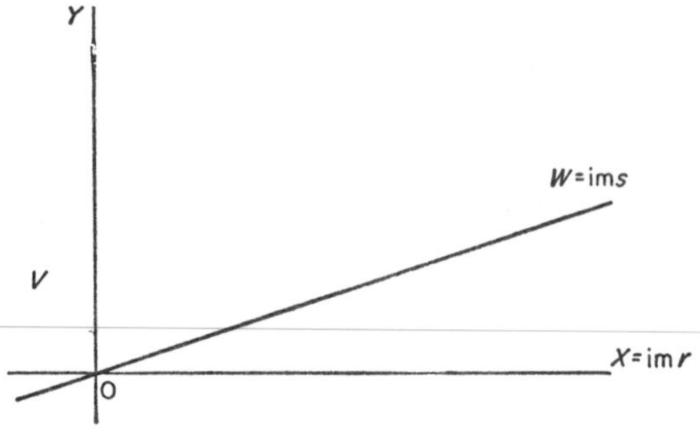

By Prop. 8.8, $X = \operatorname{im} r$ and $W = \operatorname{im} s$, where r and s are linear sections of the linear partition $\pi : V \to V/Y$. Define $t : X \to Y$ by $t(x) = s\,r^{-1}(x) - x$, for all $x \in X$. Then $W = \operatorname{im} s = \operatorname{graph} t$.

Uniqueness is by Prop. 1.22. □

Spaces of linear complements

It was proved at the end of Chapter 5 that the set of linear sections of a linear partition $\pi : V \to V/Y$ has a natural affine structure, with linear part $\mathscr{L}(V/Y, Y)$, and we have just seen that the map $s \rightsquigarrow \operatorname{im} s$ is a bijection of the set of sections of π to the set of linear complements of Y in V, so that the latter set also has a natural affine structure. The *affine space of linear complements* of Y in V, so defined, will be denoted by $\Theta(V, Y)$.

It follows at once, from Prop. 8.10, that for any $X \in \Theta(V, Y)$ the map

$$\gamma : \mathscr{L}(X, Y) \to \Theta(V, Y); \quad t \rightsquigarrow \operatorname{graph} t$$

is an affine isomorphism, sending 0 to X.

When V is finite-dimensional, the dimension of $\Theta(V, Y)$ is $k(n - k)$, where $k = \dim Y$ and $n = \dim V$.

Prop. 8.11. Let X and X' be linear complements of Y in V and let $t \in L(X, Y)$ and $t' \in L(X', Y)$. Then

$$\operatorname{graph} t' = \operatorname{graph} t \quad \Leftrightarrow \quad t' = tp - q$$

where $(p, q) : X' \to X \oplus Y$ is the inclusion map.

Proof

The result is 'obvious from the diagram', but has nevertheless to be checked.

$\Rightarrow$: Let $\operatorname{graph} t' = \operatorname{graph} t$ and let $x' \in X'$. Then there exists $x \in X$

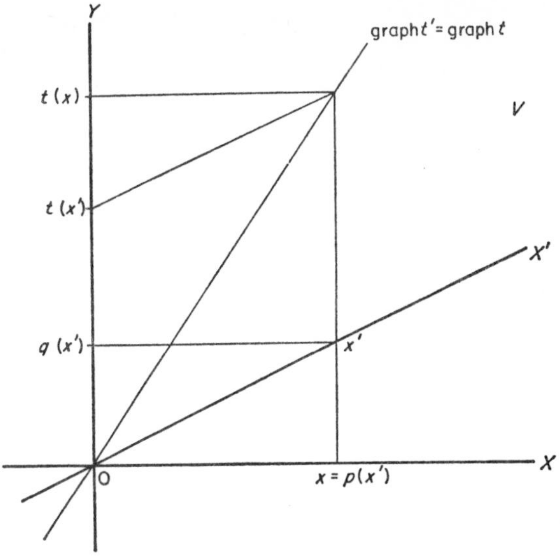

such that $x' + t(x') = x + t(x)$. Since $x' = p(x') + q(x')$ it follows that $x = p(x')$ and that $t(x) - t'(x') = q(x')$. So, for all $x' \in X'$, $t'(x') = tp(x') - q(x')$; that is, $t' = tp - q$.

$\Leftarrow$: Suppose $t'(x') = tp(x') - q(x')$, where $x' \in X'$, and let $x = p(x')$. By reversing the steps of the above argument, $x' + t(x') = x + t(x)$. Since p is bijective, it follows that graph t' = graph t. □

To put the result in another way, this says that

$$(\gamma')^{-1}\gamma(t) = tp - q$$

where $\gamma(t)$ = graph t and $\gamma'(t')$ = graph t'. The fact that the map

$$(\gamma')^{-1}\gamma : L(X,Y) \rightarrow L(X',Y); \quad t \rightsquigarrow tp - q$$

is affine is just another manifestation of the affine structure for $\Theta(V,Y)$.

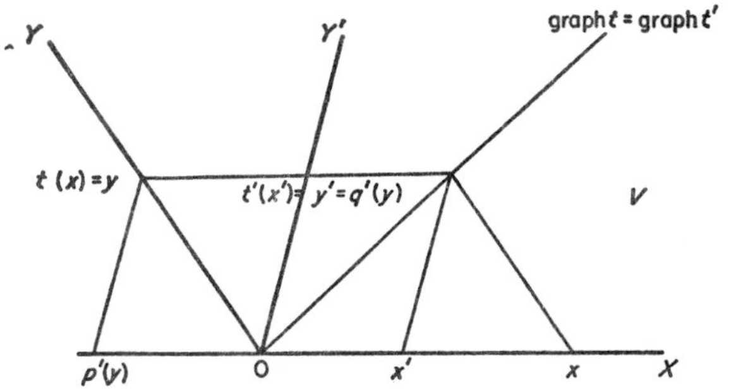

Prop. 8.12. Let Y and Y' be linear complements of the linear subspace X of the linear space V and let $t \in L(X,Y)$ and $t' \in L(X,Y')$. Then

$$\text{graph}\, t' = \text{graph}\, t \iff t' = q't(1_X + p't)^{-1}$$

where $(p',q') : Y \to V \cong X \times Y'$ is the inclusion.

Proof $\Rightarrow$: Let graph t' = graph t and let $x' \in X$. Then, for some $x \in X$, $x' + t'(x') = x + t(x)$. Since $t(x) = p't(x) + q't(x)$ it follows that $x' - x = p't(x)$ and that $t'(x') = q't(x)$. So, for all $x \in X$, $q't(x) = t'(x + p't(x))$; that is, $q't = t'(1_X + p't)$. Finally, $1_X + p't$ is invertible, with inverse $1_X + pt'$, where $p(x + y) = x$ for all $x \in X$, $y \in Y$. So $t' = q't(1_X + p't)^{-1}$.

$\Leftarrow$: Suppose $q't(x) = t'(x + p't(x))$, where $x \in X$, and let $x' = x + p't(x)$. Then $x' + t'(x') = x + t(x)$. Since $1_X + p't$ is bijective, it follows that graph t' = graph t. □

Grassmannians

The last few propositions have application to the description of the *Grassmannians* of a finite-dimensional **K**-linear space V, the set $\mathcal{G}_k(V)$, consisting of all the linear subspaces of V of a given dimension k not greater than the dimension of V, being, by definition, the *Grassmannian* of (*linear*) k-planes in V. When **K** is ordered, in particular when **K** = **R**, there is also interest in the set $\mathcal{G}_k^+(V)$, consisting of all the oriented linear subspaces of V of a given dimension k, this set being, by definition, the *Grassmannian* of *oriented* (*linear*) *k-planes* in V.

An important example is the Grassmannian $\mathcal{G}_1(V)$ of lines in V through 0, also called the *projective space* of the linear space V.

Since any two linear complements in V of a linear subspace Y have the same dimension, the set $\Theta(V,Y)$ of all the linear complements in V of Y is a subset of $\mathcal{G}_k(V)$, where k is the codimension of Y in V. By Propositions 8.6 and 5.22 or 8.11, $\mathcal{G}_k(V)$ may therefore be regarded as the union of a set of overlapping affine spaces, each of dimension $k(\dim V - k)$. The same is true of $\mathcal{G}_k^+(V)$, when **K** = **R**.

In particular, the projective space of an $(n + 1)$-dimensional linear space V is the union of a set of overlapping n-dimensional affine spaces, each of the form $\Theta(V,Y)$, where Y is a linear hyperplane of V. Such a projective space is said to be *n-dimensional*. The projective space $\mathcal{G}_1(\mathbf{K}^{n+1})$ is also denoted by $\mathbf{K}P^n$ or by $P^n(\mathbf{K})$. A zero-dimensional projective space is called a *projective point*, a one-dimensional projective space is called a *projective line* and a two-dimensional projective space is called a *projective plane*.

Proposition 8.7 may be applied in two ways to the description of

$\mathscr{G}_k(V)$. It implies, first, that if X and X' are any two points of $\mathscr{G}_k(V)$, then there is an affine subspace $\Theta(V,Y)$ of $\mathscr{G}_k(V)$ to which they both belong. Secondly, if the roles of X and Y are interchanged, the proposition implies that any two of the affine subspaces $\Theta(V,Y)$ and $\Theta(V,Y')$ intersect, Y and Y' being linear subspaces of Y of codimension k. Proposition 8.12 describes their intersection in terms of the linear structures on $\Theta(V,Y)$ and $\Theta(V,Y')$ with common origin some common point X of $\Theta(V,Y) \cap \Theta(V,Y')$.

Exercise 8.13. Apply Propositions 8.6, 8.7, 8.11 and 8.12 to the description of $\mathscr{G}_k^+(V)$. □

As we shall see in Chapters 17 and 20, it follows at once from the above remarks that any Grassmannian $\mathscr{G}_k(V)$ or $\mathscr{G}_k^+(V)$ is in a natural way a smooth manifold.

Each point of the projective space $\mathscr{G}_1(X)$ of a linear space X is a line through 0 in X, this line being uniquely determined by any one of its points x other than 0. The line, or projective point, $\mathrm{im}\{x\}$ will also be denoted by $[x]$. When $X = \mathbf{K}^{n+1}$, with $x = (x_i : i \in n+1)$, $[x]$ will also be denoted by $[x_i : i \in n+1]$, or by $[x_0, x_1, \ldots, x_n]$, these notations being particularly convenient in examples when one is working with some particular small value of n. For example, $[x_0,x_1]$ denotes a point of $\mathscr{G}_1(\mathbf{K}^2)$, namely the line in $\mathbf{K}^2$ through 0 and (x_0,x_1). (Confusion here with the closed intervals of $\mathbf{R}$, which are similarly denoted, is most unlikely in practice.)

The projective line $\mathbf{K}P^1 = \mathscr{G}_1(\mathbf{K}^2)$ is often thought of simply as the union of two copies of the field $\mathbf{K}$, glued together by the map $\mathbf{K} \rightarrowtail \mathbf{K}$; $x \rightsquigarrow x^{-1}$, for $\mathbf{K}P^1$ is the union of the images of the maps

$$i_0 : \mathbf{K} \to \mathbf{K}P^1; \quad y \rightsquigarrow [1,y]$$

and

$$i_1 : \mathbf{K} \to \mathbf{K}P^1; \quad x \rightsquigarrow [x,1],$$

with $[1,y] = [x,1]$ if, and only if, $y = x^{-1}$. In this model only one point of the first copy of $\mathbf{K}$ fails to correspond to a point of the second. This point $[1,0]$ is often denoted by ∞ and called *the point at infinity* on the projective line. Every other point $[x,y]$ of $\mathbf{K}P^1$ is represented by a unique point xy^{-1} in the second copy of $\mathbf{K}$. When we are using this representation of $\mathbf{K}P^1$ we shall simply write $K \cup \{\infty\}$ in place of $\mathbf{K}P^1$.

Example 8.14. Let $\sum_{i \in n+1} a^i x_i$ be a polynomial of positive degree n over the infinite field $\mathbf{K}$. Then the polynomial map $\mathbf{K} \to \mathbf{K}$; $x \rightsquigarrow \sum_{i \in n+1} a_i x^i$ may be regarded as the restriction to $\mathbf{K}$ with target $\mathbf{K}$ of the map

$$\mathbf{K} \cup \{\infty\} \to \mathbf{K} \cup \{\infty\}; \quad [x,y] \rightsquigarrow [\sum_{i \in n+1} a_i x^i y^{n-i}, y^n],$$

for this map sends $[x\lambda,\lambda] = [x,1] = x$ to

$$[\sum_{i \in n+1} a^i x^i \lambda^n, \lambda^n] = [\sum_{i \in n+1} a_i{}^i x, 1] = \sum_{i \in n+1} a_i x^i$$

and $[\lambda,0] = [1,0] = \infty$ to $[a_n\lambda^n,0] = [1,0] = \infty$, for any x and any non-zero λ in $\mathbf{K}$. □

This will be useful in the proof of the fundamental theorem of algebra in Chapter 19, in the particular case when $\mathbf{K} = \mathbf{C}$.

The projective plane $\mathbf{K}P^2 = \mathscr{G}_1(\mathbf{K}^3)$ may be thought of, similarly, as the union of three copies of $\mathbf{K}^2$ suitably glued together, for $\mathbf{K}P^2$ is the union of the images of the maps

$$i_0 : \mathbf{K}^2 \to \mathbf{K}P^2; \quad (y_0,z_0) \rightsquigarrow [1,y_0,z_0],$$
$$i_1 : \mathbf{K}^2 \to \mathbf{K}P^2; \quad (x_1,z_1) \rightsquigarrow [x_1,1,z_1]$$

and

$$i_2 : \mathbf{K}^2 \to \mathbf{K}P^2; \quad (x_2,y_2) \rightsquigarrow [x_2,y_2,1].$$

In this representation, (y_0,z_0) in the domain of i_0 and (x_1,z_1) in the domain of i_1 represent the same point of $\mathscr{G}_1(\mathbf{K}^3)$ if, and only if, $x_1 = y_0^{-1}$ and $z_1 = z_0 y_0^{-1}$. (Cf. the construction of the Cayley plane on page 285.)

As was the case with $\mathbf{K}P^1$, it is often convenient in working with $\mathbf{K}P^2$ to regard one of the injections, say i_2, as standard and to regard all the points of $\mathscr{G}_1(\mathbf{K}^3)$ not lying in the image of i_2 as *lying at infinity*. Observe that the set of points lying at infinity is a projective line, namely the projective space of the plane $\{(x,y,z) \in \mathbf{K}^3 : z = 0\}$ in $\mathbf{K}^3$.

Similar remarks apply to projective spaces of dimension greater than 2. The following proposition formalizes the intuition concerning 'points at infinity' in the general case.

Prop. 8.15. Any projective space of positive dimension n may be represented, as a set, as the disjoint union of an n-dimensional affine space and the $(n - 1)$-dimensional projective space of its vector space (the *hyperplane at infinity*).

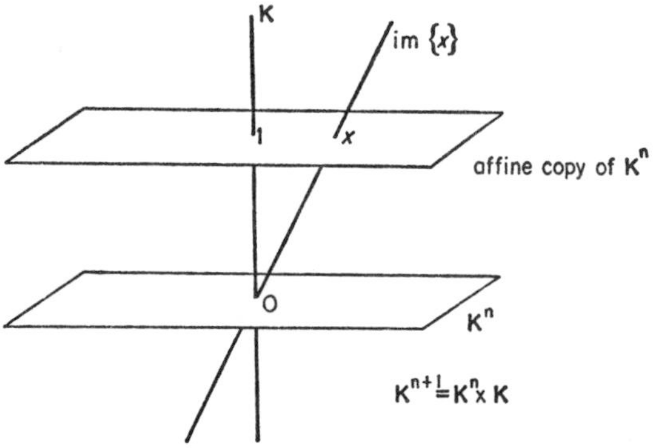

Proof It is sufficient to consider $\mathbf{K}P^n = \mathscr{G}_1(\mathbf{K}^{n+1})$ and to show that $\mathbf{K}P^n$ may be regarded as the disjoint union of an affine copy of $\mathbf{K}^n$ and $\mathbf{K}P^{n-1} = \mathscr{G}_1(\mathbf{K}^n)$.

To do so, set $\mathbf{K}^{n+1} = \mathbf{K}^n \oplus \mathbf{K}$. Then each line through 0 in $\mathbf{K}^{n+1}$ is either a line of $\mathbf{K}^n$ or a linear complement in $\mathbf{K}^{n+1}$ of $\mathbf{K}^n$, these two possibilities being mutually exclusive. Conversely, each line of $\mathbf{K}^n$ and each linear complement in $\mathbf{K}^{n+1}$ of $\mathbf{K}^n$ is a line through 0 in $\mathbf{K}^{n+1}$. That is, $\mathbf{K}P^n$ is the disjoint union of $\mathbf{K}P^{n-1} = \mathscr{G}_1(\mathbf{K}^n)$ and the affine copy of $\mathbf{K}^n$, $\Theta(\mathbf{K}^{n+1},\mathbf{K}^n)$. □

Suppose now that X and Y are finite-dimensional linear spaces. Each *injective* linear map $t: X \to Y$ induces a map

$$\mathscr{G}_k(t) = t_{\vdash} | \mathscr{G}_k(X) : \mathscr{G}_k(X) \to \mathscr{G}_k(Y)$$

for each number k not greater than the dimension of X.

In particular, if $\dim X > 0$, t induces a map

$$\mathscr{G}_1(t) = t_{\vdash} | \mathscr{G}_1(X) : \mathscr{G}_1(X) \to \mathscr{G}_1(Y).$$

Such a map is said to be a *projective map*.

Prop. 8.16. Let t and $u: X \to Y$ induce the same projective map $\mathscr{G}_1(X) \to \mathscr{G}_1(Y)$, X and Y being $\mathbf{K}$-linear spaces of positive finite dimension. Then there is a non-zero element λ of $\mathbf{K}$ such that $u = \lambda t$. □

Projective subspaces of a projective space $\mathscr{G}_1(V)$ are defined in the obvious way. Each projective subspace of a given dimension k is the projective space of a linear subspace of V of dimension $k + 1$. Conversely, the projective space of any linear subspace of V of dimension greater than zero is a projective subspace of $\mathscr{G}_1(V)$. It follows that, for any finite k, the Grassmannian $\mathscr{G}_{k+1}(V)$, the set of *linear* subspaces of dimension $k + 1$ of the *linear* space V, may be identified with the set of *projective* subspaces of dimension k of the *projective* space $\mathscr{G}_1(V)$. For example, the Grassmannian $\mathscr{G}_2(\mathbf{K}^4)$ of linear planes in $\mathbf{K}^4$ may be identified with the set of projective lines in $\mathbf{K}P^3$.

There is a projective version of Prop. 6.29.

Prop. 8.17. Let X and Y be linear subspaces of a finite-dimensional linear space V. Then

$$\dim \mathscr{G}_1(X + Y) + \dim (\mathscr{G}_1(X) \cap \mathscr{G}_1(Y)) = \dim \mathscr{G}_1(X) + \dim \mathscr{G}_1(Y),$$

where, by convention, $\dim \emptyset = \dim \mathscr{G}_1\{0\} = -1$. □

The projective subspace $\mathscr{G}_1(X + Y)$ of V is said to be the *join* of the projective subspaces $\mathscr{G}_1(X)$ and $\mathscr{G}_1(Y)$.

The Grassmannian $\mathscr{G}_k(X)$ of k-planes in a finite-dimensional $\mathbf{K}$-linear space X is related to the set of k-framings of X, $GL(\mathbf{K}^k,X)$, by the surjective map

$$h : GL(\mathbf{K}^k,X) \rightarrow \mathscr{G}_k(X); \quad t \rightsquigarrow \operatorname{im} t.$$

Now $L(\mathbf{K},X)$ is naturally isomorphic to X and in this isomorphism $GL(\mathbf{K},X)$ corresponds to $X \setminus \{0\}$. Therefore in the case that $k = 1$ the map h may be identified with the map

$$X \setminus \{0\} \rightarrow \mathscr{G}_1(X); \quad x \rightsquigarrow \operatorname{im}\{x\}.$$

The latter map is called the *Hopf map* over the projective space $\mathscr{G}_1(X)$.

It is not our purpose to develop the ideas of projective geometry. There are many excellent books which do so. See, in particular, [20], which complements in a useful way the linear algebra presented in this book.

Further Exercises

8.18. Let V be a linear space and let X, Y and Z be linear subspaces of V such that $V = (X \oplus Y) \oplus Z$. Prove that $V = X \oplus (Y \otimes Z)$. □

8.19. Let $u : X \oplus Y \rightarrow Z$ be a linear map such that $u \mid Y$ is invertible, and let α denote the isomorphism $X \times Y \rightarrow X \oplus Y$; $(x,y) \rightsquigarrow x + y$. Prove that there exists a unique linear map $t : X \rightarrow Y$ such that $\ker u = \alpha_{\vdash}(\operatorname{graph} t)$. □

8.20. Let $t : X \rightarrow X$ be a linear involution of a linear space X (that is, t is linear and $t^2 = 1_X$). Prove that

$$X = \operatorname{im}(1_X + t) \oplus \operatorname{im}(1_X - t)$$

and that t preserves this direct sum decomposition, reducing to the identity on one component and to minus the identity on the other.

Interpret this when $X = \mathbf{K}(n)$, for any finite n, and t is transposition. □

8.21. Let $t : X \rightarrow X$ be a linear map, X being a finite-dimensional linear space. Prove that $X = \ker t \oplus \operatorname{im} t$ if, and only if, $\operatorname{rk} t = \operatorname{rk} t^2$. □

8.22. Let $t : V \rightarrow V$ be a linear map, V being a finite-dimensional linear space, and suppose that, for each direct sum decomposition $X \oplus Y$ of V,

$$t_{\vdash}(V) = t_{\vdash}(X) + t_{\vdash}(Y).$$

Prove that t is an isomorphism. □

8.23. Verify that the map

$$\mathscr{G}_1(\mathbf{R}^3) \rightarrow \mathscr{G}_1(\mathbf{R}^3); \quad [x,y,z] \rightsquigarrow [yz,zx,xy]$$

is well-defined, and determine the domain, the image and the fibres of the map. □

8.24. Verify that the map

$$\mathscr{G}_1(\mathbf{R}^3) \to \mathbf{R}^3; \quad [x,y,z] \rightsquigarrow \frac{2z}{x^2 + y^2 + z^2}(x,y,z)$$

is well defined, with image

$$\{(u,v,w) \in \mathbf{R}^3 : u^2 + v^2 + (w - 1)^2 = 1\},$$

and determine the fibres of the map. □

CHAPTER 9

ORTHOGONAL SPACES

As we have already remarked in the introduction to Chapter 3, the *intuitive plane*, the plane of our intuition, has more structure than the two-dimensional affine (or linear) space which it has been used to represent in Chapters 3 to 8. For consider two intersecting oriented lines l_0, l_1 in the intuitive plane.

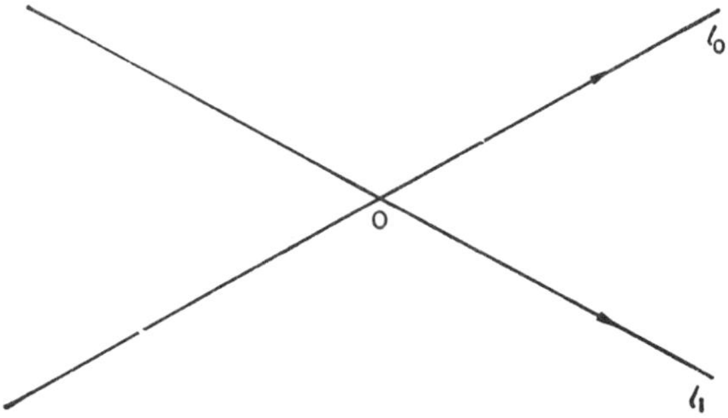

Then we 'know' how to *rotate* one line on to the other, keeping the point of intersection 0 fixed and respecting the orientations, and we 'know' how to project one line on to the other *orthogonally*. Both these maps are linear, if the point of intersection is chosen as the origin for both lines. There are two special cases, namely when $l_1 = l_0$ and when $l_1 = -l_0$ ($-l_0$ is the same line as l_0, but has the opposite orientation). If $l_1 = l_0$ each map is the identity, while if $l_1 = -l_0$ one is minus the identity (after 0 has been chosen!) and the other is the identity. If $l_1 \neq l_0$ or $-l_0$ *neither* map is defined by the affine structure alone.

The first intuition, that we can rotate one oriented line on to another, acquired by playing with rulers and so forth, is the basis of our concept of the *length* of a line-segment or, equivalently, of the *distance* between two points, since it enables us to compare line-segments on lines which need not be parallel to one another. The only additional facts (or experiences) required to set the concept up, other than the fundamental correspondence between the points of a line and the real numbers already discussed in Chapter 2, are first that the rotation of oriented lines is *translation-invariant*: that is, if l_0' and l_1' are the (oriented) images of l_0 and l_1 by some translation of the plane, then the rotation of l_0 on to l_1 followed by the translation of l_1 on to l_1' is equal to the translation of l_0 on to l_0' followed by the rotation of l_0' on to l_1'; and secondly that it is *transitive*: that is, given three oriented lines, l_0, l_1, l_2, say, with a common point of intersection, then the rotation of l_0 on to l_2 is the composite of the rotation of l_0 on to l_1 and the rotation of l_1 on to l_2.

In what follows, the oriented affine line joining two distinct points a and b of the plane, with $b \dot{-} a$ taken to be a positive tangent vector at a, will be denoted by ab. The *length* of a line segment $[a,b]$ (with respect to some *unit* line-segment assigned the length 1) will be denoted by $|a - b|$.

The second intuition, that we can project one line on to another orthogonally, is derived from the more fundamental one that we know what is meant by two lines being *at right angles* or *orthogonal* to one another. Through any point of a line in the plane there is a unique line through that point orthogonal to the given line and distinct from it. Orthogonality also is translation-invariant: if two lines of the plane are at right angles, then their images in any translation of the plane also are at right angles. The orthogonal projection of l_1 on to l_0 is then the restriction to l_1 of the projection of the plane on to l_0 with fibres the (mutually parallel) lines of the plane orthogonal to l_0.

The final observation, which leads directly to Pythagoras' 'theorem', is that if we are given two intersecting oriented lines l_0 and l_1 with $l_1 \neq l_0$ or $-l_0$, then the linear map obtained by first rotating l_0 on to l_1 and then projecting l_1 orthogonally on to l_0 is a linear *contraction* of l_0, that is, multiplication by a real number, the *cosine* of the *angle* (l_0,l_1), of absolute value <1. Moreover, the contraction coefficient, the cosine, remains the same if the roles of l_0 and l_1 are interchanged.

To deduce Pythagoras' theorem, consider three non-collinear points a, b, c of the plane, such that the lines ac and bc are at right angles,

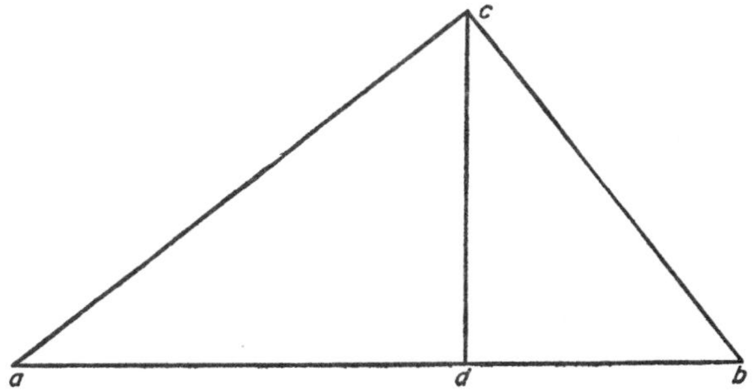

and let d be the image of c by the orthogonal projection of the plane on to the line ab.

Let λ be the cosine of the angle (ac, ad) and let μ be the cosine of the angle (bc, bd). Then

$$|a - d| = \lambda |a - c| = \lambda^2 |a - b|$$

and

$$|d - b| = \mu |c - b| = \mu^2 |a - b|.$$

Since λ^2 and μ^2 are each less than 1, d lies between a and b, and so

$$|a - b| = |a - d| + |d - b|,$$

implying that

$$|a - b|^2 = |a - c|^2 + |c - b|^2.$$

If $\mathbf{R}^2$ is taken as a model for the intuitive plane, with $\mathbf{R} \times \{0\}$ and $\{0\} \times \mathbf{R}$ representing mutually orthogonal lines, and with (1,0) and (0,1) each at unit distance from the origin, then it follows that the distance of any point (x,y) of $\mathbf{R}^2$ from the origin is $\sqrt{(x^2 + y^2)}$. The geometry therefore provides a motivation for studying the *quadratic form*

$$\mathbf{R}^2 \to \mathbf{R}; \quad (x,y) \rightsquigarrow x^2 + y^2.$$

As we shall see in detail in this chapter, we can reconstruct all the phenomena which we have just noted on the intuitive plane by starting with a real linear space and a distinguished quadratic form on the space. To begin with, we consider arbitrary real-valued quadratic forms on *real* linear spaces, positive-definite ones, such as $(x,y) \rightsquigarrow x^2 + y^2$, being considered specially later on. A final section is concerned with analogues over the *complex* field. Other generalizations are deferred until Chapter 11. Geometrical applications of both chapters will be found in Chapter 12, while some of the deeper properties of the orthogonal groups and their analogues are discussed in Chapters 13, 17 and 20.

Real orthogonal spaces

A quadratic form on a real linear space X is most conveniently introduced in terms of a *symmetric scalar product* on X. This, by definition, is a bilinear map

$$X^2 \to \mathbf{R}; \quad (a,b) \rightsquigarrow a \cdot b$$

such that, for all $a, b \in X$, $b \cdot a = a \cdot b$. The map

$$X \to \mathbf{R}; \quad a \rightsquigarrow a \cdot a$$

is called the *quadratic form* of the scalar product, $a^{(2)} = a \cdot a$ being called the *square* of a. (The notation a^2 is reserved for later use in Chapter 13.) Since, for each $a, b \in X$,

$$2a \cdot b = a^{(2)} + b^{(2)} - (a - b)^{(2)},$$

the scalar product is uniquely determined by its quadratic form. In particular, the scalar product is the zero map if, and only if, its quadratic form is the zero map.

The following are examples of scalar products on $\mathbf{R}^2$:

$$((x,y),(x',y')) \rightsquigarrow 0,\ xx',\ xx' + yy',\ -xx' + yy' \text{ and } xy' + yx',$$

their respective quadratic forms being

$$(x,y) \rightsquigarrow 0,\ x^2,\ x^2 + y^2,\ -x^2 + y^2 \text{ and } 2xy.$$

It is a consequence of the symmetry of the scalar product that, for all $a, b \in X$, $b \cdot a = 0 \Leftrightarrow a \cdot b = 0$.

When $a \cdot b = 0$ the elements a and b of X are said to be *mutually orthogonal*. Subsets A and B of X are said to be *mutually orthogonal* if, for each $a \in A$, $b \in B$, $a \cdot b = 0$.

A real linear space with scalar product will be called a (*real*) *orthogonal space*, any linear subspace W of an orthogonal space X being tacitly assigned the restriction to W^2 of the scalar product for X.

An orthogonal space X is said to be *positive-definite* if, for all non-zero $a \in X$, $a^{(2)} > 0$, and to be *negative-definite* if, for all non-zero $a \in X$, $a^{(2)} < 0$. An example of a positive-definite space is the linear space $\mathbf{R}^2$ with the scalar product

$$((x,y),(x',y')) \rightsquigarrow xx' + yy'.$$

An orthogonal space whose scalar product is the zero map is said to be *isotropic* (the term derives from its use in the special theory of relativity), and an orthogonal space that is the linear direct sum of two isotropic subspaces is said to be *neutral*.

It is convenient to have short notations for the orthogonal spaces that most commonly occur in practice. The linear space $\mathbf{R}^{p+q}$ with the scalar product

$$(a,b) \rightsquigarrow -\sum_{i \in p} a_i b_i + \sum_{j \in q} a_{p+j} b_{p+j}$$

will therefore be denoted by $\mathbf{R}^{p,q}$, while the linear space $\mathbf{R}^{2n}$ with the scalar product

$$(a,b) \rightarrow \sum_{i \in n} (a_i b_{n+i} + a_{n+i} b_i)$$

will be denoted by $\mathbf{R}^n_{\text{hb}}$, or by $\mathbf{R}_{\text{hb}}$ when $n = 1$. The letters hb are an abbreviation for *hyperbolic*, $\mathbf{R}_{\text{hb}}$ being the *standard hyperbolic plane*. The linear space underlying $\mathbf{R}^{p,q}$ will frequently be identified with $\mathbf{R}^p \times \mathbf{R}^q$ and the linear space underlying $\mathbf{R}^n_{\text{hb}}$ with $\mathbf{R}^n \times \mathbf{R}^n$. The linear subspaces $\mathbf{R}^n \times \{0\}$ and $\{0\} \times \mathbf{R}^n$ of $\mathbf{R}^n \times \mathbf{R}^n$ are isotropic subspaces of $\mathbf{R}^n_{\text{hb}}$. This orthogonal space is therefore neutral. The orthogonal spaces $\mathbf{R}^{0,n}$ and $\mathbf{R}^{n,0}$ are, respectively, positive-definite and negative-definite.

When there is only one orthogonal structure assigned to a linear space X, the dot notation for the scalar product will normally be most convenient, and will often be used without special comment. Alternative notations will be introduced later.

Invertible elements

An element a of an orthogonal space X is said to be *invertible* if $a^{(2)} \neq 0$, the element $a^{(-1)} = (a^{(2)})^{-1}a$ being called the *inverse* of a. Every non-isotropic real orthogonal space X possesses invertible elements since the quadratic form on X is zero only if the scalar product is zero.

Prop. 9.1. Let a be an invertible element of an orthogonal space X. Then, for some $\lambda \in \mathbf{R}$, $(\lambda a)^{(2)} = \pm 1$.

Proof Since $(\lambda a)\cdot(\lambda a) = \lambda^2 a^{(2)}$ and since $a^{(2)} \neq 0$ we may choose $\lambda = (\sqrt{(|a^{(2)}|)})^{-1}$. □

Prop. 9.2. If a and b are invertible elements of an orthogonal space X with $a^{(2)} = b^{(2)}$, then $a + b$ and $a - b$ are mutually orthogonal and either $a + b$ or $a - b$ is invertible.

Proof Since $(a + b)\cdot(a - b) = a^{(2)} - b^{(2)} = 0$, $a + b$ is orthogonal to $a - b$. Also

$$\begin{aligned}(a + b)^{(2)} + (a - b)^{(2)} &= a^{(2)} + 2a\cdot b + b^{(2)} + a^{(2)} - 2a\cdot b + b^{(2)} \\ &= 4a^{(2)},\end{aligned}$$

so, if $a^{(2)} \neq 0$, $(a + b)^{(2)}$ and $(a - b)^{(2)}$ are not both zero. □

The elements $a + b$ and $a - b$ need not both be invertible even when $a \neq \pm b$. Consider, for example, $\mathbf{R}^{1,2}$. In this case we have $(1,1,1)^{(2)} = (1,1,-1)^{(2)} = 1$, and $(0,0,2) = (1,1,1) - (1,1,-1)$ is invertible, since $(0,0,2)^{(2)} = 4 \neq 0$. However, $(2,2,0) = (1,1,1) + (1,1,-1)$ is non-invertible, since $(2,2,0)^{(2)} = 0$.

For a sequel to Prop. 9.2, see Prop. 9.40.

Linear correlations

Let X be any real linear space, with dual space $X^{\mathscr{L}}$. Any linear map $\xi : X \to X^{\mathscr{L}};\ x \rightsquigarrow x^{\xi} = \xi(x)$ is said to be a *linear correlation* on X. An example of a linear correlation on $\mathbf{R}^n$ is transposition:

$$\tau : \mathbf{R}^n \to (\mathbf{R}^n)^L; \quad x \rightsquigarrow x^{\tau}.$$

(In accordance with the remark on page 106, we usually write X^L in place of $X^{\mathscr{L}}$ whenever X is finite-dimensional.)

A correlation ξ is said to be *symmetric* if, for all $a, b \in X$, $a^\xi(b) = b^\xi(a)$. A symmetric correlation ξ induces a scalar product $(a,b) \rightsquigarrow a^\xi b = a^\xi(b)$. Conversely, any scalar product $(a,b) \rightsquigarrow a \cdot b$ is so induced by a unique symmetric correlation, namely the map $a \rightsquigarrow (a\,\cdot)$, where, for all $a, b \in X$, $(a\cdot)b = a\cdot b$. A real linear space X with a correlation ξ will be called a real *correlated* (*linear*) *space*. By the above remarks any real orthogonal space may be thought of as a symmetric real correlated space, and conversely.

Non-degenerate spaces

In this section we suppose, for simplicity, that X is a *finite-dimensional* real orthogonal space, with correlation ξ. For such a space, ξ is, by Cor. 6.33, injective if, and only if, it is bijective, in which case X, its scalar product, its quadratic form, and its correlation are all said to be *non-degenerate*. If, on the other hand, ξ is not injective, that is if $\ker \xi \neq \{0\}$, then X is said to be *degenerate*. The kernel of ξ, $\ker \xi$, is also called the *kernel* of X and denoted by $\ker X$. An element $a \in X$ belongs to $\ker X$ if, and only if, for all $x \in X$, $a\cdot x = a^\xi x = 0$, that is if, and only if, a is orthogonal to each element of X. From this it at once follows that a positive-definite space is non-degenerate. The rank of ξ is also called the *rank* of X and denoted by $\operatorname{rk} X$. The space X is non-degenerate if, and only if, $\operatorname{rk} X = \dim X$.

Prop. 9.3. Let A be a finite set of mutually orthogonal invertible elements of a real orthogonal space X. Then the linear image of A, $\operatorname{im} A$, is a non-degenerate subspace of X.

Proof Let λ be any set of coefficients for A such that $\sum_{a\in A} \lambda_a a \neq 0$. Then, by the orthogonality condition,

$$\Big(\sum_{a\in A} \lambda_a a\Big)^\xi\Big(\sum_{a\in A} \lambda_a a^{(-1)}\Big) = \sum_{a\in A} \lambda_a^2 > 0,$$

where ξ is the correlation on X. So $\sum_{a\in A} \lambda_a a \notin \ker(\operatorname{im} A)$. Therefore $\ker(\operatorname{im} A) = \{0\}$. That is, $\operatorname{im} A$ is non-degenerate. □

Cor. 9.4. For any finite p, q, the orthogonal space $\mathbf{R}^{p,q}$ is non-degenerate. □

Prop. 9.5. Let X be a finite-dimensional real orthogonal space and let X' be a linear complement in X of $\ker X$. Then X' is a non-degenerate subspace of X. □

Orthogonal maps

As always, there is interest in the maps preserving a given structure.

Let X and Y be real orthogonal spaces, with correlations ξ and η, respectively. A map $t : X \to Y$ is said to be a *real orthogonal map* if it is linear and, for all $a,b \in X$,

$$t(a)^{\eta}\, t(b) = a^{\xi} b$$

or, informally, in terms of the dot notation,

$$t(a) \cdot t(b) = a \cdot b.$$

This condition may be re-expressed in terms of a commutative diagram involving the linear dual $t^{\mathscr{L}}$ of t, as follows.

Prop. 9.6. Let X, Y, ξ and η be as above. Then a linear map $t : X \to Y$ is orthogonal if, and only if, $t^{\mathscr{L}} \eta t = \xi$, that is, if, and only if, the diagram

$$\begin{array}{ccc} X & \longrightarrow & Y \\ {\scriptstyle \xi}\downarrow & & \downarrow{\scriptstyle \eta} \\ X^{\mathscr{L}} & \xleftarrow{\;t^{\mathscr{L}}\;} & Y^{\mathscr{L}} \end{array}$$

commutes.

Proof $t^{\mathscr{L}} \eta t = \xi \Leftrightarrow$ for all $a, b \in X$, $t^{\mathscr{L}} \eta t(a)(b) = \xi(a)(b)$
$\Leftrightarrow$ for all $a, b \in X$, $(\eta t(a)) t(b) = \xi(a)(b)$
$\Leftrightarrow$ for all $a, b \in X$, $t(a)^{\eta}\, t(b) = a^{\xi} b$. □

Cor. 9.7. If X is non-degenerate, then any orthogonal map $t : X \to Y$ is injective.

Proof Let t be such a map. Then $(t^{\mathscr{L}} \eta) t = \xi$ is injective and so, by Prop. 1.3, t is injective. □

Prop. 9.8. Let W, X and Y be orthogonal spaces and let $t : X \to Y$ and $u : W \to X$ be orthogonal maps. Then 1_X is orthogonal, tu is orthogonal and, if t is invertible, t^{-1} is orthogonal. □

An invertible orthogonal map $t : X \to Y$ will be called an *orthogonal isomorphism*, and two orthogonal spaces X and Y so related will be said to be isomorphic.

Prop. 9.9. For any finite n the orthogonal spaces $\mathbf{R}^{n,n}$ and $\mathbf{R}^{n}_{\text{hb}}$ are isomorphic.

Proof It is convenient to identify $\mathbf{R}^{2n}$ with $\mathbf{R}^{n} \times \mathbf{R}^{n}$ and to indicate the scalar products of $\mathbf{R}^{n,n}$ and $\mathbf{R}^{0,n}$ by $\cdot$ and the scalar product of $\mathbf{R}^{n}_{\text{hb}}$ by ${}_{\text{hb}}^{\cdot}$. Then the map

$$\mathbf{R}^{n,n} \to \mathbf{R}^{n}_{\text{hb}};\quad (x,y) \rightsquigarrow (\sqrt{2})^{-1}(-x + y\,, x + y)$$

is an orthogonal isomorphism; for it is clearly a linear isomorphism, while, for any (x,y), $(x',y') \in \mathbf{R}^{n,n}$,

$$\begin{aligned}\tfrac{1}{2}(-x+y\,,x+y)_{\text{hb}}(-x'+y',x'+y')\\ &= \tfrac{1}{2}((x+y)\cdot(-x'+y')+(-x+y)\cdot(x'+y'))\\ &= -x\cdot x'+y\cdot y'\\ &= (x,y)\cdot(x',y'). \quad \square\end{aligned}$$

Any two-dimensional orthogonal space isomorphic to the standard hyperbolic plane $\mathbf{R}_{\text{hb}}$ will be called a *hyperbolic plane*.

Prop. 9.10. Let X be an orthogonal space. Then any two linear complements in X of $\ker X$ are isomorphic as orthogonal spaces. □

An invertible orthogonal map $t : X \to X$ will be called an *orthogonal automorphism* of the orthogonal space X. By Cor. 9.7 any orthogonal transformation of a non-degenerate *finite-dimensional* orthogonal space X is an orthogonal automorphism of X.

For orthogonal spaces X and Y the set of orthogonal maps $t : X \to Y$ will be denoted $O(X,Y)$ and the group of orthogonal automorphisms $t : X \to X$ will be denoted $O(X)$. For any finite p, q, n the groups $O(\mathbf{R}^{p,q})$ and $O(\mathbf{R}^{0,n})$ will also be denoted, respectively, by $O(p,q;\mathbf{R})$ and $O(n;\mathbf{R})$ or, more briefly, by $O(p,q)$ and $O(n)$.

An orthogonal transformation of a finite-dimensional orthogonal space X may or may not preserve the orientations of X. An orientation-preserving orthogonal transformation of a finite-dimensional orthogonal space X is said to be a *special* orthogonal transformation, or a *rotation*, of X. The subgroup of $O(X)$ consisting of the special orthogonal transformations of X is denoted by $SO(X)$, the groups $SO(\mathbf{R}^{p,q})$ and $SO(\mathbf{R}^{0,n})$ also being denoted, respectively, by $SO(p,q)$ and by $SO(n)$.

An orthogonal automorphism of X that reverses the orientations of X will be called an *antirotation* of X.

Prop. 9.11. For any finite p, q, the groups $O(p,q)$ and $O(q,p)$ are isomorphic, as are the groups $SO(p,q)$ and $SO(q,p)$. □

Adjoints

Suppose now that $t : X \to Y$ is a linear map of a non-degenerate finite-dimensional orthogonal space X, with correlation ξ, to an orthogonal space Y, with correlation η. Since ξ is bijective there will be a unique linear map $t^* : Y \to X$ such that $\xi t^* = t^{\mathscr{L}}\eta$, that is, such that, for any $x \in X$, $y \in Y$, $t^*(y)\cdot x = y\cdot t(x)$. The map $t^* = \xi^{-1}t^{\mathscr{L}}\eta$ is called the *adjoint* of t with respect to ξ and η.

Prop. 9.12. Let W, X and Y be non-degenerate finite-dimensional real orthogonal spaces. Then

(i) the map $L(X,Y) \rightarrow L(Y,X)$; $t \rightsquigarrow t^*$ is linear,
(ii) for any $t \in L(X,Y)$, $(t^*)^* = t$,
(iii) for any $t \in L(X,Y)$, $u \in L(W,X)$, $(tu)^* = u^*t^*$. □

Cor. 9.13. Let X be a non-degenerate finite-dimensional real orthogonal space. Then the map

$$\text{End}\, X \rightarrow \text{End}\, X; \quad t \rightsquigarrow t^*$$

is an *anti-involution* of the real algebra $\text{End}\, X = L(X,X)$. □

Prop. 9.14. Let $t : X \rightarrow Y$ be a linear map of a non-degenerate finite-dimensional orthogonal space X, with correlation ξ to an orthogonal space Y with correlation η. Then t is orthogonal if, and only if, $t^*t = 1_X$.

Proof Since ξ is bijective,

$$t^L\eta t = \xi \quad \Leftrightarrow \quad \xi^{-1}t^L\eta t = t^*t = 1_X. \qquad \square$$

Cor. 9.15. A linear automorphism $t : X \rightarrow X$ of a non-degenerate finite-dimensional orthogonal space X is orthogonal if, and only if, $t^* = t^{-1}$. □

Prop. 9.16. Let $t : X \rightarrow X$ be a linear transformation of a finite-dimensional non-degenerate orthogonal space X. Then $x \cdot t(x) = 0$, for all $x \in X$, if, and only if, $t^* = -t$.

Proof $x \cdot t(x) = 0$, for all $x \in X$,

$$\begin{aligned}
&\Leftrightarrow \quad x \cdot t(x) + x' \cdot t(x') - (x - x') \cdot t(x - x') \\
&\qquad = x \cdot t(x') + x' \cdot t(x) = 0, \text{ for all } x, x' \in X \\
&\Leftrightarrow \quad t(x') \cdot x + t^*(x') \cdot x = 0, \text{ for all } x, x' \in X \\
&\Leftrightarrow \quad (t + t^*)(x') = 0, \text{ for all } x' \in X, \text{ since } \ker X = 0 \\
&\Leftrightarrow \quad t + t^* = 0. \qquad \square
\end{aligned}$$

Cor. 9.17. Let $t : X \rightarrow X$ be an orthogonal transformation of a finite-dimensional non-degenerate orthogonal space X. Then $x \cdot t(x) = 0$, for all $x \in X$, if, and only if, $t^2 = -1_X$. □

Exercise 9.18. Let t: $X \rightarrow X$ be a linear transformation of a finite-dimensional orthogonal space X, and suppose that t^2 is orthogonal. Discuss whether or not t is necessarily orthogonal. Discuss, in particular, the case where X is positive-definite. □

Examples of adjoints

The next few propositions show what the adjoint of a linear map looks like in several important cases. It is convenient throughout these examples to use the same letter to denote not only a linear map $t: \mathbf{R}^p \longrightarrow \mathbf{R}^q$ but also its $q \times p$ matrix over $\mathbf{R}$. Elements of $\mathbf{R}^p$ are identified with column matrices and elements of $(\mathbf{R}^p)^L$ with row matrices, as at the end of Chapter 3. (We write $(\mathbf{R}^p)^L$ and not $(\mathbf{R}^p)^{\mathscr{L}}$, since $\mathbf{R}^p$ is finite-dimensional.) For any linear map $t: \mathbf{R}^p \longrightarrow \mathbf{R}^q$, t^τ denotes both the transpose of the matrix of t and also the linear map $\mathbf{R}^q \longrightarrow \mathbf{R}^p$ represented by this matrix.

Prop. 9.19. Let $t: \mathbf{R}^{0,p} \longrightarrow \mathbf{R}^{0,q}$ be a linear map. Then $t^* = t^\tau$.

Proof For any $x \in \mathbf{R}^p$, $y \in \mathbf{R}^q$,

$$y \cdot t(x) = y^\tau t x = (t^\tau y)^\tau x = t^\tau(y) \cdot x.$$

Now $\mathbf{R}^{0,p}$ is non-degenerate, implying that the adjoint of t is unique. So $t^* = t^\tau$. □

The case $p = q = 2$ is worth considering in more detail.

Example 9.20. Let t: $\mathbf{R}^{0,2} \longrightarrow \mathbf{R}^{0,2}$ be a linear map with matrix $\begin{pmatrix} a & c \\ b & d \end{pmatrix}$. Then t^* has matrix $\begin{pmatrix} a & b \\ c & d \end{pmatrix}$ and t is therefore orthogonal if, and only if,

$$\begin{pmatrix} a & b \\ c & d \end{pmatrix}\begin{pmatrix} a & c \\ b & d \end{pmatrix} = \begin{pmatrix} 1 & 0 \\ 0 & 1 \end{pmatrix},$$

that is, if, and only if, $a^2 + b^2 = c^2 + d^2 = 1$ and $ac + bd = 0$, from which it follows that the matrix is either of the form $\begin{pmatrix} a & -b \\ b & a \end{pmatrix}$ or of the form $\begin{pmatrix} a & b \\ b & -a \end{pmatrix}$, with $a^2 + b^2 = 1$. The map in the first case is a rotation and in the second case an antirotation, as can be verified by examination of the sign of the determinant. □

To simplify notations in the next two propositions $\mathbf{R}^{p,q}$ and $\mathbf{R}^n_{\text{hb}}$ are identified, as linear spaces, with $\mathbf{R}^p \times \mathbf{R}^q$ and $\mathbf{R}^n \times \mathbf{R}^n$, respectively. The entries in the matrices are linear maps.

Prop. 9.21. Let $t: \mathbf{R}^{p,q} \longrightarrow \mathbf{R}^{p,q}$ be linear, and let $t = \begin{pmatrix} a & c \\ b & d \end{pmatrix}$. Then $t^* = \begin{pmatrix} a^\tau - & b^\tau \\ -c^\tau & d^\tau \end{pmatrix}$.

Proof For all $(x,y), (x',y') \in \mathbf{R}^{p,q}$,

$$\begin{aligned}(x,y)\cdot(ax' + cy', bx' + dy') &= -x^\tau(ax' + cy') + y^\tau(bx' + dy') \\ &= -(a^\tau x)^\tau x' - (c^\tau x)^\tau y' + (b^\tau y)^\tau x' + (d^\tau y)^\tau y' \\ &= -(a^\tau x - b^\tau y)^\tau x' + (-c^\tau x + d^\tau y)^\tau y' \\ &= (a^\tau x - b^\tau y, -c^\tau x + d^\tau y)\cdot(x',y'). \quad \square\end{aligned}$$

Cor. 9.22. For such a linear map t, $\det t^* = \det t$, and, if t is orthogonal, $(\det t)^2 = 1$. □

Prop. 9.23. Let $t : \mathbf{R}^n_{\mathrm{hb}} \to \mathbf{R}^n_{\mathrm{hb}}$ be linear, where $t = \begin{pmatrix} a & c \\ b & d \end{pmatrix}$. Then $t^* = \begin{pmatrix} d^\tau & c^\tau \\ b^\tau & a^\tau \end{pmatrix}$. □

Orthogonal annihilators

Let X be a finite-dimensional real orthogonal space with correlation $\xi : X \to X^L$ and let W be a linear subspace of X. In Chapter 6 the dual annihilator $W^@$ of W was defined to be the subspace of X^L annihilating W, namely

$$\{\beta \in X^L : \text{for all } w \in W, \beta(w) = 0\}.$$

By Prop. 6.28, $\dim W^@ = \dim X - \dim W$. We now define $W^\perp = \xi^{-1}(W^@)$. That is,

$$W^\perp = \{a \in X : a\cdot w = 0, \text{ for all } w \in W\}.$$

This linear subspace of X is called the *orthogonal annihilator* of W in X. Its dimension is not less than $\dim X - \dim W$, being equal to this when ξ is bijective, that is, when X is non-degenerate.

A linear complement Y of W in X that is also a linear subspace of W is said to be *an orthogonal complement* of W in X. The direct sum decomposition $W \oplus Y$ of X is then said to be *an orthogonal decomposition* of X.

Prop. 9.24. Let W be a linear subspace of a finite-dimensional real orthogonal space X. Then $\ker W = W \cap W^\perp$. □

Prop. 9.25. Let W be a linear subspace of a non-degenerate finite-dimensional real orthogonal space X. Then $X = W \oplus W^\perp$ if, and only if W is non-degenerate, $W^\perp$, in this case, being the unique orthogonal complement of W in X. □

Proof $\Leftarrow$: Suppose W is non-degenerate. Then $W \cap W^\perp = \{0\}$. Also, since the correlation on X is injective, $\dim W^\perp = \dim W^@ = \dim X - \dim W$, implying, by Prop. 6.29, that

$$\dim(W + W^\perp) = \dim W + \dim W^\perp - \dim(W \cap W^\perp) = \dim X,$$

and therefore, by Cor. 6.17, that $W + W^{\perp} = X$. It follows that $X = W \oplus W^{\perp}$.

$\Rightarrow$: Suppose W is degenerate. Then $W \cap W^{\perp} \neq \{0\}$, implying that X is not the direct sum of W and $W^{\perp}$. □

Cor. 9.26. Let a be an element of a non-degenerate finite-dimensional real orthogonal space X. Then $X = \text{im}\{a\} \oplus \text{im}\{a\}^{\perp}$ if, and only if, a is invertible. □

Prop. 9.27. Let W be a linear subspace of a non-degenerate finite-dimensional real orthogonal space X. Then $(W^{\perp})^{\perp} = W$. □

Prop. 9.28. Let V and W be linear subspaces of a finite-dimensional orthogonal space X. Then $V \subset W^{\perp} \Leftrightarrow W \subset V^{\perp}$. □

A first application of the orthogonal annihilator is to isotropic subspaces.

Prop. 9.29. Let W be a linear subspace of a finite-dimensional real orthogonal space X. Then W is isotropic if, and only if, $W \subset W^{\perp}$. □

Cor. 9.30. Let W be an isotropic subspace of a non-degenerate finite-dimensional real orthogonal space X. Then $\dim W \leqslant \frac{1}{2} \dim X$. □

By this corollary it is only just possible for a non-degenerate finite-dimensional real orthogonal space to be neutral. As we noted earlier, $\mathbf{R}^n_{\text{hb}}$, and therefore also $\mathbf{R}^{n,n}$, is such a space.

The basis theorem

Let W be a linear subspace of a non-degenerate real orthogonal space X. Then, by Prop. 9.25, $X = W \oplus W^{\perp}$ if, and only if, W is non-degenerate. Moreover, if W is non-degenerate, then, by Prop. 9.27, $W^{\perp}$ also is non-degenerate.

These remarks lead to the basis theorem, which we take in two stages.

Theorem 9.31. An n-dimensional non-degenerate real orthogonal space, with $n > 0$, is expressible as the direct sum of n non-degenerate mutually orthogonal lines.

Proof By induction. The basis, with $n = 1$, is a tautology. Suppose now the truth of the theorem for any n-dimensional space and consider an $(n + 1)$-dimensional orthogonal space X. Since the scalar product on X is not zero, there exists an invertible element $a \in X$ and therefore a non-degenerate line, $\text{im}\{a\}$, in X. So $X = (\text{im}\{a\}) \oplus (\text{im}\{a\})^{\perp}$. By hypothesis, $(\text{im}\{a)\}^{\perp}$, being n-dimensional, is the direct sum of n non-degenerate mutually orthogonal lines; so the step is proved, and hence the theorem. □

A linearly free subset S of a real orthogonal space X, such that any two distinct elements of S are mutually orthogonal, with the square of any element of the basis equal to 0, -1 or 1, is said to be an *orthonormal subset* of X. If S also spans X, then S is said to be an *orthonormal basis* for X.

Theorem 9.32. (*Basis theorem.*) Any finite-dimensional orthogonal space X has an orthonormal basis.

Proof Let X' be a linear complement in X of $\ker X$. Then, by Prop. 9.5, X' is a non-degenerate subspace of X and so has an orthonormal basis, B, say, by Theorem 9.31 and Prop. 9.1. Let A be any basis for $\ker X$. Then $A \cup B$ is an orthonormal basis for X. □

Cor. 9.33. (The *classification theorem*, continued in Cor. 9.48.) Any non-degenerate finite-dimensional orthogonal space X is isomorphic to $\mathbf{R}^{p,q}$ for some finite p, q. □

Cor. 9.34. For any orthogonal automorphism $t : X \to X$ of a non-degenerate finite-dimensional orthogonal space X, $(\det t)^2 = 1$, and, for any rotation t of X, $\det t = 1$.

Proof Apply Cor. 9.33 and Cor. 9.22. □

Reflections

Prop. 9.35. Let $W \oplus Y$ be an orthogonal decomposition of an orthogonal space X. Then the map

$$X \to X;\quad w + y \rightsquigarrow w - y,$$

where $w \in W$ and $y \in Y$, is orthogonal. □

Such a map is said to be *a reflection* of X in W. When $Y = W^\perp$ the map is said to be *the* reflection of X in W. A reflection of X in a linear hyperplane W is said to be a *hyperplane reflection* of X. Such a reflection exists if $\dim X > 0$, for the hyperplane can be chosen to be an orthogonal complement of $\operatorname{im}\{a\}$, where a is either an element of $\ker X$ or an invertible element of X.

Prop. 9.36. A hyperplane reflection of a finite-dimensional orthogonal space X is an antirotation of X. □

Cor. 9.37. Let X be an orthogonal space of positive finite dimension. Then $SO(X)$ is a normal subgroup of $O(X)$ and $O(X)/SO(X)$ is isomorphic to the group $\mathbf{Z}_2$. □

Here $\mathbf{Z}_2$ is the additive group consisting of the set $2 = \{0,1\}$ with addition mod 2. The group is isomorphic to the multiplicative group $\{1,-1\}$, a group that will later also be denoted by S^0.

Exercise 9.38. Let $\pi : O(X) \to \mathbf{Z}_2$ be the group surjection, with kernel $SO(X)$, defined in Cor. 9.37. For which values of dim X is there a group injection $s : \mathbf{Z}_2 \to O(X)$, such that $\pi s = 1_{\mathbf{Z}_2}$ and such that im s is a normal subgroup of $O(X)$? □

Exercise 9.39. Let X be an orthogonal space of positive finite dimension. For which values of dim X, if any, is the group $O(X)$ isomorphic to the group product $SO(X) \times \mathbf{Z}_2$? (It may be helpful to refer back to the last section of Chapter 5.) □

If a is an invertible element of a finite-dimensional orthogonal space X, then the hyperplane $(\operatorname{im}\{a\})^\perp$ is the unique orthogonal complement of the line $\operatorname{im}\{a\}$. The reflection of X in this hyperplane will be denoted by ρ_a.

Prop. 9.40. Suppose that a and b are invertible elements of a finite-dimensional orthogonal space X, such that $a^{(2)} = b^{(2)}$. Then a may be mapped to b either by a single hyperplane reflection of X or by the composite of two hyperplane reflections of X.

Proof By Prop. 9.2, either $a - b$ or $a + b$ is invertible, $a - b$ and $a + b$ being in any case mutually orthogonal. In the first case, ρ_{a-b} exists and

$$\begin{aligned}\rho_{a-b}(a) &= \rho_{a-b}(\tfrac{1}{2}(a - b) + \tfrac{1}{2}(a + b)) \\ &= -\tfrac{1}{2}(a - b) + \tfrac{1}{2}(a + b) = b.\end{aligned}$$

In the second case, ρ_{a+b} exists and

$$\rho_b\rho_{a+b}(a) = \rho_b(-b) = b. \qquad \square$$

Theorem 9.41. Any orthogonal transformation $t : X \to X$ of a non-degenerate finite-dimensional orthogonal space X is expressible as the composite of a finite number of hyperplane reflections of X, the number being not greater than 2 dim X, or, if X is positive-definite, dim X.

Indication of proof This is a straightforward induction based on Prop. 9.40. Suppose the theorem true for n-dimensional spaces and let X be $(n + 1)$-dimensional, with an orthonormal basis $\{e_i : i \in n + 1\}$, say. Then, by Prop. 9.40, there is an orthogonal map $u : X \to X$, either a hyperplane reflection or the composite of two hyperplane reflections of X, such that $ut(e_n) = e_n$. The map ut induces an orthogonal transformation of the n-dimensional linear image of $\{e_i : i \in n\}$ and the inductive hypothesis is applicable to this. The details are left as an exercise. □

By Prop. 9.36 the number of reflections composing t is even when t is a rotation and odd when t is an antirotation. The following corollaries are important. To simplify notations we write $\mathbf{R}^2$ for $\mathbf{R}^{0,2}$ and $\mathbf{R}^3$ for $\mathbf{R}^{0,3}$.

Cor. 9.42. Any antirotation of $\mathbf{R}^2$ is a reflection in some line of $\mathbf{R}^2$. □

Cor. 9.43. Any rotation of $\mathbf{R}^3$ is the composite of two plane reflections of $\mathbf{R}^3$. □

Prop. 9.44. The only rotation t of $\mathbf{R}^2$ leaving a non-zero point of $\mathbf{R}^2$ fixed is the identity.

Proof Let a be such a point and let $b_0 = \lambda a$, where $\lambda^{-1} = \sqrt{(a^{(2)})}$. Then there exists a unique element $b_1 \in \mathbf{R}^2$ such that (b_0, b_1) is a positively oriented orthonormal basis for $\mathbf{R}^2$. Since t is a rotation leaving b_0 fixed, b_1 also is left fixed by t. So t is the identity. □

Prop. 9.45. Any rotation t of $\mathbf{R}^3$, other than the identity, leaves fixed each point of a unique line in $\mathbf{R}^3$.

Proof Let $t = \rho_b \rho_a$, where a and b are invertible elements of $\mathbf{R}^3$. Then either $b = \lambda a$, for some $\lambda \in \mathbf{R}$, in which case $\rho_b = \rho_a$ and $t = 1$, or there exists an element c, orthogonal to the plane spanned by a and b. Since $\rho_a(c)$ and $\rho_b(c)$ are each equal to $-c$, it follows that $t(c) = c$ and therefore that each point of the line $\operatorname{im}\{c\}$ is left fixed by t.

If each point of more than one line is left fixed then, if orientations are to be preserved, t must be the identity, by an argument similar to that used in the proof of Prop. 9.44.

Hence the result. □

The line left fixed is called the *axis* of the rotation. These special cases will be studied further in Chapter 10.

Theorem 9.41 has an important part to play in Chapter 13.

Signature

Prop. 9.46. Let $U \oplus V$ and $U' \oplus V'$ be orthogonal decompositions of an orthogonal space X such that, for all non-zero $u' \in U'$, $u'^{(2)} < 0$, and, for all $v \in V$, $v^{(2)} \geqslant 0$. Then the projection of X on to U with kernel V maps U' injectively to U.

Proof Let $u' = u + v$ be any element of U'. Then $u'^{(2)} = u^{(2)} + v^{(2)}$, so that, if $u = 0$, $u'^{(2)} = v^{(2)}$, implying that $u'^{(2)} = 0$ and therefore that $u' = 0$. □

Cor. 9.47. If, also, $u^{(2)} < 0$, for all non-zero $u \in U$, and $v'^{(2)} \geqslant 0$, for all $v' \in V'$, then $\dim U = \dim U'$.

Proof By Prop. 9.46, $\dim U' \leqslant \dim U$. By a similar argument $\dim U \leqslant \dim U'$. □

Cor. 9.48. (Continuation of the *classification theorem*, Cor. 9.33.) The orthogonal spaces $\mathbf{R}^{p,q}$ and $\mathbf{R}^{p',q'}$ are isomorphic if, and only if, $p = p'$ and $q = q'$. □

Cor. 9.49. Let $\begin{pmatrix} a & c \\ b & d \end{pmatrix} : \mathbf{R}^{p,q} \rightarrow \mathbf{R}^{p,q}$ be an orthogonal automorphism of $\mathbf{R}^{p,q}$, $\mathbf{R}^{p,q}$ being identified as usual with $\mathbf{R}^p \times \mathbf{R}^q$. Then $a : \mathbf{R}^p \rightarrow \mathbf{R}^p$ and $d : \mathbf{R}^q \rightarrow \mathbf{R}^q$ are linear isomorphisms. □

The orientations of $\mathbf{R}^p$ and $\mathbf{R}^q$ will be called the *semi-orientations* of the *orthogonal* space $\mathbf{R}^{p,q}$, and an orthogonal automorphism $\begin{pmatrix} a & c \\ b & d \end{pmatrix}$ of $\mathbf{R}^{p,q}$ will be said to *preserve* the semi-orientations of $\mathbf{R}^{p,q}$ if a preserves the orientations of $\mathbf{R}^p$ and d preserves the orientations of $\mathbf{R}^q$.

Exercise 9.50. Let $SO^+(1,1)$ denote the set of orthogonal automorphisms of $\mathbf{R}^{1,1}$ that preserve both the semi-orientations of $\mathbf{R}^{1,1}$. Prove that $SO^+(1,1)$ is a normal subgroup of $SO(1,1)$ with quotient group isomorphic to $\mathbf{Z}_2$.

(Show first that any element of $SO^+(1,1)$ may be written in the form

$$\begin{pmatrix} \cosh u & \sinh u \\ \sinh u & \cosh u \end{pmatrix}, \quad \text{where } u \in \mathbf{R}.$$

Note that it has to be proved that $SO^+(1,1)$ is a subset of $SO(1,1)$.) □

Groups $SO^+(p,q)$ analogous to $SO^+(1,1)$ exist for arbitrary finite p and q. These groups, the *proper Lorentz groups*, are discussed on pages 268 and 427 (Prop. 20.96).

It follows from Cor. 9.33 that the quadratic form $x \rightsquigarrow x^{(2)}$ of a finite-dimensional orthogonal space X may be represented as a 'sum of squares'

$$x = \sum_{i \in n} x_i e_i \rightsquigarrow \sum_{i \in n} \zeta_i x_i^2$$

with respect to some suitable orthonormal basis $\{e_i : i \in n\}$ for X, with $\zeta_i = e_i^{(2)} = 0, -1$, or 1, for each $i \in n$. By Cor. 9.48, the number, p, of negative squares and the number, q, of positive squares are each independent of the basis chosen and $\dim \ker X + p + q = \dim X$; that is, $\operatorname{rk} X = p + q$. The pair of numbers (p,q) will be called the *signature* of the quadratic form and of the orthogonal space. The number $\inf\{p,q\}$ will be called the *index* of the form and of the orthogonal space.

(The definitions of 'signature' and 'index' are not standard in the literature, and almost all possibilities occur. For example, the signature is frequently defined to be $-p + q$ and the index to be p. The number we have called the index is sometimes called the *Witt index* of the orthogonal space.)

The geometrical significance of the index is brought out in the next proposition.

Prop. 9.51. Let W be an isotropic subspace of the orthogonal space $\mathbf{R}^{p,q}$. Then $\dim W \leqslant \inf\{p,q\}$.

Proof There is an obvious orthogonal decomposition $\mathbf{R}^{p,q} = X \oplus Y$, where $X \cong \mathbf{R}^{p,0}$, and $Y \cong \mathbf{R}^{0,q}$. As in the proof of Prop. 9.46, the restrictions to W of the projections of $\mathbf{R}^{p,q}$ on to X and Y, with kernels Y and X respectively, are injective. Hence the result. □

The bound is attained since there is a subspace of $\mathbf{R}^{p,q}$ isomorphic to $\mathbf{R}^{r,r}$, where $r = \inf\{p,q\}$, and $\mathbf{R}^{r,r}$ is neutral.

Witt decompositions

A *Witt decomposition* of a non-degenerate finite-dimensional real orthogonal space X is a direct sum decomposition of X of the form $W \oplus W' \oplus (W \oplus W')^\perp$, where W and W' are isotropic subspaces of X. (Some authors restrict the use of the term to the case where $\dim W = \operatorname{index} X$.)

One application of a Witt decomposition $X = W \oplus W' \oplus (W \oplus W')^\perp$ is to the representation as linear subspaces of X of the various quotient spaces and dual spaces involving the isotropic subspace W and its annihilators $W^@$ and $W^\perp$:

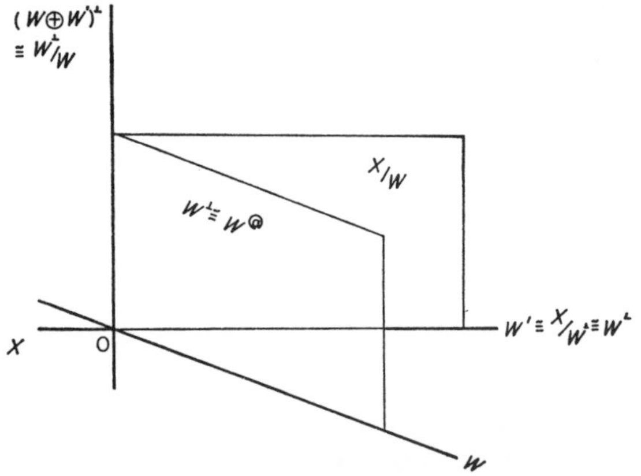

namely, $W^\perp/W$ may be represented by $(W \oplus W')^\perp$
X/W may be represented by $W' \oplus (W \oplus W')^\perp$
and $X/W^\perp$ and W^L may be represented by W'.

We shall return to this in Chapter 12.

The exact sequence of linear maps

$$\{0\} \to W^\perp/W \to X/W \to X/W^\perp \to \{0\}$$

relating the various quotient spaces has already been described in Chapter 5, where it was assumed only that $W^\perp$ was a linear subspace of X and W a linear subspace of $W^\perp$.

Prop. 9.52. Let X be a non-degenerate finite-dimensional real orthogonal space with a one-dimensional isotropic subspace W. Then there exists another one-dimensional isotropic subspace W' such that the plane spanned by W and W' is a hyperbolic plane.

Proof Let w be a non-zero element of W. Since X is non-degenerate, there exists $x \in X$ such that $w \cdot x \neq 0$ and x may be chosen so that $w \cdot x = 1$. Then for any $\lambda \in \mathbf{R}$, $(x + \lambda w)^{(2)} = x^{(2)} + 2\lambda$, this being zero if $\lambda = -\frac{1}{2}x^{(2)}$. Let W' be the line spanned by $w' = x - \frac{1}{2}x^{(2)}w$. The line is isotropic since $(w')^{(2)} = 0$, and the plane spanned by w and w' is isomorphic to $\mathbf{R}^{1,1}$ since $w \cdot w' = w' \cdot w = 1$, and therefore, for any $a, b \in \mathbf{R}$, $(aw + bw') \cdot w = b$ and $(aw + bw') \cdot w' = a$, both being zero only if $a = b = 0$. □

Cor. 9.53. Let W be an isotropic subspace of a non-degenerate finite-dimensional real orthogonal space X. Then there exists an isotropic subspace W' of X such that $X = W \oplus W' \oplus (W \oplus W')^\perp$ (a Witt decomposition of X). □

Cor. 9.54. Any non-degenerate finite-dimensional real orthogonal space may be expressed as the direct sum of a finite number of hyperbolic planes and a positive- or negative-definite subspace, any two components of the decomposition being mutually orthogonal. □

Neutral spaces

By Prop. 9.51 a non-degenerate finite-dimensional orthogonal space is neutral if, and only if, its signature is (n,n), for some finite number n, that is, if, and only if, it is isomorphic to $\mathbf{R}_{\mathrm{hb}}^n$, or, equivalently, to $\mathbf{R}^{n,n}$, for some n. The following proposition sometimes provides a quick method of detecting whether or not an orthogonal space is neutral.

Prop. 9.55. A non-degenerate finite-dimensional real orthogonal space X is neutral if, and only if, there exists a linear map $t : X \rightarrow X$ such that $t^*t = -1$.

$$(t^*t = -1 \Leftrightarrow \text{for all } x \in X,\ (t(x))^{(2)} = -x^{(2)}.) \qquad \square$$

The next will be of use in Chapter 13.

Prop. 9.56. Let W be a possibly degenerate n-dimensional real orthogonal space. Then W is isomorphic to an orthogonal subspace of $\mathbf{R}^{n,n}$. $\square$

Positive-definite spaces

By Cor. 9.33 any finite-dimensional positive-definite orthogonal space is isomorphic to $\mathbf{R}^{0,n}$ for some n. Throughout the remainder of this chapter this orthogonal space will be denoted simply by $\mathbf{R}^n$.

Let X be a positive-definite space. For all $a, b \in X$ the *norm* of a is, by definition,

$$|a| = \sqrt{(a^{(2)})},$$

defined for all $a \in X$ since $a^{(2)} \geqslant 0$, and the *distance of a from b* or the *length* of the line-segment $[a,b]$ is by definition $|a - b|$. In particular, for all $\lambda \in \mathbf{R}(= \mathbf{R}^{0,1})$, $\lambda^{(2)} = \lambda^2$, and $|\lambda| = \sqrt{(\lambda^{(2)})}$ is the usual absolute value.

Prop. 9.57. Let $a, b, c \in \mathbf{R}$. Then the quadratic form

$$f : \mathbf{R}^2 \rightarrow \mathbf{R};\quad (x,y) \rightsquigarrow ax^2 + 2bxy + cy^2$$

is positive-definite if, and only if, $a > 0$ and $ac - b^2 > 0$.

Proof $\Rightarrow$: Suppose that f is positive-definite. Then $f(1,0) = a > 0$ and $f(-b,a) = a(ac - b^2) > 0$. So $a > 0$ and $ac - b^2 > 0$.

$\Leftarrow$: If $a > 0$, then, for all $x, y \in \mathbf{R}^2$,

$$f(x,y) = a^{-1}((ax + by)^2 + (ac - b^2)y^2).$$

It follows that, if also $ac - b^2 > 0$, then $f(x \cdot y)$ is non-negative, and is zero if, and only if, $ax + by = 0$ and $y = 0$, that is, if, and only if, $x = 0$ and $y = 0$. $\square$

Prop. 9.58. Let X be a positive-definite space. Then for all $a, b \in X$, $\lambda \in \mathbf{R}$,

(1) $|a| \geqslant 0$,
(2) $|a| = 0 \Leftrightarrow a = 0$,
(3) $|a - b| = 0 \Leftrightarrow a = b$,
(4) $|\lambda a| = |\lambda|\,|a|$,
(5) a and b are collinear with 0 if, and only if, $|b|\,a = \pm|a|\,b$,

(6) a and b are collinear with 0 and not on opposite sides of 0 if, and only if, $|b|a = |a|b$,
(7) $|a \cdot b| \leqslant |a||b|$,
(8) $a \cdot b \leqslant |a||b|$,
(9) $|a + b| \leqslant |a| + |b|$ (the *triangle inequality*),
(10) $||a| - |b|| \leqslant |a - b|$,

with equality in (7) if, and only if, a and b are collinear with 0 and in (8), (9) and (10) if, and only if, a and b are collinear with 0 and not on opposite sides of 0.

Proof (1), (2) and (3) are immediate consequences of the positive-definiteness of the quadratic form and the definition of the norm.

(4) For all $a \in x$, $\lambda \in \mathbf{R}$,

$$|\lambda a|^2 = (\lambda a)^{(2)} = \lambda^2 a^{(2)} = |\lambda|^2 |a|^2.$$

So $$|\lambda a| = |\lambda||a|.$$

(5) $\Leftarrow$: If $a = b = 0$ there is nothing to prove. If either a of b is not zero, then either

$$a = \pm\frac{|a|}{|b|}b \quad \text{or} \quad b = \pm\frac{|b|}{|a|}a.$$

In either case a and b are collinear with 0.
$\Rightarrow$: If a and b are collinear with 0, then either there exists $\lambda \in \mathbf{R}$ such that $b = \lambda a$, in which case $|b| = \pm\lambda |a|$ and

$$|b|a = \pm\lambda|a|a = \pm|a|b,$$

or there exists $\mu \in \mathbf{R}$ such that $a = \mu b$, from which it follows, similarly, that $|a|b = \pm|b|a$.

(6) This follows at once from (5).

(7) For all $x, y \in \mathbf{R}$,

$$0 \leqslant |xa + yb|^2 = (xa + yb)^{(2)} = a^{(2)}x^2 + 2(a \cdot b)xy + b^{(2)}y^2.$$

Therefore, by Prop. 9.57, $(a \cdot b)^2 \leqslant a^{(2)}b^{(2)} = |a|^2|b|^2$ and so $|a \cdot b| \leqslant |a||b|$.

(8) This follows by transitivity from the inequalities $a \cdot b \leqslant |a \cdot b|$ and $|a \cdot b| \leqslant |a||b|$.

(9) $|a + b| \leqslant |a| + |b|$

$$\begin{aligned}
&\Leftrightarrow & (a+b)^{(2)} &\leqslant (|a| + |b|)^2 \\
&\Leftrightarrow & a^{(2)} + 2a \cdot b + b^{(2)} &\leqslant a^{(2)} + 2|a||b| + b^{(2)} \\
&\Leftrightarrow & a \cdot b &\leqslant |a||b|,
\end{aligned}$$

which is (8).

(10) For all $a, b \in X$,

$$|a| = |(a - b) + b| \leqslant |a - b| + |b|$$

and

$$|b| = |(b-a)+a| \leqslant |a-b| + |a|.$$

So $|a| - |b| \leqslant |a-b|$ and $|b| - |a| \leqslant |a-b|$; that is,

$$||a| - |b|| \leqslant |a-b|.$$

Finally,

$$\begin{aligned} |b|a = |a|b &\Leftrightarrow |b|a - |a|b = 0 \\ &\Leftrightarrow (|b|a - |a|b)^{(2)} = 0 \\ &\Leftrightarrow |b|^2|a|^2 - 2|a||b|a\cdot b + |a|^2|b|^2 = 0. \end{aligned}$$

So

$$\begin{aligned} |b|a = |a|b &\Leftrightarrow a\cdot b = |a||b| &(8)\\ &\Leftrightarrow |a+b| = |a| + |b|, &(9) \end{aligned}$$

while $|b|a = -|a|b \Leftrightarrow a\cdot b = -|a||b|$, so

$$|a\cdot b| = |a||b| \Leftrightarrow |b|a = \pm|a|b. \qquad (7)$$

Also $|a| = |a-b| + |b|$

$$\begin{aligned} &\Leftrightarrow |b|(a-b) = |a-b|b = (|a| - |b|)b \\ &\Leftrightarrow |b|a = |a|b \end{aligned}$$

and, similarly, $|b| = |a-b| + |a| \Leftrightarrow |b|a = |a|b$. So

$$||a| - |b|| = |a-b| \Leftrightarrow |b|a = |a|b. \quad \square \qquad (10)$$

Inequality (7) of Prop. 9.58 is known as the *Cauchy–Schwarz inequality*. It follows from this inequality that, for all non-zero $a, b \in X$,

$$-1 \leqslant \frac{a\cdot b}{|a||b|} \leqslant 1,$$

$\dfrac{a\cdot b}{|a||b|}$ being equal to 1 if, and only if, b is a positive multiple of a, and equal to -1 if, and only if, b is a negative multiple of a. The *absolute angle* between the line-segments $[0,a]$ and $[0,b]$ is defined by

$$\cos\theta = \frac{a\cdot b}{|a||b|}, \quad 0 \leqslant \theta \leqslant \pi,$$

with $a\cdot b = 0 \Leftrightarrow \cos\theta = 0 \Leftrightarrow \theta = \pi/2$, this being consistent with the ordinary usage of the word 'orthogonal'.

A map $t: X \longrightarrow Y$ between positive-definite spaces X and Y is said to *preserve scalar product* if, for all $a, b \in X$, $t(a)\cdot t(b) = a\cdot b$, to *preserve norm* if, for all $a \in X$, $|t(a)| = |a|$, to *preserve distance* if, for all $a, b \in X$, $|t(a) - t(b)| = |a-b|$, and to *preserve zero* if $t(0) = 0$.

According to our earlier definition, t is *orthogonal* if it is linear and preserves scalar product.

Of the various definitions of an orthogonal map proved equivalent in

Prop. 9.59, (iii) is probably the most natural from a practical point of view, being closest to our intuition of a rotation or antirotation.

Prop. 9.59. Let $t: X \rightarrow Y$ be a map between positive-definite spaces X and Y. Then the following are equivalent:

(i) t is orthogonal,
(ii) t is linear and preserves norm,
(iii) t preserves distance and zero,
(iv) t preserves scalar product.

Proof (i) $\Rightarrow$ (ii): Suppose t is orthogonal. Then t is linear and, for any $a, b \in X$, $t(a)\cdot t(b) = a \cdot b$. In particular, $(t(a))^{(2)} = a^{(2)}$, implying that $|t(a)| = |a|$. That is, t is linear and preserves norm.

(ii) $\Rightarrow$ (iii): Suppose t is linear and preserves norm. Then for any $a, b \in X$, $|t(a) - t(b)| = |t(a - b)| = |a - b|$, while $t(0) = 0$. That is, t preserves distance and zero.

(iii) $\Rightarrow$ (iv): Suppose t preserves distance and zero. Then, for any $a \in X$, $|t(a)| = |t(a) - 0| = |t(a) - t(0)| = |a - 0| = |a|$. So t preserves norm. It follows that, for all $a, b \in X$,

$$\begin{aligned} t(a)\cdot t(b) &= \tfrac{1}{2}(t(a)^{(2)} + t(b)^{(2)} - (t(a) - t(b))^{(2)}) \\ &= \tfrac{1}{2}(a^{(2)} + b^{(2)} - (a - b)^{(2)}) \\ &= a \cdot b. \end{aligned}$$

That is, t preserves scalar product.

(iv) $\Rightarrow$ (i): Suppose that t preserves scalar product. Then, for all $a, b \in X$ and all $\lambda \in \mathbf{R}$,

$$(t(a + b) - t(a) - t(b))^{(2)} = ((a + b) - a - b)^{(2)} = 0$$

and

$$(t(\lambda a) - \lambda t(a))^{(2)} = ((\lambda a) - \lambda a)^{(2)} = 0,$$

implying that $t(a + b) = t(a) + t(b)$ and $t(\lambda a) = \lambda t(a)$. That is, t is linear, and therefore orthogonal. □

Euclidean spaces

A real affine space X, with an orthogonal structure for the vector space X_* is said to be an *orthogonal affine space*, the orthogonal affine space being said to be *positive-definite* if its vector space is positive-definite. A finite-dimensional positive-definite real affine space is said to be a *euclidean space*.

Prop. 9.60. Let X be a euclidean space, let W be an affine subspace of X and let $a \in X$. Then the map $W \rightarrow \mathbf{R};\ w \rightsquigarrow |w - a|$ is bounded below and attains its infimum at a unique point of W.

Proof If $a \in W$ the infimum is 0, attained only at a.

Suppose now that $a \notin W$. Since the problem concerns only the affine subspace of X spanned by W and $\{a\}$ in which W is a hyperplane, we may, without loss of generality, assume that W is a hyperplane in X. Set $a = 0$ and let W have equation $b \cdot x = c$, say, where $b \in X$ and $c \in \mathbf{R}$, with $b \neq 0$ and $c \neq 0$. By Prop. 9.58 (8),

$$c = b \cdot x \leqslant |\, b \,|\, |\, x \,|, \quad \text{for all } x \in W,$$

with equality if, and only if, $|\, b \,|\, x = |\, x \,|\, b$; that is, $|\, x \,| \geqslant c \,|\, b \,|^{-1}$, for all $x \in W$, with equality if, and only if, $x = c\, b^{(-1)}$. □

The unique point p at which the infimum is attained is called the *foot of the perpendicular* from a to W, the *perpendicular* from a to W being the line-segment $[a,p]$.

Spheres

Let X be a euclidean space, and let $a \in X$ and $r \in \mathbf{R}^+$. Then the set $\{x \in X : |\, x - a \,| = r\}$ is called the *sphere* with *centre* a and *radius* r in X. When X is linear, the sphere $\{x \in X : |\, x \,| = 1\}$ is said to be *the unit sphere* in X. The unit sphere in $\mathbf{R}^{n+1}$ is usually denoted by S^n, and called the *unit n-sphere*. In particular,

$$S^0 = \{x \in \mathbf{R} : x^2 = 1\} = \{-1,1\},$$

$$S^1 = \{(x,y) \in \mathbf{R}^2 : x^2 + y^2 = 1\}, \text{ the } \textit{unit circle},$$

and $$S^2 = \{(x,y,z) \in \mathbf{R}^3 : x^2 + y^2 + z^2 = 1\}, \text{ the } \textit{unit sphere}.$$

In studying S^n it is often useful to identify $\mathbf{R}^{n+1}$ with $\mathbf{R}^n \times \mathbf{R}$. The points $(0,1)$ and $(0,-1)$ are then referred to, respectively, as the *North* and *South poles* of S^n.

Prop. 9.61. Let S be the unit sphere in a linear euclidean space X, and let $t : X \rightarrow X$ be a linear transformation of X such that $t_\vdash(S) \subset S$. Then t is orthogonal, and $t_\vdash(S) = S$. □

Prop. 9.62. Let S be a sphere in a euclidean space X, and let W be any affine subspace of X. Then $W \cap S$ is a sphere on W, or a point, or is null.

Proof Let a be the centre and r the radius of S and let $T = S \cap W$. Then $w \in T$ if, and only if, $w \in W$ and $|\, w - a \,| = r$. Let the foot of the perpendicular from a on W be 0. Then, for all $w \in W$, $w \cdot a = 0$ and $|\, w - a \,| = r$ if, and only if,

$$w \cdot w - 2w \cdot a + a \cdot a = w \cdot w + a \cdot a = r^2,$$

that is, if, and only if, $w^{(2)} = r^2 - a^{(2)}$.

The three cases then correspond to the cases

$$r^2 > a^{(2)}, \quad r^2 = a^{(2)} \quad \text{and} \quad r^2 < a^{(2)}. \qquad \square$$

In the particular case that W passes through the centre of S, the set $S \cap W$ is a sphere in W with centre the centre of S. Such a sphere is said to be a *great sphere* on S. If also $\dim W = 1$, $S \cap W$ consists of a pair of points on S that are *mutually antipodal*, that is, the centre of S is the mid-point of the line-segment joining them. Either of the points is said to be the *antipode* of the other.

Prop. 9.63. For any finite n, the map

$$S^n \rightarrowtail \mathbf{R}^n; \quad (u,v) \rightsquigarrow \frac{u}{1-v},$$

undefined only at the North pole, (0,1), is invertible.

Proof Since $(u,v) \in S^n$, $|u|^2 + v^2 = 1$. So, if $x = \dfrac{u}{1-v}$,

$$|x|^2 = \frac{1+v}{1-v} = \frac{2}{1-v} - 1,$$

and v, and therefore u, is uniquely determined by x. $\square$

The map defined in Prop. 9.63 is said to be the *stereographic projection* of S^n from the North pole on to its equatorial plane, $\mathbf{R}^n \times \{0\}$, identified with $\mathbf{R}^n$. For, since $(u,v) = (1-v)\left(\dfrac{u}{1-v},0\right) + v(0,1)$, the three points $(0,1)$, (u,v) and $\left(\dfrac{u}{1-v},0\right)$ are collinear.

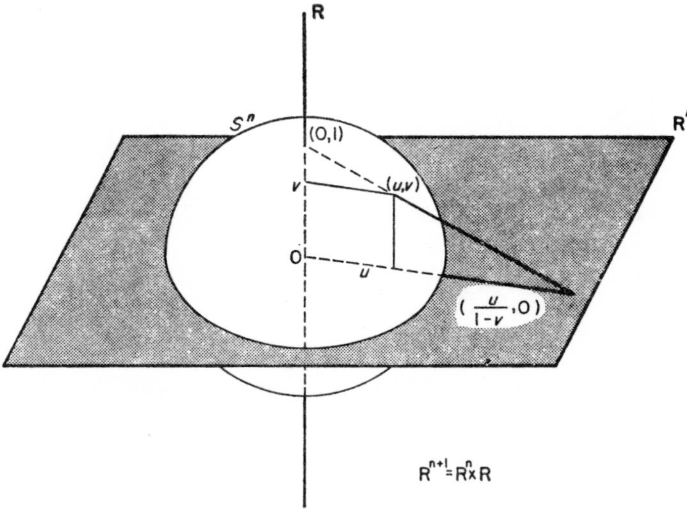

Similarly the map $S^n \to \mathbf{R}^n$; $(u,v) \rightsquigarrow \dfrac{u}{1+v}$, undefined only at the South pole, is invertible. This is stereographic projection from the South pole on to the equatorial plane.

Prop. 9.64. Let $f: S^n \to \mathbf{R}^n$ be the stereographic projection of S^n on to $\mathbf{R}^n$ from the North pole, and let T be a sphere on S^n. Then $f_\vdash(T)$ is a sphere in $\mathbf{R}^n$ if T does not pass through $(0,1)$, and is an affine subspace of $\mathbf{R}^n$ if T does pass through $(0,1)$. Conversely, every sphere or affine subspace in $\mathbf{R}^n$ is of the form $f_\vdash(T)$, where T is a sphere on S^n.

Indication of proof Let $x = \dfrac{u}{1-v}$, where $(u,v) \in \mathbf{R}^n \times \mathbf{R}$, with $v \neq 1$, and $|u|^2 + v^2 = 1$, and where $x \in \mathbf{R}^n$. Then, for any $a, c \in \mathbf{R}$ and $b \in \mathbf{R}^n$,

$$\begin{aligned} a\,|x|^2 + 2b\cdot x + c &= 0 \\ \Leftrightarrow\quad a\frac{|u|^2}{(1-v)^2} + 2b\cdot\frac{u}{1-v} + c &= 0 \\ \Leftrightarrow\quad a(1+v) + 2b\cdot u + c(1-v) &= 0 \\ \Leftrightarrow\quad 2b\cdot u + (a-c)v + (a+c) &= 0. \end{aligned}$$

The rest of the proof should now be clear. □

The following proposition will be required in Chapter 20.

Prop. 9.65. Let $t, u: \mathbf{R}^{n+1} \to \mathbf{R}^{n+1}$ be linear transformations of $\mathbf{R}^{n+1}$. Then the following statements are equivalent:

(i) For each $x \in S^n$, $(x,t(x),u(x))$ is an orthonormal 3-frame in $\mathbf{R}^{n+1}$
(ii) t and u are orthogonal, $t^2 = u^2 = -1_{n+1}$ and $ut = -tu$.

(Use Props. 9.16 and 9.61 and Cor. 9.15.) □

Complex orthogonal spaces

Much of this chapter extends at once to complex orthogonal spaces, or indeed to orthogonal spaces over any commutative field $\mathbf{K}$, $\mathbf{R}$ being replaced simply by $\mathbf{C}$, or by $\mathbf{K}$, in the definitions, propositions and theorems. Exceptions are the signature theorem, which is false, and the whole section on positive-definite spaces, which is irrelevant since positive-definiteness is not defined. The main classification theorem for complex orthogonal spaces is the following.

Theorem 9.66. Let X be a non-degenerate n-dimensional complex orthogonal space. Then X is isomorphic to $\mathbf{C}^n$ with its standard complex scalar product

$$\mathbf{C}^n \times \mathbf{C}^n \to \mathbf{C}^n : (a,b) \rightsquigarrow \sum_{i \in n} a_i b_i. \qquad \square$$

As in the real case, a neutral non-degenerate finite-dimensional orthogonal space is even-dimensional, but in the complex case we can say more.

Prop. 9.67. Let X be any non-degenerate complex orthogonal space, of even dimension $2n$. Then X is neutral, being isomorphic not only to $\mathbf{C}^{2n}$, but also to $\mathbf{C}^{n,n}$ and to $\mathbf{C}^n_{\text{hb}}$. □

Note that the analogue of Prop. 9.55, with $\mathbf{C}$ replacing $\mathbf{R}$, is false.
Finally, an exercise on adjoints.

Prop. 9.68. Let X be a non-degenerate finite-dimensional complex orthogonal space. Then the maps

$$X \times X \to \mathbf{R}; \quad (a,b) \rightsquigarrow \operatorname{re}(a \cdot b) \quad \text{and} \quad (a,b) \rightsquigarrow \operatorname{pu}(a \cdot b)$$

(cf. page 47) are symmetric scalar products on X, regarded as a real linear space. Moreover the adjoint t^* of any complex linear map $t: X \to X$ with respect to the complex scalar product, coincides with the adjoint of t with respect to either of the induced real scalar products. □

Complex orthogonal spaces are not to be confused with *unitary spaces*, complex linear spaces that carry a hermitian form. These are among the other generalizations of the material of this chapter to be discussed in Chapter 11.

Further Exercises

9.69. Let X be a finite-dimensional linear space over $\mathbf{K}$, where $\mathbf{K} = \mathbf{R}$ or $\mathbf{C}$. Prove that the map

$$(X \times X^L)^2 \to \mathbf{K}; \quad ((x,t),(y,u)) \rightsquigarrow t(y) + u(x)$$

is a neutral non-degenerate scalar product on the linear space $X \times X^L$. □

9.70. Prove that $\mathbf{R}(2)$ with the quadratic form

$$\mathbf{R}(2) \to \mathbf{R}; \quad t \rightsquigarrow \det t$$

is isomorphic as a real orthogonal space with the space $\mathbf{R}^{2,2}$, the subset $\left\{\begin{pmatrix}1 & 0\\ 0 & 1\end{pmatrix}, \begin{pmatrix}1 & 0\\ 0 & -1\end{pmatrix}, \begin{pmatrix}0 & -1\\ 1 & 0\end{pmatrix}, \begin{pmatrix}0 & 1\\ 1 & 0\end{pmatrix}\right\}$ being an orthonormal basis. Verify that $t \in \mathbf{R}(2)$ is invertible with respect to the quadratic form if, and only if, it is invertible as an element of the algebra $\mathbf{R}(2)$. □

9.71. For any $\begin{pmatrix}a & c\\ b & d\end{pmatrix} \in \mathbf{R}(2)$, define $\begin{pmatrix}a & c\\ b & d\end{pmatrix}^- = \begin{pmatrix}d & -c\\ -b & a\end{pmatrix}$, the space $\mathbf{R}(2)$ being assigned the determinant quadratic form, and let any $\lambda \in \mathbf{R}$

be identified with $\begin{pmatrix} \lambda & 0 \\ 0 & \lambda \end{pmatrix} \in \mathbf{R}(2)$. Verify that, for any $t \in \mathbf{R}(2)$, $t^- t = t^{(2)}$ and that the subset $T = \{t \in \mathbf{R}(2) : t + t^- = 0\}$ is an orthogonal subspace of $\mathbf{R}(2)$ isomorphic to $\mathbf{R}^{2,1}$. □

9.72. Let $u \in \mathbf{R}(2)$ and let $t \in T$, where T is as in Exercise 9.71. Suppose also that t is orthogonal to $u - u^-$. Show that $tu \in T$. Hence prove that any element of $\mathbf{R}(2)$ is expressible as the product of two elements of T. □

9.73. With T as in 9.71, prove that, for any invertible $u \in T$, the map $T \to T$; $t \rightsquigarrow -utu^{-1}$ is reflection in the plane $(\operatorname{im}\{u\})^\perp$. □

9.74. Prove that, for any $u \in SL(2;\mathbf{R})$, the maps $\mathbf{R}(2) \to \mathbf{R}(2)$; $t \rightsquigarrow ut$ and $t \rightsquigarrow tu$ are rotations of $\mathbf{R}(2)$. (It has to be shown not only that the quadratic form is preserved but also that orientations are preserved.) □

9.75. For any $u, v \in SL(2;\mathbf{R})$, let

$$\rho_{u,v} : \mathbf{R}(2) \to \mathbf{R}(2); \quad t \rightsquigarrow utv^{-1}$$

and let ρ_u denote the restriction of $\rho_{u,u}$ with domain and target T. Prove that the maps

$$SL(2;\mathbf{R}) \to SO(T); \quad u \rightsquigarrow \rho_u$$

and
$$SL(2;\mathbf{R}) \times SL(2;\mathbf{R}) \to SO(\mathbf{R}(2)); \quad (u,v) \rightsquigarrow \rho_{u,v}$$

are surjective group maps and that the kernel in either case is isomorphic to $\mathbf{Z}_2$. □

9.76. Let X and Y be positive-definite spaces, let Z be a linear space and let $a: X \to Z$ and $b: Y \to Z$ be linear maps such that, for all $x \in X$, $y \in Y$, $a(x) = b(y) \Rightarrow x^{(2)} = y^{(2)}$. Prove that a and b are injective and that, if X and Y are finite-dimensional,

$$\dim \{(x,y) \in X \times Y : a(x) = b(y)\} \leqslant \inf \{\dim X, \dim Y\}. \quad □$$

9.77. Prove that the graph of a linear map $t : \mathbf{R}^n \to \mathbf{R}^n$ is an isotropic subspace of $\mathbf{R}^n_{\text{hb}}$ if, and only if, $t + t^\tau = 0$. □

9.78. Find linear injections $\alpha : \mathbf{R} \to \mathbf{R}(2)$ and $\beta : \mathbf{R}^{1,1} \to \mathbf{R}(2)$ such that, for all $x \in \mathbf{R}^{1,1}$, $(\beta(x))^2 = \alpha(x^{(2)})$. □

9.79. Let $t : \mathbf{R}^2 \to \mathbf{R}^2$ be the orthogonal projection of $\mathbf{R}^2$ on to the line $\{(x,y) \in \mathbf{R}^2 : ax + by = 0\}$, where $a, b \in \mathbf{R}$, not both zero. Find the matrix of t and verify that $t^2 = t$. □

9.80. Let i and $j : \mathbf{R}^n \to S^n$ be the 'inverses' of the stereographic projection of S^n, the unit sphere in $\mathbf{R}^{n+1} = \mathbf{R}^n \times \mathbf{R}$, from its North and South poles respectively on to its equatorial plane $\mathbf{R}^n = \mathbf{R}^n \times \{0\}$. Prove that, for all $x \in \mathbf{R}^n \setminus \{0\}$,

$$i_{\text{sur}}^{-1} j(x) = j_{\text{sur}}^{-1} i(x) = x^{(-1)}. \quad \square$$

9.81. Let $\mathscr{S}(\mathbf{R}^{p,q+1}) = \{x \in \mathbf{R}^{p,q+1} : x^{(2)} = 1\}$. Prove that, for any $(x,y) \in \mathbf{R}^p \times S^q$, $(x, \sqrt{(1 + x^{(2)})}y) \in \mathscr{S}(\mathbf{R}^{p,q+1})$, and that the map

$$\mathbf{R}^p \times S^q \to \mathscr{S}(\mathbf{R}^{p,q+1}); \quad (x,y) \rightsquigarrow (x, \sqrt{(1 + x^{(2)})}y)$$

is bijective. $\square$

9.82. Determine whether or not the point-pairs (or 0-spheres)

$$\{x \in \mathbf{R} : x^2 - 8x - 12 = 0\} \quad \text{and} \quad \{x \in \mathbf{R} : x^2 - 10x + 7 = 0\}$$

are linked. $\square$

9.83. Determine whether or not the circles

$$\{(x,y,z) \in \mathbf{R}^3 : x^2 + y^2 + z^2 = 5 \quad \text{and} \quad x + y - 1 = 0\}$$

and

$$\{(x,y,z) \in \mathbf{R}^3 : x^2 + y^2 + z^2 + 2y - 4z = 0 \quad \text{and} \quad x - z + 1 = 0\}$$

are linked. $\square$

9.84. Let A and B be mutually disjoint circles which link in $\mathbf{R}^3$. Prove that the map

$$A \times B \to S^2; \quad (a,b) \rightsquigarrow (b - a)/|b - a|$$

is surjective, and describe the fibres of the map. $\square$

9.85. What was your definition of 'linked' in the preceding three exercises? Can two circles in $\mathbf{R}^4$ be linked, according to your definition? Extend your definition to cover point-pairs on S^1 or circles on S^3. Also try to extend the definition to curves other than circles, either in $\mathbf{R}^3$ or on S^3. (You will first have to decide what is meant by a curve! The problem of obtaining a good definition of *linking* is a rather subtle topological one. This is something to follow up after you have read Chapters 16 and 19. An early discussion is by Brouwer [6] and an even earlier one by Gauss [16]! See [50], page 60 and page 66, and also [13].) $\square$

9.86. Show that any two mutually disjoint great circles on S^3 are linked. $\square$

9.87. Where can one find a pair of linked 3-spheres? $\square$

9.88. [61.] Let X and Y be isomorphic non-degenerate orthogonal spaces, let U and V be orthogonal subspaces of X and Y respectively and suppose that $s : U \to V$ is an orthogonal isomorphism. Construct an orthogonal isomorphism $t : X \to Y$ such that $s = (t \mid U)_{\text{sur}}$. $\square$

CHAPTER 10

QUATERNIONS

Certain real (and complex) algebras arise naturally in the detailed study of the groups $O(p,q)$, where $p, q \in \omega$. These are the *Clifford algebras*, defined and studied in Chapter 13. As we shall see in that chapter, examples of such algebras are **R** itself, the real algebra of complex numbers **C** and the real algebra of quaternions **H**. The present chapter is mainly concerned with the algebra **H**, but it is convenient first of all to recall certain properties of **C** and to relate **C** to the group of orthogonal transformations $O(2)$ of $\mathbf{R}(2)$.

The real algebra **C** was defined at the end of Chapter 2, and its algebraic properties listed. There was further discussion in Chapter 3. In particular, it was remarked in Prop. 3.38 that there is a unique algebra involution of **C** different from the identity, namely conjugation

$$\mathbf{C} \to \mathbf{C}; \quad z = x + iy \rightsquigarrow \bar{z} = x - iy,$$

where $(x,y) \in \mathbf{R}^2$. Now since, for any $z = x + iy \in \mathbf{C}$, $|z|^2 = \bar{z}z = x^2 + y^2$, **C** can in a natural way be identified not only with the linear space $\mathbf{R}^2$ but also with the positive-definite orthogonal space $\mathbf{R}^{0,2}$. Moreover, since, for any $z \in \mathbf{C}$,

$$|z| = 1 \Leftrightarrow x^2 + y^2 = 1,$$

the subgroup of $\mathbf{C}^*$, consisting of all complex numbers of absolute value 1, may be identified with the unit circle S^1 in $\mathbf{R}^2$.

In what follows, the identification of **C** with $\mathbf{R}^2 = \mathbf{R}^{0,2}$ is taken for granted.

It was remarked in Chapter 3, Prop. 3.31, that, for any complex number $c = a + ib$, the map $\mathbf{C} \to \mathbf{C}$; $z \rightsquigarrow cz$ may be regarded as a real linear map, with matrix $\begin{pmatrix} a & -b \\ b & a \end{pmatrix}$. Moreover, as we remarked in Chapter 7, $|c|^2 = \bar{c}c = \det \begin{pmatrix} a & -b \\ b & a \end{pmatrix}$. Now, for all $c, z \in \mathbf{C}$, $|cz| = |c|\,|z|$; so multiplication by c is a rotation of $\mathbf{R}^2$ if, and only if, $|c| = 1$. Conversely, it is easy to see that any rotation of $\mathbf{R}^2$ is so induced. In fact, this statement is just Example 9.20. The following statement sums this all up.

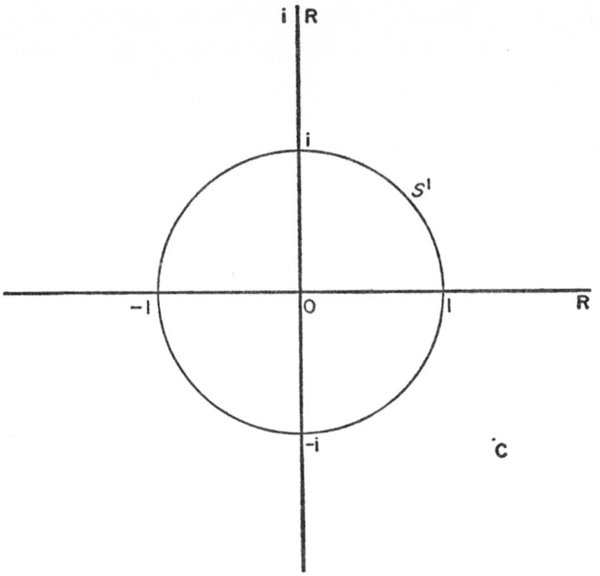

Prop. 10.1 $S^1 \cong SO(2)$. □

The group S^1 is called the *circle group*.

Antirotations of $\mathbf{R}^2$ can also be handled by $\mathbf{C}$, for conjugation is an antirotation, and any other antirotation can be regarded as the composite of conjugation with a rotation, that is, with multiplication by a complex number of absolute value 1.

By Theorem 2.69, any complex number of absolute value 1 may be expressed in the form $e^{i\theta}$. The rotation of $\mathbf{R}^2$ corresponding to the number $e^{i\theta}$ is often referred to as the rotation of $\mathbf{R}^2$ *through the angle* θ. In particular, since $-1 = e^{i\pi} = e^{-i\pi}$, the map $\mathbf{C} \to \mathbf{C}$; $z \rightsquigarrow -z$ is also referred to not only as the reflection of $\mathbf{R}^2$ in $\{0\}$ but also as the rotation of $\mathbf{R}^2$ through the angle π or, equivalently, through the angle $-\pi$.

Note that, for any $a, b \in S^1$, with $b = ae^{i\theta}$

$$a \cdot b = \tfrac{1}{2}(\bar{a}b + \bar{b}a) = \tfrac{1}{2}(e^{i\theta} + e^{-i\theta}) = \cos\theta,$$

in accordance with the remarks made on angle following Prop. 9.58.

In any complete discussion of the complex exponential map, the relationship between angle and arc length along the circle S^1 is developed. It may be shown, in particular, that 2π is the circumference of the circle S^1.

The algebra of quaternions, $\mathbf{H}$, introduced in the next section, will be seen to be analogous in many ways to the algebra of complex numbers $\mathbf{C}$. For example, it has application, as we shall see, to the description of certain groups of orthogonal transformations, namely $O(3)$ and $O(4)$. The letter H is the initial letter of the surname of Sir William Hamilton, who first studied quaternions and gave them their name [23].

The algebra H

Let 1, i, j and k denote the elements of the standard basis for $\mathbf{R}^4$. The *quaternion product* on $\mathbf{R}^4$ is then the $\mathbf{R}$-bilinear product

$$\mathbf{R}^4 \times \mathbf{R}^4 \to \mathbf{R}^4; \quad (a,b) \rightsquigarrow ab$$

with unity 1, defined by the formulae

$$\mathrm{i}^2 = \mathrm{j}^2 = \mathrm{k}^2 = -1$$

and $\quad \mathrm{ij} = \mathrm{k} = -\mathrm{ji}, \quad \mathrm{jk} = \mathrm{i} = -\mathrm{kj} \quad$ and $\quad \mathrm{ki} = \mathrm{j} = -\mathrm{ik}.$

Prop. 10.2. The quaternion product is associative. □

On the other hand the quaternion product is not commutative. For example, $\mathrm{ji} \neq \mathrm{ij}$. Moreover, it does not necessarily follow that if $a^2 = b^2$, then $a = \pm b$. For example, $\mathrm{i}^2 = \mathrm{j}^2$, but $\mathrm{i} \neq \pm\mathrm{j}$.

The linear space $\mathbf{R}^4$, with the quaternion product, is a real algebra $\mathbf{H}$ known as the *algebra of quaternions*. In working with $\mathbf{H}$ it is usual to identify $\mathbf{R}$ with $\mathrm{im}\{1\}$ and $\mathbf{R}^3$ with $\mathrm{im}\{\mathrm{i,j,k}\}$, the first identification having been anticipated by our use of the symbol 1. The subspace $\mathrm{im}\{\mathrm{i,j,k}\}$ is known as the subspace of *pure quaternions*. Each quaternion q is uniquely expressible in the form $\mathrm{re}\, q + \mathrm{pu}\, q$, where $\mathrm{re}\, q \in \mathbf{R}$ and $\mathrm{pu}\, q \in \mathbf{R}^3$, $\mathrm{re}\, q$ being called the *real part* of q and $\mathrm{pu}\, q$ the *pure part* of q.

Prop. 10.3. A quaternion is real if, and only if, it commutes with every quaternion. That is, $\mathbf{R}$ is the centre of $\mathbf{H}$.

Proof $\Rightarrow$: Clear.

$\Leftarrow$: Let $q = a + b\mathrm{i} + c\mathrm{j} + d\mathrm{k}$, where a, b, c and d are real, be a quaternion commuting with i and j. Since q commutes with i,

$$a\mathrm{i} - b + c\mathrm{k} - d\mathrm{j} = \mathrm{i}q = q\mathrm{i} = a\mathrm{i} - b - c\mathrm{k} + d\mathrm{j},$$

implying that $2(c\mathrm{k} - d\mathrm{j}) = 0$. So $c = d = 0$. Similarly, since q commutes with j, $b = 0$. So $q = a$, and is real. □

Cor. 10.4. The ring structure of $\mathbf{H}$ induces the real linear structure, and any ring automorphism or anti-automorphism of $\mathbf{H}$ is a real linear automorphism of $\mathbf{H}$, and therefore also a real algebra automorphism or anti-automorphism of $\mathbf{H}$.

Proof By Prop. 10.3 the injection of $\mathbf{R}$ in $\mathbf{H}$, and hence the real scalar multiplication $\mathbf{R} \times \mathbf{H} \to \mathbf{H}$; $(\lambda,q) \rightsquigarrow \lambda q$, is determined by the ring structure.

Also, again by Prop. 10.3, any automorphism or anti-automorphism t of **H** maps **R** to **R**, this restriction being an automorphism of **R** and therefore the identity, by Prop. 2.60. Therefore t not only respects addition and respects or reverses ring multiplication but also, for any $\lambda \in \mathbf{R}$ and $q \in \mathbf{H}$,

$$\begin{aligned} t(\lambda q) &= t(\lambda)\, t(q) \quad \text{or} \quad t(q)\, t(\lambda) \\ &= \lambda t(q) \quad \text{or} \quad t(q)\lambda \\ &= \lambda t(q). \quad \square \end{aligned}$$

This result is to be contrasted with the more involved situation for the field of complex numbers described on page 48. The automorphisms and anti-automorphisms of **H** are discussed in more detail below.

Prop. 10.5 A quaternion is pure if, and only if, its square is a non-positive real number.

Proof $\Rightarrow$: Consider $q = bi + cj + dk$, where $b, c, d \in \mathbf{R}$. Then $q^2 = -(b^2 + c^2 + d^2)$, which is real and non-positive.

$\Leftarrow$: Consider $q = a + bi + cj + dk$, where $a, b, c, d \in \mathbf{R}$. Then

$$q^2 = a^2 - b^2 - c^2 - d^2 + 2a(bi + cj + dk).$$

If q^2 is real, either $a = 0$ and q is pure, or $b = c = d = 0$ and $a \neq 0$, in which case q^2 is positive. So, if q^2 is real and non-positive, q is pure. $\square$

Cor. 10.6. The direct sum decomposition of **H**, with components the real and pure subspaces, is induced by the ring structure for **H**. $\square$

The *conjugate* $\bar{q}$ of a quaternion q is defined to be the quaternion $\operatorname{re} q - \operatorname{pu} q$.

Prop. 10.7. Conjugation: $\mathbf{H} \to \mathbf{H};\ q \rightsquigarrow \bar{q}$ is an algebra anti-involution. That is, for all $a, b \in \mathbf{H}$ and all $\lambda \in \mathbf{R}$,

$$\begin{aligned} \overline{a + b} = \bar{a} + \bar{b},\ \overline{\lambda a} &= \lambda \bar{a}, \\ \bar{\bar{a}} = a \quad \text{and} \quad \overline{ab} &= \bar{b}\bar{a}. \end{aligned}$$

Moreover, $a \in \mathbf{R} \Leftrightarrow \bar{a} = a$ and $a \in \mathbf{R}^3 \Leftrightarrow \bar{a} = -a$, while $\operatorname{re} a = \frac{1}{2}(a + \bar{a})$ and $\operatorname{pu} a = \frac{1}{2}(a - \bar{a})$. $\square$

Now let **H** be assigned the standard positive-definite scalar product on $\mathbf{R}^4$, denoted as usual by $\cdot$.

Prop. 10.8. For all $a, b \in \mathbf{H}$, $a \cdot b = \frac{1}{2}(\bar{a}b + \bar{b}a)$, with $a^{(2)} = \bar{a}a$.
In particular, $\bar{a}a$ is non-negative. In particular, also, for all $a, b \in \mathbf{R}^3$,

$$a \cdot b = -\tfrac{1}{2}(ab + ba) = -\operatorname{re}(ab),$$

with $a^{(2)} = -a^2$, and with $a \cdot b = 0$ if, and only if, a and b anti-commute. $\square$

The non-negative number $|a| = \sqrt{(\bar{a}a)}$ is called the ***norm*** or ***absolute value*** of the quaternion a.

Prop. 10.9. Let $x \in \mathbf{H}$. Then

$$x^2 + bx + c = 0,$$

where $b = -(x + \bar{x})$, and $c = \bar{x}x$, b and c both being real. □

Cor. 10.10. Let x be a non-real element of $\mathbf{H}$. Then $\operatorname{im}\{1,x\}$ is a subalgebra of $\mathbf{H}$ isomorphic with $\mathbf{C}$. In particular, $\operatorname{im}\{1,x\}$ is commutative. □

Prop. 10.11. Each non-zero $a \in \mathbf{H}$ is invertible, with $a^{-1} = |a|^{-2}\,\bar{a}$ and with $|a^{-1}| = |a|^{-1}$. □

Note that the quaternion inverse of a is the conjugate of the scalar product inverse of a, $a^{(-1)} = |a|^{-2}\,a$.

By Prop. 10.11, $\mathbf{H}$ may be regarded as a non-commutative field. The group of non-zero quaternions will be denoted by $\mathbf{H}^*$. □

Prop. 10.12. For all $a, b \in \mathbf{H}$, $|ab| = |a|\,|b|$.

Proof For all $a, b \in \mathbf{H}$,

$$\begin{aligned} |ab|^2 &= \overline{ab}\,ab = \bar{b}\bar{a}ab \\ &= \bar{a}a\bar{b}b, \quad \text{since } \bar{a}a \in \mathbf{R}, \\ &= |a|^2\,|b|^2. \end{aligned}$$

Therefore, since $|q| \geqslant 0$ for all $q \in \mathbf{H}$,

$$|ab| = |a|\,|b|. \quad \square$$

A quaternion q is said to be a *unit quaternion* if $|q| = 1$.

Prop. 10.13. The set of unit quaternions coincides with the unit sphere S^3 in $\mathbf{R}^4$ and is a subgroup of $\mathbf{H}^*$. □

Prop. 10.14. Let $q \in \mathbf{H}$ be such that $q^2 = -1$. Then $q \in S^2$, the unit sphere in $\mathbf{R}^3$.

Proof Since q^2 is real and non-positive, $q \in \mathbf{R}^3$, and, since $q^2 = -1$, $|q| = 1$. So $q \in S^2$. □

The *vector product* $a \times b$ of a pair (a,b) of pure quaternions is defined by the formula

$$a \times b = \operatorname{pu}(ab).$$

Prop. 10.15. For all $a, b \in \mathbf{R}^3$,

$$ab = -a\cdot b + a \times b, \quad \text{and} \quad a \times b = \tfrac{1}{2}(ab - ba) = -(b \times a),$$

while $a \times a = a\cdot(a \times b) = b\cdot(a \times b) = 0$.

If a and b are mutually orthogonal elements of $\mathbf{R}^3$, a and b anti-commute, that is, $ba = -ab$, and $a \times b = ab$. In particular,

$$\mathrm{i} \cdot (\mathrm{j} \times \mathrm{k}) = \mathrm{i} \cdot (\mathrm{jk}) = \mathrm{i} \cdot \mathrm{i} = -\mathrm{i}^2 = 1. \quad \square$$

Prop. 10.16. Let q be a quaternion. Then there exists a non-zero pure quaternion b such that qb also is a pure quaternion.

Proof Let b be any non-zero element of $\mathbf{R}^3$ orthogonal to the pure part of q. Then

$$qb = (\mathrm{re}\, q)b + (\mathrm{pu}\, q) \times b \in \mathbf{R}^3. \quad \square$$

Cor. 10.17. Each quaternion is expressible as the product of a pair of pure quaternions. □

Prop. 10.18. For all $a, b, c \in \mathbf{R}^3$,

$$a \times (b \times c) = (a \cdot c)b - (a \cdot b)c,$$

and $$a \times (b \times c) + b \times (c \times a) + c \times (a \times b) = 0.$$

Proof For all $a, b, c \in \mathbf{R}^3$,

$$\begin{aligned} &4a \times (b \times c) - 4(a \cdot c)b + 4(a \cdot b)c \\ &= 2a \times (bc - cb) - 2(a \cdot c)b - 2b(a \cdot c) + 2(a \cdot b)c + 2c(a \cdot b) \\ &= a(bc - cb) - (bc - cb)a + (ac + ca)b + b(ac + ca) \\ &\qquad - (ab + ba)c - c(ab + ba) \\ &= 0. \quad \square \end{aligned}$$

Prop. 10.19. The map

$$\mathbf{R}^3 \times \mathbf{R}^3 \to \mathbf{R}^3; \quad (a,b) \rightsquigarrow a \times b$$

is alternating bilinear, and the map

$$(\mathbf{R}^3)^3 \to \mathbf{R}^3; \quad (a,b,c) \rightsquigarrow a \cdot (b \times c)$$

is alternating trilinear, with

$$a \cdot (b \times c) = b \cdot (c \times a) = c \cdot (a \times b) = \det \mathrm{col}^{-1}(a,b,c). \quad \square$$

Automorphisms and anti-automorphisms of H

By Cor. 10.4 the field automorphisms and anti-automorphisms of **H** coincide with the real algebra automorphisms and anti-automorphisms of **H**. In this section we show that they are also closely related to the orthogonal automorphisms of $\mathbf{R}^3$. The relationship one way is given by the next proposition.

Prop. 10.20. Any automorphism or anti-automorphism u of **H** is of the form $\mathbf{H} \to \mathbf{H}$; $a \rightsquigarrow \mathrm{re}\, a + t(\mathrm{pu}\, a)$, where t is an orthogonal automorphism of $\mathbf{R}^3$.

Proof By Cor. 10.4, Prop. 2.60 and Prop. 10.5, u is a linear map leaving each real quaternion fixed and mapping $\mathbf{R}^3$ to itself. Also, for each $x \in \mathbf{R}^3$, $(u(x))^2 = u(x^2) = x^2$, since $x^2 \in \mathbf{R}$, while $|x|^2 = -x^2$. So

$$t : \mathbf{R}^3 \to \mathbf{R}^3; \quad x \rightsquigarrow u(x)$$

is linear and respects norm. Therefore, by Prop. 9.59, it is an orthogonal automorphism of $\mathbf{R}^3$. □

In the reverse direction we have the following fundamental result.

Prop. 10.21. Let q be an invertible pure quaternion. Then, for any pure quaternion x, qxq^{-1} is a pure quaternion, and the map

$$-\rho_q : \mathbf{R}^3 \to \mathbf{R}^3; \quad x \rightsquigarrow -q\,x\,q^{-1}$$

is reflection in the plane $(\operatorname{im}\{q\})^\perp$.

Proof Since $(qxq^{-1})^2 = x^2$, which is real and non-positive, qxq^{-1} is pure, by Prop. 10.5. Also $-\rho_q$ is linear, and $-\rho q(q) = -q$, while, for any $r \in (\operatorname{im}\{q\})^\perp$, $-\rho_q(r) = -qrq^{-1} = rqq^{-1} = r$. Hence the result. □

Proposition 10.21 is used twice in the proof of Prop. 10.22.

Prop. 10.22. Each rotation of $\mathbf{R}^3$ is of the form ρ_q for some non-zero quaternion q, and every such map is a rotation of $\mathbf{R}^3$.

Proof Since, by Prop. 9.43, any rotation of $\mathbf{R}^3$ is the composite of two plane reflections it follows, by Prop. 10.21, that the rotation can be expressed in the given form. The converse is by Cor. 10.17 and Prop. 10.21. □

In fact, each rotation of $\mathbf{R}^3$ can be so represented by a unit quaternion, unique up to sign. This follows from Prop. 10.23.

Prop. 10.23. The map $\rho : \mathbf{H}^* \to SO(3)$; $q \rightsquigarrow \rho_q$ is a group surjection, with kernel $\mathbf{R}^*$, the restriction of ρ to S^3 also being surjective, with kernel $S^0 = \{1, -1\}$.

Proof The map ρ is surjective, by Prop. 10.22, and is a group map since, for all $q, r \in \mathbf{H}^*$, and all $x \in \mathbf{R}^3$,

$$\rho_{qr}(x) = qrx(qr)^{-1} = \rho_q\rho_r(x).$$

Moreover, $q \in \ker \rho$ if, and only if, $q\,x\,q^{-1} = x$, for all $x \in \mathbf{R}^3$, that is, if, and only if, $qx = xq$, for all $x \in \mathbf{R}^3$. Therefore, by Prop. 10.3, $\ker \rho = \mathbf{R} \cap \mathbf{H}^* = \mathbf{R}^*$.

The restriction of ρ to S^3 also is surjective simply because, for any $\lambda \in \mathbf{R}^*$ and for any $q \in \mathbf{H}^*$, $\rho_{\lambda q} = \rho_q$, and λ may be chosen so that $|\lambda q| = 1$. Finally, $\ker(\rho \mid S^3) = \ker \rho \cap S^3 = \mathbf{R}^* \cap S^3 = S^0$. □

Prop. 10.24. Any unit quaternion q is expressible in the form $aba^{-1}b^{-1}$, where a and b are non-zero quaternions.

Proof By Prop. 10.16 there is, for any unit quaternion q, a non-zero pure quaternion b such that qb is a pure quaternion. Since $|q| = 1$, $|qb| = |b|$. There is therefore, by Prop. 10.22, a non-zero quaternion a such that $qb = aba^{-1}$, that is, such that $q = aba^{-1}b^{-1}$. □

Proposition 10.22 also leads to the following converse to Prop. 10.20.

Prop. 10.25. For each $t \in O(3)$, the map

$$u : \mathbf{H} \to \mathbf{H}; \quad a \rightsquigarrow \operatorname{re} a + t(\operatorname{pu} a)$$

is an automorphism or anti-automorphism of **H**, u being an automorphism if t is a rotation and an anti-automorphism if t is an antirotation of $\mathbf{R}^3$.

Proof For each $t \in SO(3)$, the map u can, by Prop. 10.22, be put in the form

$$\mathbf{H} \to \mathbf{H}; \quad a \rightsquigarrow qaq^{-1} = \operatorname{re} a + q(\operatorname{pu} a)q^{-1},$$

where $q \in \mathbf{H}^*$, and such a map is an automorphism of **H**.

Also $-1_{\mathbf{R}^3}$ is an antirotation of $\mathbf{R}^3$, and if $t = -1_{\mathbf{R}^3}$, u is conjugation, which is an anti-automorphism of **H**. The remainder of the proposition follows at once, since any anti-automorphism of **H** can be expressed as the composite of any particular anti-automorphism, for example, conjugation, with some automorphism. □

Cor. 10.26. An involution of **H** either is the identity or corresponds to the rotation of $\mathbf{R}^3$ through π about some axis, that is, reflection in some line through 0. Any anti-involution of **H** is conjugation composed with such an involution and corresponds either to the reflection of $\mathbf{R}^3$ in the origin or to the reflection of $\mathbf{R}^3$ in some plane through 0. □

It is convenient to single out one of the non-trivial involutions of **H** to be typical of the class. For technical reasons we choose the involution $\mathbf{H} \to \mathbf{H}$; $a \rightsquigarrow \mathrm{j}a\mathrm{j}^{-1}$, corresponding to the reflection of $\mathbf{R}^3$ in the line $\operatorname{im}\{\mathrm{j}\}$. This will be called the *main involution* of **H** and, for each $a \in \mathbf{H}$, $\hat{a} = \mathrm{j}a\mathrm{j}^{-1}$ will be called the *involute* of a. The main involution commutes with conjugation. The composite will be called *reversion* and, for each $a \in \mathbf{H}$, $\tilde{a} = \overline{\hat{a}} = \hat{\bar{a}}$ will be called the *reverse* of a. A reason for this is that **H** may be regarded as being generated as an algebra by i and k, and reversion sends i to i and k to k but sends ik to ki, reversing the multiplication. (Cf. page 252.)

It is a further corollary of Prop. 10.25 that the basic frame (i,j,k) for

$\mathbf{R}^3$ does not hold a privileged position in $\mathbf{H}$. This can also be shown directly as follows.

Prop. 10.27. Let a be any orthogonal basic framing for $\mathbf{R}^3$, inducing an orthonormal basis $\{a_0,a_1,a_2\}$ for $\mathbf{R}^3$. Then, for all $i \in 3$, $a_i^2 = -1$, and, for any distinct $i, j \in 3$, $a_j a_i = -a_i a_j$. Also

$$a_0 a_1 a_2 = \det a = +1 \text{ or } -1,$$

according as a respects or reverses the orientations of $\mathbf{R}^3$. If the framing a is positively oriented, then

$$a_0 = a_1 a_2, \quad a_1 = a_2 a_0 \quad \text{and} \quad a_2 = a_0 a_1,$$

while if the framing a is negatively oriented, then

$$a_0 = a_2 a_1, \quad a_1 = a_0 a_2 \quad \text{and} \quad a_2 = a_1 a_0. \qquad \square$$

The following proposition is required in the proof of Prop. 11.24.

Prop. 10.28. The map $\mathbf{H} \to \mathbf{H};\ x \rightsquigarrow \tilde{x}\,x$ has as image the three-dimensional real linear subspace $\{y \in \mathbf{H} : \tilde{y} = y\} = \operatorname{im}\{1,\mathrm{i},k\}$.

Proof It is enough to prove that the map $S^3 \to S^3;\ x \rightsquigarrow \tilde{x}\,x = \hat{x}^{-1}\,x$ has as image the unit sphere in $\operatorname{im}\{1,\mathrm{i},\mathrm{k}\}$.

So let $y \in \mathbf{H}$ be such that $\bar{y}y = 1$ and $\tilde{y} = \hat{y}$. Then

$$1 + y = \bar{y}y + y = (\hat{y} + 1)y.$$

So, if $y \neq -1$, $y = \hat{x}^{-1}\,x$, where

$$x = (1 + y)(|\,1 + y\,|)^{-1}. \quad \text{Finally, } -1 = \tilde{\mathrm{i}}\,\mathrm{i}. \qquad \square$$

Rotations of $\mathbf{R}^4$

Quaternions may also be used to represent rotations of $\mathbf{R}^4$.

Prop. 10.29. Let q be a unit quaternion. Then the map $q_L : \mathbf{R}^4 \to \mathbf{R}^4;\ x \rightsquigarrow qx$, where $\mathbf{R}^4$ is identified with $\mathbf{H}$, is a rotation of $\mathbf{R}^4$, as is the map $q_R : \mathbf{R}^4 \to \mathbf{R}^4;\ x \rightsquigarrow xq$.

Proof The map q_L is linear, and preserves norm by Prop. 10.12; so it is orthogonal, by Prop. 9.59. That it is a rotation follows from Prop. 10.24 which states that there exist non-zero quaternions a and b such that $q = aba^{-1}b^{-1}$, and therefore such that $q_L = a_L b_L (a_L)^{-1}(b_L)^{-1}$, implying that $\det_{\mathbf{R}}(q_L) = 1$. Similarly for q_R. $\square$

Prop. 10.30. The map

$$\rho : S^3 \times S^3 \to SO(4); \quad (q,r) \rightsquigarrow q_L \bar{r}_R$$

is a group surjection with kernel $\{(1,1),(-1,-1)\}$.

Proof For any $q, q', r, r' \in S^3$ and any $x \in \mathbf{H}$,

$$\rho(q'q,r'r)(x) = (q'q)_L\,(\overline{r'r})_R\,x$$
$$= q'qx\bar{r}\bar{r}' = \rho(q',r')\,\rho(q,r)(x)$$

Therefore, for any $(q,r), (q',r') \in S^3 \times S^3$,

$$\rho((q',r')(q,r)) = \rho(q',r')\,\rho(q,r);$$

that is, ρ is a group map. That it has the stated kernel follows from the observation that if q and r are unit quaternions such that $qx\bar{r} = x$ for all $x \in \mathbf{H}$, then, by choosing $x = 1$, $q\bar{r} = 1$, from which it follows that $qxq^{-1} = x$ for all $x \in \mathbf{H}$, or, equivalently, that $qx = xq$ for all $x \in \mathbf{H}$. This implies, by Prop. 10.3, that $q \in \{1, -1\} = \mathbf{R} \cap S^3$.

To prove that ρ is surjective, let t be any rotation of $\mathbf{R}^4$ and let $s = t(1)$. Then $|s| = 1$ and the map $\mathbf{R}^4 \to \mathbf{R}^4$; $x \rightsquigarrow \bar{s}(t(x))$ is a rotation of $\mathbf{R}^4$ leaving 1 and therefore each point of $\mathbf{R}$ fixed. So, by Prop. 10.22, there exists a unit quaternion r such that, for all $x \in \mathbf{R}^4$,

$$\bar{s}(t(x)) = rxr^{-1}$$

or, equivalently, $t(x) = qx\bar{r}$, where $q = sr$. □

Antirotations also are easily represented by quaternions, since conjugation is an antirotation and since any antirotation is the composite of any given antirotation and a rotation.

Linear spaces over H

Much of the theory of linear spaces and linear maps developed for commutative fields in earlier chapters extends over $\mathbf{H}$. Because of the non-commutativity of $\mathbf{H}$ it is, however, necessary to distinguish two types of linear space over $\mathbf{H}$, namely *right* linear spaces and *left* linear spaces.

A *right* linear space over $\mathbf{H}$ consists of an additive group X and a map

$$X \times \mathbf{H} \to X; \quad (x,\lambda) \rightsquigarrow x\lambda$$

such that the usual distributivity and unity axioms hold and such that, for all $x \in X$, $\lambda, \lambda' \in \mathbf{H}$,

$$(x\lambda)\lambda' = x(\lambda\lambda').$$

A *left* linear space over $\mathbf{H}$ consists of an additive group X and a map

$$\mathbf{H} \times X \to X; \quad (\mu,x) \rightsquigarrow \mu x$$

such that the usual distributivity and unity axioms hold and such that, for all $x \in X$, $\mu, \mu' \in \mathbf{H}$,

$$\mu'(\mu x) = (\mu'\mu)x.$$

The additive group $\mathbf{H}^n$, for any finite n, and in particular $\mathbf{H}$ itself, can be assigned either a right or a left $\mathbf{H}$-linear structure in an obvious way.

Unless there is explicit mention to the contrary, it will normally be assumed that the *right* **H**-linear structure has been chosen. (As we shall see below, a natural notation for $\mathbf{H}^n$ with the obvious left **H**-linear structure would be $(\mathbf{H}^n)^{\mathrm{L}}$ or $(\mathbf{H}^{\mathrm{L}})^n$).

Linear maps $t : X \to Y$, where X and Y are **H**-linear spaces, may be defined, provided that each of the spaces X and Y is a right linear space or that each is a left linear space. For example, if X and Y are both *right* linear spaces, then t is said to be *linear* (or *right linear*) if it respects addition and, for all $x \in X$, $\lambda \in \mathbf{H}$, $t(x\lambda) = (t(x))\lambda$, an analogous definition holding in the *left* case.

The set of linear maps $t : X \to Y$ between right, or left, linear spaces X and Y over **H** will be denoted in either case by $\mathscr{L}(X,Y)$, or by $L(X,Y)$ when X and Y are finite-dimensional (see below). However, the usual recipe for $\mathscr{L}(X,Y)$ to be a linear space fails. For suppose we define, for any $t \in \mathscr{L}(X,Y)$ and $\lambda \in \mathbf{H}$, a map $t\lambda : X \to Y$ by the formula $(t\lambda)x = t(x)\lambda$, X and Y being right **H**-linear spaces. Then, for any $t \in \mathscr{L}(X,Y)$ and any $x \in X$,

$$t(x)\mathrm{k} = (t\mathrm{ij})(x) = (t\mathrm{i})(x\mathrm{j}) = t(x\mathrm{j})\mathrm{i} = -t(x)\mathrm{k},$$

leading at once to a contradiction if $t \neq 0$, as is possible. Normally $\mathscr{L}(X,Y)$ is regarded as a linear space over the *centre* of **H**, namely **R**. In particular, for any right **H**-linear space X, the set End $X = \mathscr{L}(X,X)$ is normally regarded as a *real* algebra.

On the other hand, for any right linear space X over **H**, a left **H**-linear structure can be assigned to $\mathscr{L}(X,\mathbf{H})$ by setting $(\mu t)(x) = \mu(t(x))$, for all $t \in \mathscr{L}(X,\mathbf{H})$, $x \in \mathbf{H}$ and $\mu \in \mathbf{H}$. This left linear space is called the *linear dual* of X and is also denoted by $X^{\mathscr{L}}$. The linear dual of a left **H**-linear space is analogously defined. It is a right **H**-linear space.

Each right **H**-linear map $t: X \to Y$ induces a left linear map $t^{\mathscr{L}} : Y^{\mathscr{L}} \to X^{\mathscr{L}}$ by the formula

$$t^{\mathscr{L}}(\gamma) = \gamma t, \quad \text{for each } \gamma \in Y^{\mathscr{L}},$$

and if $t \in \mathscr{L}(X,Y)$ and $u \in \mathscr{L}(W,X)$, W, X and Y all being right **H**-linear spaces, then

$$(tu)^{\mathscr{L}} = u^{\mathscr{L}} t^{\mathscr{L}}.$$

The section on matrices in Chapter 3 generalizes at once to **H**-linear maps. For example, any right **H**-linear map $t: \mathbf{H}^n \to \mathbf{H}^m$ may be represented in the obvious way by an $m \times n$ matrix $\{t_{ij} : (i,j) \in m \times n\}$ over **H**. In particular, any element of the right **H**-linear space $\mathbf{H}^m$ may be represented by a column matrix. Scalar multipliers have, however, to be written on the right and not on the left as has been our custom hitherto.

For example, suppose that $t \in \text{End } \mathbf{H}^2$, and let $x, y \in \mathbf{H}^2$ be such that

$y = t(x)$. Then this statement may be written in matrix notations in the form

$$\begin{pmatrix} y_0 \\ y_1 \end{pmatrix} = \begin{pmatrix} t_{00} & t_{01} \\ t_{10} & t_{11} \end{pmatrix} \begin{pmatrix} x_0 \\ x_1 \end{pmatrix}.$$

The statement that, for any $x \in \mathbf{H}^2$ and any $\lambda \in \mathbf{H}$, $t(x\lambda) = (t(x))\lambda$, becomes, in matrix notations,

$$\begin{pmatrix} t_{00} & t_{01} \\ t_{10} & t_{11} \end{pmatrix} \begin{pmatrix} x_0\lambda \\ x_1\lambda \end{pmatrix} = \left(\begin{pmatrix} t_{00} & t_{01} \\ t_{10} & t_{11} \end{pmatrix} \begin{pmatrix} x_0 \\ x_1 \end{pmatrix} \right) \lambda.$$

The left **H**-linear space $(\mathbf{H}^n)^{\mathrm{L}}$ dual to the right **H**-linear space $\mathbf{H}^n$ may be identified with the additive group $\mathbf{H}^n$ assigned its left **H**-linear structure. Elements of this space may be represented by row matrices. A left **H**-linear map $u : (\mathbf{H}^m)^{\mathrm{L}} \longrightarrow (\mathbf{H}^n)^{\mathrm{L}}$ is then represented by an $m \times n$ matrix that multiplies elements of $(\mathbf{H}^m)^{\mathrm{L}}$ on the right.

$\mathbf{H}(n)$ will be a notation for the real algebra of $n \times n$ matrices over **H**.

Subspaces of right or left **H**-linear spaces and products of such spaces are defined in the obvious way. The material of Chapters 4 and 5 also goes over without essential change, as does the material of Chapter 6 on linear independence and the basis theorem for finite-dimensional spaces and its corollaries, except that care must be taken to put scalar multipliers on the correct side. Any right linear space X over **H** with a finite basis is isomorphic to $\mathbf{H}^n$ as a right linear space, n, the number of elements in the basis, being uniquely determined by X. This number n is called the *quaternionic dimension*, $\dim_{\mathbf{H}} X$ of X. Analogous remarks apply in the left case. For any finite-dimensional **H**-linear space X,

$$\dim_{\mathbf{H}} X^{\mathrm{L}} = \dim_{\mathbf{H}} X.$$

Any quaternionic linear space X may be regarded as a real linear space and, if X is finite-dimensional, $\dim_{\mathbf{R}} X = 4 \dim_{\mathbf{H}} X$. Such a space may also be regarded as a complex linear space, once some representation of **C** as a subalgebra of **H** has been chosen, with $\dim_{\mathbf{C}} X = 2 \dim_{\mathbf{H}} X$ when X is finite-dimensional. In the following discussion **C** is identified with $\operatorname{im}\{1,i\}$ in **H**, and, for each $n \in \omega$, $\mathbf{C}^{2n} = \mathbf{C}^n \times \mathbf{C}^n$ is identified with $\mathbf{H}^n$ by the (right) complex linear isomorphism

$$\mathbf{C}^n \times \mathbf{C}^n \longrightarrow \mathbf{H}^n; \quad (u,v) \rightsquigarrow u + jv.$$

Prop. 10.31. Let $a + jb \in \mathbf{H}(n)$, where a and $b \in \mathbf{C}(n)$. Then the corresponding element of $\mathbf{C}(2n)$ is $\begin{pmatrix} a & -\bar{b} \\ b & \bar{a} \end{pmatrix}$.

Proof For any u, v, u', $v' \in \mathbf{C}^n$ and any $a, b \in \mathbf{C}(n)$, the equation $u' + jv' = (a + jb)(u + jv)$ is equivalent to the pair of equations

$$u' = au - \bar{b}v$$

and

$$v' = bu + \bar{a}v. \qquad \square$$

In particular, when $n = 1$, this becomes an explicit representation of $\mathbf{H}$ as a subalgebra of $\mathbf{C}(2)$, analogous to the representation of $\mathbf{C}$ as a subalgebra of $\mathbf{R}(2)$ given in Prop. 3.31.

Notice that, for any $q = a + jb \in \mathbf{H}$, with $a, b \in \mathbf{C}$,

$$|q|^2 = \bar{q}q = \bar{a}a + \bar{b}b = \det \begin{pmatrix} a & -\bar{b} \\ b & \bar{a} \end{pmatrix}.$$

This remark is a detail in the proof of Prop. 10.33 below.

The lack of commutativity in $\mathbf{H}$ is most strongly felt when one tries to introduce products, as the following proposition shows.

Prop. 10.32. Let X, Y and Z be right linear spaces over $\mathbf{H}$ and let $t : X \times Y \to Z;\ (x,y) \rightsquigarrow x \cdot y$ be a right bilinear map. Then $t = 0$.

Proof For any $(x,y) \in X \times Y$,

$$\begin{aligned}(x \cdot y)\mathrm{k} &= (x \cdot y)\mathrm{ij} = (x \cdot y\mathrm{i})\mathrm{j} = x\mathrm{j} \cdot y\mathrm{i} = (x\mathrm{j} \cdot y)\mathrm{i} \\ &= (x \cdot y)\mathrm{ji} = -(x \cdot y)\mathrm{k}.\end{aligned}$$

Since $\mathrm{k} \neq 0$, $x \cdot y = 0$. So $t = 0$. □

It follows from this, *a fortiori*, that there is no non-trivial n-linear map $X^n \to \mathbf{H}$ for a right $\mathbf{H}$-linear space X for any $n > 1$. In particular, there is no direct analogue of the determinant for the algebra of endomorphisms of a finite-dimensional quaternionic linear space, in particular the right $\mathbf{H}$-linear space $\mathbf{H}^n$. There is, however, an analogue of the *absolute* determinant.

In fact the material of Chapter 7 holds for the non-commutative field $\mathbf{H}$ up to and including Theorem 7.8. This theorem can then be used to prove the following analogue, for quaternionic linear endomorphisms, of Cor. 7.33 for complex linear endomorphisms.

Prop. 10.33. Let X be a finite-dimensional right $\mathbf{H}$-linear space and let $t : X \to X$ be an $\mathbf{H}$-linear map. Then $\det_{\mathbf{C}} t$ is a non-negative real number. □

(Here, as above, $\mathbf{C}$ is identified with the subalgebra $\operatorname{im}\{1,\mathrm{i}\}$ of $\mathbf{H}$.)

Theorem 10.34 is the analogue, for $\mathbf{H}$, of Theorem 7.28 for $\mathbf{R}$ and $\mathbf{C}$.

Theorem 10.34. Let $n \in \omega$. Then there exists a unique map

$$\Delta : \mathbf{H}(n) \to \mathbf{R};\quad a \rightsquigarrow \Delta(a)$$

such that

(i) for each $a \in \mathbf{H}(n)$, $i \in n$ and $\lambda \in \mathbf{H}$, $\Delta(a({}^{\lambda}e_i)) = \Delta(a)\,|\,\lambda\,|$

(ii) for each $a \in \mathbf{H}(n)$, $i, j \in n$ with $i \neq j$, $\Delta(a({}^{1}e_{ij})) = \Delta(a)$

and (iii) $\Delta({}^{n}1) = 1$.

Proof The existence of such a map Δ follows from the remark that, by identifying $\mathbf{C}$ with $\operatorname{im}\{1,i\}$ in $\mathbf{H}$, any element $a \in \mathbf{H}(n)$ may be regarded as an endomorphism of the *complex* linear space $\mathbf{H}^n$. As such it has a determinant $\det_{\mathbf{C}} a \in \mathbf{C}$, and, by Prop. 10.33, this is a non-negative real number. Now define $\Delta(a) = \sqrt{(\det_{\mathbf{C}} a)}$, for all $a \in \mathbf{H}(n)$. Then it may be readily verified that conditions (i), (ii) and (iii) are satisfied. Moreover, for any $a, b \in \mathbf{H}(n)$, $\Delta(ba) = \Delta(b)\,\Delta(a)$.

The uniqueness of Δ follows easily from the analogue of Theorem 7.8 for $\mathbf{H}$, by the argument hinted at in the sketch of the proof of Theorem 7.28. □

Prop. 10.35. Let $t \in \mathbf{H}(n)$, for some finite n. Then t is invertible if, and only if, $\Delta(t) \neq 0$. □

It is usual to write simply $\det t$ for $\Delta(t)$, for any $n \in \omega$ and any $t \in \mathbf{H}(n)$. The subgroup $\{t \in \mathbf{H}(n) : \det t = 1\}$ of $GL(n;\mathbf{H})$ is denoted by $SL(n;\mathbf{H})$.

Right and left $\mathbf{H}$-linear spaces are examples of *right* and *left* Λ-modules, where Λ is a not necessarily commutative ring with unity, the ring Λ simply replacing the field $\mathbf{H}$ in all the definitions. One can, for example, consider right and left modules over the ring ${}^s\mathbf{H}$, for any positive s, and extend to the quaternionic case the appropriate part of Chapter 8.

The remainder of Chapter 8 also extends to the quaternionic case, including the definitions of Grassmannians and projective spaces and their properties. The only point to note is that, if Y is a subspace of a finite-dimensional right $\mathbf{H}$-linear space V, then $\Theta(V,Y)$, the set of linear complements of Y in V, has in a natural way a *real* affine structure, with vector space the real linear space $L(V/Y, Y)$, but it has not, in general, a useful quaternionic affine structure.

Generalizing the ideas of Chapter 9 to the quaternionic case is a bigger problem. This is discussed in Chapter 11.

Tensor product of algebras

Certain algebras over a *commutative* field $\mathbf{K}$ admit a decomposition somewhat analogous to the direct sum decompositions of a linear space, but involving the multiplicative structure rather than the additive structure.

Suppose B and C are subalgebras of a finite-dimensional algebra A over $\mathbf{K}$, the algebra being associative and with unity, such that

(i) A is generated as an algebra by B and C
(ii) $\dim A = \dim B \dim C$
and (iii) for any $b \in B$, $c \in C$, $cb = bc$.

Then we say that A is the *tensor product* $B \otimes_{\mathbf{K}} C$ of B and C over $\mathbf{K}$, the abbreviation $B \otimes C$ being used in place of $B \otimes_{\mathbf{K}} C$ when the field $\mathbf{K}$ is not in doubt.

Prop. 10.36. Let B and C be subalgebras of a finite-dimensional algebra A over $\mathbf{K}$, such that $A = B \otimes C$, the algebra A being associative and with unity. Then $B \cap C = \mathbf{K}$ (the field $\mathbf{K}$ being identified with the set of scalar multiples of $1_{(A)}$). □

It is tempting to suppose that this proposition can be used as an alternative to condition (ii) in the definition. That this is not so is shown by the following example.

Example 10.37.

Let
$$A = \left\{ \begin{pmatrix} a & b & c \\ 0 & a & 0 \\ 0 & 0 & a \end{pmatrix} \in \mathbf{R}(3) : a, b, c \in \mathbf{R} \right\},$$
let
$$B = \left\{ \begin{pmatrix} a & b & 0 \\ 0 & a & 0 \\ 0 & 0 & a \end{pmatrix} \in \mathbf{R}(3) : a, b \in \mathbf{R} \right\}$$
and let
$$C = \left\{ \begin{pmatrix} a & 0 & c \\ 0 & a & 0 \\ 0 & 0 & a \end{pmatrix} \in \mathbf{R}(3) : a, c \in \mathbf{R} \right\}.$$
Then A is generated as an algebra by B and C, $B \cap C = \mathbf{R}$, and any element of B commutes with any element of C. But $\dim A = 3$, while $\dim B = \dim C = 2$, so that $\dim A \neq \dim B \dim C$. □

Condition (ii) is essential to the proof of the following proposition.

Prop. 10.38. Let A be a finite-dimensional associative algebra with unity over $\mathbf{K}$ and let B and C be subalgebras of A such that $A = B \otimes C$. Also let $\{e_i : i \in \dim B\}$ and $\{f_j : j \in \dim C\}$ be bases for the linear spaces B and C respectively. Then the set
$$\{e_i f_j : i \in \dim B, j \in \dim C\}$$
is a basis for the linear space A. □

This can be used in the proof of the next proposition.

Prop. 10.39. Let A and A' be finite-dimensional associative algebras with unity over $\mathbf{K}$ and let B and C be subalgebras of A, and B' and C' be subalgebras of A' such that $A = B \otimes C$ and $A' = B' \otimes C'$. Then if $B \cong B'$ and if $C \cong C'$, it follows that $A \cong A'$. □

Proposition 10.39 encourages various extensions and abuses of the notation $\otimes$. In particular, if A, B, C, B' and C' are associative algebras with unity over $\mathbf{K}$ such that

$$A = B \otimes C, \quad B' \cong B \quad \text{and} \quad C' \cong C,$$

one frequently writes $A \cong B' \otimes C'$, even though there is no unique construction of $B' \otimes C'$. The precise meaning of such a statement will always be clear from the context.

The tensor product of algebras is a special case and generalization of the tensor product of linear spaces. We have chosen not to develop the theory of tensor products in general, as we have no essential need of the more general concept.

The following propositions involving the tensor product of algebras will be of use in determining the table of Clifford algebras in Chapter 13.

Prop. 10.40. Let A be an associative algebra with unity over a commutative field $\mathbf{K}$ and let B, C and D be subalgebras of A. Then

$$A = B \otimes C \;\Leftrightarrow\; A = C \otimes B$$

and

$$A = B \otimes (C \otimes D) \;\Leftrightarrow\; A = (B \otimes C) \otimes D. \qquad \square$$

In the latter case it is usual to write, simply, $A = B \otimes C \otimes D$.

Prop. 10.41. For any commutative field $\mathbf{K}$, and for any $p, q \in \omega$,

$$\mathbf{K}(pq) \cong \mathbf{K}(p) \otimes \mathbf{K}(q).$$

Proof Let $\mathbf{K}^{pq}$ be identified as a linear space with $\mathbf{K}^{p \times q}$, the linear space of $p \times q$ matrices over $\mathbf{K}$. Then the maps $\mathbf{K}(p) \to \mathbf{K}(pq)$; $a \rightsquigarrow a_{\mathrm{L}}$ and $\mathbf{K}(q) \to \mathbf{K}(pq)$; $b \rightsquigarrow (b^{\tau})_{\mathrm{R}}$ are algebra injections whose images in $\mathbf{K}(pq)$ satisfy conditions (i)–(iii) for $\otimes$, a_{L} and $(b^{\tau})_{\mathrm{R}}$ being defined, for each $a \in \mathbf{K}(p)$ and $b \in \mathbf{K}(q)$, and for each $c \in \mathbf{K}^{p \times q}$, by the formulae

$$a_{\mathrm{L}}(c) = ac \quad \text{and} \quad (b^{\tau})_{\mathrm{R}}(c) = cb^{\tau}.$$

For example, the commutativity condition (iii) follows directly from the associativity of matrix multiplication. $\square$

In particular, for any $p, q \in \omega$,

$$\mathbf{R}(pq) \cong \mathbf{R}(p) \otimes_{\mathbf{R}} \mathbf{R}(q).$$

In this case we can say slightly more.

Prop. 10.42. For any $p, q \in \omega$, let $\mathbf{R}^p$, $\mathbf{R}^q$ and $\mathbf{R}^{pq}$ be regarded as positive-definite orthogonal spaces in the standard way, and let $\mathbf{R}^{p \times q}$ be identified with $\mathbf{R}^{pq}$. Then the algebra injections

$$\mathbf{R}(p) \to \mathbf{R}(pq); \quad a \rightsquigarrow a_{\mathrm{L}}$$

and

$$\mathbf{R}(q) \to \mathbf{R}(pq); \quad b \rightsquigarrow (b^{\tau})_{\mathrm{R}}$$

send the orthogonal elements of $\mathbf{R}(p)$ and $\mathbf{R}(q)$, respectively, to orthogonal elements of $\mathbf{R}(pq)$. □

Cor. 10.43. The product of any finite ordered set of elements belonging either to the copy of $O(p)$ or to the copy of $O(q)$ in $\mathbf{R}(pq)$ is an element of $O(pq)$. □

In what follows, $\mathbf{C}$ and $\mathbf{H}$ will both be regarded as *real* algebras, of dimensions 2 and 4, respectively, and $\otimes = \otimes_{\mathbf{R}}$.

Prop. 10.44. $\mathbf{R} \otimes \mathbf{R} = \mathbf{R}$, $\mathbf{C} \otimes \mathbf{R} = \mathbf{C}$, $\mathbf{H} \otimes \mathbf{R} = \mathbf{H}$, $\mathbf{C} \otimes \mathbf{C} \cong {}^2\mathbf{C}$, $\mathbf{H} \otimes \mathbf{C} \cong \mathbf{C}(2)$ and $\mathbf{H} \otimes \mathbf{H} \cong \mathbf{R}(4)$.

Proof The first three of these statements are obvious. To prove that $\mathbf{C} \otimes \mathbf{C} = {}^2\mathbf{C}$ it is enough to remark that ${}^2\mathbf{C}$ is generated as a real algebra by the subalgebras $\left\{\begin{pmatrix} z & 0 \\ 0 & z \end{pmatrix} : z \in \mathbf{C}\right\}$ and $\left\{\begin{pmatrix} z & 0 \\ 0 & \bar{z} \end{pmatrix} : z \in \mathbf{C}\right\}$, each isomorphic to $\mathbf{C}$, conditions (i)–(iii) being readily verified.

To prove that $\mathbf{H} \otimes \mathbf{C} = \mathbf{C}(2)$, let $\mathbf{C}^2$ be identified with $\mathbf{H}$ as a right complex linear space by the map $\mathbf{C}^2 \to \mathbf{H};\ (z,w) \rightsquigarrow z + jw$, as before. Then, for any $q \in \mathbf{H}$ and any $c \in \mathbf{C}$, the maps

$$q_L : \mathbf{H} \to \mathbf{H};\quad x \rightsquigarrow qx \quad \text{and} \quad c_R : \mathbf{H} \to \mathbf{H};\quad x \rightsquigarrow xc$$

are complex linear, and the maps

$$\mathbf{H} \to \mathbf{C}(2);\quad q \rightsquigarrow q_L \quad \text{and} \quad \mathbf{C} \to \mathbf{C}(2);\quad c \rightsquigarrow c_R$$

are algebra injections. Conditions (ii) and (iii) are obviously satisfied by the images of these injections. To prove (i) it is enough to remark that the matrices

$$\begin{pmatrix} 1 & 0 \\ 0 & 1 \end{pmatrix}, \begin{pmatrix} i & 0 \\ 0 & -i \end{pmatrix}, \begin{pmatrix} 0 & -1 \\ 1 & 0 \end{pmatrix}, \begin{pmatrix} 0 & -i \\ -i & 0 \end{pmatrix},$$
$$\begin{pmatrix} i & 0 \\ 0 & i \end{pmatrix}, \begin{pmatrix} -1 & 0 \\ 0 & 1 \end{pmatrix}, \begin{pmatrix} 0 & -i \\ i & 0 \end{pmatrix}, \begin{pmatrix} 0 & 1 \\ 1 & 0 \end{pmatrix},$$

representing

$$\begin{matrix} 1, & i_L, & j_L, & k_L \\ i_R, & i_L i_R, & j_L i_R, & k_L i_R, \end{matrix}$$

respectively, span $\mathbf{C}(2)$ linearly.

The proof that $\mathbf{H} \otimes \mathbf{H} \cong \mathbf{R}(4)$ is similar, the maps

$$q_L : \mathbf{H} \to \mathbf{H};\quad x \rightsquigarrow qx \quad \text{and} \quad \bar{r}_R : \mathbf{H} \to \mathbf{H};\quad x \rightsquigarrow x\bar{r}$$

being real linear, for any $q, r \in \mathbf{H}$ and the maps

$$\mathbf{H} \to \mathbf{R}(4);\quad q \rightsquigarrow q_L \quad \text{and} \quad \mathbf{H} \to \mathbf{R}(4);\quad r \rightsquigarrow \bar{r}_R$$

being algebra injections whose images satisfy conditions (i)–(iii). □

In this last case it is worth recalling Prop. 10.29, which states that the image, by either of these injections, of a quaternion of absolute value 1 is an orthogonal element of **R**(4). At the end of Chapter 11 we make a similar remark about the isomorphism of **H** ⊗ **C** with **C**(2) and draw an analogy between them both and Prop. 10.42.

It is an advantage to be able to detect quickly whether or not a subalgebra of a given real associative algebra A is isomorphic to one of the algebras

$$\mathbf{R},\ \mathbf{C},\ \mathbf{H},\ {}^2\mathbf{R},\ {}^2\mathbf{C},\ {}^2\mathbf{H},\ \mathbf{R}(2),\ \mathbf{C}(2)\ \text{or}\ \mathbf{H}(2),$$

or whether a subalgebra of a given complex associative algebra A is isomorphic to one of the algebras **C**, ^{2}**C** or **C**(2). The following proposition is useful in this context.

Prop. 10.45. Let A be a real associative algebra with unity 1. Then

1 generates **R**;

any two-dimensional subalgebra generated by an element e_0 of A such that $e_0^2 = -1$ is isomorphic to **C**;

any two-dimensional subalgebra generated by an element e_0 of A such that $e_0^2 = 1$ is isomorphic to ^{2}**R**;

any four-dimensional subalgebra generated by a set $\{e_0,e_1\}$ of mutually anticommuting elements of A such that $e_0^2 = e_1^2 = -1$ is isomorphic to **H**;

any four-dimensional subalgebra generated by a set $\{e_0,e_1\}$ of mutually anticommuting elements of A such that $e_0^2 = e_1^2 = 1$ is isomorphic to **R**(2);

any eight-dimensional subalgebra generated by a set $\{e_0,e_1,e_2\}$ of mutually anticommuting elements of A such that $e_0^2 = e_1^2 = e_2^2 = -1$ is isomorphic to ^{2}**H**;

any eight-dimensional subalgebra generated by a set $\{e_0,e_1,e_2\}$ of mutually anticommuting elements of A such that $e_0^2 = e_1^2 = e_2^2 = 1$ is isomorphic to **C**(2).

(Sets of elements meeting the required conditions include

$$\left\{\begin{pmatrix}0 & 1\\ 1 & 0\end{pmatrix}, \begin{pmatrix}1 & 0\\ 0 & -1\end{pmatrix}\right\} \text{ for } \mathbf{R}(2),$$

$$\left\{\begin{pmatrix}\mathrm{i} & 0\\ 0 & -\mathrm{i}\end{pmatrix}, \begin{pmatrix}\mathrm{j} & 0\\ 0 & -\mathrm{j}\end{pmatrix}, \begin{pmatrix}\mathrm{k} & 0\\ 0 & -\mathrm{k}\end{pmatrix}\right\} \text{ or } \left\{\begin{pmatrix}\mathrm{i} & 0\\ 0 & \mathrm{i}\end{pmatrix}, \begin{pmatrix}\mathrm{j} & 0\\ 0 & -\mathrm{j}\end{pmatrix}, \begin{pmatrix}\mathrm{k} & 0\\ 0 & \mathrm{k}\end{pmatrix}\right\}$$

$$\left(\text{but not} \left\{\begin{pmatrix}\mathrm{i} & 0\\ 0 & \mathrm{i}\end{pmatrix}, \begin{pmatrix}\mathrm{j} & 0\\ 0 & \mathrm{j}\end{pmatrix}, \begin{pmatrix}\mathrm{k} & 0\\ 0 & \mathrm{k}\end{pmatrix}\right\} \text{ nor } \left\{\begin{pmatrix}\mathrm{i} & 0\\ 0 & -\mathrm{i}\end{pmatrix}, \begin{pmatrix}\mathrm{j} & 0\\ 0 & \mathrm{j}\end{pmatrix}, \begin{pmatrix}\mathrm{k} & 0\\ 0 & -\mathrm{k}\end{pmatrix}\right\}\right)$$

for ^{2}**H** and $\left\{\begin{pmatrix}0 & 1\\ 1 & 0\end{pmatrix}, \begin{pmatrix}1 & 0\\ 0 & -1\end{pmatrix}, \begin{pmatrix}0 & -\mathrm{i}\\ \mathrm{i} & 0\end{pmatrix}\right\}$ for **C**(2).) □

In particular we have the following results, including several we have had before.

Prop. 10.46. The subset of matrices of the real algebra $\mathbf{K}(2)$ of the form

(i) $\begin{pmatrix} a & b \\ b & a \end{pmatrix}$ is a subalgebra isomorphic to ${}^2\mathbf{R}$, ${}^2\mathbf{C}$ or ${}^2\mathbf{H}$,

(ii) $\begin{pmatrix} a & -b \\ b & a \end{pmatrix}$ is a subalgebra isomorphic to $\mathbf{C}$, ${}^2\mathbf{C}$ or $\mathbf{C}(2)$,

(iii) $\begin{pmatrix} a & b' \\ b & a' \end{pmatrix}$ is a subalgebra isomorphic to ${}^2\mathbf{R}$, $\mathbf{R}(2)$ or $\mathbf{C}(2)$,

(iv) $\begin{pmatrix} a & -b' \\ b & a' \end{pmatrix}$ is a subalgebra isomorphic to $\mathbf{C}$, $\mathbf{H}$ or ${}^2\mathbf{H}$,

according as $\mathbf{K} = \mathbf{R}$, $\mathbf{C}$ or $\mathbf{H}$, respectively, where for any $a \in \mathbf{K}$, $a' = a$, $\bar{a}$ or $\hat{a}$, respectively. □

Each of the algebra injections listed in Prop. 10.46 is induced by a (non-unique) real linear injection. For example, those of the form (iii) may be regarded as being the injections of the appropriate endomorphism algebras induced by the real linear injections

$$ {}^2\mathbf{R} \rightarrow \mathbf{R}^2; \quad (x,y) \rightsquigarrow (x + y, x - y),$$
$$ \mathbf{R}^2 \rightarrow \mathbf{C}^2; \quad (x,y) \rightsquigarrow (x + iy, x - iy)$$

and

$$ \mathbf{C}^2 \rightarrow \mathbf{H}^2; \quad (z,w) \rightsquigarrow (z + jw, \bar{z} + j\bar{w}).$$

Real algebras, A, B, C and D, say, frequently occur in a commutative square of algebra injections of the form

$$\begin{array}{ccc} A & \longrightarrow & C \\ \downarrow & & \downarrow \\ B & \longrightarrow & D \end{array}$$

the algebra D being generated by the images of B and C. Examples of such squares, which may easily be constructed using the material of Prop. 10.46, include

$$\begin{array}{ccc} \mathbf{R} & \longrightarrow & {}^2\mathbf{R} \\ \downarrow & & \downarrow \\ \mathbf{C} & \longrightarrow & \mathbf{R}(2) \end{array}, \qquad \begin{array}{ccc} {}^2\mathbf{R} & \longrightarrow & \mathbf{R}(2) \\ \downarrow & & \downarrow \\ \mathbf{R}(2) & \longrightarrow & {}^2\mathbf{R}(2) \end{array},$$

$$\begin{array}{ccc} \mathbf{C} & \longrightarrow & \mathbf{R}(2) \\ \downarrow & & \downarrow \\ \mathbf{H} & \longrightarrow & \mathbf{C}(2) \end{array}, \qquad \begin{array}{ccc} \mathbf{C} & \longrightarrow & {}^2\mathbf{C} \\ \downarrow & & \downarrow \\ {}^2\mathbf{C} & \longrightarrow & \mathbf{C}(2) \end{array}, \qquad \begin{array}{ccc} {}^2\mathbf{C} & \longrightarrow & \mathbf{C}(2) \\ \downarrow & & \downarrow \\ \mathbf{C}(2) & \longrightarrow & {}^2\mathbf{C}(2) \end{array},$$

$$\begin{array}{ccc} \mathbf{H} & \longrightarrow & \mathbf{C}(2) \\ \downarrow & & \downarrow \\ {}^2\mathbf{H} & \longrightarrow & \mathbf{H}(2) \end{array} \quad \text{and} \quad \begin{array}{ccc} {}^2\mathbf{H} & \longrightarrow & \mathbf{H}(2) \\ \downarrow & & \downarrow \\ \mathbf{H}(2) & \longrightarrow & {}^2\mathbf{H}(2) \end{array}.$$

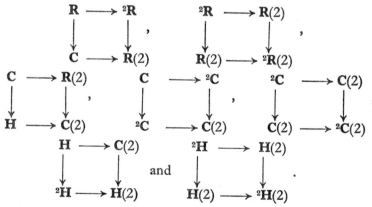

The emphasis, in these last few pages, has been on algebras over **R**. The reader is invited to consider how much of what has been said holds for the complex field **C**.

Automorphisms and anti-automorphisms of $^s\mathbf{K}$

Some knowledge of the real algebra automorphisms and anti-automorphisms of $^s\mathbf{K}$, where $\mathbf{K} = \mathbf{R}$, **C** or **H** and s is positive, is required in Chapter 11.

The problem is reduced by considering the primitive idempotents of $^s\mathbf{K}$. An *idempotent* of an algebra A is an element a of A such that $a^2 = a$, and a *primitive* idempotent of A is an idempotent of A that is not the sum of two non-zero idempotents of A.

Prop. 10.47. Any automorphism or anti-automorphism of a real algebra A permutes the primitive idempotents of A. □

Prop. 10.48. For any positive s the elements of the standard basis for $^s\mathbf{K}$ are the primitive idempotents of $^s\mathbf{K}$. □

A permutation π of a finite set S is said to be *reducible* if there is a proper subset T of S such that $\pi_\vdash(T) \subset T$, and an automorphism or anti-automorphism of a real algebra A is said to be *reducible* if the induced permutation of the primitive idempotents of A is reducible. A permutation or automorphism or anti-automorphism that is not reducible is said to be *irreducible*.

Prop. 10.49. Let s be a positive number such that $^s\mathbf{K}$ admits an irreducible involution or anti-involution. Then $s = 1$ or 2. □

Any automorphism or anti-automorphism of **K** is irreducible. By Prop. 2.60 the only automorphism of **R** is $1_\mathbf{R}$ and, by Prop. 3.38, the only (real algebra) automorphisms of **C** are $1_\mathbf{C}$ and conjugation, both of which are involutions. The automorphisms of **H** are represented, by Prop. 10.22, by the rotations of the space of pure quaternions, while the anti-automorphisms of **H** are represented, similarly, by the antirotations of that space. By Cor. 10.26 the involutions of **H** are $1_\mathbf{H}$ and those corresponding to reflection of the space of pure quaternions in any line through 0, while the anti-involutions are conjugation, which corresponds to reflection in 0, and those corresponding to reflection of the space of pure quaternions in any plane through 0. Notations for certain involutions and anti-involutions of **H** have been given earlier in the chapter.

Prop. 10.50. An automorphism or anti-automorphism of $^2\mathbf{K}$ is reducible if, and only if, it is of the form

$$^2\mathbf{K} \to {}^2\mathbf{K};\quad (\lambda,\mu) \rightsquigarrow (\lambda^\chi,\mu^\psi),$$

where χ, $\psi : \mathbf{K} \to \mathbf{K}$ are, respectively, both automorphisms or anti-automorphisms of $\mathbf{K}$. It is an involution or anti-involution of ${}^2\mathbf{K}$ if, and only if, both χ and ψ are, respectively, involutions or anti-involutions of $\mathbf{K}$. □

Such an automorphism or anti-automorphism is denoted by $\chi \times \psi$. More interesting are the irreducible automorphisms and anti-automorphisms of ${}^2\mathbf{K}$.

Prop. 10.51. An automorphism or anti-automorphism of ${}^2\mathbf{K}$ is irreducible if, and only if, it is of the form

$$ {}^2\mathbf{K} \to {}^2\mathbf{K}; \quad (\lambda,\mu) \rightsquigarrow (\mu^{\psi},\lambda^{\chi}), $$

where χ and ψ are, respectively, both automorphisms or anti-automorphisms of ${}^2\mathbf{K}$.

An involution or anti-involution of ${}^2\mathbf{K}$ is irreducible if, and only if, it is of the form

$$ {}^2\mathbf{K} \to {}^2\mathbf{K}; \quad (\lambda,\mu) \rightsquigarrow (\mu^{\psi},\lambda^{\psi^{-1}}), $$

where ψ is an automorphism or anti-automorphism (not necessarily an involution or anti-involution) of $\mathbf{K}$. □

The involution

$$ {}^2\mathbf{K} \to {}^2\mathbf{K}; \quad (\lambda,\mu) \rightsquigarrow (\mu,\lambda) $$

will be denoted by hb, the involution $(\psi \times \psi^{-1})\,\text{hb} = \text{hb}\,(\psi^{-1} \times \psi)$ being denoted also, more briefly, by hb ψ. The letters hb are an abbreviation for the word *hyperbolic*, the choice of word being suggested by the observation that, when $\mathbf{K} = \mathbf{R}$, the set $\{(\lambda,\mu) \in {}^2\mathbf{K} : (\lambda,\mu)^{\text{hb}}(\lambda,\mu) = 1\}$ is just the rectangular hyperbola $\{(\lambda,\mu) \in \mathbf{R}^2 : \lambda\mu = 1\}$.

The symbols $\mathbf{R}$, $\mathbf{C}$, $\bar{\mathbf{C}}$, $\mathbf{H}$, $\hat{\mathbf{H}}$, $\bar{\mathbf{H}}$ and $\tilde{\mathbf{H}}$ denote $\mathbf{R}$, $\mathbf{C}$ or $\mathbf{H}$ with the indicated involution or anti-involution distinguished, this being taken to be the identity when there is no indication to the contrary. The symbols hb $\mathbf{K}^{\psi}$ and $({}^2\mathbf{K})^{\text{hb}\,\psi}$ will both denote the algebra ${}^2\mathbf{K}$ with the involution or anti-involution hb ψ.

Two automorphisms or anti-automorphisms β, γ of an algebra A are said to be *similar* if there is an automorphism α of A such that $\gamma\alpha = \alpha\beta$. If no such automorphism exists, β and α are said to be *dissimilar*.

Prop. 10.52. The two involutions of $\mathbf{C}$ are dissimilar. □

Prop. 10.53. The sets of involutions and anti-involutions of $\mathbf{H}$ each divide into two equivalence classes with respect to similarity, the identity, or conjugation, being the sole element in one class, and all the rest belonging to the other class. □

Prop. 10.54. For any automorphisms or anti-automorphisms ψ and χ of $\mathbf{K}$ the involutions or anti-involutions hb ψ and hb χ of ${}^2\mathbf{K}$ are similar.

Proof Let $\alpha = \chi^{-1}\psi$. Then

$$(\text{hb}\,\chi)(1 \times \alpha) = (1 \times \alpha)(\text{hb}\,\psi),$$

since, for any $(\lambda,\mu) \in {}^2\mathbf{K}$,

$$(\text{hb}\,\chi)(1 \times \alpha)(\lambda,\mu) = (\text{hb}\,\chi)(\lambda,\mu^\alpha) = (\mu^{\chi\alpha},\lambda^{\chi^{-1}})$$

and $$(1 \times \alpha)(\text{hb}\,\psi)(\lambda,\mu) = (1 \times \alpha)(\mu^\psi,\lambda^{\psi^{-1}}) = (\mu^\psi,\lambda^{\alpha\psi^{-1}}).$$ □

Cor. 10.55. Let ψ be an irreducible anti-involution of ${}^s\mathbf{K}$, for some positive s. Then ψ is similar to one and only one of the following:

$$1_{\mathbf{R}},\quad 1_{\mathbf{C}},\quad \mathbf{C} \to \mathbf{C};\ \lambda \rightsquigarrow \bar{\lambda},$$
$$\mathbf{H} \to \mathbf{H};\ \lambda \rightsquigarrow \bar{\lambda},\quad \mathbf{H} \to \mathbf{H};\ \lambda \rightsquigarrow \tilde{\lambda},$$
$${}^2\mathbf{R} \to {}^2\mathbf{R};\ (\lambda,\mu) \rightsquigarrow (\mu,\lambda),\quad {}^2\mathbf{C} \to {}^2\mathbf{C};\ (\lambda,\mu) \rightsquigarrow (\bar{\mu},\bar{\lambda})$$

or $${}^2\mathbf{H} \to {}^2\mathbf{H};\ (\lambda,\mu) \rightsquigarrow (\bar{\mu},\bar{\lambda}),$$

eight in all. □

Further Exercises

10.56. Prove that the complex numbers $9 + i$, $4 + 13i$, $-8 + 8i$ and $-3 - 4i$ are the vertices of a square, when $\mathbf{C}$ is identified with the orthogonal space $\mathbf{R}^2$. □

10.57. Show that any circle in $\mathbf{C}$ is of the form

$$\{z \in \mathbf{C} : z\bar{z} + \bar{a}z + a\bar{z} + c = 0\}$$

for some $a \in \mathbf{C}$ and $c \in \mathbf{R}$, a and c not being both zero. □

10.58. Find the greatest and least values of $|\,2z + 2 + 3i\,|$ if $z \in \mathbf{C}$ and $|\,z - i\,| \leqslant 1$. □

10.59. The point w describes the circle

$$\{z \in \mathbf{C} : |\,z - 2 - i\,| = 2\}$$

in an anticlockwise direction, starting at $w = i$. Describe the motion of $1/z$. □

10.60. Find the domain and image of the map

$$\mathbf{C} \to \mathbf{C};\quad z \rightsquigarrow \frac{z + i}{iz + 1}$$

and find the image by the map of the interior of the unit circle. □

10.61. Let $\mathbf{C}$ be identified with $\mathbf{R} \times \mathbf{R} \times \{0\}$ in $\mathbf{R}^3$ and let $f : \mathbf{C} \to S^2$ be the inverse of stereographic projection from the South pole. So f maps 0 to the North pole and maps the unit circle to the equator of the sphere. In terms of this representation, describe the map of Exercise 10.60. (A ping-pong ball with some complex numbers marked on it may be of assistance!) If two points on the sphere are antipodal to one another, how are the corresponding complex numbers related?

(The map $\bar{f}: \mathbf{C} \cup \{\infty\} \to S^2$, which agrees with f on $\mathbf{C}$ and which sends ∞ to the South pole, is called the *Riemann representation* of $\mathbf{C}P^1 = \mathbf{C} \cup \{\infty\}$ on S^2. Cf. page 141.) □

10.62. Let g be a map of the form

$$\mathbf{C} \cup \{\infty\} \to \mathbf{C} \cup \{\infty\}; \quad z \rightsquigarrow \frac{az + c}{bz + d},$$

where $a, b, c, d \in \mathbf{C}$ and where the conventions governing the symbol ∞ are the obvious extensions of those introduced on page 141. Verify that g corresponds by the Riemann representation of Exercise 10.61 to a rotation of S^2 if, and only if, $c = -\bar{b}$ and $d = \bar{a}$, and show that every rotation of S^2 may be so represented. □

10.63. Verify that the matrix $\begin{pmatrix} 1 & \mathrm{i} \\ \mathrm{j} & \mathrm{k} \end{pmatrix}$ is invertible in $\mathbf{H}(2)$, but that the matrix $\begin{pmatrix} 1 & \mathrm{j} \\ \mathrm{i} & \mathrm{k} \end{pmatrix}$ is not. □

10.64. Does a *real* algebra involution of $\mathbf{H}(2)$ necessarily map a matrix of the form $\begin{pmatrix} a & 0 \\ 0 & a \end{pmatrix}$, where $a \in \mathbf{H}$, to one of the same form? □

10.65. Verify that, for any pair of invertible quaternions (a,b),

$$\begin{pmatrix} aba^{-1}b^{-1} & 0 \\ 0 & 1 \end{pmatrix} = \begin{pmatrix} a & 0 \\ 0 & a^{-1} \end{pmatrix}\begin{pmatrix} b & 0 \\ 0 & b^{-1} \end{pmatrix}\begin{pmatrix} (ba)^{-1} & 0 \\ 0 & ba \end{pmatrix}$$

and, for any invertible quaternion c,

$$\begin{pmatrix} c & 0 \\ 0 & c^{-1} \end{pmatrix}\begin{pmatrix} 1 & c^{-1} \\ 0 & 1 \end{pmatrix}\begin{pmatrix} 1 & 0 \\ -c & 1 \end{pmatrix}\begin{pmatrix} 1 & c^{-1} \\ 0 & 1 \end{pmatrix} = \begin{pmatrix} 1 & 1 \\ 0 & 1 \end{pmatrix}\begin{pmatrix} 1 & 0 \\ -1 & 1 \end{pmatrix}\begin{pmatrix} 1 & 1 \\ 0 & 1 \end{pmatrix}.$$

Hence, and by Prop. 10.24 and the analogue for $\mathbf{K} = \mathbf{H}$ of Theorem 7.8, prove that, for any $n \geqslant 2$, an element t of $GL(n;\mathbf{H})$ is unimodular if, and only if, $\det t = 1$. □

10.66. Verify that the map $\alpha: \mathbf{C}^2 \to \mathbf{H};\ x \rightsquigarrow x_0 + \mathrm{j}x_1$ is a right $\mathbf{C}$-linear isomorphism, and compute

$$\alpha^{-1}(\widetilde{\alpha(x)}\,\alpha(y)), \quad \text{for any } x, y \in \mathbf{C}^2.$$

Let $Q = \{(x,y) \in (\mathbf{C}^2)^2 : x_0y_0 + x_1y_1 = 1\}$.

Prove that, for any $(a,b) \in \mathbf{H}^* \times \mathbf{C}$,

$$(\alpha^{-1}(\tilde{a}), \alpha^{-1}(a^{-1}(1 + \mathrm{j}b))) \in Q$$

and that the map

$$\mathbf{H}^* \times \mathbf{C} \to \mathbf{Q}; \quad (a,b) \rightsquigarrow (\alpha^{-1}(\tilde{a}), \alpha^{-1}(a^{-1}(1 + jb)))$$

is bijective. □

10.67. Extend Chapter 8 to the quaternionic case. □

10.68. Show that the fibres of the restriction of the Hopf map

$$\mathbf{C}^2 \to \mathbf{C}P^1; \quad (z_0,z_1) \rightsquigarrow [z_0,z_1]$$

to the sphere $S^3 = \{(z_0,z_1) \in \mathbf{C}^2 : \bar{z}_0 z_0 + \bar{z}_1 z_1 = 1\}$ are circles, any two of which link. (Cf. page 144 and Exercise 9.85.) □

10.69. Show that the fibres of the restriction of the Hopf map

$$\mathbf{H}^2 \to \mathbf{H}P^1; \quad (q_0,q_1) \rightsquigarrow [q_0,q_1]$$

to the sphere $S^7 = \{(q_0,q_1) \in \mathbf{H}^2 : \bar{q}_0 q_0 + \bar{q}_1 q_1 = 1\}$ are 3-spheres, any two of which link. □

(The map of Exercise 10.68 is discussed in [28]. Analogues, such as the maps of Exercise 10.69 and Exercise 14.22, are discussed in [29]. The major topological problem raised by these papers was solved by Adams [1].)

10.70. Reread Chapter 0. □

CHAPTER 11

CORRELATIONS

Throughout this chapter $\mathbf{A}$ denotes some positive power ${}^s\mathbf{K}$ of $\mathbf{K} = \mathbf{R}$, $\mathbf{C}$ or $\mathbf{H}$. One of the main theorems of the chapter is Theorem 11.32, which indicates how any real algebra anti-involution of the algebra $\mathbf{A}(n)$, where n is finite, may be regarded as the adjoint involution induced by some appropriate product on the right $\mathbf{A}$-linear space $\mathbf{A}^n$. Theorem 11.32 and Theorem 11.25 together classify the irreducible anti-involutions of $\mathbf{A}(n)$ into ten classes. The ten types of product, which include the real and complex symmetric scalar products of Chapter 9, are extensively studied, the analogies with the theorems and techniques of Chapter 9 being close, in every case. For example, there are in each case groups analogous to the orthogonal and special orthogonal groups $O(n)$ and $SO(n)$.

It is convenient here, and in Chapter 12, to make a distinction between right and left linear spaces over $\mathbf{A}$ even when $\mathbf{A}$ is commutative and, in particular, to regard the dual of a right $\mathbf{A}$-linear space always as a left $\mathbf{A}$-linear space and conversely.

Since one of our aims is to introduce various products on (right) $\mathbf{A}$-linear spaces analogous to the scalar products of Chapter 9, and since, in the particular case that $\mathbf{A}$ is a power of $\mathbf{H}$, no non-zero bilinear maps exist, it is necessary at an early stage in the discussion to generalize the concept of bilinearity suitably.

First, we generalize the concept of linearity by introducing semi-linear maps.

Semi-linear maps

Let X and Y each be a right or a left linear space over $\mathbf{A}$. An $\mathbf{R}$-linear map $t : X \to Y$ is said to be *semi-linear* over $\mathbf{A}$ if there is an automorphism or anti-automorphism $\psi : \mathbf{A} \to \mathbf{A};\ \lambda \rightsquigarrow \lambda^\psi$ such that, for all $x \in X$, and all $\lambda \in \mathbf{A}$,

$$\begin{aligned} t(x\lambda) &= t(x)\lambda^\psi, & t(x\lambda) &= \lambda^\psi t(x), \\ t(\lambda x) &= \lambda^\psi t(x) \quad \text{or} & t(\lambda x) &= t(x)\lambda^\psi \end{aligned}$$

as the case may be, ψ being an automorphism if $\mathbf{A}$ operates on X and Y on the same side and an anti-automorphism if $\mathbf{A}$ operates on X and Y on opposite sides. The terms *right*, *right-to-left*, *left* and *left-to-right* semi-linear maps over $\mathbf{A}$ have the obvious meanings.

The semi-linear map t determines the automorphism or anti-automorphism ψ uniquely, unless $t = 0$. On occasions it is convenient to refer directly to ψ, the map t being said to be ***semi-linear over* A *with respect to* ψ** or, briefly, $\mathbf{A}^\psi$-linear (*not* '$\mathbf{A}^\psi$ semi-linear', since, when $\psi = 1_\mathbf{A}$, $\mathbf{A}^\psi$ is usually abbreviated to $\mathbf{A}$ and the term '$\mathbf{A}$ semi-linear' could therefore be ambiguous).

Examples 11.1. The following maps are invertible right semi-linear maps over $\mathbf{H}$:

$$\mathbf{H} \to \mathbf{H};\quad x \rightsquigarrow x,$$
$$x \rightsquigarrow ax, \quad \text{for any non-zero } a \in \mathbf{H},$$
$$x \rightsquigarrow xb, \quad \text{for any non-zero } b \in \mathbf{H},$$
$$x \rightsquigarrow axb, \quad \text{for any non-zero } a, b \in \mathbf{H},$$
$$x \rightsquigarrow \hat{x}\ (= \mathrm{j}x\mathrm{j}^{-1})$$

and
$$\mathbf{H}^2 \to \mathbf{H}^2;\quad (x,y) \rightsquigarrow (ax,by), \quad \text{for any non-zero } a, b \in \mathbf{H},$$
$$(x,y) \rightsquigarrow (by,ax), \quad \text{for any non-zero } a, b \in \mathbf{H},$$

the corresponding automorphisms of $\mathbf{H}$ being, respectively,

$$1_\mathbf{H},\ 1_\mathbf{H},\ \lambda \rightsquigarrow b\lambda b^{-1},\ \lambda \rightsquigarrow b\lambda b^{-1},\ \lambda \rightsquigarrow \hat{\lambda} \text{ and } 1_\mathbf{H},\ 1_\mathbf{H}.$$

By contrast, the map

$$\mathbf{H}^2 \to \mathbf{H}^2;\quad (x,y) \rightsquigarrow (xa,yb), \quad \text{with } a, b \in \mathbf{H},$$

is *not* right semi-linear over $\mathbf{H}$, unless $\lambda a = \mu b$, with $\lambda, \mu \in \mathbf{R}$.

The maps

$$\mathbf{H}^2 \to \mathbf{H}^2;\quad (x,y) \rightsquigarrow (\bar{x},\bar{y})$$
$$(x,y) \rightsquigarrow (\bar{y},\bar{x})$$

are invertible right-to-left $\bar{\mathbf{H}}$-linear maps. The first of these is also a right-to-left ${}^2\bar{\mathbf{H}}$-linear map, and the second a right-to-left hb $\bar{\mathbf{H}}$-linear map. (See Chapter 10 for the notations.) □

Semi-linear maps over ${}^2\mathbf{K}$ are classified by the following proposition.

Prop. 11.2. Let X and Y be ${}^2\mathbf{K}$-linear spaces. Then any $({}^2\mathbf{K})^{\chi\times\psi}$-linear map $X \to Y$ is of the form

$$X_0 \oplus X_1 \to Y_0 \oplus Y_1;\quad (x_0,x_1) \rightsquigarrow (r(x_0),s(x_1)),$$

where $r : X_0 \to Y_0$ is $\mathbf{K}^\chi$-linear and $s : X_1 \to Y_1$ is $\mathbf{K}^\psi$-linear, while any $({}^2\mathbf{K})^{\mathrm{hb}\,(\chi\times\psi)}$-linear map $X \to Y$ is of the form

$$X_0 \oplus X_1 \to Y_0 \oplus Y_1;\quad (x_0,x_1) \rightsquigarrow (s(x_1),r(x_0)),$$

where $r : X_0 \to Y_1$ is $\mathbf{K}^\chi$-linear and $s : X_1 \to Y_0$ is $\mathbf{K}^\psi$-linear.

Proof We indicate the proof for a $({}^2\mathbf{K})^{\mathrm{hb}(\chi\times\psi)}$-linear map $t : X \to Y$, assuming, for the sake of definiteness, that X is a right ${}^2\mathbf{K}$-linear space and Y a left ${}^2\mathbf{K}$-linear space. Then, for all $a \in X_0$, $b \in X_1$,

$$t(a,0) = t((a,0)(1,0)) = (0,1)t(a,0)$$

and

$$t(0,b) = t((0,b)(0,1)) = (1,0)t(0,b).$$

So maps $r : X_0 \to Y_1$ and $s : X_1 \to Y_0$ are defined by

$$(0,r(a)) = t(a,0) \quad\text{and}\quad (s(b),0) = t(0,b), \quad\text{for all } (a,b) \in X.$$

It is then a straightforward matter to check that these maps r and s have the required properties.

The proofs in the other cases are similar. □

The first of the two maps described in Prop. 11.2 will be denoted by $r \times s$ and the second by hb $(r \times s)$.

A particular case that will occur is when $Y = X^{\mathscr{L}}$. In this case Y_0 and Y_1 are usually identified, in the obvious ways (cf. Prop. 8.4), with $X_0^{\mathscr{L}}$ and $X_1^{\mathscr{L}}$.

An $\mathbf{A}^{\psi}$-linear map $t : X \to Y$ is said to be *irreducible* if ψ is irreducible. Otherwise, it is said to be *reducible*. If t is irreducible, and if ψ is an involution or anti-involution then, by Prop. 10.49, $\mathbf{A} = \mathbf{K}$ or ${}^2\mathbf{K}$. The map $r \times s$ in Prop. 11.2 is reducible, while the map hb $(r \times s)$ is irreducible.

Prop. 11.3. The composite of a pair of composable semi-linear maps is semi-linear and the inverse of an invertible semi-linear map is semi-linear. □

An invertible semi-linear map is said to be a *semi-linear isomorphism.*

Prop. 11.4. Let X be a right $\mathbf{A}$-linear space, let α be an automorphism of $\mathbf{A}$ and let X^{α} consist of the set X with addition defined as before, but with a new scalar multiplication namely,

$$X^{\alpha} \times \mathbf{A} \to X^{\alpha}; \quad (x,\lambda) \rightsquigarrow x\lambda^{\alpha^{-1}}.$$

Then X^{α} is a right $\mathbf{A}$-linear space and the set identity $X \to X^{\alpha}$; $x \rightsquigarrow x$ is an $\mathbf{A}^{\alpha}$-linear isomorphism. □

Prop. 11.5. Let $t : X \to Y$ be a semi-linear map over $\mathbf{K}$. Then $\operatorname{im} t$ is a $\mathbf{K}$-linear subspace of Y and $\ker t$ is a $\mathbf{K}$-linear subspace of X. □

The analogue of this proposition for a power of $\mathbf{K}$ greater than 1 is false. The image and kernel of a semi-linear map over ${}^s\mathbf{K}$ are ${}^s\mathbf{K}$-modules but not, in general, ${}^s\mathbf{K}$-linear spaces, if $s > 1$. (Cf. foot of page 135.)

Rank and *kernel rank* are defined for semi-linear maps as for linear maps.

Prop. 11.6. Let $t : X \longrightarrow Y$ be an $\mathbf{A}^\psi$-linear map. Then, for any $\gamma \in Y^{\mathscr{L}}$, $\psi^{-1}\gamma t \in X^{\mathscr{L}}$.

Proof The map $\psi^{-1}\gamma t$ is certainly $\mathbf{R}$-linear. It remains to consider its interaction with $\mathbf{A}$-multiplication. There are four cases, of which we consider only one, namely the case in which X and Y are each right $\mathbf{A}$-linear. In this case, for each $x \in X$, $\lambda \in \mathbf{A}$,

$$\begin{aligned} \psi^{-1}\gamma t(x\lambda) &= \psi^{-1}\gamma(t(x)\lambda^\psi) = \psi^{-1}((\gamma t(x))\lambda^\psi) \\ &= (\psi^{-1}\gamma t(x))\lambda. \end{aligned}$$

The proofs in the other three cases are similar. □

The map $t^{\mathscr{L}} : Y^{\mathscr{L}} \longrightarrow X^{\mathscr{L}}$, defined, for all $\gamma \in X^{\mathscr{L}}$, by the formula $t^{\mathscr{L}}(\gamma) = \psi^{-1}\gamma t$, is called the *dual* of t. This definition is more vividly displayed by the diagram.

$$\begin{array}{ccc} X & \longrightarrow & Y \\ {\scriptstyle t^{\mathscr{L}}(\gamma) = \psi^{-1}\gamma t}\downarrow & \overset{t^{\mathscr{L}}}{\leftsquigarrow} & \downarrow{\scriptstyle \gamma} \\ \mathbf{A} & \xrightarrow{\psi} & \mathbf{A} \end{array}$$

Prop. 11.7. The dual $t^{\mathscr{L}}$ of an $\mathbf{A}^\psi$-linear map $t : X \longrightarrow Y$ is $\mathbf{A}^{\psi^{-1}}$-linear. □

Many properties of the duals of $\mathbf{R}$-linear maps carry over to semi-linear maps over $\mathbf{A}$.

Correlations

A *correlation* on a right $\mathbf{A}$-linear space X is an $\mathbf{A}$-semi-linear map $\xi : X \longrightarrow X^{\mathscr{L}}$; $x \rightsquigarrow x^\xi = \xi(x)$. The map $X \times X \longrightarrow A$; $(a,b) \rightsquigarrow a^\xi b = a^\xi(b)$ is the *product* induced by the correlation, and the map $X \longrightarrow \mathbf{A}$; $a \rightsquigarrow a^\xi a$ the *form* induced by the correlation. Such a product is $\mathbf{R}$-bilinear, but not, in general, $\mathbf{A}$-bilinear, for although the map

$$X \longrightarrow \mathbf{A}; \quad x \rightsquigarrow a^\xi x$$

is $\mathbf{A}$-linear, for any $a \in X$, the map

$$X \longrightarrow \mathbf{A}; \quad x \rightsquigarrow x^\xi b,$$

for any $b \in X$, is, in general, not linear but only (right-to-left) semi-linear over $\mathbf{A}$. Products of this kind are said to be *sesqui-linear*, the prefix being derived from a Latin word meaning 'one and a half'.

An $\mathbf{A}^\psi$-correlation $\xi : X \longrightarrow X^{\mathscr{L}}$ and the induced product on the right $\mathbf{A}$-linear space X are said to be, respectively, *symmetric* or *skew with respect* to ψ or *over* $\mathbf{A}^\psi$ according as, for each $a, b \in X$,

$$b^\xi a = (a^\xi b)^\psi \quad \text{or} \quad -(a^\xi b)^\psi.$$

Symmetric products over $\bar{\mathbf{C}}$ or $\bar{\mathbf{H}}$ are called *hermitian products*, and their forms are called *hermitian* forms.

Examples 11.8. Any anti-involution ψ of $\mathbf{A}$ may be regarded as a symmetric $\mathbf{A}^\psi$-correlation on $\mathbf{A} = \mathbf{A}^{\mathscr{L}}$.

2. The product

$$\mathbf{R}^2 \times \mathbf{R}^2 \to \mathbf{R};\quad ((a,b),(a',b')) \rightsquigarrow ba' - ab'$$

is skew over $\mathbf{R}$, with zero form.

3. The product

$$^2\mathbf{R} \times {}^2\mathbf{R} \to {}^2\mathbf{R};\quad ((a,b),(a',b')) \rightsquigarrow (ba',ab')$$

is symmetric over hb $\mathbf{R}$.

4. The correlations previously studied in Chapter 9 were symmetric correlations over $\mathbf{R}$ or $\mathbf{C}$.

5. The product

$$\mathbf{C}^2 \times \mathbf{C}^2 \to \mathbf{C};\quad ((a,b),(a',b')) \rightsquigarrow \bar{a}a' + \bar{b}b',$$

is hermitian. □

Prop. 11.9. Let ξ be a non-zero symmetric or skew $\mathbf{A}^\psi$-correlation on a right $\mathbf{A}$-linear space X. Then ψ is an anti-involution of $\mathbf{A}$. □

Symmetric and skew correlations are particular examples of reflexive correlations, a correlation ξ being said to be *reflexive* if, for all $a, b \in X$,

$$b^\xi a = 0 \Leftrightarrow a^\xi b = 0.$$

Not all correlations are reflexive.

Example 11.10. The $\mathbf{R}$-bilinear product on $\mathbf{R}^2$:

$$\mathbf{R}^2 \times \mathbf{R}^2 \to \mathbf{R};\quad ((a,b),(a',b')) \rightsquigarrow aa' + ab' + bb'$$

is induced by a correlation that is not reflexive. □

The next proposition is in contrast to Example 11.8, 2 above.

Prop. 11.11. Let ξ be a non-zero irreducible reflexive correlation on a right $^2\mathbf{K}$-linear space X. Then, for some $x \in X$, $x^\xi x$ is invertible.

Proof Since ξ is irreducible, $\xi = \text{hb}\,(\eta \times \zeta)$, where $\zeta : X_1 \to X_0^{\mathscr{L}}$ and $\eta : X_0 \to X_1^{\mathscr{L}}$ are semi-linear. Since ξ is reflexive,

$$(a,0)^\xi(0,b) = 0 \Leftrightarrow (0,b)^\xi(a,0) = 0.$$

That is,

$$(0,a^\eta b) = 0 \Leftrightarrow (b^\zeta a,0) = 0$$

or

$$a^\eta b \neq 0 \Leftrightarrow b^\zeta a \neq 0.$$

Since ξ is non-zero, η or ζ is non-zero. Suppose η is non-zero. Then there exists $(a,b) \in X_0 \times X_1$ such that $a^\eta b \neq 0$ and $b^\zeta a \neq 0$, that is, such that $(a,b)^\xi(a,b)$ is invertible. □

If ξ is symmetric we can say more.

Prop. 11.12. Let ξ be an hb $\mathbf{K}^\psi$ symmetric correlation on a right ${}^2\mathbf{K}$-linear space X and suppose that, for some $x \in X$, $x^\xi x$ is invertible. Then there exists $\lambda \in {}^2\mathbf{K}$, such that $(x\lambda)^\xi(x\lambda) = 1$.

Proof As in the proof of Prop. 11.11, $\xi = \text{hb}\,(\eta \times \zeta)$. Now ξ is symmetric, so, for all $(a,b) \in X$,

$$((a,b)^\xi(a,b))^{\text{hb}\psi} = (a,b)^\xi(a,b),$$

that is,
$$((a^\eta b)^\psi, (b^\zeta a)^{\psi^{-1}}) = (b^\zeta a, a^\eta b).$$

In particular, $a^\eta b = 1$ if, and only if, $b^\zeta a = 1$. Now, if $x = (a,b)$ is invertible, $b^\zeta a \neq 0$. Choose $\lambda = ((b^\zeta a)^{-1}, 1)$. □

An invertible correlation is said to be *non-degenerate.*

Prop. 11.13. Let ξ be a non-degenerate correlation on a finite-dimensional right $\mathbf{K}$-linear space X, and let x be any element of X. Then there exists $x' \in X$ such that $x^\xi x' = 1$. □

Prop. 11.14. Let ξ be a non-degenerate irreducible correlation on a finite-dimensional right ${}^2\mathbf{K}$-linear space X, and let x be a regular element of X. Then there exists $x' \in X$ such that $x^\xi x' = 1\ (=(1,1))$. □

Equivalent correlations

Theorem 11.25 below classifies irreducible reflexive correlations with respect to an equivalence which will now be defined.

Semi-linear correlations $\xi, \eta : X \longrightarrow X^{\mathscr{L}}$ on a right $\mathbf{A}$-linear space X are said to be *equivalent* if, for some invertible $\lambda \in \mathbf{A}$, $\eta = \lambda\xi$. This is clearly an equivalence on any set of semi-linear correlations on X. Several important cases are discussed in the following four propositions.

Prop. 11.15. Any skew $\bar{\mathbf{C}}$-correlation on a (right) $\mathbf{C}$-linear space X is equivalent to a symmetric $\bar{\mathbf{C}}$-correlation on X, and conversely.

Proof Let ξ be a skew $\bar{\mathbf{C}}$-correlation on X. Then $\mathrm{i}\xi$ is a $\bar{\mathbf{C}}$-correlation on X since, for all $x \in X$, $\lambda \in \mathbf{C}$,

$$(\mathrm{i}\xi)(x\lambda) = \mathrm{i}(\xi(x\lambda)) = \mathrm{i}\bar{\lambda}\xi(x) = \bar{\lambda}(\mathrm{i}\xi)(x).$$

Moreover, for all $a, b \in X$,

$$b^{\mathrm{i}\xi}a = \mathrm{i}b^\xi a = (-\bar{\mathrm{i}})(-\overline{a^\xi b}) = \overline{a^{\mathrm{i}\xi}b}.$$

That is, $\mathrm{i}\xi$ is symmetric over $\bar{\mathbf{C}}$.

Similarly, if ξ is symmetric over $\bar{\mathbf{C}}$, then i^ξ is skew over $\bar{\mathbf{C}}$. □

Prop. 11.16. Let ψ be an anti-involution of $\mathbf{H}$ other than conjugation. Then any skew $\mathbf{H}^\psi$-correlation on a right $\mathbf{H}$-linear space X is equivalent to a symmetric $\bar{\mathbf{H}}$-correlation on X, and conversely.

Proof We give the proof for the case $\mathbf{H}^\psi = \tilde{\bar{\mathbf{H}}}$. Let ξ be a skew $\tilde{\bar{\mathbf{H}}}$-correlation on X. Then $\mathrm{j}\xi$ is an $\bar{\mathbf{H}}$-correlation on X, since, for all $x \in X$, $\lambda \in \mathbf{H}$,

$$(\mathrm{j}\xi)(x\lambda) = \mathrm{j}\tilde{\lambda}\xi(x) = \bar{\lambda}(\mathrm{j}\xi)(x).$$

Moreover, for all $a, b \in X$,

$$b^{\mathrm{j}\xi}a = \mathrm{j}b^\xi a = -\mathrm{j}\widetilde{a^\xi b} = \overline{a^\xi b}\,\bar{\mathrm{j}} = \overline{a^{\mathrm{j}\xi}b}.$$

That is, $\mathrm{j}\xi$ is symmetric over $\bar{\mathbf{H}}$.

Similarly, if ξ is symmetric over $\bar{\mathbf{H}}$, then $\mathrm{j}\xi$ is skew over $\tilde{\bar{\mathbf{H}}}$. ☐

Prop. 11.17. Let ψ be as in Prop. 11.16. Then any symmetric $\mathbf{H}^\psi$-correlation on a right $\mathbf{H}$-linear space X is equivalent to a skew $\tilde{\bar{\mathbf{H}}}$-correlation on X, and conversely. ☐

Prop. 11.18. Any irreducible skew correlation on a right ${}^2\mathbf{K}$-linear space X is equivalent to an irreducible symmetric correlation on X, and conversely.

Proof Let ξ be an irreducible skew $({}^2\mathbf{K})^\psi$ correlation on X. Then $(1,-1)\xi$ also is an irreducible correlation on X, and, for all $a, b \in X$,

$$\begin{aligned} b^{(1,-1)\xi}a &= (1,-1)b^\xi a = -(a^\xi b)^\psi(1,-1) \\ &= (a^\xi b)^\psi(1,-1)^\psi, \quad \text{since } \psi \text{ is irreducible,} \\ &= a^{(1,-1)\xi}b. \end{aligned}$$

That is, $(1,-1)\xi$ is symmetric.

Similarly, if ξ is symmetric, $(1,-1)\xi$ is skew. ☐

A correlation that is equivalent to a symmetric or skew correlation will be called a *good* correlation. Good correlations are almost the same as reflexive correlations, as the next theorem shows.

Theorem 11.19. Any non-zero reflexive irreducible correlation on a finite-dimensional right $\mathbf{A}$-linear space X of dimension greater than one is a good correlation.
(A counter-example in the one-dimensional case is the correlation on $\mathbf{H}$ with product $\mathbf{H}^2 \to \mathbf{H}$; $(a,b) \rightsquigarrow \bar{a}(1 + \mathrm{j})b$.)

Proof By Prop. 10.49, $\mathbf{A} = \mathbf{K}$ or ${}^2\mathbf{K}$.

Consider first a reflexive $\mathbf{K}^\psi$-correlation ξ on a finite-dimensional $\mathbf{K}$-linear space X. Then, for all $a,b \in X$, $(b^\xi a)^{\psi^{-1}} = 0 \Leftrightarrow b^\xi a = 0 \Leftrightarrow a^\xi b = 0$. That is, for any non-zero $a \in X$, the kernel of the (surjective) $\mathbf{K}$-linear map $X \to \mathbf{K}$; $b \rightsquigarrow (b^\xi a)^{\psi^{-1}}$ coincides with $\ker a^\xi$. Therefore, by the $\mathbf{K}$-analogue of Prop. 5.15, there exists $\lambda_a \in \mathbf{K}$, non-zero, such that, for all $b \in X$,

$$(b^\xi a)^{\lambda^{-1}} = \lambda_a a^\xi b.$$

Now λ_a does not depend on a. For let a' be any other non-zero element of X. Since $\dim X > 1$, there exists $c \in X$, linearly free both of a and of a' (separately!). Then, since

$$b^\xi a + b^\xi c = b^\xi(a + c),$$

it follows that

$$\lambda_a\, a^\xi b + \lambda_c\, c^\xi b = \lambda_{a+c}(a + c)^\xi b,$$

for all $b \in X$. So

$$a\lambda_a^{\psi^{-1}} + c\lambda_c^{\psi^{-1}} = (a + c)\lambda_{a+c}^{\psi^{-1}}.$$

But a and c are free of each other. So

$$\lambda_a^{\psi^{-1}} = \lambda_{a+c}^{\psi^{-1}} = \lambda_c^{\psi^{-1}},$$

implying that $\lambda_a = \lambda_c$. Similarly $\lambda_{a'} = \lambda_c$. So $\lambda_{a'} = \lambda_a$. That is, there exists $\lambda \in \mathbf{K}$, non-zero, such that, for all $a, b \in X$,

$$(b^\xi a)^{\psi^{-1}} = \lambda a^\xi b.$$

There are two cases.

Suppose first that $a^\xi a = 0$, for all $a \in X$. Then, since $2(a^\xi b + b^\xi a) = a^\xi a + b^\xi b - (a - b)^\xi(a - b)$,

$$(b^\xi a)^{\psi^{-1}} = \lambda a^\xi b = -\lambda b^\xi a,$$

for all $a, b \in X$. Now any element of $\mathbf{K}$ is expressible in the form $b^\xi a$, for suitable a and b. In particular, for suitable a and b, $b^\xi a = 1$. So $\lambda = -1$ and $\psi = 1_\mathbf{K}$. That is, the correlation is a skew $\mathbf{K}$-correlation, with $\mathbf{K} = \mathbf{R}$ or $\mathbf{C}$.

The alternative is that, for some $x \in X$, $x^\xi x \neq 0$, implying that, for some invertible $\mu \in \mathbf{K}$, $(\mu^{-1})^{\psi^{-1}} = \lambda\mu^{-1}$ or, equivalently, $\mu^{-1} = (\mu^{-1})^\psi \lambda^\psi$. Then, for all $a, b \in x$,

$$b^{\mu\xi} a = \mu(\lambda a^\xi b)^\psi = \mu(\mu a^\xi b)^\psi(\mu^{-1})^\psi\lambda^\psi = \mu(\mu a^\xi b)^\psi \mu^{-1} = \mu(a^{\mu\xi} b)^\psi \mu^{-1}.$$

Moreover, for all $\lambda \in \mathbf{K}$, $(b\lambda)^{\mu\xi} a = (\mu\lambda^\psi\mu^{-1}) b^{\mu\xi} a$. The correlation $\mu\xi$, equivalent to ξ, is therefore a symmetric $\mathbf{K}^{\psi'}$-correlation, where, for any $\nu \in \mathbf{K}$,

$$\nu^{\psi'} = \mu\nu^\psi\mu^{-1}.$$

The proof for an irreducible $({}^2\mathbf{K})^\psi$ reflexive correlation ξ on a ${}^2\mathbf{K}$-linear space X is basically the same, but care has to be taken, since a non-zero element of ${}^2\mathbf{K}$ or of X is not necessarily regular. One proves first that there exists an invertible $\lambda \in {}^2\mathbf{K}$ such that, for all *regular* $a \in X$ and all $b \in X$,

$$(b^\xi a)^{\psi^{-1}} = \lambda a^\xi b.$$

It is then easy to deduce that this formula also holds for all $a \in X$. Next, by Prop. 11.11, $x^\xi x$ is invertible, for some $x \in X$. The remainder of the proof is as before. The conclusion is that any such correlation is equivalent to a symmetric correlation. □

It is easy to verify that skew **R**- or **C**-correlations are *essentially skew*.

Prop. 11.20. Let ξ be a skew **K**-correlation on a finite-dimensional **K**-linear space X, **K** being **R** or **C**, and let η be any correlation on X equivalent to ξ. Then η also is a skew **K**-correlation. □

The following corollary of Theorem 11.19 and Prop. 11.20 complements Prop. 11.11, both being required in the proof of the basis theorem for symmetric correlated spaces (Theorem 11.40).

Cor. 11.21. Let ξ be a non-zero symmetric correlation on a finite-dimensional **K**-linear space X. Then there exists $x \in X$ such that $x^\xi x \neq 0$. □

Corollary 11.21 may also be regarded as a corollary of the following proposition, which may be proved, for example, by case examination.

Prop. 11.22. Let ξ be a symmetric $\mathbf{K}^\psi$-linear correlation on a right **K**-linear space X. Then ξ is uniquely determined by its form $X \to \mathbf{K}$; $x \rightsquigarrow x^\xi x$. □

The next proposition complements Prop. 11.12.

Prop. 11.23. Let X be a right **K**-linear space, let ξ be a symmetric $\mathbf{K}^\psi$-correlation and suppose that $x \in X$ is such that $x^\xi x \neq 0$. Then, if $\mathbf{K}^\psi = \mathbf{R}, \bar{\mathbf{C}}$ or $\bar{\mathbf{H}}$, there exists $\lambda \in \mathbf{K}$ such that $(x\lambda)^\xi(x\lambda) = 1$ or -1, while if $\mathbf{K}^\psi = \mathbf{C}$ or $\tilde{\mathbf{H}}$, there exists $\lambda \in \mathbf{K}$ such that $(x\lambda)^\xi(x\lambda) = 1$.

Proof Since, for all $\lambda \in \mathbf{K}$, $(x\lambda)^\xi(x\lambda) = \lambda^\psi(x^\xi x)\lambda$, it is enough to prove that, for some $\lambda \in \mathbf{K}$, $(\lambda^{-1})^\psi\, \lambda^{-1} = (\lambda^\psi)^{-1}\, \lambda^{-1} = \pm x^\xi x$, as the case may be. Now, when ψ is conjugation, that is, when $\mathbf{K}^\psi = \mathbf{R}, \bar{\mathbf{C}}$ or $\bar{\mathbf{H}}$, $\overline{x^\xi x} = x^\xi x$, by the symmetry of ψ, and $x^\xi x$ is therefore real. So in these cases λ^{-1} may be taken to be the square root of $|\, x^\xi x \,|$. When $\mathbf{K}^\psi = \mathbf{C}$, $x^\xi x \in \mathbf{C}$ and we may take λ^{-1} to be the square root of $x^\xi x$. Finally, if $\mathbf{K}^\psi = \tilde{\mathbf{H}}$, $\widetilde{x^\xi x} = x^\xi x$ and so, by Prop. 10.28, $x^\xi x$ belongs to the image of the map $\mathbf{H} \to \mathbf{H}$; $\mu \rightsquigarrow \tilde{\mu}\mu$. □

Methods similar to those used in the proof of Theorem 11.19 may be used to prove the following.

Theorem 11.24. Let ξ and η be non-degenerate correlations on a finite-dimensional right **K**-linear space X of dimension greater than one and suppose that the induced *projective correlations* $\mathscr{G}_1(\xi)$ and $\mathscr{G}_1(\eta)$ are equal. Then ξ and η are equivalent. □

Because of Theorem 11.24, equivalent correlations are sometimes said to be ***projectively equivalent***. Note, however, the slight dimensional restriction.

The final theorem of this section is a crude classification of good irreducible correlations with respect to equivalence. There are ten classes (assigned code numbers later, on page 270).

Theorem 11.25. Let ξ be a good irreducible correlation on a right $\mathbf{A}$-linear space X. Then ξ is equivalent to one of the following:

a symmetric $\mathbf{R}$-correlation;
a skew $\mathbf{R}$-correlation;
a symmetric $\mathbf{C}$-correlation;
a skew $\mathbf{C}$-correlation;
a symmetric or, equivalently, a skew $\bar{\mathbf{C}}$-correlation;
a symmetric $\bar{\mathbf{H}}$- or, equivalently, a skew $\tilde{\bar{\mathbf{H}}}$-correlation;
a symmetric $\tilde{\bar{\mathbf{H}}}$- or, equivalently, a skew $\bar{\mathbf{H}}$-correlation;
a symmetric or, equivalently, a skew hb $\mathbf{R}$-correlation;
a symmetric or, equivalently, a skew hb $\mathbf{C}^{\psi}$-correlation,

where $\mathbf{C}^{\psi} = \mathbf{C}$ or $\bar{\mathbf{C}}$; or, finally,

a symmetric or, equivalently, a skew hb $\mathbf{H}^{\psi}$-correlation,

where ψ is an anti-automorphism of $\mathbf{H}$.

These ten possibilities are mutually exclusive.

Proof This is an immediate corollary of Cor. 10.55 and Theorem 11.19, together with Props. 11.15, 11.16, 11.17 and 11.18. □

Algebra anti-involutions

In Chapter 9 we noted how any non-degenerate real symmetric scalar product on a finite-dimensional real linear space X induces an anti-involution, the adjoint anti-involution, of the real algebra End X. In a similar way any non-degenerate good correlation on a finite-dimensional right $\mathbf{A}$-linear space X induces an anti-involution of the *real* algebra End X of $\mathbf{A}$-linear automorphisms of X.

It is convenient to begin by considering several spaces. In the next few propositions X, Y and Z will all be finite-dimensional right $\mathbf{A}$-linear spaces.

Prop. 11.26. Let ξ be a non-degenerate $\mathbf{A}^{\psi}$-correlation on X, let η be a non-degenerate $\mathbf{A}^{\psi}$-correlation on Y and let $t: X \rightarrow Y$ be an $\mathbf{A}$-linear map. Then there exists a unique map $t^*: Y \rightarrow X$, namely the $\mathbf{A}$-linear map $\xi^{-1}t^L\eta$, such that, for all $a \in X$, $b \in Y$,

$$b^{\eta}t(a) = t^*(b)^{\xi}a. \quad \square$$

The map t^* is called the *adjoint* of t with respect to ξ and η. The adjoint of a linear map $u : X \longrightarrow X$ with respect to ξ will be denoted by u^ξ. The map u is said to be *self-adjoint* if $u^\xi = u$ and *skew-adjoint* if $u^\xi = -u$. The real linear subspaces $\{u \in \operatorname{End} X : u^\xi = u\}$ and $\{u \in \operatorname{End} X : u^\xi = -u\}$ of the real linear space $\operatorname{End} X = L(X,X)$ will be denoted by $\operatorname{End}_+(X,\xi)$ and $\operatorname{End}_-(X,\xi)$, respectively.

Prop. 11.27. Let $t : X \longrightarrow Y$ and $u : Y \longrightarrow Z$ be $\mathbf{A}$-linear maps and let ξ, η and ζ be non-degenerate $\mathbf{A}^\varphi$-correlations on X, Y and Z respectively. Then $(1_X)^\xi = 1_X$ and $(ut)^* = t^* u^*$, where t^* is the adjoint of t with respect to ξ and η, u^* the adjoint of u with respect to η and ζ, and $(ut)^*$ the adjoint of ut with respect to ξ and ζ. □

Prop. 11.28. Let ξ and η be equivalent non-degenerate correlations on X. Then, for each $\mathbf{A}$-linear map $u : X \longrightarrow X$, $u^\eta = u^\xi$.

Proof For some $\lambda \in \mathbf{A}$, $\eta = \lambda\xi$. Therefore, for any $u \in \operatorname{End} X$,

$$\eta u^\xi = \lambda \xi u^\xi = \lambda u^L \xi = u^L(\lambda\xi) = u^L\eta,$$

since u^L is left $\mathbf{A}$-linear. So

$$u^\eta = \eta^{-1} u^L \eta = u^\xi. \quad \square$$

Prop. 11.29. Let ξ be an irreducible good correlation on X. Then, for any $\mathbf{A}$-linear map $u : X \longrightarrow X$,

$$(u^\xi)^\xi = u. \quad \square$$

Prop. 11.30. Let ξ be as in Prop. 11.29. Then the map

$$\operatorname{End} X \longrightarrow \operatorname{End} X; \quad t \rightsquigarrow t^\xi$$

is a real algebra anti-involution. □

There is an important converse to Prop. 11.30. The following proposition is required early in the proof.

Prop. 11.31. Let X be a finite-dimensional $\mathbf{A}$-linear space, let α be an anti-automorphism of the real algebra $\operatorname{End} X$ and let $t \in \operatorname{End} X$, with $\operatorname{rk} t = 1$. Then $\operatorname{rk} t^\alpha = 1$.

Proof Consider first the case that $\mathbf{A} = \mathbf{K}$, where $\mathbf{K} = \mathbf{R}$, $\mathbf{C}$ or $\mathbf{H}$. Then, by Theorem 6.39, which holds also when $\mathbf{K} = \mathbf{H}$, t generates a minimal left ideal of $\operatorname{End} X$. Since α is an anti-automorphism of $\operatorname{End} X$, the image of this ideal by α is a minimal right ideal of $\operatorname{End} X$. This ideal is generated by t^α; so, by Theorem 6.39 again, or, rather, its analogue for right ideals, $\operatorname{rk} t^\alpha = 1$.

The case $\mathbf{A} = {}^2\mathbf{K}$ is slightly trickier. In this case the left ideal generated by an element t of $\operatorname{End} X$, with $\operatorname{rk} t = 1$, is not minimal, but has exactly two minimal left ideals as proper subideals. Moreover, this can only occur if rk $t = 1$. Otherwise the proof goes as before. The details are left to the reader. □

Now the converse to Prop. 11.30.

Theorem 11.32. Let X be a finite-dimensional right $\mathbf{A}$-linear space. Then any anti-involution α of the real algebra End X is representable as the adjoint anti-involution induced by a non-degenerate reflexive correlation on X.

Proof The case dim $X = 0$ is trivial; so there is no loss of generality in supposing that $X = \mathbf{A}^n \times \mathbf{A}$, for some $n \in \omega$.

Let $u = \begin{pmatrix} 0 & 0 \\ 0 & 1 \end{pmatrix}^\alpha$. Then $u^2 = u$, while, by Prop. 11.31, u has rank 1. Let $v : \operatorname{im} u \longrightarrow \mathbf{A}$ be an $\mathbf{A}$-linear isomorphism and let $s = vu_{\text{sur}}$, $i = u_{\text{inc}}v^{-1}$. Then by the analogue of Prop. 3.20 for $\mathbf{A}$-linear maps with image an $\mathbf{A}$-linear space,

$$si = 1_{\mathbf{A}} = sui,$$

while, for all $(c, d) \in \mathbf{A}^n \times \mathbf{A}$,

$$i\,s\begin{pmatrix} 0 & c \\ 0 & d \end{pmatrix}^\alpha = \begin{pmatrix} 0 & 0 \\ 0 & 1 \end{pmatrix}^\alpha \begin{pmatrix} 0 & c \\ 0 & d \end{pmatrix}^\alpha = \begin{pmatrix} 0 & c \\ 0 & d \end{pmatrix}^\alpha.$$

The map

$$\psi : \mathbf{A} \longrightarrow \mathbf{A}; \quad \lambda \rightsquigarrow s\begin{pmatrix} 0 & 0 \\ 0 & \lambda \end{pmatrix}^\alpha i$$

is a ring anti-automorphism of $\mathbf{A}$; for it respects addition, while, for any $\lambda, \mu \in \mathbf{A}$,

$$\begin{aligned} (\lambda\mu)^\psi &= s\begin{pmatrix} 0 & 0 \\ 0 & \lambda\mu \end{pmatrix}^\alpha i \\ &= s\begin{pmatrix} 0 & 0 \\ 0 & \mu \end{pmatrix}^\alpha i\,s\begin{pmatrix} 0 & 0 \\ 0 & \lambda \end{pmatrix}^\alpha i \\ &= \mu^\psi\lambda^\psi, \end{aligned}$$

with $$1^\psi = sui = 1.$$

Now define

$$\xi : \mathbf{A}^n \times \mathbf{A} \longrightarrow (\mathbf{A}^n \times \mathbf{A})^L; \quad (c,d) \rightsquigarrow s\begin{pmatrix} 0 & c \\ 0 & d \end{pmatrix}^\alpha.$$

Then ξ is $\mathbf{A}^\psi$-linear; for it respects addition, while, for any $(c,d) \in \mathbf{A}^n \times \mathbf{A}$ and any $\lambda \in \mathbf{A}$,

$$\begin{aligned} \xi(c\lambda, d\lambda) &= s\begin{pmatrix} 0 & c\lambda \\ 0 & d\lambda \end{pmatrix}^\alpha \\ &= s\left(\begin{pmatrix} 0 & c \\ 0 & d \end{pmatrix}\begin{pmatrix} 0 & 0 \\ 0 & \lambda \end{pmatrix}\right)^\alpha \end{aligned}$$

$$= s\begin{pmatrix}0 & 0\\ 0 & \lambda\end{pmatrix}^{\alpha} i\, s\begin{pmatrix}0 & c\\ 0 & d\end{pmatrix}^{\alpha}$$
$$= \lambda^{\psi}\xi(c,d).$$

Moreover ξ is injective; for if $s\begin{pmatrix}0 & c\\ 0 & d\end{pmatrix}^{\alpha} = 0$, for any $(c,d) \in \mathbf{A}^n \times \mathbf{A}$, then $\begin{pmatrix}0 & c\\ 0 & d\end{pmatrix}^{\alpha} = i\, s\begin{pmatrix}0 & c\\ 0 & d\end{pmatrix}^{\alpha} = 0$, implying that $(c,d) = 0$, since $\alpha^2 = 1$. So ξ is a non-degenerate $\mathbf{A}^{\psi}$-linear correlation on $\mathbf{A}^n \times \mathbf{A}$.

This correlation is reflexive; since, for all $(c,d), (c',d') \in \mathbf{A}^n \times \mathbf{A}$,

$$\left(\begin{pmatrix}c'\\ d'\end{pmatrix}^{\xi}\begin{pmatrix}c\\ d\end{pmatrix}\right)^{\psi} = s\left(\begin{pmatrix}0\\ s\end{pmatrix}\begin{pmatrix}0 & c'\\ 0 & d'\end{pmatrix}^{\alpha}\begin{pmatrix}0 & c\\ 0 & d\end{pmatrix}\right)^{\alpha}_{i}$$
$$= s\begin{pmatrix}0 & c\\ 0 & d\end{pmatrix}^{\alpha}\begin{pmatrix}c'\\ d'\end{pmatrix}\begin{pmatrix}0 & 1\end{pmatrix}\begin{pmatrix}0\\ s\end{pmatrix}^{\alpha}_{i}$$
$$= \begin{pmatrix}c\\ d\end{pmatrix}^{\xi}\begin{pmatrix}c'\\ d'\end{pmatrix}\mu,$$

where $\mu \in \mathbf{A}$, from which it follows that if $(c,d)^{\xi}(c',d') = 0$, then $(c',d')^{\xi}(c,d) = 0$, ψ being an anti-automorphism of $\mathbf{A}$.

Finally, for any $(c,d), (c',d') \in \mathbf{A}^n \times \mathbf{A}$, and any $t \in \operatorname{End}(\mathbf{A}^n \times \mathbf{A})$,

$$(t^{\alpha}(c',d'))^{\xi}(c,d) = (c',d')^{\xi}\, t(c,d),$$

each side being equal to $s\begin{pmatrix}0 & c'\\ 0 & d'\end{pmatrix}^{\alpha} t\begin{pmatrix}c\\ d\end{pmatrix}$, since $t^{\alpha\alpha} = t$. That is, t^{α} is the adjoint of t with respect to the correlation ξ. □

Correlated spaces

An $\mathbf{A}^{\psi}$-correlated space (X,ξ) consists of a right $\mathbf{A}$-linear space X and an $\mathbf{A}^{\psi}$-correlation ξ on X. Such a space is said to be *non-degenerate*, *irreducible*, *reflexive*, *good*, *symmetric* or *skew* if its correlation ξ is, respectively, non-degenerate, irreducible, reflexive, good, symmetric or skew, to be *isotropic* if its correlation is zero, and *neutral* if X is the direct sum of two isotropic subspaces, each linear subspace of X being tacitly assigned the correlation induced on it by ξ in the obvious way.

Example 11.33. The right $\mathbf{A}$-linear space $\mathbf{A}^2$ with the $\mathbf{A}^{\psi}$ sesquilinear product

$$\mathbf{A}^2 \times \mathbf{A}^2 \to \mathbf{A}; \quad ((a,b),(a',b')) \rightsquigarrow b^{\psi}a' + a^{\psi}b'$$

is a symmetric neutral non-degenerate $\mathbf{A}^{\psi}$-correlated space. □

This space is denoted here by $\mathbf{A}^{\psi}_{\mathrm{hb}}$, and called the *standard* $\mathbf{A}^{\psi}$-*hyperbolic plane*.

Example 11.34. The right **A**-linear space $\mathbf{A}^2$ with the $\mathbf{A}^\psi$ sesquilinear product

$$\mathbf{A}^2 \times \mathbf{A}^2 \to \mathbf{A}; \quad ((a,b),(a',b')) \rightsquigarrow b^\psi a' - a^\psi b'$$

is a skew neutral non-degenerate $\mathbf{A}^\psi$-correlated space. □

This space is denoted here by $\mathbf{A}^\psi_{\mathrm{sp}}$, and called the *standard* $\mathbf{A}^\psi$-*symplectic plane.*

By analogy with Chapter 9, points a and b of a reflexive correlated space (X, ξ) are said to be *mutually orthogonal* if $a^\xi b = 0$, this being equivalent to the condition that $b^\xi a = 0$. *Orthogonal annihilators* of subspaces are then defined just as before, a subspace of a non-degenerate correlated space being isotropic if, and only if, it is a subspace of its annihilator and a non-degenerate subspace of a non-degenerate correlated space having a unique *orthogonal complement.*

The $\mathbf{A}^\psi$-*product* of two $\mathbf{A}^\psi$-correlated spaces (X,ξ) and (Y,η) is the $\mathbf{A}^\psi$-correlated space $(X \times Y, \zeta)$ where, for all (a,b), $(a',b') \in X \times Y$,

$$(a,b)^\zeta(a',b') = a^\xi a' + b^\eta b'.$$

Such a product of two non-degenerate, isotropic or neutral correlated spaces is, respectively, non-degenerate, isotropic or neutral. The subspaces $X \times \{0\}$ and $\{0\} \times Y$ of $X \times Y$ are orthogonal complements of each other in $(X \times Y, \zeta)$. The *negative* of a correlated space (X,ξ) is the correlated space $(X, -\xi)$.

A *correlated map* $t : (X,\xi) \to (Y,\eta)$ is a (right) $\mathbf{A}^\alpha$-linear map, where α is an automorphism of $\mathbf{A}$, such that, for all $a, b \in X$,

$$t(a)^\eta\, t(b) = (a^\xi b)^\alpha,$$

an invertible map of this type being a *correlated isomorphism.* (We omit the usual routine remarks.)

Prop. 11.35. Let ξ be an $\mathbf{A}^\psi$-correlation on a right **A**-linear space X and let χ be any anti-automorphism of **A** similar to ψ, in the sense of page 194. Then there exists a right **A**-linear space Y and an $\mathbf{A}^\chi$-correlation η on Y such that $(Y,\eta) \cong (X,\zeta)$.

Proof Since χ and ψ are similar, there exists an automorphism α of **A** such that $\alpha\psi = \chi\alpha$. Let $Y = X^\alpha$ (cf. Prop. 11.4) and let $\eta : Y \to Y^{\mathscr{L}}$ be defined for all $a, b \in Y$, by the formula

$$a^\eta b = (a^\xi b)^\alpha.$$

The image of η is genuinely in $Y^{\mathscr{L}}$, since for any $\mu \in \mathbf{A}$,

$$a^\eta(b\mu^{\alpha^{-1}}) = (a^\xi b\mu^{\alpha^{-1}})^\alpha = (a^\xi b)^\alpha \mu.$$

Moreover, for any $\lambda \in \mathbf{A}$,

$$(a\lambda^{\alpha^{-1}})^{\eta}b = ((a\lambda^{\alpha^{-1}})^{\xi}b)^{\alpha} = (\lambda^{\psi\alpha^{-1}}a^{\xi}b)^{\alpha}$$
$$= \lambda^{\chi}a^{\xi}b, \quad \text{since } \chi = \alpha\psi\alpha^{-1}.$$

That is, η is $\mathbf{A}^{\chi}$-linear.

Finally, the set identity $(X,\xi) \longrightarrow (Y,\eta)$ is a correlated isomorphism, since it is a semi-linear isomorphism and, from its very definition,

$$a^{\eta}b = (a^{\xi}b)^{\alpha}, \quad \text{for all } a, b \in X. \qquad \square$$

Prop. 11.36. Let (X,ξ) and (Y,η) be non-degenerate finite-dimensional $\mathbf{A}^{\psi}$-correlated spaces. Then an $\mathbf{A}$-linear map $t : (X,\xi) \longrightarrow (Y,\eta)$ is correlated if, and only if, $t^*t = 1_X$, where t^* denotes the adjoint of t with respect to ξ and η. $\square$

Cor. 11.37. Let (X,ξ) and (Y,η) be as in Prop. 11.36. Then any correlated map $t : (X,\xi) \longrightarrow (Y,\eta)$ is injective. $\square$

Cor. 11.38. Let (X,ξ) be as in Prop. 11.36, and let $t \in \operatorname{End} X$. Then t is a correlated automorphism of (X,ξ) if, and only if, $t^{\xi}t = 1_X$. $\square$

Prop. 11.39. Let (X,ξ) and (Y,η) be as in Prop. 11.36 and suppose, further, that ξ and η are each symmetric, or skew. Then, for any $t \in L(X, Y)$,

$$(t^*t)^{\xi} = \pm t^*t,$$

the $+$ sign applying if ξ and η are both symmetric or both skew, and the $-$ sign if one is symmetric and the other skew. $\square$

Detailed classification theorems

By Theorem 11.25 and Prop. 11.35 any irreducible good correlated space is equivalent, up to isomorphism, either to a symmetric $\mathbf{A}^{\psi}$-correlated space, where $\mathbf{A}^{\psi} = \mathbf{R}, \mathbf{C}, \bar{\mathbf{C}}, \tilde{\mathbf{H}}, \bar{\mathbf{H}}$, hb $\mathbf{R}$, hb $\bar{\mathbf{C}}$ or hb $\bar{\mathbf{H}}$, or to a skew $\mathbf{R}$- or $\mathbf{C}$-correlated space. In each of these cases there are classification theorems analogous to the classification theorems of Chapter 9. We state them without proof whenever the proof is the obvious analogue of a proof in that chapter.

Theorem 11.40. (The *basis theorem* for symmetric correlated spaces.) Each irreducible symmetric finite-dimensional $\mathbf{A}^{\psi}$-correlated space has an orthonormal basis.

(Cf. Cor. 11.21, Prop. 11.11, Prop. 11.23 and Prop. 11.12.) $\square$

Theorem 11.41. (*Classification theorem.*)

(i) Let (X,ξ) be a non-degenerate symmetric $\bar{\mathbf{K}}$-correlated space of finite dimension n over $\mathbf{K}$, where $\bar{\mathbf{K}} = \mathbf{R}$, $\bar{\mathbf{C}}$ or $\bar{\mathbf{H}}$. Then there exists a unique pair of natural numbers (p,q), with $p + q = n$, such that (X,ξ) is isomorphic to $\bar{\mathbf{K}}^{p,q}$, this being the right $\mathbf{K}$-linear space $\mathbf{K}^{p+q}$, with the hermitian product

$$(\mathbf{K}^{p+q})^2 \longrightarrow \mathbf{K}; \quad (a,b) \rightsquigarrow -\sum_{i\in p} \bar{a}_i b_i + \sum_{j\in q} \bar{a}_{p+j} b_{p+j}.$$

(ii) Let (X,ξ) be a non-degenerate symmetric $\mathbf{A}^\varphi$-correlated space of finite dimension n over $\mathbf{A}$, where $\mathbf{A}^\varphi = \mathbf{C}$, $\tilde{\mathbf{H}}$, hb $\mathbf{R}$, hb $\bar{\mathbf{C}}$, or hb $\bar{\mathbf{H}}$. Then (X,ξ) is isomorphic to $(\mathbf{A}^\varphi)^n$, this being the right $\mathbf{A}$-linear space $\mathbf{A}^n$, with the product

$$(\mathbf{A}^n)^2 \longrightarrow \mathbf{A}; \quad (a,b) \rightsquigarrow \sum_{i\in n} a_i^\varphi b_i. \qquad \square$$

In (i) the pair of numbers (p,q) is called the *signature* of the correlated space (X,ξ).

The following proposition concerns powers of hyperbolic planes.

Prop. 11.42. For any $n \in \omega$ there are the following isomorphisms of correlated spaces:

$$(\bar{\mathbf{K}}_{\mathrm{hb}})^n \cong \bar{\mathbf{K}}^{n,n}, \quad \text{where } \bar{\mathbf{K}} = \mathbf{R}, \bar{\mathbf{C}} \text{ or } \bar{\mathbf{H}}$$

and $(\mathbf{A}^\varphi_{\mathrm{hb}})^n = (\mathbf{A}^\varphi)^{2n}$, where $\mathbf{A}^\varphi = \mathbf{C}$, $\tilde{\mathbf{H}}$, hb $\mathbf{R}$, hb $\bar{\mathbf{C}}$ or hb $\bar{\mathbf{H}}$. $\square$

The Witt construction of Prop. 9.52 generalizes to each of the ten classes of correlated space as follows.

Prop. 11.43. Let (X,ξ) be a non-degenerate irreducible finite-dimensional symmetric or skew $\mathbf{A}^\varphi$-correlated space, and suppose that W is a one-dimensional isotropic subspace of X. Then there exists another one-dimensional isotropic subspace W' distinct from W such that the plane spanned by W and W' is, respectively, a hyperbolic or symplectic $\mathbf{A}^\varphi$-plane, that is, isomorphic to $\mathbf{A}^\varphi_{\mathrm{hb}}$ or to $\mathbf{A}^\varphi_{\mathrm{sp}}$.

Proof In the argument which follows, the upper of two alternative signs refers to the symmetric case and the lower to the skew case.

Let w be a regular element of W. Since X is non-degenerate there exists, by Prop. 11.13 or Prop. 11.14, an element $x \in X$ such that $w^\xi x = 1$. Then, for any $\lambda \in \mathbf{A}$,

$$(x + w\lambda)^\xi(x + w\lambda) = x^\xi x + \lambda^\varphi \pm \lambda,$$

this being zero if $\lambda = \mp\frac{1}{2}x^\xi x$, since $x^\xi x = \pm(x^\xi x)^\varphi$. Let $w' = x \mp \frac{1}{2}w\, x^\xi x$. Then $w^\xi w' = 1$, $w'^\xi w = \pm 1$ and $w'^\xi w' = 0$. Now let $W' = \operatorname{im}\{w'\}$, a ${}^2\mathbf{K}$-line in the ${}^2\mathbf{K}$ case, since $w'^\xi w = \pm 1$ $(= \pm(1,1))$. Then the plane spanned by W and W' is, respectively, a hyperbolic or symplectic $\mathbf{A}^\varphi$-plane. $\square$

Cor. 11.44. Let W be an isotropic subspace of a non-degenerate irreducible finite-dimensional symmetric or skew $\mathbf{A}^{\varphi}$-correlated space X. Then there exists an isotropic subspace W' of X such that $X = W \oplus W' \oplus (W \oplus W')^{\perp}$. □

Such a decomposition of X will be called a *Witt decomposition* of X with respect to the isotropic subspace W.

Cor. 11.45. Let X be a non-degenerate irreducible finite-dimensional symmetric or skew $\mathbf{A}^{\varphi}$-correlated space. Then there is a unique number k such that X is isomorphic either to $(\mathbf{A}^{\varphi}_{\mathrm{hb}})^k \times Y$ in the symmetric case, or to $(\mathbf{A}^{\varphi}_{\mathrm{sp}})^k \times Y$ in the skew case, where in either case Y is a subspace of X admitting no non-zero isotropic subspace. □

Cor. 11.46. (*Classification theorem* for skew $\mathbf{R}$- or $\mathbf{C}$-correlated spaces.)

Let X be a non-degenerate finite-dimensional skew $\mathbf{R}$- or $\mathbf{C}$-correlated space. Then X is isomorphic to $\mathbf{R}^k_{\mathrm{sp}}$ or to $\mathbf{C}^k_{\mathrm{sp}}$, where $2k = \dim X$, $\dim X$ necessarily being even. □

Cor. 11.47. (*Classification theorem* for neutral correlated spaces.)

Any neutral non-degenerate irreducible finite-dimensional symmetric or skew $\mathbf{A}^{\varphi}$-correlated space X is isomorphic either to $(\mathbf{A}^{\varphi}_{\mathrm{hb}})^k$ or to $(\mathbf{A}^{\varphi}_{\mathrm{sp}})^k$, where $2k = \dim X$. Typical spaces of each of the ten types are

$$\mathbf{R}^n_{\mathrm{hb}} \cong \mathbf{R}^{n,n}, \quad \mathbf{R}^n_{\mathrm{sp}},$$
$$\mathbf{C}^n_{\mathrm{hb}} \cong \mathbf{C}^{2n}, \quad \mathbf{C}^n_{\mathrm{sp}}, \quad \bar{\mathbf{C}}^n_{\mathrm{hb}} \cong \bar{\mathbf{C}}^{n,n} \approx \bar{\mathbf{C}}^n_{\mathrm{sp}}$$
$$\mathbf{H}^n_{\mathrm{hb}} \approx \bar{\mathbf{H}}_{\mathrm{sp}}, \quad \mathbf{H}^n_{\mathrm{sp}} \approx \bar{\mathbf{H}}^n_{\mathrm{hb}} \cong \bar{\mathbf{H}}^{n,n},$$
$$(\mathrm{hb}\mathbf{R})^n_{\mathrm{hb}} \cong (\mathrm{hb}\mathbf{R})^{2n} \approx (\mathrm{hb}\mathbf{R})^n_{\mathrm{sp}},$$
$$\begin{array}{ccccccccc} (\mathrm{hb}\mathbf{C})^n_{\mathrm{hb}} & \cong & (\mathrm{hb}\mathbf{C})^{2n} & \approx & (\mathrm{hb}\mathbf{C})^n_{\mathrm{sp}} & & (\mathrm{hb}\mathbf{H})^n_{\mathrm{hb}} \cong & (\mathrm{hb}\mathbf{H})^{2n} & \approx (\mathrm{hb}\mathbf{H})^n_{\mathrm{sp}}, \\ \wr\Vert & & \wr\Vert & & \wr\Vert & \text{and} & \wr\Vert & \wr\Vert & \wr\Vert \\ (\mathrm{hb}\bar{\mathbf{C}})^n_{\mathrm{hb}} & \cong & (\mathrm{hb}\bar{\mathbf{C}})^{2n} & \approx & (\mathrm{hb}\bar{\mathbf{C}})^n_{\mathrm{sp}} & & (\mathrm{hb}\bar{\mathbf{H}})^n_{\mathrm{hb}} \cong & (\mathrm{hb}\mathbf{H})^{2n} & \approx (\mathrm{hb}\mathbf{H})^n_{\mathrm{sp}} \end{array}$$

where $\approx$ denotes isomorphism up to equivalence. □

The *index* of a non-degenerate finite-dimensional $\mathbf{A}^{\varphi}$-correlated space (X,ξ) is the dimension of the isotropic subspace of greatest dimension in (X,ξ).

Prop. 11.48. The index of a non-degenerate finite-dimensional $\mathbf{A}^{\varphi}$-correlated space (X,ξ) is at most half the dimension of X. □

Prop. 11.49. The correlated spaces $\mathbf{R}^{n,n+k}$, $\mathbf{R}^n_{\mathrm{sp}}$, $\mathbf{C}^{2n+1}$, $\bar{\mathbf{C}}^{n,n+k}$, $\mathbf{C}^n_{\mathrm{sp}}$, $\tilde{\mathbf{H}}^{2n}$, $\tilde{\mathbf{H}}^{2n+1}$, $\bar{\mathbf{H}}^{n,n+k}$, $(\mathrm{hb}\mathbf{K}^{\varphi})^{2n}$ and $(\mathrm{hb}\mathbf{K}^{\varphi})^{2n+1}$ all have index n, for any finite n and k. □

Positive-definite spaces

A $\mathbf{K}^\psi$-correlated space (X,ξ) is said to be *positive-definite* if, for each non-zero $a \in X$, $a^\xi a$ is a positive real number, and to be *negative-definite* if its negative $(X,-\xi)$ is positive-definite.

Prop. 11.50. An n-dimensional $\mathbf{K}^\psi$-correlated space (X,ξ) is positive-definite if, and only if, (X,ξ) is isomorphic to $\mathbf{R}^n$, $\bar{\mathbf{C}}^n$ or $\bar{\mathbf{H}}^n$. □

Prop. 11.51. Every non-zero linear subspace of a finite-dimensional positive-definite correlated space is non-degenerate and has a unique orthogonal complement. □

Particular adjoint anti-involutions

The following information on various adjoint anti-involutions will be useful later. The information is given in tabular form. The notations are all as before, with the additional convention that if, for any finite m and n, $a \in L(\mathbf{K}^n,\mathbf{K}^m)$ and if ψ is any anti-involution of $\mathbf{K}$, then a^ψ denotes the element of $L(\mathbf{K}^n,\mathbf{K}^m)$ whose matrix is obtained from the matrix of a by applying the anti-involution ψ to each term of the matrix. The map $a^{\psi\tau}$ is then the transpose of a^ψ.

Table 11.52.

Linear space, X	*Correlated space, (X,ξ)*	$t \in \operatorname{End} X$	t^ξ
$\mathbf{K}^n$	$(\mathbf{K}^\psi)^n$ $(\mathbf{K}^\psi = \mathbf{C}$ or $\bar{\mathbf{H}})$	t	$t^{\psi\tau}$
$\mathbf{K}^p \times \mathbf{K}^q$	$\bar{\mathbf{K}}^{p,q}$ $(\bar{\mathbf{K}} = \mathbf{R}, \bar{\mathbf{C}}$ or $\bar{\mathbf{H}})$	$\begin{pmatrix} a & c \\ b & d \end{pmatrix}$	$\begin{pmatrix} \bar{a}^\tau & -\bar{b}^\tau \\ -\bar{c}^\tau & \bar{d}^\tau \end{pmatrix}$
$\mathbf{K}^n \times \mathbf{K}^n$	$(\mathbf{K}^\psi)^n_{\text{hb}}$	$\begin{pmatrix} a & c \\ b & d \end{pmatrix}$	$\begin{pmatrix} d^{\psi\tau} & c^{\psi\tau} \\ b^{\psi\tau} & a^{\psi\tau} \end{pmatrix}$
$\mathbf{K}^n \times \mathbf{K}^n$	$(\mathbf{K}^\psi)^n_{\text{sp}}$	$\begin{pmatrix} a & c \\ b & d \end{pmatrix}$	$\begin{pmatrix} d^{\psi\tau} & -c^{\psi\tau} \\ -b^{\psi\tau} & a^{\psi\tau} \end{pmatrix}$
${}^2\mathbf{K}^n$	$(hb\mathbf{K}^\psi)^n$	$\begin{pmatrix} a & 0 \\ 0 & d \end{pmatrix}$	$\begin{pmatrix} d^{\psi\tau} & 0 \\ 0 & a^{\psi\tau} \end{pmatrix}$

It follows, for example, that the correlated automorphisms of $(\text{hb}\,\mathbf{K}^\psi)^n$ are the endomorphisms of ${}^2\mathbf{K}^n$ of the form $\begin{pmatrix} a & 0 \\ 0 & (a^{\psi\tau})^{-1} \end{pmatrix}$, where a is any automorphism of $\mathbf{K}^n$. In that case the group of correlated automorphisms is isomorphic to $GL(n,\mathbf{K})$.

Groups of correlated automorphisms

By analogy with the real orthogonal case, the group of correlated automorphisms of a correlated space (X,ξ) will be denoted by $O(X,\xi)$.

There are various (nearly) standard notations for the groups of correlated automorphisms of each of the standard correlated spaces. The table is:

Table 11.53.

Correlated space	*Group of correlated automorphisms*	*Subgroup* $\{t : \det t = 1\}$
$\mathbf{R}^{p,q}$	$O(p,q;\mathbf{R})$ or $O(p,q)$ with $O(n) = O(0,n) \cong O(n,0)$	$SO(p,q)$
$\mathbf{R}^n_{sp}$	$Sp(2n;\mathbf{R})$	$Sp(2n;\mathbf{R})$
$\mathbf{C}^n$	$O(n;\mathbf{C})$	$SO(n;\mathbf{C})$
$\mathbf{C}^n_{sp}$	$Sp(2n;\mathbf{C})$	$Sp(2n;\mathbf{C})$
$\bar{\mathbf{C}}^{p,q}$	$U(p,q)$ with $U(n) = U(0,n) \cong U(n,0)$	$SU(p,q)$
$\tilde{\mathbf{H}}^n$	$O(n;\mathbf{H})$	$O(n;\mathbf{H})$
$\bar{\mathbf{H}}^{p,q}$	$Sp(p,q;\mathbf{H})$ or $Sp(p,q)$ with $Sp(n) = Sp(0,n) \cong Sp(n,0)$	$Sp(p,q)$
hb$\mathbf{R}^n$	$GL(n;\mathbf{R})$	$SL(n;\mathbf{R})$
hb$\mathbf{C}^n \cong$ hb$\bar{\mathbf{C}}^n$	$GL(n;\mathbf{C})$	$SL(n;\mathbf{C})$
hb$\tilde{\mathbf{H}}^n \cong$ hb$\bar{\mathbf{H}}^n$	$GL(n;\mathbf{H})$	$SL(n;\mathbf{H})$

The letter O stands for *orthogonal*, the letter U for *unitary* and the letters Sp for *symplectic*. The rather varied uses of the word 'symplectic' tend to be a bit confusing at first. It is to be noted that when one speaks of the symplectic group, of a given degree n, one is normally referring to the group $Sp(n;\mathbf{H})$. This usage tends to give the word quaternionic overtones. In fact the word was first used to describe the groups $Sp(2n;\mathbf{R})$ and $Sp(2n;\mathbf{C})$ and to indicate their connection with the sets of isotropic planes in the correlated spaces $\mathbf{R}^n_{sp}$ and $\mathbf{C}^n_{sp}$. Such a set of isotropic planes, regarded as a set of projective lines in the associated projective space $\mathscr{G}_1(\mathbf{R}^{2n})$ or $\mathscr{G}_1(\mathbf{C}^{2n})$, is known to projective geometers as a *line complex*. The groups were therefore originally called *complex groups*.

This was leading to hopeless confusion when H. Weyl [58] coined the word 'symplectic', derived from the Greek synonym of the Latin word 'complex'! Whether the situation is any less complicated now is a matter of dispute.

A final warning: some authors write $Sp(n;\mathbf{R})$ and $Sp(n;\mathbf{C})$ where we have written $Sp(2n;\mathbf{R})$ and $Sp(2n;\mathbf{C})$.

There are numerous relationships between the different groups. Various group injections, namely those induced by the injections

$$\mathbf{R} \to \mathbf{C};\quad \lambda \rightsquigarrow \lambda,\quad \mathbf{H} \to \mathbf{C}^2;\quad z + jw \rightsquigarrow (z,w)$$

and $$\mathbf{K} \to {}^2\mathbf{K};\quad \lambda \rightsquigarrow (\lambda,\lambda)\quad \text{where } \mathbf{K} = \mathbf{R}, \mathbf{C} \text{ or } \mathbf{H},$$

are so standard as usually to be regarded as inclusions. In particular, the groups $O(n;\mathbf{C})$, $U(p,q)$, with $p + q = n$, and $GL(n;\mathbf{R})$ are all regarded as subgroups of $GL(n;\mathbf{C})$, for any n, while $Sp(2n;\mathbf{C})$ and $GL(n;\mathbf{H})$ are regarded as subgroups of $GL(2n;\mathbf{C})$.

Prop. 11.54. For any finite n,

$$O(n) = O(n;\mathbf{C}) \cap GL(n;\mathbf{R}) = O(n;\mathbf{C}) \cap U(n)$$
$$Sp(n) = Sp(2n;\mathbf{C}) \cap GL(n;\mathbf{H}) = Sp(2n;\mathbf{C}) \cap U(2n)$$
$$O(n;\mathbf{H}) = O(2n;\mathbf{C}) \cap GL(n;\mathbf{H})$$

and $$Sp(2n;\mathbf{R}) = Sp(2n;\mathbf{C}) \cap GL(2n;\mathbf{R}),$$

while, with rather obvious definitions of $O(p,q;\mathbf{C})$ and $Sp(2p,2q,\mathbf{C})$, isomorphic respectively to $O(n;\mathbf{C})$ and $Sp(2n,\mathbf{C})$,

$$O(p,q) = O(p,q;\mathbf{C}) \cap U(p,q)$$

and $$Sp(p,q) = Sp(2p,2q;\mathbf{C}) \cap U(p,q).$$

(The equation $Sp(n) = Sp(2n;\mathbf{C}) \cap U(2n)$, for example, follows readily from the observation that, for all $z + jw,\ z' + jw' \in \mathbf{H}$,

$$(\overline{z + jw})(z' + jw') = (\bar{z}z' + \bar{w}w') - j(wz' - zw').) \qquad \square$$

There are analogues of the unit sphere S^n in $\mathbf{R}^{n+1}$ for the other non-degenerate finite-dimensional correlated spaces. Suppose first that (X,ξ) is a symmetric correlated space. Then

$$\mathscr{S}(X,\xi) = \{x \in X : x^\xi x = 1\}$$

is defined to be the *unit quasi-sphere* in (X,ξ). In particular, $\mathscr{S}(\mathbf{R}^{n+1}) = S^n$, while $\mathscr{S}(\bar{\mathbf{C}}^{n+1})$ and $\mathscr{S}(\bar{\mathbf{H}}^{n+1})$ are identifiable in an obvious way with S^{2n+1} and S^{4n+3}, respectively, for any number n. Note also that, for $(X,\xi) = \text{hb}\tilde{\mathbf{K}}^{n+1}$, with $\tilde{\mathbf{K}} = \mathbf{R}, \mathbf{C}$ or $\tilde{\mathbf{H}}$,

$$\begin{aligned}\mathscr{S}(X,\xi) &= \{x \in {}^2\mathbf{K}^{n+1} : x^\xi x = 1\}\\ &= \{x \in {}^2\mathbf{K}^{n+1} : (\tilde{x}_1{}^\tau x_0, \tilde{x}_0{}^\tau x_1) = (1,1)\}\\ &= \{x \in {}^2\mathbf{K}^{n+1} : \tilde{x}_0{}^\tau x_1 = 1\},\end{aligned}$$

since $\tilde{x}_1{}^\tau x_0 = 1$ if, and only if, $\tilde{x}_0{}^\tau x_1 = 1$.

A slightly different definition is necessary in the essentially skew cases. The appropriate definition is

$$\mathscr{S}(\mathbf{K}^n_{\mathrm{sp}}) = \{(x,y) \in (\mathbf{K}^n_{\mathrm{sp}})^2 : x \cdot y = 1\},$$

where $\cdot$ denotes the product on $\mathbf{K}^n_{\mathrm{sp}}$, or, equivalently,

$$\mathscr{S}(\mathbf{K}^n_{\mathrm{sp}}) = \left\{\begin{pmatrix} a & c \\ b & d \end{pmatrix} \in (\mathbf{K}^n)^{2\times 2} : a \cdot d - b \cdot c = 1\right\},$$

where $\cdot$ denotes the standard scalar product on $\mathbf{K}^n$, $\mathbf{K}$ being $\mathbf{R}$ or $\mathbf{C}$.

The verification of the following theorem is a straightforward check!

Theorem 11.55. For any $p, q, n \in \omega$, let $\mathbf{R}^{p,q+1}$, $\mathbf{C}^{n+1}$, $\bar{\mathbf{C}}^{p,q+1}$, $\tilde{\mathbf{H}}^{n+1}$, $\bar{\mathbf{H}}^{p,q+1}$, hb $\mathbf{R}^{n+1}$, hb $\mathbf{C}^{n+1}$, hb$\tilde{\mathbf{H}}^{n+1}$, $\mathbf{R}^{n+1}_{\mathrm{sp}}$ and $\mathbf{C}^{n+1}_{\mathrm{sp}}$ be identified with $\mathbf{R}^{p,q} \times \mathbf{R}$, $\mathbf{C}^n \times \mathbf{C}$, $\bar{\mathbf{C}}^{p,q} \times \bar{\mathbf{C}}$, $\tilde{\mathbf{H}}^n \times \tilde{\mathbf{H}}$, $\bar{\mathbf{H}}^{p,q} \times \bar{\mathbf{H}}$, hb$\mathbf{R}^n \times$ hb$\mathbf{R}$, hb$\mathbf{C}^n \times$ hb$\mathbf{C}$, hb$\tilde{\mathbf{H}}^n \times$ hb$\tilde{\mathbf{H}}$, $\mathbf{R}^n_{\mathrm{sp}} \times \mathbf{R}_{\mathrm{sp}}$ and $\mathbf{C}^n_{\mathrm{sp}} \times \mathbf{C}_{\mathrm{sp}}$, respectively, in the obvious ways. Then the pairs of maps

$$\begin{aligned}
O(p,q) &\to O(p,q+1) \to \mathscr{S}(\mathbf{R}^{p,q+1}),\\
SO(p,q) &\to SO(p,q+1) \to \mathscr{S}(\mathbf{R}^{p,q+1}), \quad p+q > 0,\\
O(n;\mathbf{C}) &\to O(n+1;\mathbf{C}) \to \mathscr{S}(\mathbf{C}^{n+1}),\\
SO(n;\mathbf{C}) &\to SO(n+1;\mathbf{C}) \to \mathscr{S}(\mathbf{C}^{n+1}), \quad n > 0,\\
U(p,q) &\to U(p,q+1) \to \mathscr{S}(\bar{\mathbf{C}}^{p,q+1}),\\
SU(p,q) &\to SU(p,q+1) \to \mathscr{S}(\bar{\mathbf{C}}^{p,q+1}), \quad p+q > 0,\\
O(n;\tilde{\mathbf{H}}) &\to O(n+1;\tilde{\mathbf{H}}) \to \mathscr{S}(\tilde{\mathbf{H}}^{n+1}),\\
Sp(p,q) &\to Sp(p,q+1) \to \mathscr{S}(\bar{\mathbf{H}}^{p,q+1}),\\
GL(n;\mathbf{R}) &\to GL(n+1;\mathbf{R}) \to \mathscr{S}(\mathrm{hb}\mathbf{R}^{n+1}),\\
SL(n;\mathbf{R}) &\to SL(n+1;\mathbf{R}) \to \mathscr{S}(\mathrm{hb}\mathbf{R}^{n+1}), \quad n > 0,\\
GL(n;\mathbf{C}) &\to GL(n+1;\mathbf{C}) \to \mathscr{S}(\mathrm{hb}\mathbf{C}^{n+1}),\\
SL(n;\mathbf{C}) &\to SL(n+1;\mathbf{C}) \to \mathscr{S}(\mathrm{hb}\mathbf{C}^{n+1}), \quad n > 0,\\
GL(n;\mathbf{H}) &\to GL(n+1;\mathbf{H}) \to \mathscr{S}(\mathrm{hb}\tilde{\mathbf{H}}^{n+1}),\\
SL(n;\mathbf{H}) &\to SL(n+1;\mathbf{H}) \to \mathscr{S}(\mathrm{hb}\tilde{\mathbf{H}}^{n+1}), \quad n > 0,\\
Sp(2n;\mathbf{R}) &\to Sp(2n+2;\mathbf{R}) \to \mathscr{S}(\mathbf{R}^{n+1}_{\mathrm{sp}}),\\
\text{and} \quad Sp(2n;\mathbf{C}) &\to Sp(2n+2;\mathbf{C}) \to \mathscr{S}(\mathbf{C}^{n+1}_{\mathrm{sp}})
\end{aligned}$$

are each left-coset exact (page 97), the first map in each case being the injection $s \rightsquigarrow \begin{pmatrix} s & 0 \\ 0 & 1 \end{pmatrix}$ and the second being, in all but the last two cases, the map $t \rightsquigarrow t(0,1)$, the last column of t, and, in the last two cases, the map $t \rightsquigarrow (t(0,(1,0)), t(0,(0,1)))$, the last two columns of t. □

Note, in particular, the left-coset exact pairs of maps

$$O(n) \to O(n+1) \to S^n$$
$$SO(n) \to SO(n+1) \to S^n \quad (n > 0)$$
$$U(n) \to U(n+1) \to S^{2n+1}$$
$$SU(n) \to SU(n+1) \to S^{2n+1} \quad (n > 0)$$

and
$$Sp(n) \to Sp(n+1) \to S^{4n+3}.$$

For applications of Theorem 11.55, see Cor. 20.83 and Cor. 20.85.

Finally, some simple observations concerning the groups $O(n)$, $U(n)$ and $Sp(n)$ for small values of n.

It is clear, first of all, that $U(1) = S^1$, the group of complex numbers of absolute value 1, and therefore that $U(1) \cong SO(2)$. Also $Sp(1) = S^3$, the group of quaternions of absolute value 1.

Now consider $\mathbf{C}(2)$. Here there is an analogue to Prop. 10.30.

Prop. 11.56. For any $q \in S^3$ and any $c \in S^1$ the map

$$\bar{\mathbf{C}}^2 \to \bar{\mathbf{C}}^2; \quad x \rightsquigarrow qxc$$

is unitary, $\mathbf{C}^2$ being identified with $\mathbf{H}$ in the usual way. Moreover, any element of $U(2)$ can be so represented, two distinct elements (q,c) and (q',c') of $S^3 \times S^1$ representing the same unitary map if, and only if, $(q',c') = -(q,c)$.

Proof The map $x \rightsquigarrow qxc$ is complex linear, for any $(q,c) \in S^3 \times S^1$, since it clearly respects addition, while, for any $\lambda \in \mathbf{C}$, $q(x\lambda)c = (qxc)\lambda$, since $\lambda c = c\lambda$. To prove that it is unitary, it then is enough to show that it respects the hermitian form

$$\mathbf{C}^2 \to \mathbf{R}; \quad x \rightsquigarrow \bar{x}^\tau x.$$

However, since, for all $(x_0,x_1) \in \mathbf{C}^2$,

$$\bar{x}_0 x_0 + \bar{x}_1 x = (\bar{x}_0 - \bar{x}_1 \mathrm{j})(x_0 + \mathrm{j}x_1) = |\, x_0 + \mathrm{j}x_1 \,|^2$$

it is enough to verify instead that the map, regarded as a map from $\mathbf{H} \to \mathbf{H}$, preserves the norm on $\mathbf{H}$, and this is obvious.

Conversely, let $t \in U(2)$ and let $r = t(1)$. Then $|\, r \,| = 1$ and the map

$$\mathbf{C}^2 \to \mathbf{C}^2; \quad x \rightsquigarrow \bar{r}t(x)$$

is an element of $U(2)$ leaving 1, and therefore every point of $\mathbf{C}$ fixed, and mapping the orthogonal complement in $\bar{\mathbf{C}}^2$ of $\mathbf{C}$, the complex line $\mathrm{j}\mathbf{C} = \{\mathrm{j}z : z \in \mathbf{C}\}$, to itself. It follows that there is an element c of S^1, defined uniquely up to sign, such that, for all $x \in \mathbf{C}^2$, $\bar{r}t(x) = \bar{c}xc$ or, equivalently, $t(x) = qxc$, where $q = r\bar{c}$, Finally, since c is defined uniquely up to sign, the pair (q,c) also is defined uniquely up to sign.

An alternative proof of the converse goes as follows.

Let $t \in U(2)$. Then $t = c_{\mathrm{R}}u$, where $c^2 = \det t$ and $u \in SU(2)$. Now the matrix of u can readily be shown to be of the form $\begin{pmatrix} a & -\bar{b} \\ b & \bar{a} \end{pmatrix}$, from which it follows that $u = q_{\mathrm{L}}$, where $q = a + jb$. The result follows at once. □

Cor. 11.57. The following is a commutative diagram of exact sequences of group maps:

$$\begin{array}{ccccccccc}
 & & & & \{1\} & & \{1\} & & \\
 & & & & \downarrow & & \downarrow & & \\
 & & \{1\} & \longrightarrow & S^0 & \xrightarrow{1} & S^0 & \longrightarrow & \{1\} \\
 & & \downarrow & & \downarrow{\scriptstyle \text{incl.}} & & \downarrow{\scriptstyle \text{incl.}} & & \\
\{1\} & \longrightarrow & S^3 & \xrightarrow{\text{incl.}} & S^3 \times S^1 & \xrightarrow{\text{proj.}} & S^1 & \longrightarrow & \{1\} \\
 & & \downarrow & & \downarrow{\scriptstyle f} & & \downarrow{\scriptstyle \text{squaring}} & & \\
\{1\} & \longrightarrow & SU(2) & \xrightarrow{\text{incl.}} & U(2) & \xrightarrow{\text{det.}} & S^1 & \longrightarrow & \{1\} \\
 & & \downarrow & & \downarrow & & \downarrow & & \\
 & & \{1\} & & \{1\} & & \{1\} & &
\end{array}$$

the map f being defined by the formula

$$f(q,c) = q_{\mathrm{L}}c_{\mathrm{R}}, \quad \text{for all } (q,c) \in S^3 \times S^1.$$

In particular, $Sp(1) = S^3 \cong SU(2)$. □

Now, by Prop. 10.44, $\mathbf{C}(2) = \mathbf{H} \otimes \mathbf{C}$, the representative of any $q \in \mathbf{H}$ being q_{L} and the representative of any $c \in \mathbf{C}$ being c_{R}. It follows, by Prop. 11.56, that the product of any finite ordered set of elements belonging either to the copy of $Sp(1) = S^3$ or to the copy of $U(1) = S^1$ in $\mathbf{C}(2)$ is an element of $U(2)$.

This result is to be compared with Prop. 10.42 and the remarks following Prop. 10.44. Note also that in the standard inclusion of $\mathbf{C}$ in $\mathbf{R}(2)$ the elements representing the elements of $U(1) = S^1$ are all orthogonal.

For an important application of these remarks, see Prop. 13.27.

Further Exercises

11.58. Classify semi-linear maps over ${}^s\mathbf{K}$, for any positive s. (Cf. Prop. 11.2.) □

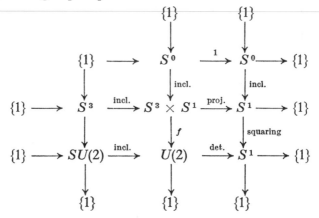

11.59. State and prove the analogues of Props. 3.32 and 3.33 for semi-linear maps. □

11.60 Let $t \rightsquigarrow t^\alpha$ be a real algebra automorphism of $\mathbf{A}(n)$, where $\mathbf{A} = \mathbf{K}$ or ${}^2\mathbf{K}$. Prove that there is a semi-linear isomorphism $\phi : \mathbf{A}^n \to \mathbf{A}^n$ such that, for all $t \in \mathbf{A}(n)$, $t^\alpha = \phi t \phi^{-1}$. □

11.61. Let tw**K** denote the *twisted square* of **K**, that is, $\mathbf{K} \times \mathbf{K}$ with the product $(\mathbf{K} \times \mathbf{K})^2 \to (\mathbf{K} \times \mathbf{K});\ (\lambda,\mu)(\lambda',\mu') \rightsquigarrow (\lambda\lambda',\mu'\mu)$. Show that, for any finite-dimensional **K**-linear space X, the **K**-linear space $X \times X^L$ may be regarded as a tw**K**-linear space by defining scalar multiplication by the formula

$$(x,\omega)(\lambda,\mu) = (x\lambda,\mu\omega), \quad \text{for any } (x,\omega) \in X \times X^L,\ (\lambda,\mu) \in \mathrm{tw}\mathbf{K}.$$

Develop the theory of tw**K**-correlated spaces. Show, in particular, that for any finite-dimensional **K**-linear space X, the map

$$(X \times X^L)^2 \to \mathrm{tw}\mathbf{K};\ ((a,\alpha),(b,\beta)) \rightsquigarrow (\alpha(b),\beta(a))$$

is the product of a non-degenerate symmetric $(\mathrm{tw}\mathbf{K})^{\mathrm{hb}}$-correlation on $X \times X^L$ (hb being an *anti*-involution of tw**K**). □

11.62. Let ξ be a symmetric $\bar{\mathbf{C}}$-linear correlation on a **C**-linear space X such that, for all $x \in X$, $x^\xi x = 0$. Prove that $\xi = 0$. □

11.63. Show

(i) that there is a group map $SU(2) \to S^3 \times S^3$ making the diagram

$$\begin{array}{ccc} & & S^3 \times S^3 \\ & \nearrow & \downarrow \\ SU(2) & \longrightarrow & SO(4) \end{array}$$

commute,

and (ii) that there is no group map $U(2) \to S^3 \times S^3$ making the diagram

$$\begin{array}{ccc} & & S^3 \times S^3 \\ & \nearrow & \downarrow \\ U(2) & \longrightarrow & SO(4) \end{array}$$

commute,

the vertical map in either case being the group map defined in Prop. 10.29 and the horizontal maps being the standard group injections induced by the usual identification of $\mathbf{C}^2$ with $\mathbf{R}^4$. □

11.64. Let $t \in U(3)$. Show that

$$\overline{t_{02}} = \det\begin{pmatrix} t_{10} & t_{11} \\ t_{20} & t_{21} \end{pmatrix}, \quad \overline{t_{12}} = -\det\begin{pmatrix} t_{00} & t_{01} \\ t_{20} & t_{21} \end{pmatrix}$$

and

$$\overline{t_{22}} = \det\begin{pmatrix} t_{00} & t_{01} \\ t_{10} & t_{11} \end{pmatrix}. \quad □$$

11.65. Show that the diagram of maps

$$\begin{array}{ccccc}
Sp(1) = SU(2) & \longrightarrow & SU(3) & \longrightarrow & S^5 \\
\downarrow & & \downarrow & & \\
Sp(2) & \longrightarrow & SU(4) & \xrightarrow{\pi} & T \\
\downarrow & & \downarrow & & \\
S^7 & \xrightarrow{1} & S^7 & &
\end{array}$$

where any $z + jw$ in $Sp(1)$ is identified with $\begin{pmatrix} z & -\bar{w} \\ w & \bar{z} \end{pmatrix}$ in $SU(2)$, is commutative, the top row and the two columns being special cases of the left-coset exact pairs defined in Theorem 11.55, and the map π being the surjection, with image in $SU(4)$, defined, for all $t \in SU(4)$, by the formula $\pi(t) = tt^{\sim}$, where

$$t^{\sim} = \begin{pmatrix} t_{11} & -t_{01} & t_{31} & -t_{21} \\ -t_{10} & t_{00} & -t_{30} & t_{20} \\ t_{13} & -t_{03} & t_{33} & -t_{23} \\ -t_{12} & t_{02} & -t_{32} & t_{22} \end{pmatrix}.$$

Hence construct a bijection $S^5 \longrightarrow T$ that makes the square

$$\begin{array}{ccc}
SU(3) & \longrightarrow & S^5 \\
\downarrow & & \downarrow \\
SU(4) & \xrightarrow{\pi} & T
\end{array}$$

commute and show that this bijection is the restriction to S^5 with target T of an injective real linear map

$$\gamma : \mathbf{C}^3 \longrightarrow \mathbf{C}(4). \qquad \square$$

11.66. Let X and Y be isomorphic non-degenerate symmetric or skew $\mathbf{A}^{\varphi}$-correlated spaces, let U and V be correlated subspaces of X and Y, respectively, and suppose that $s : U \longrightarrow V$ is a correlated isomorphism. Construct a correlated isomorphism $t : X \longrightarrow Y$ such that $s = (t \mid U)_{\text{sur}}$. (Cf. Exercise 9.88.)

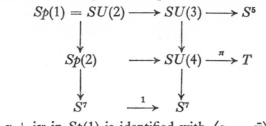

CHAPTER 12

QUADRIC GRASSMANNIANS

The central objects of study in this chapter are the quadric Grassmannians of finite-dimensional correlated spaces. Particular topics include affine quadrics and their classification, parabolic charts on a quadric Grassmannian and various coset space representations of quadric Grassmannians.

There is no attempt to be exhaustive. The purpose of the chapter, rather, is to provide a fund of examples that will illustrate the material of later chapters, in particular Chapters 17 and 20.

All linear spaces will be finite-dimensional linear spaces over $\mathbf{A} = \mathbf{K}$ or ${}^2\mathbf{K}$, where $\mathbf{K} = \mathbf{R}$, $\mathbf{C}$ or $\mathbf{H}$. On a first reading one should assume that $\mathbf{A} = \mathbf{R}$ or $\mathbf{C}$ and that ψ is the identity, ignoring references to the more complicated cases.

Grassmannians

Grassmannians of linear spaces have already been introduced in Chapters 8 and 10, but it is convenient to recall the definitions here, varying the notations slightly.

Let X be a right $\mathbf{A}$-linear space. Then, for any finite k, the set $\mathscr{G}_k(X)$ of linear subspaces of X of dimension k over $\mathbf{A}$ is the *Grassmannian of linear k-planes* in X, the Grassmannian $\mathscr{G}_1(X)$ of lines in X through 0 being called also the *projective space* of X. In the real case there are also the Grassmannians $\mathscr{G}_k^+(X)$ of *oriented* linear k-planes in X.

As we saw in Chapter 8, and again in Chapter 10 in the quaternionic case, various subsets of the Grassmannian $\mathscr{G}_k(X)$ may be regarded in a natural way as affine spaces. For any linear subspace Y of codimension k in X, the inclusion $\Theta(X,Y) \longrightarrow \mathscr{G}_k(X)$ will be called a *natural chart* on $\mathscr{G}_k(X)$, $\Theta(X,Y)$ being, as before, the affine space of linear complements of Y in X. In this context it is convenient to regard as *equivalent* injections $A \longrightarrow \mathscr{G}_k(X)$ and $B \longrightarrow \mathscr{G}_k(X)$, where A and B are affine spaces over $\mathbf{A}$ (or the centre of $\mathbf{A}$), whenever there is an affine isomorphism $A \to B$ such that the diagram

$$\begin{array}{ccc} A & \searrow & \\ \downarrow & & \mathscr{G}_k(X) \\ B & \nearrow & \end{array}$$

is commutative. A *standard chart* on $\mathscr{G}_k(X)$ is then defined to be an injection $A \to \mathscr{G}_k(X)$, with A affine, equivalent to one of the natural charts.

Suppose, for example, that $X = W \oplus Y (\cong W \times Y)$ with $W \in \mathscr{G}_k(X)$. Then the map

$$L(W, Y) \to \mathscr{G}_k(X); \quad t \rightsquigarrow \text{graph } t$$

is a standard chart on $\mathscr{G}_k(X)$, the map

$$L(W, Y) \to \Theta(X, Y); \quad t \rightsquigarrow \text{graph } t$$

being an affine isomorphism.

Standard charts on the projective space $\mathscr{G}_1(X)$ include maps of the form

$$H \to \mathscr{G}_1(X); \quad x \rightsquigarrow [x] = \text{im}\{x\},$$

where H is an affine hyperplane of X not passing through 0, and maps of the form

$$Y \to \mathscr{G}_1(X); \quad y \rightsquigarrow [b + y]$$

where Y is a linear hyperplane of X and $b \in X \setminus Y$.

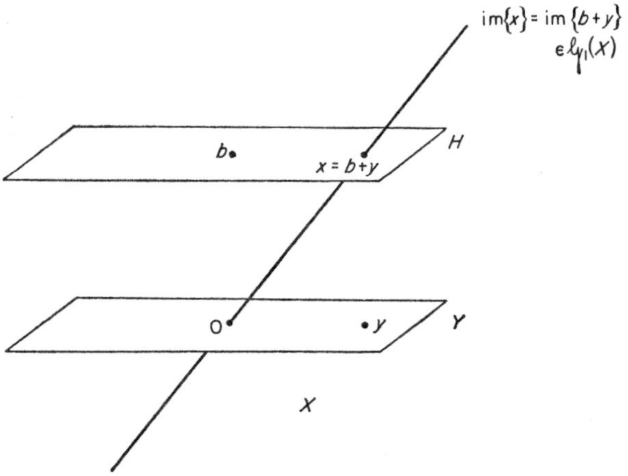

A standard atlas on $\mathscr{G}_k(X)$ is a set of standard charts on $\mathscr{G}_k(X)$ such that every point of $\mathscr{G}_k(X)$ is in the image of at least one chart.

Example 12.1. The set of maps $\mathbf{R}^2 \to \mathscr{G}_1(\mathbf{R}^3)$; $(x,y) \rightsquigarrow [x,y,1]$, $(x,z) \rightsquigarrow [x,1,z]$ and $(y,z) \rightsquigarrow [1,y,z]$ is a standard atlas for $\mathscr{G}_1(\mathbf{R}^3)$. □

Quadric Grassmannians

Now let ξ be an irreducible symmetric or skew correlation on the right **A**-linear space X. The kth *quadric Grassmannian* of the correlated space is, by definition, the subset $\mathscr{I}_k(X,\xi)$ of $\mathscr{G}_k(X)$ consisting of the k-dimensional isotropic subspaces of (X,ξ).

Prop. 12.2. Let ξ be such a correlation on X and let η be any correlation equivalent to ξ. Then, for each k, $\mathscr{I}_k(X,\eta) = \mathscr{I}_k(X,\xi)$. In particular, $\mathscr{I}_k(X,-\xi) = \mathscr{I}_k(X,\xi)$, for each k. □

The counterimage of $\mathscr{I}_k(X,\xi)$ by any one of the standard charts of $\mathscr{G}_k(X)$ will be called an *affine form* of $\mathscr{I}_k(X,\xi)$ or simply an *affine* quadric Grassmannian.

We shall mainly be concerned with the case when ξ is non-degenerate. In this case the dimension of an isotropic subspace is at most half the dimension of X. When (X,ξ) is a non-degenerate neutral space, necessarily of even dimension, isotropic subspaces of half the dimension of X exist. Such subspaces will be termed *semi-neutral* subspaces, and the set of semi-neutral subspaces of (X,ξ) will be called a *semi-neutral* quadric Grassmannian.

There are isotropic lines in (X,ξ) unless (X,ξ) is positive- or negative-definite. The subset $\mathscr{I}_1(X,\xi)$ of the projective space $\mathscr{G}_1(X)$ is called the *projective quadric* of (X,ξ). The counterimage of $\mathscr{I}_1(X,\xi)$ by any one of the standard charts of $\mathscr{G}_1(X)$ will be called an *affine form* of $\mathscr{I}_1(X,\xi)$ or simply an *affine quadric*. When ξ is non-degenerate, an affine form of $\mathscr{I}_1(X,\xi)$ will be called a *non-degenerate* affine quadric.

A line W in X is isotropic with respect to the correlation ξ on X or, equivalently, is a point of the projective quadric if, and only if, for every $x \in W$, $x^\xi x = 0$. This equation is frequently referred to as the *equation of the quadric* $\mathscr{I}_1(X,\xi)$.

Just as the elements of $\mathscr{G}_k(X)$ may, when $k \geqslant 1$, be interpreted as $(k-1)$-dimensional projective subspaces of the projective space $\mathscr{G}_1(X)$ rather than as k-dimensional linear subspaces of X, so the elements of $\mathscr{I}_k(X,\xi)$ may when $k \geqslant 1$, be interpreted as $(k-1)$-dimensional projective spaces lying on the projective quadric $\mathscr{I}_1(X,\xi)$ rather than as k-dimensional isotropic subspaces of (X,ξ). We shall refer to this as the *projective interpretation* of the quadric Grassmannians.

When (X,ξ) is isomorphic either to $\mathbf{R}^n_{sp}$ or to $\mathbf{C}^n_{sp}$, every line in (X,ξ) is isotropic. In these cases, therefore, the first interesting quadric Grassmannian is not $\mathscr{I}_1(X,\xi)$ but $\mathscr{I}_2(X,\xi)$, the set of isotropic planes in (X,ξ). This set is usually called the (*projective*) *line complex* of (X,ξ), the terminology reflecting the projective rather than the linear interpretation of $\mathscr{I}_2(X,\xi)$. (See also page 216.)

Affine quadrics

Let ξ denote a symmetric or skew $\mathbf{K}^{\varphi}$-correlation on the right $\mathbf{K}$-linear space X and let Y be a linear hyperplane of X (the ${}^2\mathbf{K}$ case being excluded from the discussion because of complications when ξ is degenerate). Then the natural chart $\Theta(X,Y) \to \mathscr{G}_1(X)$ determines an affine form of the quadric $\mathscr{I}_1(X,\xi)$.

There are various possibilities, which may be conveniently grouped into four types, namely

(i) ξ non-degenerate, Y a non-degenerate subspace of X,
(ii) ξ non-degenerate, Y a degenerate subspace,
(iii) ξ degenerate, $\ker(X,\xi) \subset Y$,
(iv) ξ degenerate, $\ker(X,\xi) \not\subset Y$.

We consider the various types in turn.

Type (i)—ξ non-degenerate, Y non-degenerate.

Since Y is non-degenerate, $X = Y^{\perp} \oplus Y$, the line $W = Y^{\perp}$ also being a non-degenerate subspace. We may suppose, without loss of generality, that $W \cong \mathbf{K}^{\varphi}$, even in the cases where signature is relevant, since $\mathscr{I}_1(X,-\xi) = \mathscr{I}_1(X,\xi)$, and we choose such an isomorphism. Let η be the correlation induced on Y by ξ.

There is then an isomorphism $\mathbf{K} \times Y \to X$ determining an affine isomorphism

$$Y \to \Theta(X,Y)\,; \quad y \rightsquigarrow \operatorname{im}\{(1,y)\}.$$

The equation of $\mathscr{I}_1(X,\xi)$ with respect to the former isomorphism is

$$w^{\varphi}w + y^{\eta}y = 0, \quad \text{with} \quad w \in \mathbf{K} \cong W \quad \text{and} \quad y \in Y,$$

and the equation of the affine form of $\mathscr{I}_1(X,\xi)$ is

$$1 + y^{\eta}y = 0,$$

this being obtained by setting $w = 1$ in the previous equation.

Such an affine quadric is said to be *central*, with *centre* 0, for if y is a point of the affine quadric, so is $-y$.

Type (ii)—ξ non-degenerate, Y degenerate.

In this case $\dim(\ker Y) = 1$, by Prop. 9.24. Let $W' = \ker Y$ and let W be any isotropic line not lying in Y. Then since X is non-degenerate, $W \oplus W'$ is a hyperbolic or symplectic plane, for otherwise $W' = \ker X$. The line W therefore determines a Witt decomposition $W \oplus W' \oplus Z$ of X in which $Z = (W \oplus W')^{\perp}$ also is non-degenerate. Choose an isomorphism $W \oplus W' \to \mathbf{K}^{\varphi}_{hb}$ or $\mathbf{K}^{\varphi}_{sp}$ and let ζ be the correlation induced on Z by ξ.

The isomorphism $\mathbf{K} \times Y = \mathbf{K} \times \mathbf{K} \times Z \to X$ determines an affine isomorphism

$$Y \to \Theta(X, Y);\; y \rightsquigarrow \operatorname{im}\{(1,y)\}$$

as before. The equation of $\mathscr{I}_1(X,\xi)$ with respect to the former isomorphism is

$$w^{\psi}w' \pm w'^{\psi}w + z^{\zeta}z = 0,$$

with $w \in \mathbf{K} \cong W$, $w' \in \mathbf{K} \cong W'$ and $z \in Z$, the sign being $+$ if ξ is symmetric and $-$ if ξ is skew.

The equation of the affine form of $\mathscr{I}_1(X,\xi)$ is therefore

$$w' \pm w'^{\psi} + z^{\zeta}z = 0$$

obtained, as before, by setting $w = 1$ in the previous equation.

Such an affine quadric is said to be *parabolic*.

Type (iii)—ξ degenerate, $\ker(X,\xi) \subset Y$.

Two subspaces Y_0 and Y_1 of an affine space Y with vector space Y_* are said to be *complementary* if $Y_* = (Y_0)_* \oplus (Y_1)_*$. A subset Q of Y is said to be a (non-degenerate) *quadric cylinder* in Y if there are complementary subspaces Y_0 and Y_1 of Y and a non-degenerate affine quadric Q_1 in Y_1 such that Q is the union of the set of affine subspaces in Y parallel to Y_0, one through each point of Q_1.

Exercise 12.3. Show that any affine quadric of type (iii) is a quadric cylinder. □

Type (iv)—ξ degenerate, $\ker(X,\xi) \not\subset Y$.

A subset Q of an affine space Y is said to be a *quadric cone* in Y if there is a hyperplane Y' of Y, a point $v \in Y \setminus Y'$, and a not necessarily non-degenerate affine quadric Q' in Y' such that Q is the union of the set of affine lines joining v to each point of Q'. The point v is said to be a *vertex* of the cone. It need not be unique.

Exercise 12.4. Show that any affine quadric of type (iv) is a quadric cone. □

Real affine quadrics

In this section we consider briefly the various types of non-degenerate affine quadrics in the case that $X \cong \mathbf{R}^{p,q}$. As before, Y denotes a linear hyperplane of X. There are two types.

Type (i)—In this case Y is non-degenerate, so isomorphic either to $\mathbf{R}^{p,q-1}$ or to $\mathbf{R}^{p-1,q}$. Without loss of generality we may suppose as before that the former is the case, and that the affine form of $\mathscr{I}_1(X,\xi)$ has the equation

$$1 - \sum_{i \in p} y_i^2 + \sum_{j \in q-1} y_{j+p}^2 = 0$$

or, equivalently,

$$\sum_{i \in p} y_i^2 - \sum_{j \in q-1} y_{j+p}^2 = 1.$$

When $n = p + q - 1 = 1$, there is only one type of affine central quadric, whose equation may be taken to be $y^2 = 1$. This is a *pair* of *points*.

When $n = 2$ there are two types, the *ellipse* and the *hyperbola* with equations

$$y_0^2 + y_1^2 = 1 \quad \text{and} \quad y_0^2 - y_1^2 = 1,$$

respectively. Each is an affine form of the projective quadric in $\mathscr{G}_1(X)$ with equation

$$-x_0^2 + x_1^2 + x_2^2 = 0.$$

When $n = 3$ there are three types, the *ellipsoid*, the *hyperboloid of one sheet*, and the *hyperboloid of two sheets*, with equations

$$y_0^2 + y_1^2 + y_2^2 = 1$$

$$y_0^2 + y_1^2 - y_2^2 = 1$$

and

$$y_0^2 - y_1^2 - y_2^2 = 1,$$

respectively.

The phrases 'one sheet' and 'two sheets' refer to the fact that the one hyperboloid is in one piece and the other in two pieces. The subject of connectedness is one which is discussed in more detail later, in Chapters 16 and 17.

Type (*ii*)—In this case $W + W'$ is isomorphic to $\mathbf{R}_{\text{hb}}$ and Z to $\mathbf{R}^{p-1,q-1}$. The equation of the affine form of $\mathscr{I}_1(X,\xi)$ may therefore be taken to be

$$2w' - \sum_{i \in p-1} z_i^2 - \sum_{j \in q-1} z_{j+p-1}^2 = 0,$$

or, equivalently,

$$\sum_{i \in p-1} z_i^2 - \sum_{j \in q-1} z_{j+p-1}^2 = 2w'.$$

When $n = p + q - 1 = 1$, there is one type, with equation $w' = 0$. This is a *single point*.

When $n = 2$, there is again one type, the *parabola*, with equation $z^2 = 2w'$, this being a third affine form of the projective quadric with equation

$$-x_0^2 + x_1^2 + x_2^2 = 0,$$

When $n = 3$, there are two parabolic quadrics, the *elliptic paraboloid* with equation

$$z_0^2 + z_1^2 = 2w',$$

this being an affine form of the projective quadric

$$-x_0^2 + x_1^2 + x_2^2 + x_3^2 = 0$$

whose other affine manifestations are the ellipsoid and the hyperboloid of two sheets, and the *hyperbolic paraboloid*, with equation

$$z_0^2 - z_1^2 = 2w',$$

this being an affine form of the quadric

$$-x_0^2 - x_1^2 + x_2^2 + x_3^2 = 0$$

whose other affine manifestation is the hyperboloid of one sheet.

Exercise 12.5. Since $\mathbf{R}^{2,2}$ is neutral, there are isotropic planes in $\mathbf{R}^{2,2}$, projective lines on $\mathscr{I}_1(\mathbf{R}^{2,2})$ and affine lines on each of its affine forms. Find the affine lines on the hyperboloid of one sheet and on the hyperbolic paraboloid and show that in each case there are, in some natural sense, two families of lines on the quadric. (The existence of such lines is one reason for the popularity amongst architects of the hyperbolic paraboloid roof.)

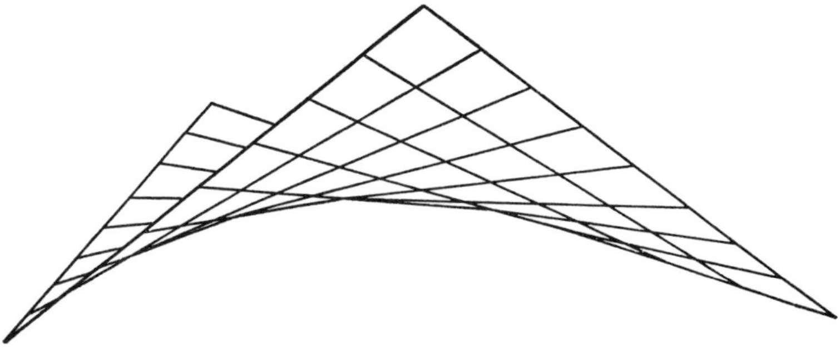

□

Charts on quadric Grassmannians

Now let (X,ξ) be any non-degenerate irreducible symmetric or skew $\mathbf{A}^\varphi$-correlated space, and consider the quadric Grassmannian $\mathscr{I}_k(X,\xi)$.

By Cor. 11.44 there is, for any $W \in \mathscr{I}_k(X,\xi)$ a Witt decomposition $W \oplus W' \oplus Z$ of X, where $W' \in \mathscr{I}_k(X,\xi)$ and $Z = (W \oplus W')^\perp$. There are, moreover, linear isomorphisms $\mathbf{A}^k \longrightarrow W$ and $\mathbf{A}^k \longrightarrow W'$ such that the product on X induced by ξ is given with respect to these framings by the formula

$$(a,b,c)^\xi(a',b',c') = b^\eta a' \pm a^\eta b' + c^\zeta c'$$

where η is the (symmetric) correlation on $(\mathbf{A}^{\varphi})^k$ and ζ is the correlation induced on Z by ξ, and where X has been identified with $W \times W' \times Z$ to simplify notations.

Both here and in the subsequent discussion, where there is a choice of sign the upper sign applies when ξ (and therefore ζ) is symmetric and the lower sign when ξ (and ζ) is skew.

Now let $Y = W' \oplus Z \cong W \times Z$ and consider the standard chart on $\mathscr{G}_k(X)$

$$L(W,Y) \to \mathscr{G}_k(X); \quad (s,t) \rightsquigarrow \text{graph}\,(s,t).$$

The counterimage by this chart of $\mathscr{I}_k(X,\xi)$ is given by the following proposition.

Prop. 12.6. Let $(s,t) \in L(W,Y)$, the notations and sign convention being those just introduced. Then

$$\text{graph}\,(s,t) \in \mathscr{I}_k(X,\xi) \iff s \pm s^{\eta} + t^*t = 0,$$

where t^* is the adjoint of t with respect to the correlations η on $\mathbf{A}^k$ and ζ on Z.

In particular, when $Z = \{0\}$, that is, when $\mathscr{I}_k(X,\xi)$ is semi-neutral, the counterimage of $\mathscr{I}_k(X,\xi)$ by the chart

$$L(W,Y) \to \mathscr{G}_k(X); \quad (s,t) \rightsquigarrow \text{graph}\,(s,t)$$

is a real linear subspace of its source.

Proof For all $a, b \in W$,

$$\begin{aligned}(a,s(a),t(a))^{\xi}\,(b,s(b),t(b)) &= s(a)^{\eta}b \pm a^{\eta}s(b) + t(a)^{\zeta}t(b)\\ &= (s(a) \pm s^{\eta}(a) + t^*t(a))^{\eta}b\end{aligned}$$

by Prop. 11.26. Therefore

$$\text{graph}\,(s,t) \in \mathscr{I}_k(X,\xi) \iff s \pm s^{\eta} + t^*t = 0.$$

The second part of the proposition follows from the remark that $\text{End}_+(\mathbf{A}^k,\eta)$ and $\text{End}_-(\mathbf{A}^k,\eta)$ (cf. page 208) are real linear subspaces of $\text{End}\,\mathbf{A}^k$, while $t = 0$ when $Z = \{0\}$. □

Prop. 12.7. Let the notations be as above. Then the map

$$f\colon \text{End}_{\mp}(\mathbf{A}^k,\eta) \times L(\mathbf{A}^k,Z) \to L(\mathbf{A}^k,Y); \quad (s,t) \rightsquigarrow (s - \tfrac{1}{2}t^*t\,,\,t)$$

is injective, with image the affine form of $\mathscr{I}_k(X,\xi)$ in $L(\mathbf{A}^k,Y)$ $(= L(W,Y))$.

Proof That the map f is injective is obvious. That its image is as stated follows from the fact that, for any $t \in L(\mathbf{A}^k,Z)$, $(t^*t)^{\eta} = \pm t^*t$, by Prop. 11.39. Therefore, for any (s,t),

$$(s - \tfrac{1}{2}t^*t) \pm (s - \tfrac{1}{2}t^*t)^{\eta} + t^*t = 0$$

That is, the image of f is a subset of $\mathscr{I}_k(X,\xi)$, by Prop. 12.6. Conversely, if graph $(s',t') \in \mathscr{I}_k(X,\xi)$, let $s = s' + \frac{1}{2}t'^*t'$ and let $t = t'$. Then $s \pm s^\eta = 0$; so $s \in \mathrm{End}_\mp(\mathbf{A}^k,\eta)$ and $t \in L(\mathbf{A}^k,Z)$, while $s' = s - \frac{1}{2}t^*t$ and $t' = t$.

That is, the affine form of $\mathscr{I}_k(X,\xi)$ is a subset of $\operatorname{im} f$. The image of f and the affine quadric Grassmannian therefore coincide. □

The composite of the map f_{sur} with the inclusion of $\operatorname{im} f$ in $\mathscr{I}_k(X,\xi)$ will be called a *parabolic chart* on $\mathscr{I}_k(X,\xi)$ *at* W. A set of parabolic charts on $\mathscr{I}_k(X,\xi)$, one at each point of $\mathscr{I}_k(X,\xi)$, will be called a *parabolic atlas* for $\mathscr{I}_k(X,\xi)$.

Grassmannians as coset spaces

Let X be a finite-dimensional real linear space. Then, as was noted in Chapter 8, there is a surjective map

$$h : GL(\mathbf{R}^k,X) \longrightarrow \mathscr{G}_k(X); \quad u \rightsquigarrow \operatorname{im} u$$

associating to each k-frame on X its linear image.

Prop. 12.8. Let ξ be a positive-definite correlation on X. Then $h \mid O(\mathbf{R}^k,X)$ is surjective. □

The map introduced in the next proposition is a slight variant of this.

Prop. 12.9. Let $\mathbf{R}^n$ be identified with $\mathbf{R}^k \times \mathbf{R}^{n-k}$. Then the map

$$f : O(\mathbf{R}^n) \longrightarrow \mathscr{G}_k(\mathbf{R}^n); \quad t \rightsquigarrow t_\vdash(\mathbf{R}^k \times \{0\})$$

is surjective, its fibres being the left cosets in $O(\mathbf{R}^n)$ of the subgroup $O(\mathbf{R}^k) \times O(\mathbf{R}^{n-k})$.

Proof With $\mathbf{R}^n$ identified with $\mathbf{R}^k \times \mathbf{R}^{n-k}$, any map $t : \mathbf{R}^n \longrightarrow \mathbf{R}^n$ is of the form $\begin{pmatrix} a & c \\ b & d \end{pmatrix}$ where $a \in L(\mathbf{R}^k,\mathbf{R}^k)$, $b \in L(\mathbf{R}^k,\mathbf{R}^{n-k})$, $c \in L(\mathbf{R}^{n-k}, \mathbf{R}^k)$ and $d \in L(\mathbf{R}^{n-k},\mathbf{R}^{n-k})$. Since the first k columns of the matrix span $t_\vdash(\mathbf{R}^k \times \{0\})$, $t_\vdash(\mathbf{R}^k \times \{0\}) = \mathbf{R}^k \times \{0\}$ if, and only if, $b = 0$. However, if t is orthogonal with $b = 0$, then c also is zero, since any two columns of the matrix are mutually orthogonal. The subgroup $O(\mathbf{R}^k) \times O(\mathbf{R}^{n-k})$, consisting of all $\begin{pmatrix} a & 0 \\ 0 & d \end{pmatrix} \in O(\mathbf{R}^k \times \mathbf{R}^{n-k})$, is therefore the fibre of f over $\mathbf{R}^k \times \{0\}$.

The map f is surjective by Prop. 9.25. (Any element of $\mathscr{G}_k(\mathbf{R}^n)$ is a non-degenerate subspace of $\mathbf{R}^n$ and so has an orthonormal basis that extends to an orthonormal basis for the whole of $\mathbf{R}^n$.)

Finally, if t and $u \in O(\mathbf{R}^n)$ are such that $t_\vdash(\mathbf{R}^k \times \{0\}) = u_\vdash(\mathbf{R}^k \times \{0\})$, then $(u^{-1}t)_{\vdash}(\mathbf{R}^k \times \{0\}) = \mathbf{R}^k \times \{0\}$, from which it follows directly that the fibres of the map f are the left cosets in $O(\mathbf{R}^n)$ of the subgroup $O(\mathbf{R}^k) \times O(\mathbf{R}^{n-k})$.. □

In the terminology of Chapter 5, page 97, the pair of maps

$$O(k) \times O(n-k) \xrightarrow{\text{inc}} O(n) \xrightarrow{f} \mathcal{G}_k(\mathbf{R}^n)$$

is left-coset exact, and $f_{\text{inj}} : O(n)/(O(k) \times O(n-k)) \longrightarrow \mathcal{G}_k(\mathbf{R}^n)$ is a coset space representation of $\mathcal{G}_k(\mathbf{R}^n)$.

Prop. 12.10. For each finite n, k, with $k \leqslant n$, there are coset space representations

$$U(n)/(U(k) \times U(n-k)) \longrightarrow \mathcal{G}_k(\mathbf{C}^n)$$

$$Sp(n)/(Sp(k) \times Sp(n-k)) \longrightarrow \mathcal{G}_k(\mathbf{H}^n)$$

and

$$SO(n)/(SO(k) \times SO(n-k)) \longrightarrow \mathcal{G}_k^+(\mathbf{R}^n),$$

analogous to the coset space representation

$$O(n)/(O(k) \times O(n-k)) \longrightarrow \mathcal{G}_k(\mathbf{R}^n)$$

constructed in Prop. 12.9. □

Quadric Grassmannians as coset spaces

Coset space representations analogous to those of Prop. 12.10 exist for each of the quadric Grassmannians.

We begin by considering a particular case, the semi-neutral Grassmannian $\mathcal{I}_n(\mathbf{C}_{\text{hb}}^n)$ of the neutral $\mathbf{C}$-correlated space $\mathbf{C}_{\text{hb}}^n$.

Prop. 12.11. There exists a bijection

$$O(2n)/U(n) \longrightarrow \mathcal{I}_n(\mathbf{C}_{\text{hb}}^n),$$

where $O(2n)/U(n)$ denotes the set of left cosets in $O(2n)$ of the standard image of $U(n)$ in $O(2n)$.

Proof This bijection is constructed as follows.

The linear space underlying the correlated space $\mathbf{C}_{\text{hb}}^n$ is $\mathbf{C}^n \times \mathbf{C}^n$, and this same linear space also underlies the positive-definite correlated space $\bar{\mathbf{C}}^n \times \bar{\mathbf{C}}^n$. Any linear map $t: \mathbf{C}^n \times \mathbf{C}^n \longrightarrow \mathbf{C}^n \times \mathbf{C}^n$ that respects both correlations is of the form $\begin{pmatrix} a & \bar{b} \\ b & \bar{a} \end{pmatrix}$, with $a^\tau b + b^\tau a = 0$ and $\bar{a}^\tau b + \bar{b}^\tau b = 1$, for, by Table 11.52, the respective adjoints of any such map $t = \begin{pmatrix} a & c \\ b & d \end{pmatrix}$ are $\begin{pmatrix} d^\tau & c^\tau \\ b^\tau & a^\tau \end{pmatrix}$ and $\begin{pmatrix} \bar{a}^\tau & \bar{c}^\tau \\ \bar{b}^\tau & \bar{d}^\tau \end{pmatrix}$, and these are equal if, and only if, $d = \bar{a}$ and $c = \bar{b}$. By Prop. 10.46 such a map may be identified with an element of $O(2n)$ or, when $b = 0$, with an element of $U(n)$, the injection $U(n) \longrightarrow O(2n)$ being the standard one.

Suppose that W is any n-dimensional isotropic subspace of $\mathbf{C}^n_{\text{hb}}$. A positive-definite orthonormal basis may be chosen for W as a subspace of $\bar{\mathbf{C}}^n \times \bar{\mathbf{C}}^n$. Suppose this is done, and the basis elements arranged in some order to form the columns of a $2n \times n$ matrix $\begin{pmatrix} a \\ b \end{pmatrix}$. Then W is the image of the isotropic subspace $\mathbf{C}^n \times \{0\}$ by the map $\begin{pmatrix} a & \bar{b} \\ b & \bar{a} \end{pmatrix}$. Moreover, $a^\tau b + b^\tau a = 0$, since W is isotropic for the hyperbolic correlation, while $\bar{a}^\tau a + \bar{b}^\tau b = 1$, since the basis chosen for W is orthonormal with respect to the positive-definite correlation.

Now let f be the map

$$O(2n) \to \mathscr{I}_n(\mathbf{C}^n_{\text{hb}}); \quad \begin{pmatrix} a & \bar{b} \\ b & \bar{a} \end{pmatrix} \rightsquigarrow \operatorname{im} \begin{pmatrix} a \\ b \end{pmatrix}.$$

The map is clearly surjective; so none of the fibres is null. Secondly, $f^{-1}(\mathbf{C}^n \times \{0\}) = U(n)$. Finally, by an argument similar to that used in the proof of Prop. 12.9, the remaining fibres of f are the left cosets in $O(2n)$ of the subgroup $U(n)$. □

Theorem 12.12. Let $(X,\xi) = (\mathbf{A}^\psi_{\text{hb}})^n$ or $(\mathbf{A}^\psi_{\text{sp}})^n$, where ψ is irreducible and n is finite. Then in each of the ten standard cases there is a coset space representation of the semi-neutral Grassmannian $\mathscr{I}_n(X,\xi)$, as follows:

$$(O(n) \times O(n))/O(n) \to \mathscr{I}_n(\mathbf{R}^n_{\text{hb}})$$
$$U(n)/O(n) \to \mathscr{I}_n(\mathbf{R}^n_{\text{sp}})$$
$$O(2n)/U(n) \to \mathscr{I}_n(\mathbf{C}^n_{\text{hb}})$$
$$(U(n) \times U(n))/U(n) \to \mathscr{I}_n(\bar{\mathbf{C}}^n_{\text{hb}}) = \mathscr{I}_n(\bar{\mathbf{C}}^n_{\text{sp}})$$
$$Sp(n)/U(n) \to \mathscr{I}_n(\mathbf{C}^n_{\text{sp}})$$
$$U(2n)/Sp(n) \to \mathscr{I}_n(\tilde{\mathbf{H}}^n_{\text{hb}}) = \mathscr{I}_n(\bar{\mathbf{H}}^n_{\text{sp}})$$
$$(Sp(n) \times Sp(n))/Sp(n) \to \mathscr{I}_n(\tilde{\mathbf{H}}^n_{\text{sp}}) = \mathscr{I}_n(\bar{\mathbf{H}}^n_{\text{hb}})$$
$$O(2n)/(O(n) \times O(n)) \to \mathscr{I}_n(\text{hb}\,\mathbf{R})^n_{\text{hb}}$$
$$U(2n)/(U(n) \times U(n)) \to \mathscr{I}_n(\text{hb}\,\bar{\mathbf{C}})^n_{\text{hb}}$$
$$Sp(2n)/(Sp(n) \times Sp(n)) \to \mathscr{I}_n(\text{hb}\,\bar{\mathbf{H}})^n_{\text{hb}}.$$

Proof The third of these is the case considered in Prop. 12.11. The details in each of the other cases follow the details of this case, but using the appropriate part of Prop. 10.46. □

This is a theorem to return to after one has studied Tables 13.66.

Cayley charts

The first of the cases listed in Theorem 12.12 merits further discussion in view of the following remark.

Prop. 12.13. Let f be the map

$$O(n) \times O(n) \longrightarrow O(n); \quad (a,b) \rightsquigarrow ab^{-1}.$$

Then $f^{-1}\{{}^{n}1\}$ is the image of $O(n)$ by the injective group map

$$O(n) \longrightarrow O(n) \times O(n); \quad a \rightsquigarrow (a,a)$$

and the map

$$f_{\text{inj}} : (O(n) \times O(n))/O(n) \longrightarrow O(n).$$

is bijective. □

It follows from this that $O(n)$ may be represented as the semi-neutral Grassmannian $\mathscr{I}_n(\mathbf{R}^{n}_{\text{hb}})$. The charts on $O(n)$ corresponding to the parabolic charts on $\mathscr{I}_n(\mathbf{R}^{n}_{\text{hb}})$ will be called the *Cayley charts* on $O(n)$. The following is an independent account of this case.

Let $(X,\xi) \cong \mathbf{R}^{n,n} \cong \mathbf{R}^{n}_{\text{hb}}$, and consider the quadric $\mathscr{I}_1(X,\xi)$. Its equation may be taken to be either

$$x^{\tau}x = y^{\tau}y, \quad \text{where } (x,y) \in \mathbf{R}^n \times \mathbf{R}^n$$

or

$$u^{\tau}v = 0, \quad \text{where } (u,v) \in \mathbf{R}^n \times \mathbf{R}^n,$$

according to the isomorphism chosen, the two models being related, for example, by the equations

$$u = x + y, \quad v = -x + y.$$

Now any n-dimensional subspace of $\mathbf{R}^n \times \mathbf{R}^n$ may be represented as the image of an injective linear map

$$(a,b) = \begin{pmatrix} a \\ b \end{pmatrix} : \mathbf{R}^n \longrightarrow \mathbf{R}^n \times \mathbf{R}^n.$$

Prop. 12.14. The linear space $\operatorname{im}(a,b)$, where (a,b) is an injective element of $L(\mathbf{R}^n, \mathbf{R}^n \times \mathbf{R}^n)$, is an isotropic subspace of $\mathbf{R}^{n,n}$ if, and only if, a and b are bijective and $ba^{-1} \in O(n)$.

Proof $\Rightarrow$: Let $\operatorname{im}(a,b) \in \mathscr{I}_n(\mathbf{R}^{n,n})$, let $w \in \mathbf{R}^n$ be such that $x = a(w) = 0$ and let $y = b(w)$. Since (x,y) belongs to an isotropic subspace of $\mathbf{R}^{n,n}$, $x^{\tau}x = y^{\tau}y$, but $x = 0$, so that $y = 0$. Since (a,b) is injective, it follows that $w = 0$ and therefore that a is injective. So a is bijective, by Cor. 6.33. Similarly b is bijective.

Since a is bijective, a^{-1} exists; so, for any $(x,y) = (a(w),b(w))$, $y = ba^{-1}(x)$. But $y^{\tau}y = x^{\tau}x$. So $ba^{-1} \in O(n)$.

$\Leftarrow$: Suppose that a and b are bijective; then, as above, for any $(x,y) \in \operatorname{im}(a,b)$ $y = ba^{-1}(x)$. If also $ba^{-1} \in O(n)$, then $y^{\tau}y = x^{\tau}x$. □

Cor. 12.15. Any n-dimensional isotropic subspace of $\mathbf{R}^{n,n}$ has an equation of the form $y = t(x)$, where $t \in O(n)$, and any n-plane with such an equation is isotropic. □

Prop. 12.16. Any element of $\mathscr{I}_n(\mathbf{R}^{n,n})$ may be represented as the image of a linear map $(a,b): \mathbf{R}^n \to \mathbf{R}^n \times \mathbf{R}^n$ with a and b each orthogonal. □

This leads at once to the coset space representation for $\mathscr{I}_n(\mathbf{R}^{n,n})$ whose existence is asserted in Theorem 12.12.

Note that $\mathscr{I}_n(\mathbf{R}^{n,n}) = \{\text{graph } t : t \in O(n)\}$ divides into two disjoint classes, according as t preserves or reverses orientation.

So far we have considered the projective quadric $\mathscr{I}_1(\mathbf{R}^{n,n})$. We now consider the quadric $\mathscr{I}_1(\mathbf{R}^n_{\text{hb}})$. Let $s \in \text{End}(\mathbf{R}^n)$ be such that graph $s \in \mathscr{I}_n(\mathbf{R}^n_{\text{hb}})$. Then, for all $u, u' \in \mathbf{R}^n$,

$$s(u')^\tau u + u'^\tau s(u) = 0,$$

implying that $s + s^\tau = 0$, that is, that $s \in \text{End}_-(\mathbf{R}^n)$, this being a particular case of Prop. 12.6.

Now graph $s = \text{im}\,(1,s)$. We can transfer to $\mathscr{I}_1(\mathbf{R}^{n,n})$ by the map $\frac{1}{\sqrt{2}}\begin{pmatrix}1 & -1\\ 1 & 1\end{pmatrix} : \mathbf{R}^n_{\text{hb}} \to \mathbf{R}^{n,n}$. Then the image of graph s in $\mathbf{R}^{n,n}$, namely

$$\text{im}\,\frac{1}{\sqrt{2}}\begin{pmatrix}1 & -1\\ 1 & 1\end{pmatrix}\begin{pmatrix}1\\ s\end{pmatrix} = \text{im}\begin{pmatrix}1-s\\ 1+s\end{pmatrix},$$

is an element of $\mathscr{I}_n(\mathbf{R}^{n,n})$. So, by Prop. 12.14, $1 - s$ is invertible and $(1+s)(1-s)^{-1} \in O(n)$. In fact, since $(1+s)(1-s)^{-1} = (1+s)(1+s^\tau)^{-1}$, and since $\det\,(1+s^\tau) = \det\,(1+s)$, $(1+s)(1-s)^{-1} \in SO(n)$. (Note, in passing, that, since $(1+s)(1-s) = (1-s)(1+s)$,

$$(1+s)(1-s)^{-1} = (1-s)^{-1}(1+s).)$$

The following proposition sums this all up.

Prop. 12.17. For any $s \in \text{End}_-(\mathbf{R}^n)$, the endomorphism $1 - s$ is invertible, and $(1+s)(1-s)^{-1} \in SO(n)$. Moreover, the map

$$\text{End}_-(\mathbf{R}^n) \to SO(n);\quad s \rightsquigarrow (1+s)(1-s)^{-1}$$

is injective. □

The map given in this proposition is the Cayley chart on $SO(n)$ (or $O(n)$) at ${}^n1$. It is not surjective. For example, $-{}^n1$ does not lie in its image.

The direct analogue of Prop. 12.17, with $\mathbf{R}^{p,q}$ in place of $\mathbf{R}^n$ and $SO(p,q)$ in place of $SO(n)$, is not true when both p and q are non-zero; for $\begin{pmatrix}0 & 1\\ 1 & 0\end{pmatrix} \in \text{End}_-(\mathbf{R}^{1,1})$, but $\begin{pmatrix}1 & -1\\ -1 & 1\end{pmatrix}$ is not invertible. There is, however, the following partial analogue.

Prop. 12.18. For any $s \in \text{End}_-(\mathbf{R}^{p,q})$ for which $1 - s$ is invertible, $(1 + s)(1 - s)^{-1} \in SO(p,q)$. Moreover, the map

$$\text{End}_-(\mathbf{R}^{p,q}) \rightarrowtail SO(p,q); \quad s \rightsquigarrow (1 + s)(1 - s)^{-1}$$

is injective. □

The map given in Prop. 12.18 is, by definition, the *Cayley chart* on $SO(p,q)$ (or $O(p,q)$) at ${}^n1$.

An entirely analogous discussion to that given above for the orthogonal group $O(n)$ can be given also both for the unitary group $U(n)$ and the symplectic group $Sp(n)$.

It was remarked above that the semi-neutral Grassmannian $\mathscr{I}_n(\mathbf{R}^n_{\text{hb}})$ divides into two parts, the parts corresponding to the orientations of $\mathbf{R}^n$. The semi-neutral Grassmannian $\mathscr{I}_n(\mathbf{C}^n_{\text{hb}})$ divides similarly into two parts, the parts corresponding, in the coset space representation

$$O(2n)/U(n) \rightarrow \mathscr{I}_n(\mathbf{C}^n_{\text{hb}})$$

to the two orientations of $\mathbf{R}^{2n}$. (By Cor. 7.33, any element of $U(n)$ preserves the orientation of $\mathbf{R}^{2n}$.)

Grassmannians as quadric Grassmannians

Another case from the list in Theorem 12.12 that merits further discussion is $\mathscr{I}_n(\text{hb}\,\mathbf{R})^n_{\text{hb}}$. It has already been remarked in Chapter 11 that

$$(\text{hb}\,\mathbf{R})^n_{\text{hb}} \cong (\text{hb}\,\mathbf{R})^{2n}.$$

The space $(\text{hb}\,\mathbf{R})^{2n}$ may be thought of as the $\mathbf{R}^2$-linear space $\mathbf{R}^{2n} \times \mathbf{R}^{2n}$ with the product

$$(\mathbf{R}^{2n} \times \mathbf{R}^{2n})^2 \rightarrow \mathbf{R}^2; \quad (a,b), (a',b') \rightsquigarrow (b \cdot a', a \cdot b'),$$

where $\cdot$ is the standard scalar product on $\mathbf{R}^{2n}$.

Now it is easily verified that the isotropic subspaces of this space, of dimension n over $\mathbf{R}^2$, are the $\mathbf{R}^2$-linear subspaces of $\mathbf{R}^{2n} \times \mathbf{R}^{2n}$ of the form $V \times V^\perp$, where V is a subspace of $\mathbf{R}^{2n}$ of dimension n over $\mathbf{R}$ and $V^\perp$ is the orthogonal complement of V in $\mathbf{R}^{2n}$ with respect to the standard scalar product on $\mathbf{R}^{2n}$. This provides a bijection between $\mathscr{G}_n(\mathbf{R}^{2n})$ and $\mathscr{I}_n(\text{hb}\,\mathbf{R})^n_{\text{hb}}$. A coset space representation

$$O(2n)/(O(n) \times O(n)) \rightarrow \mathscr{G}_n(\mathbf{R}^{2n})$$

was constructed in Prop. 12.9. The induced representation for $\mathscr{I}_n(\text{hb}\,\mathbf{R})^n_{\text{hb}}$ can be made to coincide with the representation given in Theorem 12.12, by choosing the various isomorphisms appropriately.

A similar discussion can be carried through for the final two cases on the list, $\mathscr{I}_n(\text{hb}\,\bar{\mathbf{C}})^n_{\text{hb}}$ and $\mathscr{I}_n(\text{hb}\,\bar{\mathbf{H}})^n_{\text{hb}}$.

Further coset space representations

Coset space representations analogous to those listed above for the semi-neutral quadric Grassmannians exist for all the quadric Grassmannians. The results are summarized in the following theorem.

Theorem 12.19. Let (X,ξ) be a non-degenerate n-dimensional irreducible symmetric or skew $\mathbf{A}^\varphi$-correlated space. Then, for each k, in each of the ten standard cases, there is a coset space decomposition of the quadric Grassmannian $\mathscr{I}_k(X,\xi)$ as follows:

$$(O(p) \times O(q))/(O(k) \times O(p-k) \times O(q-k)) \to \mathscr{I}_k(\mathbf{R}^{p,q})$$
$$U(n)/(O(k) \times U(n-k)) \to \mathscr{I}_k(\mathbf{R}^n_{\mathrm{sp}})$$
$$O(n)/(U(k) \times O(n-2k)) \to \mathscr{I}_k(\mathbf{C}^n)$$
$$(U(p) \times U(q))/(U(k) \times U(p-k) \times U(q-k)) \to \mathscr{I}_k(\mathbf{C}^n)$$
$$Sp(n)/(U(k) \times Sp(n-k)) \to \mathscr{I}_k(\mathbf{C}^n_{\mathrm{sp}})$$
$$U(n)/(Sp(k) \times U(n-2k)) \to \mathscr{I}_k(\mathbf{H}^n)$$
$$(Sp(p) \times Sp(q))/(Sp(k) \times Sp(p-k) \times Sp(q-k)) \to \mathscr{I}_k(\bar{\mathbf{H}}^{p,q})$$
$$O(n)/(O(k) \times O(k) \times O(n-2k)) \to \mathscr{I}_k(\mathrm{hb}\ \mathbf{R})^n$$
$$U(n)/(U(k) \times U(k) \times U(n-2k)) \to \mathscr{I}_k(\mathrm{hb}\ \bar{\mathbf{C}})^n$$
$$Sp(n)/(Sp(k) \times Sp(k) \times Sp(n-2k)) \to \mathscr{I}_k(\mathrm{hb}\ \bar{\mathbf{H}})^n.$$

The resourceful reader will be able to supply the proof! □

Certain of the cases where $k = 1$ are of especial interest, and we conclude by considering several of these.

Consider first the real projective quadric $\mathscr{I}_1(\mathbf{R}^{p,q})$, where $p \geqslant 1$ and $q \geqslant 1$.

Prop. 12.20. The map

$$S^{p-1} \times S^{q-1} \to \mathscr{I}_1(\mathbf{R}^{p,q}); \quad (x,y) \rightsquigarrow \operatorname{im}\{(x,y)\}$$

is surjective, the fibre over $\operatorname{im}\{(x,y)\}$ being the set $\{(x,y), (-x,-y)\}$. □

That is, there is a bijection

$$(S^{p-1} \times S^{q-1})/S^0 \to \mathscr{I}_1(\mathbf{R}^{p,q}),$$

where the action of S^0 on $S^{p-1} \times S^{q-1}$ is defined by the formula

$$(x,y)(-1) = (-x,-y),$$

for all $(x,y) \in S^{p-1} \times S^{q-1}$.

This result is in accord with the representation of $\mathscr{I}_1(\mathbf{R}^{p,q})$ given in Theorem 12.19, in view of the familiar coset space representations

$$O(p)/O(p-1) \to S^{p-1} \quad \text{and} \quad O(q)/O(q-1) \to S^{q-1}$$

of Theorem 11.55.

The complex projective quadric $\mathcal{I}_1(\mathbf{C}^n)$ handles rather differently.

Lemma 12.21. For any n, let $z = x + iy \in \mathbf{C}^n$ where $x, y \in \mathbf{R}^n$. Then $z^{(2)} = 0$ if, and only if, $x^{(2)} = y^{(2)}$ and $x \cdot y = 0$. □

Now let $\mathrm{im}^+(x,y)$ denote the oriented plane spanned by any orthonormal pair (x,y) of elements of $\mathbf{R}^n$.

Prop. 12.22. For any orthonormal pair (x,y) of elements of $\mathbf{R}^n$, $\mathrm{im}\{x + iy\} \in \mathcal{I}_1(\mathbf{C}^n)$ and the map

$$\mathcal{G}_2^+(\mathbf{R}^n) \to \mathcal{I}_1(\mathbf{C}^n);\quad \mathrm{im}^+(x,y) \rightsquigarrow \mathrm{im}\{x + iy\}$$

is well defined and bijective. □

The coset space representation

$$SO(n)/(SO(2) \times SO(n-2)) \to \mathcal{G}_2^+(\mathbf{R}^n)$$

given in Prop. 12.10 is in accord with the coset space representation

$$O(n)/(U(1) \times O(n-2)) \to \mathcal{I}_1(\mathbf{C}^n)$$

given in Theorem 12.19, since $SO(2) \cong S^1 \cong U(1)$.

Now consider $\mathcal{I}_1(\mathbf{R}^n_{\mathrm{sp}})$. In this case every line is isotropic; so $\mathcal{I}_1(\mathbf{R}^n_{\mathrm{sp}})$ coincides with $\mathcal{G}_1(\mathbf{R}^{2n})$, for which we already have a coset space representation $O(2n)/(O(1) \times O(2n-1))$, equivalent, by Theorem 11.55, to S^{2n-1}/S^0, where the action of -1 on S^{2n-1} is the antipodal map. By Theorem 12.19 there is also a representation $U(n)/(O(1) \times U(n-1))$. This also is equivalent to S^{2n-1}/S^0 by the standard representation (Theorem 11.55 again)

$$U(n)/U(n-1) \to S^{2n-1}.$$

Finally, the same holds for $\mathcal{I}_1(\mathbf{C}^n_{\mathrm{sp}})$, which coincides with $\mathcal{G}_1(\mathbf{C}^{2n})$, for which we already have a representation $U(2n)/(U(1) \times U(2n-1))$, equivalent to S^{4n-1}/S^1. Here the action of S^1 is right multiplication, S^{4n-1} being identified with the quasi-sphere $\mathcal{S}(\bar{\mathbf{C}}^{2n})$ in $\mathbf{C}^{2n}$. Theorem 12.19 provides the alternative representation $Sp(n)/(U(1) \times Sp(n-1))$, also equivalent to S^{4n-1}/S^1 via the standard representation (Theorem 11.55 yet again)

$$Sp(n)/Sp(n-1) \to S^{4n-1}.$$

Further Exercises

12.23. Let $s \in \mathrm{End}_-(\mathbf{R}^n)$, for some finite number n. Prove that the kernel of s coincides with the linear subspace of $\mathbf{R}^n$ left fixed by the rotation $(1 - s)^{-1}(1 + s)$. Deduce that kr s is even or odd according as n is odd or even. □

12.24. Let $s \in \text{End}_-(\mathbf{C}^n)$, for some finite number n. Prove that, for some non-zero $\lambda \in \mathbf{C}$, $1 - \lambda s$ is invertible and that, for such a λ, $\ker s$ is the linear subspace of $\mathbf{C}^n$ left fixed by the rotation $(1 - \lambda s)^{-1}(1 + \lambda s)$. Hence show that $\text{kr}\, s$ is even or odd according as n is odd or even. (Cf. Prop. 2.18.) □

12.25. Let X be a four-dimensional real or complex linear space, and let Q be the projective quadric of a non-degenerate neutral quadratic form on X. Verify that the set of projective lines on Q divides into two families such that two distinct projective lines on Q intersect (necessarily in a single point) if, and only if, they belong to opposite families. Show also that any point on Q lies on exactly one projective line on Q of each family. □

12.26. Let X be a six-dimensional real or complex linear space and let Q be the projective quadric of a non-degenerate neutral quadratic form on X. Verify that the set of projective planes on Q divides into two families such that two distinct planes of the same family intersect in a point, while planes of opposite families either intersect in a line or do not intersect at all. Show also that any projective line on Q lies in exactly one projective plane on Q of each family. (Cf. Exercise 13.81.) □

12.27. Consider, for any finite n, the map

$$O(2n) \longrightarrow O(2n); \quad t \rightsquigarrow t^{-1} j\, t,$$

with j defined by the formula $j(x,y) = (-y,x)$, for any $(x,y) \in \mathbf{R}^n \times \mathbf{R}^n$. Verify that the fibres of this map are the left cosets in $O(2n)$ of a subgroup isomorphic to $U(n)$ and that each element in the image of the map is skew-symmetric. Determine whether or not every skew-symmetric orthogonal automorphism of $\mathbf{R}^{2n}$ is in the image of the map. □

12.28. Consider, for any finite k, n with $k \leqslant n$, the map

$$f : \mathbf{R}P^n \longrightarrow O(\mathbf{R}^k, \mathbf{R}^{n+1}); \ \text{im}\{a\} \rightsquigarrow \rho_a | \mathbf{R}^k,$$

where ρ_a is the reflection of $\mathbf{R}^{n+1}$ in the hyperplane $(\text{im}\{a\})^\perp$. Show that each fibre of f, with the exception of the fibre over the inclusion $\mathbf{R}^k \longrightarrow \mathbf{R}^{n+1}$, consists of a single point, and determine the exceptional fibre.

Discuss, in particular, the case that $k = 1$, $O(\mathbf{R}, \mathbf{R}^{n+1})$ being identifiable with the sphere S^n. Show that in this case f is surjective.

(This exercise played an important part in the solution of the vector fields on spheres problem. See page 420 and the review by Prof. M. F. Atiyah of Adams's paper [2].) □

CHAPTER 13

CLIFFORD ALGEBRAS

We saw in Chapter 10 how well adapted the algebra of quaternions is to the study of the groups $O(3)$ and $O(4)$. In either case the centre of interest is a real orthogonal space X, in the one case $\mathbf{R}^3$ and in the other $\mathbf{R}^4$. There is also a real associative algebra, $\mathbf{H}$ in either case. The algebra contains both $\mathbf{R}$ and X as linear subspaces, and there is an anti-involution, namely conjugation, of the algebra, such that, for all $x \in X$,

$$\bar{x}x = x^{(2)}.$$

In the former case, when $\mathbf{R}^3$ is identified with the subspace of pure quaternions, this formula can also be written in the simpler form

$$x^2 = -x^{(2)}.$$

In an analogous, but more elementary way, the algebra of complex numbers $\mathbf{C}$ may be used in the study of the group $O(2)$.

The aim of the present chapter is to put these rather special cases into a wider context. To keep the algebra simple, the emphasis is laid at first on generalizing the second of the two displayed formulae. It is shown that, for any finite-dimensional real orthogonal space X, there is a real associative algebra, A say, with unity 1, containing isomorphic copies of $\mathbf{R}$ and X as linear subspaces in such a way that, for all $x \in X$,

$$x^2 = -x^{(2)}.$$

If the algebra A is also generated as a ring by the copies of $\mathbf{R}$ and X or, equivalently, as a real algebra by $\{1\}$ and X, then A is said to be a (*real*) *Clifford algebra* for X (Clifford's term was *geometric algebra* [11]). It is shown that such an algebra can be chosen so that there is also on A an algebra anti-involution

$$A \longrightarrow A; \quad a \rightsquigarrow a^-$$

such that, for all $x \in X$, $x^- = -x$.

To simplify notations in the above definitions, $\mathbf{R}$ and X have been identified with their copies in A. More strictly there are linear injections $\alpha : \mathbf{R} \longrightarrow A$ and $\beta : X \longrightarrow A$ such that, for all $x \in X$,

$$(\beta(x))^2 = -\alpha(x^{(2)}),$$

unity in A being $\alpha(1)$.

The minus sign in the formula $x^2 = -x^{(2)}$ can be a bit of a nuisance at times. One could get rid of it at the outset simply by replacing the orthogonal space X by its negative. However, it turns up anyway in applications, and so we keep it in.

Prop. 13.1. Let A be a Clifford algebra for a real orthogonal space X and let W be a linear subspace of X. Then the subalgebra of A generated by W is a Clifford algebra for W. □

By Prop. 9.56 and Prop. 13.1 the existence of a Clifford algebra for an arbitrary n-dimensional orthogonal space X is implied by the existence of a Clifford algebra for the neutral non-degenerate space $\mathbf{R}^{n,n}$. Such an algebra is constructed below in Cor. 13.18. (An alternative construction of a Clifford algebra for an orthogonal space X depends on the prior construction of the *tensor algebra* of X, regarded as a linear space. The Clifford algebra is then defined as a quotient algebra of the (infinite-dimensional) tensor algebra. For details see, for example, [4].)

Examples of Clifford algebras are easily given for small-dimensional non-degenerate orthogonal spaces. For example, $\mathbf{R}$ itself is a Clifford algebra both for $\mathbf{R}^{0,0}$ and for $\mathbf{R}^{1,0}$, $\mathbf{C}$, regarded as a real algebra, is a Clifford algebra for $\mathbf{R}^{0,1}$, and $\mathbf{H}$, regarded as a real algebra, is a Clifford algebra both for $\mathbf{R}^{0,2}$ and for $\mathbf{R}^{0,3}$, it being usual, in the former case, to identify $\mathbf{R}^{0,2}$ with the linear image in $\mathbf{H}$ of {i,k}, while, in the latter case, $\mathbf{R}^{0,3}$ has necessarily to be identified with the linear image of {i,j,k}, the space of pure quaternions. Moreover, it follows easily from Exercises 9.71 and 9.78 that $\mathbf{R}(2)$ is a Clifford algebra for each of the spaces $\mathbf{R}^{2,0}$, $\mathbf{R}^{1,1}$ and $\mathbf{R}^{2,1}$. It is provocative to arrange these examples in a table as follows:

Table 13.2.

Clifford algebras for $\mathbf{R}^{p,q}$, *for small values of* p *and* q

$p+q$ \ $-p+q$	-4	-3	-2	-1	0	1	2	3	4
0					$\mathbf{R}$				
1				$\mathbf{R}$		$\mathbf{C}$			
2			$\mathbf{R}(2)$		$\mathbf{R}(2)$		$\mathbf{H}$		
3		?		$\mathbf{R}(2)$		?		$\mathbf{H}$	
4	?		?		?		?		?

A complete table of Clifford algebras for the non-degenerate orthogonal spaces $\mathbf{R}^{p,q}$ will be found on page 250. As can be seen from that table, one can always choose as Clifford algebra for such a space the space of endomorphisms of some finite-dimensional linear space over **R**, **C**, **H**, $^2\mathbf{R}$ or $^2\mathbf{H}$, the endomorphism space being regarded as a real algebra.

Later in the chapter we examine in some detail how a Clifford algebra A for an orthogonal space X may be used in the study of the group of orthogonal automorphisms of X. Here we only make two preliminary remarks.

Prop. 13.3. Let $a, b \in X$. Then, in A,

$$a \cdot b = -\tfrac{1}{2}(ab + ba).$$

In particular, a and b are mutually orthogonal if, and only if, a and b anticommute.

Proof

$$\begin{aligned} 2a \cdot b &= a \cdot a + b \cdot b - (a - b) \cdot (a - b) \\ &= -a^2 - b^2 + (a - b)^2 \\ &= -ab - ba. \qquad \square \end{aligned}$$

Prop. 13.4. Let $a \in X$. Then a is invertible in A if, and only if, it is invertible with respect to the scalar product, when $a^{-1} = -a^{(-1)}$.

Proof $\Rightarrow$: Let $b = a^{-1}$, in A. Then $a^{(2)}b = -a^2 b = -a$, implying that $a^{(2)} \neq 0$ and that $b = -a^{(-1)}$).

$\Leftarrow$: Let $b = a^{(-1)} = -(a^{(2)})^{-1}\, a$. Then $ba = -(a^{(2)})^{-1}\, a^2 = 1$. Similarly, $ab = 1$. That is, $b = a^{-1}$. $\square$

Notice that the inverse in A of an element of X is also an element of X.

Orthonormal subsets

One of the characteristic properties of a Clifford algebra may be re-expressed in terms of an orthonormal basis as follows.

Prop. 13.5. Let X be a finite-dimensional real orthogonal space with an orthonormal basis $\{e_i : i \in n\}$, where $n = \dim X$, and let A be a real associative algebra with unity 1 containing **R** and X as linear subspaces. Then $x^2 = -x^{(2)}$, for all $x \in X$, if, and only if,

$$e_i^{\,2} = -e_i^{(2)}, \quad \text{for all } i \in n,$$

and

$$e_i e_j + e_j e_i = 0, \quad \text{for all distinct } i \text{ and } j \in n. \qquad \square$$

This prompts the following definition.

An *orthonormal subset* of a real associative algebra A with unity 1 is a linearly free subset S of mutually anticommuting elements of A, the square a^2 of any element $a \in S$ being 0, 1 or -1.

Prop. 13.6. Let S be a subset of mutually anticommuting elements of the algebra A such that the square a^2 of any element $a \in S$ is 1 or -1. Then S is an orthonormal subset in A.

(All that has to be verified is the linear independence of A.) □

An orthonormal subset S each of whose elements is invertible, as in Prop. 13.6, is said to be *non-degenerate*. If p of the elements of S have square $+1$ and if the remaining q have square -1, then S is said to be of *type* (p,q).

Prop. 13.7. Let X be the linear image of an orthonormal subset S of the real associative algebra A. Then there is a unique orthogonal structure for X such that, for all $a \in S$, $a^{(2)} = -a^2$, and, if S is of type (p,q), X with this structure is isomorphic to $\mathbf{R}^{p,q}$. If S also generates A, then A is a Clifford algebra for the orthogonal space X. □

The dimension of a Clifford algebra

There is an obvious upper bound to the linear dimension of a Clifford algebra for a finite-dimensional orthogonal space.

It is convenient first of all to introduce the following notation. Suppose that $(e_i : i \in n)$ is an n-tuple of elements of an associative algebra A. Then, for each naturally ordered subset I of n, $\prod e_I$ will denote the product $\prod_{i \in I} e_i$, with $\prod e_\emptyset = 1$. In particular $\prod e_n = \prod_{i \in n} e_i$.

Prop. 13.8. Let A be a real associative algebra with unity 1 (identified with $1 \in \mathbf{R}$) and suppose that $(e_i : i \in n)$ is an n-tuple of elements of A generating A such that, for any $i, j \in n$,

$$e_i e_j + e_j e_i \in \mathbf{R}.$$

Then the set $\{\prod e_I : I \subset n\}$ spans A linearly. □

Cor. 13.9. Let A be a Clifford algebra for an n-dimensional orthogonal space X. Then $\dim A \leqslant 2^n$. □

The following theorem gives the complete set of possible values for $\dim A$, when X is non-degenerate.

Theorem 13.10. Let A be a Clifford algebra for an n-dimensional non-degenerate orthogonal space X of signature (p,q). Then $\dim A = 2^n$ or 2^{n-1}, the lower value being a possibility only if $p - q - 1$ is divisible by 4, in which case n is odd and $\prod e_n = +1$ or -1 for any basic orthonormal frame $(e_i : i \in n)$ for X.

Proof Let $(e_i : i \in n)$ be a basic orthonormal frame for X. Then, for each $I \subset n$, $\prod e_I$ is invertible in A and so is non-zero.

To prove that the set $\{\prod e_I : I \subset n\}$ is linearly free, it is enough to prove that if there are real numbers λ_I, for each $I \subset n$, such that $\sum_{I \subset n} \lambda_I(\prod e_I) = 0$, then, for each $J \subset n$, $\lambda_J = 0$. Since, for any $J \subset n$,

$$\sum_{I \subset n} \lambda_I(\prod e_I) = 0 \Leftrightarrow \sum_{I \subset n} \lambda_I(\prod e_I)(\prod e_J)^{-1} = 0,$$

thus making λ_J the coefficient of $e_\emptyset$, it is enough to prove that

$$\sum_{I \subset n} \lambda_I(\prod e_I) = 0 \Rightarrow \lambda_\emptyset = 0.$$

Suppose, therefore, that $\sum_{I \subset n} \lambda_I(\prod e_I) = 0$. We assert that this implies either that $\lambda_\emptyset = 0$, or, if n is odd, that $\lambda_\emptyset + \lambda_n(\prod e_n) = 0$. This is because, for each $i \in n$ and each $I \subset n$, e_i either commutes or anti-commutes with $\prod e_I$. So

$$\sum_{I \subset n} \lambda_I(\prod e_I) = 0 \Rightarrow \sum_{I \subset n} \lambda_I e_i(\prod e_I)e_i^{-1} = \sum_{J \subset n} \zeta_{I,i}\, \lambda_I(\prod e_I) = 0$$

where $\zeta_{I,i} = 1$ or -1 according as e_i commutes or anticommutes with $\prod e_I$. It follows that $\sum_I \lambda_I(\prod e_I) = 0$, where the summation is now over all I such that $\prod e_I$ commutes with e_i. After introducing each e_i in turn, we find eventually that $\sum_I \lambda_I(\prod e_I) = 0$, with the summation over all I such that $\prod e_I$ commutes with *each* e_i. Now there are at most only two such subsets of n, namely $\emptyset$, since $\prod e_\emptyset = 1$, and, when n is odd, n itself. This proves the assertion.

From this it follows that the subset $\{\prod e_I : I \subset n, \#I \text{ even}\}$ is linearly free in A for all n and that the subset $\{\prod e_I : I \subset n\}$ is free for all even n. For n odd, either $\{\prod e_I : I \subset n\}$ is free or $\prod e_n$ is real.

To explore this last possibility further let $n = p + q = 2k + 1$. Then $(\prod e_n)^2 = (\prod e_{2k+1})^2 = (-1)^{k(2k+1)+q}$. But, since $\prod e_n$ is real, $(\prod e_n)^2$ is positive. Therefore $(\prod e_n)^2 = 1$, implying that $\prod e_n = +1$ and that $k(2k + 1) + q$ is divisible by 2, that is, $4k^2 + p + 3q - 1$, or, equivalently, $p - q - 1$, is divisible by 4. Conversely, if $p - q - 1$ is divisible by 4, n is odd.

Finally, if $\prod e_n = \pm 1$, n being odd, then, for each $I \subset n$ with $\#I$ odd, $\prod e_I = \pm \prod e_{n \setminus I}$. Since, as has already been noted, the subset $\{\prod e_I : I \subset n, \#I \text{ even}\}$ is free in A, it follows in this case that $\dim A = 2^{n-1}$.

This completes the proof. □

The lower value for the dimension of a Clifford algebra of a non-degenerate finite-dimensional orthogonal space does occur; for, as has already been noted, **R** is a Clifford algebra for $\mathbf{R}^{1,0}$ and **H** is a Clifford algebra for $\mathbf{R}^{0,3}$,

The following corollary indicates how Theorem 13.10 is used in practice.

Cor. 13.11. Let A be a real associative algebra with an orthonormal subset $\{e_i : i \in n\}$ of type (p,q), where $p + q = n$. Then, if $\dim A = 2^{n-1}$, A is a Clifford algebra for $\mathbf{R}^{p,q}$ while, if $\dim A = 2^n$ and if $\prod e_n \neq \pm 1$, then A is again a Clifford algebra for $\mathbf{R}^{p,q}$, it being necessary to check that $\prod e_n \neq \pm 1$ only when $p - q - 1$ is divisible by 4. □

For example, $\mathbf{R}(2)$ is now seen to be a Clifford algebra for $\mathbf{R}^{2,0}$ simply because $\dim \mathbf{R}(2) = 2^2$ and because the set $\left\{\begin{pmatrix} 0 & 1 \\ 1 & 0 \end{pmatrix}, \begin{pmatrix} 1 & 0 \\ 0 & -1 \end{pmatrix}\right\}$ is an orthonormal subset of $\mathbf{R}(2)$ of type (2,0).

Prop. 13.12. The real algebra ${}^2\mathbf{R}$ is a Clifford algebra for $\mathbf{R}^{1,0}$. □

Universal Clifford algebras

The special role played by a Clifford algebra of dimension 2^n for an n-dimensional real orthogonal space X is brought out by the following theorem.

Theorem 13.13. Let A be a Clifford algebra for an n-dimensional real orthogonal space X, with $\dim A = 2^n$, let B be a Clifford algebra for a real orthogonal space Y, and suppose that $t : X \to Y$ is an orthogonal map. Then there is a unique algebra map $t_A : A \to B$ sending $1_{(A)}$ to $1_{(B)}$ and a unique algebra-reversing map $t_A^{\sim} : A \to B$ sending $1_{(A)}$ to $1_{(B)}$ such that the diagrams

$$\begin{array}{ccc} X & \xrightarrow{t} & Y \\ \downarrow{\scriptstyle \text{inc}} & & \downarrow{\scriptstyle \text{inc}} \\ A & \xrightarrow{t_A} & B \end{array} \quad \text{and} \quad \begin{array}{ccc} X & \xrightarrow{t} & Y \\ \downarrow{\scriptstyle \text{inc}} & & \downarrow{\scriptstyle \text{inc}} \\ A & \xrightarrow{t_A^{\sim}} & B \end{array}$$

commute.

Proof We construct t_A, the construction of $t_A^{\sim}$ being similar.

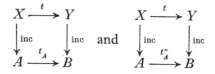

Let $(e_i : i \in n)$ be a basic orthonormal frame for X. Then if t_A exists, $t_A(\prod e_I) = \prod_{i \in I} t(e_i)$, for each non-null $I \subset n$, while $t_A(1_{(A)}) = 1_{(B)}$, by hypothesis. Conversely, since the set $\{e_I : I \subset n\}$ is a basis for A, there is a *unique linear* map $t_A : A \to B$ such that, for each $I \subset n$, $t_A(\prod e_I) = \prod_{i \in I} t(e_i)$. In particular, since, for each $i \in n$, $t_A(e_i) = t(e_i)$, the diagram

$$\begin{array}{ccc} X & \xrightarrow{t} & Y \\ \downarrow{\scriptstyle \text{inc}} & & \downarrow{\scriptstyle \text{inc}} \\ A & \xrightarrow{t_A} & B \end{array}$$

commutes. It only remains to check that t_A respects products, and for this it is enough to check that, for any $I, J \subset n$,

$$t_A((\prod e_I)(\prod e_J)) = t_A(\prod e_I)t_A(\prod e_J).$$

The verification is straightforward, if slightly tedious and depends on the fact that, since t is orthogonal, $(t(e_i))^2 = e_i^2$, for any $i \in n$, and $t(e_j)t(e_i) = -t(e_i)t(e_j)$, for any distinct $i, j \in n$. The final details are left as an exercise. □

The uniqueness is useful in several ways. For example, suppose that $Y = X$, $B = A$ and $t = 1_X$. Then $t_A = 1_A$, since the diagram

$$\begin{array}{ccc} X & \xrightarrow{1_X} & X \\ \downarrow & & \downarrow \\ A & \xrightarrow{1_A} & A \end{array}$$

commutes.

Theorem 13.13 is amplified and extended in Theorem 13.31 in the particular case that $Y = X$ and $B = A$. Immediate corollaries of 13.13 include the following.

Cor. 13.14. Let A and B be 2^n-dimensional Clifford algebras for an n-dimensional real orthogonal space X. Then $A \cong B$.

Proof Theorem 13.13 applied to the identity map 1_X in four different ways produces the commutative prisms

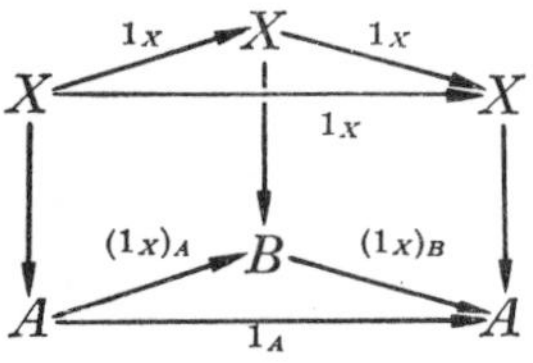

and

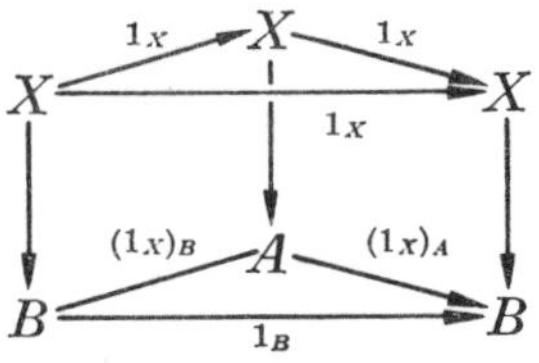

These show that $(1_X)_A : A \to B$ is an algebra isomorphism (with inverse $(1_X)_B$). □

Cor. 13.15. Any Clifford algebra B for an n-dimensional orthogonal space X is isomorphic to some quotient of any given 2^n-dimensional Clifford algebra A for X.

(What remains to be proved is that the map $(1_X)_A : A \to B$ is a *surjective* algebra map.) □

A 2^n-dimensional real Clifford algebra for an n-dimensional orthogonal space X is said to be a *universal* real Clifford algebra for X. Since any two universal Clifford algebras for X are isomorphic, and since the isomorphism between them is essentially unique, one often speaks loosely of *the* universal Clifford algebra for X. The existence of such an algebra for any X has, of course, still to be proved.

It will be convenient to denote the universal real Clifford algebra for the orthogonal space $\mathbf{R}^{p,q}$ by the symbol $\mathbf{R}_{p,q}$.

Construction of the algebras

Corollary 13.11 may now be applied to the construction of universal Clifford algebras for each non-degenerate orthogonal space $\mathbf{R}^{p,q}$. The following elementary proposition is used frequently.

Prop. 13.16. Let a and b be elements of an associative algebra A with unity 1. Then, if a and b commute, $(ab)^2 = a^2b^2$, so that, in particular,

$$a^2 = b^2 = -1 \Rightarrow (ab)^2 = 1,$$
$$a^2 = -1 \text{ and } b^2 = 1 \Rightarrow (ab)^2 = -1,$$
and
$$a^2 = b^2 = 1 \Rightarrow (ab)^2 = 1,$$

while, if a and b anticommute, $(ab)^2 = -a^2b^2$, and

$$a^2 = b^2 = -1 \Rightarrow (ab)^2 = -1,$$
$$a^2 = -1 \text{ and } b^2 = 1 \Rightarrow (ab)^2 = 1,$$
and
$$a^2 = b^2 = 1 \Rightarrow (ab)^2 = -1. \quad \square$$

The first stage in the construction is to show how to construct the universal Clifford algebra $\mathbf{R}_{p+1,q+1}$ for $\mathbf{R}^{p+1,q+1}$, given $\mathbf{R}_{p,q}$, the universal Clifford algebra for $\mathbf{R}^{p,q}$. This leads directly to the existence theorem.

Prop. 13.17. Let X be an $\mathbf{A}$-linear space, where $\mathbf{A} = \mathbf{K}$ or ${}^2\mathbf{K}$ and $\mathbf{K} = \mathbf{R}$, $\mathbf{C}$ or $\mathbf{H}$, and let S be an orthonormal subset of End X of type (p,q), generating End X as a real algebra. Then the set of matrices

$$\left\{\begin{pmatrix} a & 0 \\ 0 & -a \end{pmatrix} : a \in S\right\} \cup \left\{\begin{pmatrix} 0 & 1 \\ 1 & 0 \end{pmatrix}, \begin{pmatrix} 0 & -1 \\ 1 & 0 \end{pmatrix}\right\}$$

is an orthonormal subset of End X^2 of type $(p+1, q+1)$, generating End X^2 as a real algebra. $\square$

Cor. 13.18. For each finite n, the endomorphism algebra $\mathbf{R}(2^n)$ is a universal Clifford algebra for the neutral non-degenerate space $\mathbf{R}^{n,n}$. That is, $\mathbf{R}_{n,n} \cong \mathbf{R}(2^n)$.

Proof By induction. The basis is that $\mathbf{R}$ is a universal Clifford algebra for $\mathbf{R}^{0,0}$, and the step is Prop. 13.17. $\square$

Theorem 13.19. (*Existence theorem.*)
Every finite-dimensional orthogonal space has a universal Clifford algebra.

Proof This follows at once from the remarks following Prop. 13.1, from Prop. 13.8 and from Cor. 13.18. □

Prop. 13.20. Let S be an orthonormal subset of type $(p+1,q)$ generating an associative algebra A. Then, for any $a \in S$ with $a^2 = 1$, the set

$$\{ba : b \in S \backslash \{a\}\} \cup \{a\}$$

is an orthonormal subset of type $(q+1,p)$ generating A. □

Cor. 13.21. The universal Clifford algebras $\mathbf{R}_{p+1,q}$ and $\mathbf{R}_{q+1,p}$ are isomorphic. □

The next proposition completes the construction of the algebras $\mathbf{R}_{p,q}$, for $p + q \leqslant 4$.

Prop. 13.22. For $q \in 5$, $\mathbf{R}_{0,q}$ is isomorphic, respectively, to $\mathbf{R}$, $\mathbf{C}$, $\mathbf{H}$, ${}^2\mathbf{H}$, or $\mathbf{H}(2)$.

Proof By Cor. 13.11 it is enough, in each case, to exhibit an orthonormal subset of the appropriate type with the product of its members, in any order, not equal to 1 or -1, for each algebra has the correct real dimension, namely 2^p. Appropriate orthonormal subsets are

$$\emptyset \quad \text{for} \quad \mathbf{R}$$
$$\{\mathrm{i}\} \quad \text{for} \quad \mathbf{C}$$
$$\{\mathrm{i}, \mathrm{k}\} \quad \text{for} \quad \mathbf{H}$$
$$\left\{\begin{pmatrix} \mathrm{i} & 0 \\ 0 & -\mathrm{i} \end{pmatrix}, \begin{pmatrix} \mathrm{j} & 0 \\ 0 & -\mathrm{j} \end{pmatrix}, \begin{pmatrix} \mathrm{k} & 0 \\ 0 & -\mathrm{k} \end{pmatrix}\right\} \quad \text{for} \quad {}^2\mathbf{H}$$

and $$\left\{\begin{pmatrix} \mathrm{i} & 0 \\ 0 & -\mathrm{i} \end{pmatrix}, \begin{pmatrix} \mathrm{j} & 0 \\ 0 & -\mathrm{j} \end{pmatrix}, \begin{pmatrix} \mathrm{k} & 0 \\ 0 & -\mathrm{k} \end{pmatrix}, \begin{pmatrix} 0 & -1 \\ 1 & 0 \end{pmatrix}\right\} \quad \text{for} \quad \mathbf{H}(2).$$ □

Finally, a more sophisticated result incorporating the 'periodicity theorem'.

Prop. 13.23. Let $S = \{e_i : i \in 4\}$ be an orthonormal subset of type (0,4) of an associative algebra A with unity 1 and let R be an orthonormal subset of type (p,q) of A such that each element of S anticommutes with every element of R. Then there exists an orthonormal subset R' of type (p,q) such that each element of S commutes with every element of R'. Conversely, the existence of R' implies the existence of R.

Proof Let $a = e_0e_1e_2e_3$ and let $R' = \{ab : b \in R\}$. Since a commutes with every element of R and anticommutes with every element of S and since $a^2 = 1$, it follows at once that R' is of the required form. The converse is similarly proved. □

Cor. 13.24. For all finite p, q,

$$\mathbf{R}_{p,q+4} \cong \mathbf{R}_{p,q} \otimes \mathbf{R}_{0,4} \cong \mathbf{R}_{p,q} \otimes \mathbf{H}(2). \qquad \square$$

(The notation $\otimes$, as used in this context, was introduced on page 188.)

For example, by Prop. 10.44,

$$\mathbf{R}_{0,5} \cong \mathbf{C} \otimes \mathbf{H}(2) \cong \mathbf{C}(4),$$
$$\mathbf{R}_{0,6} \cong \mathbf{H} \otimes \mathbf{H}(2) \cong \mathbf{R}(8),$$
$$\mathbf{R}_{0,7} \cong {}^2\mathbf{H} \otimes \mathbf{H}(2) \cong {}^2\mathbf{R}(8),$$

and

$$\mathbf{R}_{0,8} \cong \mathbf{H}(2) \otimes \mathbf{H}(2) \cong \mathbf{R}(16).$$

Cor. 13.25. (The *periodicity theorem*.)

For all finite p, q,

$$\mathbf{R}_{p,q+8} \cong \mathbf{R}_{p,q} \otimes \mathbf{R}(16) \cong \mathbf{R}_{p,q}(16). \qquad \square$$

By putting together Prop. 13.22, Prop. 13.12, Prop. 13.17, Prop. 13.20, and these last two corollaries, we can construct any $\mathbf{R}_{p,q}$. Table 13.26 shows them all, for $p + q \leqslant 8$. The vertical pattern is derived from Prop. 13.17, and the horizontal symmetry about the line with equation $-p + q = -1$ is derived from Prop. 13.20.

Squares like those in the table have already made a brief appearance at the end of Chapter 10. There are clearly (non-unique) algebra injections $\mathbf{R}_{p,q} \to \mathbf{R}_{p+1,q}$ and $\mathbf{R}_{p,q} \to \mathbf{R}_{p,q+1}$, for any p, q, such that the squares commute.

Table 13.26 exhibits each of the universal Clifford algebras $\mathbf{R}_{p,q}$ as the real algebra of endomorphisms of a right $\mathbf{A}$-linear space V of the form $\mathbf{A}^m$, where $\mathbf{A} = \mathbf{R}$, $\mathbf{C}$, $\mathbf{H}$, ${}^2\mathbf{R}$ or ${}^2\mathbf{H}$. This space is called the (*real*) *spinor space* or *space of* (*real*) *spinors* of the orthogonal space $\mathbf{R}^{p,q}$.

Prop. 13.27. Let $\mathbf{R}_{p,q} = \mathbf{A}(m)$, according to Table 13.26, or its extension by Cor. 13.25. Then the representative in $\mathbf{A}(m)$ of any element of the standard orthonormal basis for $\mathbf{R}^{p+q}$ is orthogonal with respect to the standard positive-definite correlation on $\mathbf{A}^m$.

Proof This follows from Prop. 10.42, and the remarks following Prop. 10.44 and Cor. 11.57, and its truth, readily checked, for small values of p and q. □

When $\mathbf{K}$ is a double field (${}^2\mathbf{R}$ or ${}^2\mathbf{H}$), the $\mathbf{K}$-linear spaces $V(1,0)$ and $V(0,1)$ are called the (*real*) *half-spinor spaces* or *spaces of* (*real*)

Table 13.26.

The algebras $\mathbf{R}_{p,q}$, for $p + q \leqslant 8$

$p+q$ \ $-p+q$	-8	-7	-6	-5	-4	-3	-2	-1	0	1	2	3	4	5	6	7	8
0									$\mathbf{R}$								
1								${}^2\mathbf{R}$		$\mathbf{C}$							
2							$\mathbf{R}(2)$		$\mathbf{R}(2)$		$\mathbf{H}$						
3						$\mathbf{C}(2)$		${}^2\mathbf{R}(2)$		$\mathbf{C}(2)$		${}^2\mathbf{H}$					
4					$\mathbf{H}(2)$		$\mathbf{R}(4)$		$\mathbf{R}(4)$		$\mathbf{H}(2)$		$\mathbf{H}(2)$				
5				${}^2\mathbf{H}(2)$		$\mathbf{C}(4)$		${}^2\mathbf{R}(4)$		$\mathbf{C}(4)$		${}^2\mathbf{H}(2)$		$\mathbf{C}(4)$			
6			$\mathbf{H}(4)$		$\mathbf{H}(4)$		$\mathbf{R}(8)$		$\mathbf{R}(8)$		$\mathbf{H}(4)$		$\mathbf{H}(4)$		$\mathbf{R}(8)$		
7		$\mathbf{C}(8)$		${}^2\mathbf{H}(4)$		$\mathbf{C}(8)$		${}^2\mathbf{R}(8)$		$\mathbf{C}(8)$		${}^2\mathbf{H}(4)$		$\mathbf{C}(8)$		${}^2\mathbf{R}(8)$	
8	$\mathbf{R}(16)$		$\mathbf{H}(8)$		$\mathbf{H}(8)$		$\mathbf{R}(16)$		$\mathbf{R}(16)$		$\mathbf{H}(8)$		$\mathbf{H}(8)$		$\mathbf{R}(16)$		$\mathbf{R}(16)$

half-spinors, the endomorphism algebra of either being a non-universal Clifford algebra of the appropriate orthogonal space.

Complex Clifford algebras

The real field may be replaced throughout the above discussion by any commutative field—in particular by the field **C**. The notation $\mathbf{C}_n$ will denote the universal complex Clifford algebra for $\mathbf{C}^n$ unique up to isomorphism.

Prop. 13.28. For any $n, p, q \in \omega$ with $n = p + q$, $\mathbf{C}_n \cong \mathbf{R}_{p,q} \otimes_{\mathbf{R}} \mathbf{C}$, $\cong$ denoting a real algebra isomorphism. □

Cor. 13.29. For any $k \in \omega$, $\mathbf{C}_{2k} \cong \mathbf{C}(2^k)$ and $\mathbf{C}_{2k+1} \cong {}^2\mathbf{C}(2^k)$. □

The *complex spinor* and *half-spinor spaces* are defined analogously to their real counterparts.

Involuted fields

A further generalization of the concept of a Clifford algebra involves the concept of an involuted field. An *involuted field*, $\mathbf{L}^\alpha$, with fixed field **K**, consists of a commutative **K**-algebra **L** with unity 1 over a commutative field **K** and an involution α of **L**, whose set of fixed points is the set of scalar multiples of 1, identified as usual with **K**. (The algebra **L** need not be a field.) Examples include **R**, $\bar{\mathbf{C}}$ and hb **R**, each with fixed field **R**, and **C** and hb **C**, each with fixed field **C**.

Let X be a finite-dimensional orthogonal space over a commutative field **K**, let $\mathbf{L}^\alpha$ be an involuted field with fixed field **K** and let A be an associative **L**-algebra with unity, the algebra **L** being identified with the subalgebra generated by unity. Then A is said to be an *$\mathbf{L}^\alpha$-Clifford algebra* for X if it contains X as a **K**-linear subspace in such a way that, for all $x \in X$, $x^2 = -x^{(2)}$, provided also that A is generated as a ring by **L** and X or, equivalently, as an **L**-algebra by 1 and X.

All that has been said before about real Clifford algebras generalizes to $\mathbf{L}^\alpha$-Clifford algebras also. The notations $\bar{\mathbf{C}}_{p,q}$ and $(\text{hb}\,\mathbf{R})_{p,q}$ will denote the universal $\bar{\mathbf{C}}$- and hb **R**-Clifford algebras for $\mathbf{R}^{p,q}$, and the notation $(\text{hb}\,\mathbf{C})_n$ the universal hb **C**-Clifford algebra for $\mathbf{C}^n$, for any finite p, q, n.

Prop. 13.30. Let $\mathbf{L}^a$ be an involuted field with fixed field **K**, let X be a **K**-orthogonal space and let A and B be universal **K**- and $\mathbf{L}^\alpha$-Clifford algebras, respectively, for X. Then, as **K**- algebras, $B = A \otimes_{\mathbf{K}} \mathbf{L}$. □

Note that, as complex algebras, $\bar{\mathbf{C}}_{p,q}$ and $\mathbf{C}_n$ are isomorphic, for any finite n, p, q such that $n = p + q$. The detailed construction of the various tables of $\mathbf{L}^\alpha$-Clifford algebras is left to the reader.

Involutions and anti-involutions

The following theorem amplifies and extends Theorem 13.13 in various ways, in the particular case that $Y = X$ and $B = A$.

Theorem 13.31. Let A be a universal $\mathbf{L}^\alpha$-Clifford algebra for a finite-dimensional $\mathbf{K}$-orthogonal space X, $\mathbf{L}^\alpha$ being an involuted field with involution α and fixed field $\mathbf{K}$. Then, for any orthogonal automorphism $t : X \to X$, there is a unique $\mathbf{L}$-algebra automorphism $t_A : A \to A$, sending any $\lambda \in \mathbf{L}$ to λ, and a unique $\mathbf{K}$-algebra anti-automorphism $t_A^\sim : A \to A$, sending any λ to λ^α, such that the diagrams

$$\begin{array}{ccc} X & \xrightarrow{t} & X \\ \Big\downarrow{\scriptstyle \text{inc}} & & \Big\downarrow{\scriptstyle \text{inc}} \\ A & \xrightarrow{t_A} & A \end{array} \qquad \text{and} \qquad \begin{array}{ccc} X & \xrightarrow{t} & X \\ \Big\downarrow{\scriptstyle \text{inc}} & & \Big\downarrow{\scriptstyle \text{inc}} \\ A & \xrightarrow{t_A^\sim} & A \end{array}$$

commute. Moreover, $(1_X)_A = 1_A$ and, for any $t, u \in O(X)$,

$$(u\,t)_A = u_A\,t_A = u_A^\sim\,t_A^\sim.$$

If t is an orthogonal involution of X, then t_A is an algebra involution of A and $t_A^\sim$ is an algebra anti-involution of A. □

The involution of A induced by the orthogonal involution -1_X will be denoted by $a \rightsquigarrow \hat{a}$ and called the *main involution* of A, $\hat{a}$ being called the *involute* of a.

The anti-involutions of A induced by the orthogonal involutions 1_X and -1_X will be denoted by $a \rightsquigarrow a^\sim$ and $a \rightsquigarrow a^-$ and called, respectively, *reversion* and *conjugation*, $a^\sim$ being called the *reverse* of a and a^- the *conjugate* of a. (The reason for preferring $a^\sim$ to $\tilde{a}$ and a^- to $\bar{a}$ will become apparent on page 266.) Reversion takes its name from the fact that the reverse of a product of a finite number of elements of X is just their product in the reverse order.

For example, consider $a = 1 + e_0 + e_1e_2 + e_0e_1e_2 \in \mathbf{R}_{0,3}$.

$$\begin{aligned} \text{Then}\quad a &= 1 - e_0 + e_1e_2 - e_0e_1e_2, \\ a^\sim &= 1 + e_0 + e_2e_1 + e_2e_1e_0 = 1 + e_0 - e_1e_2 - e_0e_1e_2, \\ \text{while}\quad a^- &= 1 - e_0 + e_2e_1 - e_2e_1e_0 = 1 - e_0 - e_1e_2 + e_0e_1e_2. \end{aligned}$$

Prop. 13.32. Let A be a universal $\mathbf{L}^\alpha$-Clifford algebra for a finite-dimensional $\mathbf{K}$-orthogonal space X, $\mathbf{L}^\alpha$ being an involuted field with fixed field $\mathbf{K}$. Then, for any $a \in A$, $a^- = (\hat{a})^\sim = \widehat{(a^\sim)}$.

Proof Each of the anti-involutions $a \rightsquigarrow a^-$, $(\hat{a})^\sim$ and $\widehat{(a^\sim)}$ is the unique anti-involution of A induced by -1_X. □

The main involution induces, by Prop. 8.2, a direct sum decomposition $A^0 \oplus A^1$ of A, where

$$A^0 = \{a \in A : \hat{a} = a\} \quad \text{and} \quad A^1 = \{a \in A : \hat{a} = -a\}.$$

Clearly A^0 is an **L**-subalgebra of A. This subalgebra is called the *even Clifford algebra* for X. It is unique up to isomorphism. Any element $a \in A$ may be uniquely expressed as the sum of its *even part* $a^0 \in A^0$ and its *odd part* $a^1 \in A^1$. In the example above,

$$a^0 = 1 + e_1e_2 \quad \text{and} \quad a^1 = e_0 + e_0e_1e_2.$$

The even Clifford algebras for the non-degenerate real or complex finite-dimensional orthogonal spaces are determined by the next proposition.

Prop. 13.33. Let A be a universal $\mathbf{L}^\alpha$-Clifford algebra for a non-degenerate finite-dimensional **K**-orthogonal space X, $\mathbf{L}^\alpha$ being an involuted field with fixed field **K**, and let S be an orthonormal basis for X of type (p,q). Then, for any $a \in S$, the set $\{ab : b \in S \setminus \{a\}\}$ is an orthonormal subset of A^0 generating A^0, and of type $(p\,,q-1)$ or $(q\,,p-1)$, according as $a^2 = -1$ or 1. In either case, moreover, the induced isomorphism of A^0 with the universal $\mathbf{L}^\alpha$-Clifford algebra of a $(p + q - 1)$-dimensional orthogonal space respects conjugation, but not reversion.

Proof The first part is clear, by Prop. 13.16. For the last part it is enough to consider generators and to remark that if a and b are anti-commuting elements of an algebra sent to $-a$ and $-b$, respectively, by an anti-involution of the algebra, then, again by Prop. 13.16, ab is sent to $-ab$. On the other hand, if a and b are sent to a and b, respectively, by the anti-involution, then ab is not sent to ab. □

Cor. 13.34. For any finite p, q, n,

$$\mathbf{R}_{p,q+1}^{\ 0} \cong \mathbf{R}_{p,q}, \qquad \mathbf{R}_{p+1,q}^{\ 0} \cong \mathbf{R}_{q,p},$$
$$\bar{\mathbf{C}}_{p,q+1}^{\ 0} \cong \bar{\mathbf{C}}_{p,q}, \qquad \bar{\mathbf{C}}_{p+1,q}^{\ 0} \cong \bar{\mathbf{C}}_{q,p},$$
$$(\text{hb}\,\mathbf{R})_{p,q+1}^{\ 0} \cong (\text{hb}\,\mathbf{R})_{p,q}, \quad (\text{hb}\,\mathbf{R})_{p+1,q}^{\ 0} \cong (\text{hb}\,\mathbf{R})_{q,p},$$
$$\mathbf{C}_{n+1}^{\ 0} \cong \mathbf{C}_n \quad \text{and} \quad (\text{hb}\,\mathbf{C})_{n+1}^{\ 0} \cong (\text{hb}\,\mathbf{C})_n. \qquad \square$$

It follows from Cor. 13.34, in particular, that the table of the even Clifford algebras $\mathbf{R}_{p,q}^{\ 0}$, with $p + q > 0$, is, apart from relabelling, the same as the table of the Clifford algebras $\mathbf{R}_{p,q}$, except that there is an additional line of entries down the left-hand side matching the existing line of entries down the right-hand side. The symmetry about the central vertical line in the table of even Clifford algebras expresses the fact that the even Clifford algebras of a finite-dimensional non-degenerate orthogonal space and of its negative are mutually isomorphic.

So far we have considered only the universal Clifford algebras. The usefulness of the non-universal Clifford algebras is limited by the following proposition.

Prop. 13.35. Let A be a non-universal Clifford algebra for a non-degenerate finite-dimensional orthogonal space X. Then either 1_X or -1_X induces an anti-involution of A, but not both. □

If 1_X induces an anti-involution of A, we say that A is a non-universal Clifford algebra *with reversion* for X, and if -1_X induces an anti-involution, we say that A is a non-universal Clifford algebra *with conjugation* for X.

Prop. 13.36. The non-universal Clifford algebras for the orthogonal spaces $\mathbf{R}^{0,4k+3}$ have conjugation, but not reversion. □

The Clifford group

We turn to applications of the Clifford algebras to groups of orthogonal automorphisms and to the rotation groups in particular. The letter X will denote a finite-dimensional real orthogonal space and A will normally denote a universal real Clifford algebra for X—we shall make some remarks at the end about the case where A is non-universal. For each $x \in X$, $x^{(2)} = x^- x = \hat{x}\, x = -x^2$. Also, since A is universal, $\mathbf{R} \cap X = \{0\}$. The subspace $\mathbf{R} \oplus X$ will be denoted by Y and the letter y will be reserved as a notation for a point of Y. The space Y will be assigned the quadratic form

$$Y \to \mathbf{R}; \quad y \rightsquigarrow y^- y.$$

It is then the orthogonal direct sum of the orthogonal spaces $\mathbf{R}$ and X. If $X \cong \mathbf{R}^{p,q}$, then $Y \cong \mathbf{R}^{p,q+1}$.

The first proposition singles out a certain subset of A that turns out to be a subgroup of A.

Prop. 13.37. Let g be an invertible element of A such that, for all $x \in X$, $g\, x\, \hat{g}^{-1} \in X$. Then the map

$$\rho_{X,g} : X \to X; \quad x \rightsquigarrow g\, x\, \hat{g}^{-1}$$

is an orthogonal automorphism of X.

Proof For each $x \in X$,

$$(\rho_{X,g}(x))^{(2)} = \overline{g\,x\,\hat{g}^{-1}}\,g\,x\,\hat{g}^{-1} = \hat{g}\,\bar{x}\,g^{-1}\,g\,x\,g^{-1} = \bar{x}\,x = x^{(2)},$$

since $\bar{x}\,x \in \mathbf{R}$. So $\rho_{X,g}$ is an orthogonal map. Moreover, it is injective since $g\,x\,\hat{g}^{-1} = 0 \Rightarrow x = 0$ (this does not follow from the orthogonality of $\rho_{X,g}$ if X is degenerate). Finally, since X is finite-dimensional, $\rho_{X,g}$ must also be surjective. □

The element g will be said to *induce* or *represent* the orthogonal transformation $\rho_{X,g}$ and the set of all such elements g will be denoted by $\Gamma\,(X)$ or simply by Γ.

Prop. 13.38. The subset Γ is a subgroup of A.

Proof The closure of Γ under multiplication is obvious. That Γ is also closed with respect to inversion follows from the remark that, for any $g \in \Gamma$, the inverse of $\rho_{X,g}$, is $\rho_{X,g^{-1}}$. Of course $1_{(A)} \in \Gamma$. So Γ is a group. □

The group Γ is called the *Clifford group* for X in the Clifford algebra A. Since the universal algebra A is uniquely defined up to isomorphism, Γ is also uniquely defined up to isomorphism.

Prop. 13.39. $\mathbf{R}^+ = \{\lambda \in \mathbf{R} : \lambda > 0\}$ and $\mathbf{R}^* = \{\lambda \in \mathbf{R} : \lambda \neq 0\}$ are normal subgroups of Γ. □

By analogy with the notations for Grassmannians in Chapter 8, the quotient groups $\Gamma/\mathbf{R}^+$ and $\Gamma/\mathbf{R}^*$ may conveniently be denoted by $\mathscr{G}_1{}^+(\Gamma)$ and $\mathscr{G}_1(\Gamma)$, respectively. The group $\mathscr{G}_1{}^+(\Gamma)$ is also called Pin (X) for a comical reason which will be hinted at later, while the group $\mathscr{G}_1(\Gamma)$ is called the *projective Clifford group*.

Following the same analogy, the image of an element g of Γ in $\mathscr{G}_1{}^+(\Gamma)$ will be denoted by $\text{im}^+\{g\}$, while its image in $\mathscr{G}_1(\Gamma)$ will be denoted by $\text{im}\{g\}$.

There are similar propositions concerning the action of A on Y.

Prop. 13.40. Let g be an invertible element of A such that, for all $y \in Y$, $g\,y\,\hat{g}^{-1} \in Y$, Then the map

$$\rho'_{Y,g} : Y \rightarrow Y; \quad y \rightsquigarrow g\,y\,\hat{g}^{-1}$$

is an orthogonal automorphism of Y. □

Prop. 13.41. The subset $\Omega = \{g \in A : y \in Y \Rightarrow g\,y\,\hat{g}^{-1} \in Y\}$ is a subgroup of A. □

From now on we suppose that X is *non-degenerate*, and prove that in this case every orthogonal automorphism of X is represented by an element of Γ. Recall that, by Theorem 9.41, every orthogonal automorphism of X is the composite of a finite number of hyperplane reflections.

Prop. 13.42. Let a be an invertible element of X. Then $a \in \Gamma$, and the map $\rho_{X,a}$ is a reflection in the hyperplane $(\operatorname{im}\{a\})^{\perp}$.

Proof By Prop. 9.24, $X = \operatorname{im}\{a\} \oplus (\operatorname{im}\{a\})^{\perp}$, so any point of X is of the form $\lambda a + b$, where $\lambda \in \mathbf{R}$ and $b \cdot a = 0$. By Prop. 13.3, $ba = -ab$. Therefore, since $\hat{a} = -a$.

$$\rho_{X,a}(\lambda a + b) = -a(\lambda a + b)a^{-1} = -\lambda a + b.$$

Hence the result. □

Prop. 13.43. Let $a \in A$ be such that $ax = x\hat{a}$, for all $x \in X$, A being a universal Clifford algebra for X. Then $a \in \mathbf{R}$.

Proof Let $a = a^0 + a^1$, where $a^0 \in A^0$ and $a^1 \in A^1$. Then, since $ax = x\hat{a}$,

$$a^0 x = x a^0 \quad \text{and} \quad a^1 x = -x a^1$$

for all $x \in X$, in particular for each element e_i of some orthonormal basis $\{e_i : i \in n\}$ for X.

Now, by an argument used in the proof of Theorem 13.10, a^0 commutes with each e_i if, and only if, $a^0 \in \mathbf{R}$, and by a similar argument a^1 anticommutes with each e_i if, and only if, $a^1 = 0$. So $a \in \mathbf{R}$. □

Theorem 13.44. The map

$$\rho_X : \Gamma \longrightarrow O(X); \quad g \rightsquigarrow \rho_{X,g}$$

is a surjective group map with coimage the projective Clifford group $\mathscr{G}_1(\Gamma)$. That is, $\mathscr{G}_1(\Gamma)$ and $O(X)$ are isomorphic.

Proof To prove that ρ_X is a group map, let $g, g' \in \Gamma$. Then for all $x \in X$,

$$\begin{aligned}\rho_{X,gg'}(x) &= gg'\, x\, (\widehat{gg'})^{-1} \\ &= gg'\, x\, \hat{g}'^{-1}\, \hat{g}^{-1} \\ &= \rho_{Xg,}\rho_{X,g'}.\end{aligned}$$

So $\rho_{X,gg'} = \rho_{X,g}\rho_{X,g'}$, which is what had to be proved.

The surjectivity of ρ_X is an immediate corollary of Theorem 9.41 and Prop. 13.42.

Finally, suppose that $\rho_{X,g} = \rho_{X,g'}$, for $g, g' \in \Gamma$. Then, for all $x \in X$, $g\, x\, \hat{g}^{-1} = g'\, x\, \hat{g}'^{-1}$, implying that $(g^{-1} g')x = x\, \widehat{g^{-1} g'}$, and therefore that $g^{-1} g' \in \mathbf{R}$, by Prop. 13.43. Moreover, $g^{-1} g'$ is invertible and is therefore non-zero. So coim $\rho_X = \mathscr{G}_1(\Gamma)$. □

An element g of Γ represents a rotation of X if, and only if, g is the product of an even number of elements of X. The set of such elements will be denoted by Γ^0. An element g of Γ represents an antirotation of X if, and only if, g is the product of an odd number of elements of X. The set of such elements will be denoted by Γ^1. Clearly, $\Gamma^0 = \Gamma \cap A^0$ and $\Gamma^1 = \Gamma \cap A^1$.

Prop. 13.45. Let X be a non-degenerate orthogonal space of positive finite dimension. Then Γ^0 is a normal subgroup of Γ, with $\Gamma/\Gamma^0 \cong \mathbf{Z}_2$. □

Since, for any $a \in A^0$, $\hat{a} = a$, the rotation induced by an element g of Γ^0 is of the form

$$X \to X; \quad g \rightsquigarrow g\,x\,g^{-1}.$$

Similarly since, for any $a \in A^1$, $\hat{a} = -a$, the rotation induced by an element g of Γ^1 is of the form

$$X \to X; \quad g \rightsquigarrow -g\,x\,g^{-1}.$$

The quotient groups $\Gamma^0/\mathbf{R}^+$ and $\Gamma^0/\mathbf{R}^*$ will be denoted by $\mathscr{G}_1{}^+(\Gamma^0)$ and $\mathscr{G}_1(\Gamma^0)$ respectively. The group $\mathscr{G}_1{}^+(\Gamma^0)$ is also called Spin X, this name being somewhat older than the name Pin X for $\mathscr{G}_1{}^+(\Gamma)$! The use of the word 'spin' in this context is derived from certain quantum-mechanical applications of the Spin groups. The group $\mathscr{G}_1(\Gamma^0)$ is called the *even* projective Clifford group.

Prop. 13.46. The map $\Gamma^0 \to SO(X)$; $g \rightsquigarrow \rho_{X,g}$ is a surjective group map with coimage $\mathscr{G}_1(\Gamma^0)$. That is, $\mathscr{G}_1(\Gamma^0)$ and $SO(X)$ are isomorphic. □

Prop. 13.47. The groups $\mathscr{G}_1{}^+(\Gamma^0)$ and $\mathscr{G}_1(\Gamma^0)$ are normal subgroups of $\mathscr{G}_1{}^+(\Gamma)$ and $\mathscr{G}_1(\Gamma)$, respectively, the quotient group in either case being isomorphic to $\mathbf{Z}_2$, if $\dim X > 0$. □

Prop. 13.48. Let X be a non-degenerate orthogonal space of positive finite dimension. Then the maps

$$\text{Pin } X \to O(X); \quad \text{im}^+\{g\} \rightsquigarrow \rho_{X,g}$$

and

$$\text{Spin } X \to SO(X); \quad \text{im}^+\{g\} \rightsquigarrow \rho_{X,g}$$

are surjective, the kernel in each case being isomorphic to $\mathbf{Z}_2$. □

When $X = \mathbf{R}^{p,q}$, the standard notations for Γ, Γ^0, Pin X and Spin X will be $\Gamma(p,q)$, $\Gamma^0(p,q)$, Pin (p,q) and Spin (p,q). Since $\mathbf{R}_{q,p}^{\,0} \cong \mathbf{R}_{p,q}^{\,0}$, $\Gamma^0(q,p) \cong \Gamma^0(p,q)$ and Spin $(q,p) \cong$ Spin (p,q). Finally, $\Gamma^0(0,n)$ is often abbreviated to $\Gamma^0(n)$ and Spin $(0,n)$ to Spin (n).

An analogous discussion to that just given for the group Γ can be given for the subgroup Ω of the Clifford algebra A consisting of those invertible elements g of A such that, for all $y \in Y$, $g\,y\,\hat{g}^{-1} \in Y$. However, the properties of this group are deducible directly from the preceding discussion, by virtue of the following proposition.

The notations are as follows. As before, X will denote an n-dimensional non-degenerate real orthogonal space, of signature (p,q), say. This can be considered as the subspace of $\mathbf{R}^{p,q+1}$ consisting of those elements of $\mathbf{R}^{p,q+1}$ whose last co-ordinate, labelled the nth, is zero. The subalgebra $\mathbf{R}_{p,q}$ of $\mathbf{R}_{p,q+1}$ generated by X is a universal Clifford algebra for X, as also is the even Clifford algebra $\mathbf{R}_{p,q+1}^{\,0}$, by the linear injection

$$X \longrightarrow \mathbf{R}_{p,q+1}^{\,0}; \quad x \rightsquigarrow x\,e_n.$$

(Cf. Prop. 13.33.) The linear space $Y = \mathbf{R} \oplus X$ is assigned the quadratic form $y \rightsquigarrow y^- y$.

Prop. 13.49. Let $\theta : \mathbf{R}_{p,q} \longrightarrow \mathbf{R}_{p,q+1}^{\,0}$ be the isomorphism of universal Clifford algebras induced, according to Theorem 13.13, by 1_X. Then

(i) the map

$$u : Y \longrightarrow \mathbf{R}^{p,q+1}; \quad y \rightsquigarrow \theta(y)e_n^{-1}$$

is an orthogonal isomorphism,

(ii) for any $g \in \Omega$, $\theta(g) \in \Gamma^0(p,q+1)$ and the diagram

$$\begin{array}{ccc} Y & \xrightarrow{\rho'_g} & Y \\ \downarrow u & & \downarrow u \\ \mathbf{R}^{p,q+1} & \xrightarrow{\rho_{\theta(g)}} & \mathbf{R}^{p,q+1} \end{array}$$

commutes,

(iii) the map $\Omega \longrightarrow \Gamma^0(p,q+1);\ g \rightsquigarrow \theta(g)$ is a group isomorphism.

Proof (i) Since θ respects conjugation, and since $e_n^- e_n = 1$,

$$(\theta(y)e_n^{-1})^-(\theta(y)e_n^{-1}) = y^- y, \quad \text{for any } y \in Y.$$

(ii) First observe that, for any $g \in \mathbf{R}_{p,q}$, $\theta(g)e_n = e_n\theta(\hat{g})$, for the isomorphism θ and the isomorphism $g \rightsquigarrow e_n\theta(\hat{g})e_n^{-1}$ agree on X. Now let $g \in \Omega$. Then, for any $u(y) \in \mathbf{R}^{p,q+1}$, where $y \in Y$,

$$\begin{aligned} \theta(g)(\theta(y)e_n^{-1})\theta(g)^{-1} &= \theta(g)\theta(y)\theta(\hat{g})^{-1}e_n^{-1} \\ &= \theta(g\,y\,\hat{g}^{-1})e_n^{-1} = u\,\rho'_g(y). \end{aligned}$$

So $\theta(g) \in \Gamma^0(p,q+1)$, and the diagram commutes.

(iii) The map is clearly a group map, since θ is an algebra isomorphism. One proves that it is invertible by showing, by essentially the same argument as in (ii), that, for any $h \in \Gamma^0(p,q+1)$, $\theta^{-1}(h) \in \Omega$. □

Cor. 13.50. The orthogonal transformations of Y represented by the elements of Ω are the rotations of Y. □

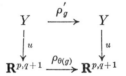

Since conjugation, restricted to Y, is an antirotation of Y, the antirotations of Y also are representable by elements of Ω in a simple manner.

It remains to make a few remarks about the non-universal case. We suppose that A is a non-universal Clifford algebra for X. Since the main involution is not now defined, we cannot proceed exactly as before. However, in the case we have just been discussing, $\hat{g} = g$ or $-g$, for any $g \in \Gamma$, according as $g \in \Gamma^0$ or Γ^1. What is true in the present case is the following.

Prop. 13.51. Let g be an invertible element of the non-universal Clifford algebra A for X such that, for all $x \in X$, $g\,x\,g^{-1} \in X$. Then the map $X \to X$; $x \rightsquigarrow g\,x\,g^{-1}$ is a rotation of X, while the map $X \to X$; $x \rightsquigarrow -g\,x\,g^{-1}$ is an antirotation of X. $\square$

In this case $\Gamma = \Gamma^0 = \Gamma^1$.

The discussion involving $Y = \mathbf{R} \oplus X$ requires that conjugation be defined, but if this is met by the non-universal Clifford algebra A, then A may be used also to describe the rotations of Y. The restriction to Y of conjugation is, as before, an antirotation of Y.

The uses of conjugation

It could be argued that until now we have not made essential use of the conjugation anti-involution on a universal Clifford algebra. This omission will be rectified in the remaining sections of this chapter.

First we introduce a chart, the Pfaffian chart, on Spin (n) for any finite n, analogous to the Cayley chart for $SO(n)$.

Secondly, we show that the groups Pin X and Spin X associated to a non-degenerate finite-dimensional real orthogonal space X may be regarded as normal subgroups of Γ and Γ^0, respectively, rather than as quotient groups. We determine explicitly, as a subset of the Clifford algebra, the groups Spin (p,q) for small p and q, considering several cases of importance in some detail. The knowledge we have gained in Chapter 11 concerning anti-involutions of endomorphism algebras is of value here.

Finally, Table 13.26 is put to work to produce a curious sequence of numbers, the Radon–Hurwitz sequence, on which we shall have more to say in Chapter 20.

The map N is a useful tool in these applications.

The map N

Let $N: A \longrightarrow A$ be defined, by the formula

$$N(a) = a^- a, \quad \text{for any } a \in A,$$

A denoting, as before, the universal Clifford algebra of the non-degenerate finite-dimensional real orthogonal space X.

Prop. 13.52.

(i) For any $g \in \Gamma$, $N(g) \in \mathbf{R}$,
(ii) $N(1) = 1$,
(iii) for any $g, g' \in \Gamma$, $N(gg') = N(g)\, N(g')$,
(iv) for any $g \in \Gamma$, $N(g) \neq 0$ and $N(g^{-1}) = (N(g))^{-1}$,
(v) for any $g \in \Gamma$, there exists a unique positive real number λ such that $|\, N(\lambda g) \,| = 1$, namely $\lambda = \sqrt{(\,|\, N(g) \,|\,)^{-1}}$.

Proof That $N(1) = 1$ is obvious. All the other statements follow directly from the observation that, by Theorem 13.44, any $g \in \Gamma$ is expressible (not necessarily uniquely) in the form

$$\prod_{i \in k} x_i = x_0 x_1 \ldots x_{k-2} x_{k-1}$$

where, for all $i \in k$, $x_i \in X$, k being finite; for it follows that

$$g^- = \prod_{i \in k} x_{k-1-i}^- = x_{k-1}^-\, x_{k-2}^- \ldots x_1^-\, x_0^-,$$

and that

$$N(g) = g^- g = \prod_{i \in k} N(x_i),$$

where, for each $i \in k$, $N(x_i) = x_i^2 \in \mathbf{R}$. $\square$

The Pfaffian chart

The Cayley chart at ${}^n1$ for the group $SO(n)$ was introduced in Chapter 12. It is the injective map

$$\mathrm{End}_-(\mathbf{R}^n) \longrightarrow SO(n); \quad s \rightsquigarrow (1 + s)(1 - s)^{-1}.$$

The analogous chart on Spin (n) is the *Pfaffian* (or *Lipschitz* [62]) chart.

Let $s \in \mathrm{End}_-(\mathbf{R}^n)$, for any finite n; that is, $s \in \mathbf{R}(n)$ and $s^\tau = -s$. The *Pfaffian* of s, pf s, is defined to be 0 if n is odd and to be the real number

$$\sum_{\pi \in P} \operatorname{sgn} \pi \prod_{k \in m} s_{\pi(2k),\pi(2k+1)}$$

if $n = 2m$ is even, P being the set of all permutations π of $2m$ for which

(i) for any $h, k \in 2m$, $h < k \Rightarrow \pi(2h) < \pi(2k)$,

and (ii) for any $k \in 2m$, $\pi(2k) < \pi(2k + 1)$.

For example, if $n = 4$, $\operatorname{pf} s = s_{01}s_{23} - s_{02}s_{13} + s_{03}s_{12}$. By convention $\operatorname{pf} s = 1$ if $n = 0$, in which case $s = {}^01 = 0$.

For any $I \subset n$, let s_I denote the matrix $(s_{ij} : i,j \in I)$. Then $s_I \in \operatorname{End}_-(\mathbf{R}^k)$, where $k = \#I$. The *complete Pfaffian* of s, $\operatorname{Pf} s$, is, by definition, the element

$$\sum_{I \subset n} \operatorname{pf} s_I \prod e_I$$

of the Clifford algebra $\mathbf{R}_{0,n}$. Since $\operatorname{pf} s_I = 0$ for $\#I$ odd, $\operatorname{Pf} s \in \mathbf{R}_{0,n}^0$. In fact $\operatorname{Pf} s \in \Gamma^0(n)$. To see this we require the following lemma.

Lemma 13.53. For any finite k and for any $g \in \Gamma^0(k+1)$, let g be expressed in the form $g_0 + g_1 e_k$, where $g_0 \in \Gamma^0(k)$ and $g_1 \in \Gamma^1(k)$, and suppose that $\rho_g = (1 + s)(1 - s)^{-1}$, where $s \in \operatorname{End}_-(\mathbf{R}^{k+1})$. Then

(i) g_0 is invertible,
(ii) there exists $a \in \mathbf{R}^k$ such that $g_1 = ag_0$,
(iii) $g_0 \in \Gamma^0(k)$,
(iv) for any $\lambda \in \mathbf{R}$, $g_0 + \lambda g_1 e_k \in \Gamma^0(k+1)$.

Proof (i) Since $e_k g e_k = -g_0 + g_1 e_k$,

$$2g_0 = g - e_k g e_k = -e_k(e_k + g e_k g^{-1})g.$$

Now e_k and g are invertible and

$$\begin{aligned} e_k + g e_k g^{-1} &= e_k + (1+s)(1-s)^{-1} e_k \\ &= 2(1-s)^{-1} e_k, \end{aligned}$$

which is a non-zero element of $\mathbf{R}^k$ and is therefore invertible, since $\mathbf{R}^k$ is positive-definite. So g_0 is invertible.

(ii) Since $2g_0 = g - e_k g e_k$ and $2g_1 = -g e_k + e_k g$ and since, by the proof of (i), g_0^{-1} is a non-zero real multiple of $g^{-1}(e_k + g e_k g^{-1})e_k$, it follows that $g_1 g_0^{-1}$ is a non-zero real multiple of

$$(-g e_k + e_k g)g^{-1}(e_k + g e_k g^{-1})e_k = g e_k g^{-1} - e_k g e_k g^{-1} e_k^{-1},$$

which is an element of $\mathbf{R}^k$.

(iii) For any $x \in \mathbf{R}^k$ there exists $x' \in \mathbf{R}^k$ and $\lambda \in \mathbf{R}$ such that

$$(g_0 + g_1 e_k)x = (x' + \lambda e_k)(g_0 + g_1 e_k).$$

This equation implies, in particular, that

$$g_0 x = x' g_0 + \lambda g_1.$$

By (ii), there exists $a \in \mathbf{R}^k$ such that $g_1 = ag_0$. Therefore, for any $x \in \mathbf{R}^k$, there exists $x'' \in \mathbf{R}^k$, namely $x'' = x' + \lambda a$, such that $g_0 x = x'' g_0$.

(iv) The proof is similar to the proof of (iii) and is left as an exercise. □

Theorem 13.54. Let $s \in \text{End}_-(\mathbf{R}^n)$. Then $\text{Pf}\, s \in \Gamma^0(n)$ and is the unique element of $\Gamma^0(n)$ inducing the rotation $(1 + s)(1 - s)^{-1}$ and with real part 1.

Proof Let $g \in \Gamma^0(n)$ be such that $\rho_g = (1 + s)(1 - s)^{-1}$. After n applications of Lemma 13.53, discarding one of the e_i each time, the real part of g is found to be invertible, and therefore non-zero. There is, therefore, a unique element of $\Gamma^0(n)$ inducing the rotation $(1 + s)(1 - s)^{-1}$ and with real part 1. We may suppose g to be this element.

Now suppose that, for each $i, j \in n$, the coefficient of e_ie_j in g is r_{ij}. Then $r \in \text{End}_-(\mathbf{R}^n)$. Since $g \in \Gamma^0(n)$, there exists, for any $x \in \mathbf{R}^n$, $x' \in \mathbf{R}^n$ such that $x'g = gx$. The coefficients of e_i on either side of this equation are equal. So, for all $i \in n$,

$$x_i' - \sum_{j\in n} r_{ij}x_j' = x_i + \sum_{j\in n} r_{ij}x_j.$$

That is,
$$(1 - r)x' = (1 + r)x.$$

So $x' = (1 - r)^{-1}(1 + r)x$, since $1 - r$ is invertible, by Prop. 12.17. Therefore $(1 - r)^{-1}(1 + r) = (1 + s)(1 - s)^{-1}$, which implies that $r = s$.

Next, by Prop. 13.52(i) and by Lemma 13.53(iii), by equating to zero the coefficients of highest degree either of $g_Ig_I^-$ or of $g_Ie_0g_I^-$, where g_I is obtained from g by omitting all terms in the expansion of g involving any e_i for which $i \notin I$, it follows, for all $I \subset n$ with $\#I > 2$, that the coefficient in the expansion of g of $\prod e_I$ is a polynomial in the terms of the matrix s_I.

By Lemma 13.53(iv) each term of the polynomial coefficient of $\prod e_I$ contains exactly one term from each row and exactly one term from each column of s_I, so that the terms of the polynomial are, up to real multiples, the terms of pf s_I.

Finally, consider any one such term,

$$\lambda s_{01}s_{23}s_{45}\, e_0e_1e_2e_3e_4e_5, \quad \text{for example.}$$

This term will be equal to the corresponding term in $\text{Pf}\, s'$ where $s' \in \text{End}_-(\mathbf{R}^n)$ is defined by

$$s'_{01} = s_{01} = -s'_{10}, \quad s'_{23} = s_{23} = -s'_{32} \quad \text{and} \quad s'_{45} = s_{45} = -s'_{54},$$

all the other terms being zero. However,

$$\begin{aligned}\text{Pf}\, s' &= (1 + s_{01}\, e_0e_1)(1 + s_{23}\, e_2e_3)(1 + s_{45}\, e_4e_5)\\ &= 1 + s_{01}\, e_0e_1 + s_{23}\, e_2e_3 + s_{45}\, e_4e_5 + \ldots\\ &\qquad \ldots + s_{01}s_{23}s_{45}\, e_0e_1e_2e_3e_4e_5,\end{aligned}$$

since each of the factors is in $\Gamma^0(6)$, the real part is 1 and the coefficients of the terms $e_i e_j$ are correct. So, in this case, $\lambda = 1$ in accordance with the theorem. The other terms are handled analogously. □

The map

$$\text{End}_-(\mathbf{R}^n) \to \text{Spin}(n); \quad s \rightsquigarrow \text{im}\{\text{Pf}\, s\}$$

will be called the *Pfaffian chart at* 1 on Spin (n).

The above account extends to a certain extent to the indefinite case, as will be seen in Chapter 20, where the relationship between the Cayley and the Pfaffian charts is studied further.

The following property of the Pfaffian is sometimes used to characterize it. (Cf. for example [3].)

Theorem 13.55. For any $s \in \text{End}_-(\mathbf{R}^n)$, $(\text{pf}\, s)^2 = \det s$.

Proof Let $s \in \text{End}_-(\mathbf{R}^n)$. Then, for any $t \in \mathbf{R}(n)$, $t^\tau s t \in \text{End}_-(\mathbf{R}^n)$. Now, for any such s and t,

$$\text{pf}\,(t^\tau s t) = \det t \text{ pf}\, s.$$

To show this it is enough, by Theorem 7.8 and Cor. 7.21, to verify that, for any $i \in n$ and $\lambda \in \mathbf{R}$,

$$\text{pf}\,(({}^\lambda e_i)^\tau s({}^\lambda e_i)) = \lambda \text{ pf}\, s$$

and that, for any $i, j \in n$ with $i \neq j$,

$$\text{pf}\,(({}^1 e_{ij})^\tau s({}^1 e_{ij})) = \text{pf}\, s,$$

where ${}^\lambda e_i$ and ${}^1 e_{ij}$ are the elementary framings of $\mathbf{R}^n$ defined on page 117. These verifications are left as exercises.

The matrix s induces a skew-symmetric correlation on $\mathbf{R}^n$ with product

$$\mathbf{R}^n \times \mathbf{R}^n \to \mathbf{R}; \quad (x,x') \rightsquigarrow x^\tau s x'.$$

Let $2m$ be the rank of this correlation. Then, by a slight extension of Cor. 11.46 to include the degenerate case, there exists $u \in GL(n;\mathbf{R})$ such that

$$(u^\tau s u)_{2k,2k+1} = 1 = -(u^\tau s u)_{2k+1,2k}$$

for all $k \in m$, and $(u^\tau s u)_{ij} = 0$ otherwise. It follows from this that $\text{Pf}\,(u^\tau s u) = \prod_{k \in m} (1 + e_{2k} e_{2k+1})$.

These are two cases. If $2m < n$, pf $u^\tau s u = 0$, implying that pf $s = 0$, since $\det u \neq 0$, while $\det u^\tau s u = 0$, implying that $\det s = 0$. If $2m = n$, pf $u^\tau s u = 1$ and $\det u^\tau s u = 1$, implying that $(\det u)^2 (\text{pf}\, s)^2 = 1 = (\det u)^2 \det s$.

In either case, $(\text{pf}\, s)^2 = \det s$. □

Spin groups

The groups Pin X and Spin X for a non-degenerate finite-dimensional real orthogonal space X are commonly regarded not as quotient groups of Γ but as normal subgroups of Γ. This is made possible by the following proposition.

Prop. 13.56. The maps

$$\{g \in \Gamma : |N(g)| = 1\} \to \text{Pin } X;\ g \rightsquigarrow \text{im}^+\{g\}$$

and $$\{g \in \Gamma^0 : |N(g)| = 1\} \to \text{Spin } X;\ g \rightsquigarrow \text{im}^+\{g\}$$

are group isomorphisms. □

The groups Pin X and Spin X will henceforth be identified with these subgroups of Γ and the maps of Prop. 13.48 identified with the maps

$$\rho_X \mid \{g \in \Gamma : |N(g)| = 1\} \quad \text{and} \quad \rho_X \mid \{g \in \Gamma^0 : |N(g)| = 1\}.$$

These maps also will be denoted loosely, from now on, by ρ.

Prop. 13.57. As subgroups of Γ and Γ^0 respectively, the groups Pin X and Spin X are normal subgroups, the quotient groups $\Gamma/\text{Pin } X$ and $\Gamma^0/\text{Spin } X$ each being isomorphic to $\mathbf{R}^+$. □

That Prop. 13.52(i) is a genuine restriction on g is illustrated by the element $1 + \prod e_4 \in \mathbf{R}_{0,4}$, since

$$N(1 + \textstyle\prod e_4) = 2(1 + \prod e_4) \notin \mathbf{R}.$$

That the same proposition does not, in general, provide a sufficient condition for g to belong to Γ is illustrated by the element $1 + \prod e_6 \in \mathbf{R}_{0,6}$, for, since

$$N(1 + \textstyle\prod e_6) = (1 - \prod e_6)(1 + \prod e_6) = 2,$$

the element is invertible, but either by explicit computation of $(1 + \prod e_6)e_0(1 + \prod e_6)^{-1}$, or by applying Theorem 13.54, it can be seen that the element does not belong to Γ. However, the condition is sufficient when $p + q \leqslant 5$, as the following proposition shows.

Prop. 13.58. Let dim $X \leqslant 5$. Then

$$\text{Spin } X = \{g \in A^0 : N(g) = \pm 1\}.$$

Proof The proof is given in full for the hardest case, namely when dim $X = 5$. The proofs in the other cases may be obtained from this one simply by deleting the irrelevant parts of the argument.

From the definition there is inclusion one way ($\subset$). What has to be proved, therefore, is that, for all $g \in A^0$ such that $N(g) = \pm 1$,

$$x \in X \Rightarrow g x g^{-1} \in X.$$

Let $\{e_i : i \in 5\}$ be an orthonormal basis for X. Then, since $X \subset A^1$ and $g \in A^0$, $x' = g\,x\,g^{-1} \in A^1$, for any $x \in X$. So there are real numbers a_i, b_{jkl}, c such that

$$x' = \sum_{i \in 5} a_i e_i + \sum_{j \in k \in l \in 5} b_{jkl} e_j e_k e_l + c \prod e_5.$$

Now $(x')^- = (g\,x\,g^{-1})^- = -x'$, since $g^{-1} = \pm g^-$, while $(e_i)^- = -e_i$, $(e_j e_k e_l)^- = e_j e_k e_l$, and $(\prod e_5)^- = -\prod e_5$. So, for all $i \in k \in l \in 5$, $b_{jkl} = 0$. That is,

$$x' = x'' + c \prod e_5, \quad \text{for some } x'' \in X.$$

The argument ends at this point if $n < 5$. In the present case it remains to prove that $c = 0$.

Now $x'^2 = x^2 \in \mathbf{R}$. So

$$x''^2 + 2cx''(\prod e_5) + c^2(\prod e_5)^2 \in \mathbf{R}.$$

Since x''^2 and $c^2(\prod e_5)^2 \in \mathbf{R}$, and $\prod e_5 \notin \mathbf{R}$, either $c = 0$ or $x'' = 0$. Whichever is the correct alternative it is the same for every x, for, if there were an element of each kind, their sum would provide a contradiction. Since the map

$$X \to A; \quad x \rightsquigarrow g\,x\,g^{-1}$$

is injective, it follows that $c = 0$. Therefore $g\,x\,g^{-1} \in X$, for each $x \in X$. □

To use Prop. 13.58 we need to know the form that conjugation takes on the Clifford algebra. Now the Clifford algebra itself is representable as an endomorphism algebra, according to Table 13.26. Also by Chapter 11, any correlation on the spinor space induces an anti-involution of the Clifford algebra, namely the appropriate adjoint involution, and conversely, by Theorem 11.32 and Theorem 11.26, any anti-involution of the Clifford algebra is so induced by a symmetric or skew correlation on the spinor space. So the problem reduces to determining in each case which anti-involution it is out of a list which we essentially already know. The job of identification is made easier by the fact that an anti-involution of an algebra is uniquely determined, by Prop. 6.38, by its restriction to any subset that generates the algebra.

For the Clifford algebras $\mathbf{R}_{0,n}$ the determination is made easy by Prop. 13.59.

Prop. 13.59. Conjugation on $\mathbf{R}_{0,n}$ is the adjoint anti-involution induced by the standard positive-definite correlation on the spinor space $\mathbf{A}^m$.

Proof By Prop. 13.27, $\bar{e}_i^{\tau} e_i = 1$, for any element e_i of the standard orthonormal basis for $\mathbf{R}^{0,n}$, here identified with its image in the Clifford algebra $\mathbf{A}(m)$. Also, by the definition of conjugacy on $\mathbf{R}_{0,n}$, $e_i^- = -e_i$. But $e_i^2 = -1$. So, for all $i \in n$, $e_i^- = \bar{e}_i^{\sim}$, from which the result follows at once, by Prop. 6.38. □

This indicates, incidentally, why we wrote a^-, and not $\bar{a}$, for the conjugate of an element a of a Clifford algebra A, the reason for writing $a^{\sim}$ and not $\tilde{a}$, for the reverse of a, being similar. The notation $\hat{a}$ is less harmful in practice, for, in the context of Prop. 13.59 at least, $\hat{a}$ in either of its senses coincides with $\hat{a}$ in its other sense.

Cor. 13.60.

Spin (1) $\cong O(1) \cong S^0$, Spin (2) $\cong U(1) \cong S^1$,
Spin (3) $\cong Sp(1) \cong S^3$, Spin (4) $\cong Sp(1) \times Sp(1) \cong S^3 \times S^3$,
Spin (5) $\cong Sp(2)$ and Spin (6) is a subgroup of $U(4)$ □.

In the case of Spin (n), for $n = 1, 2, 3, 4$, what this corollary does is to put into a wider relationship with each other various results which we have had before. It may be helpful to look at some of these cases in turn.

$\mathbf{R}^2$: The universal Clifford algebra $\mathbf{R}_{0,2}$ is $\mathbf{H}$, while the universal Clifford algebra $\mathbf{R}_{2,0}$ is $\mathbf{R}(2)$, the even Clifford algebras $\mathbf{R}_{0,2}^0$ and $\mathbf{R}_{2,0}^0$ each being isomorphic to $\mathbf{C}$.

Suppose we use $\mathbf{R}_{0,2} = \mathbf{H}$ to describe the rotations of $\mathbf{R}^2$, $\mathbf{R}^2$ being identified with im{i,k} and $\mathbf{R}_{0,2}^0 = \mathbf{C}$ being identified with im{1,i}. Then the rotation of $\mathbf{R}^2$ represented by $g \in$ Spin (2) $= U(1)$ is the map

$$x \rightsquigarrow g\,x\,g^{-1} = g\,x\,\bar{g},$$

that is, the map

$$\begin{aligned}(x_0 + \mathrm{i}x_1)\mathrm{j} = (x_0\mathrm{i} + x_1\mathrm{k}) &\rightsquigarrow (a + ib)^2(x_0 + \mathrm{i}x_1)\mathrm{j} \\ &= (a + ib)(x_0\mathrm{i} + x_1\mathrm{k})(a - ib),\end{aligned}$$

where $x = x_0\mathrm{i} + x_1\mathrm{k}$ and $g = a + ib$.

On the other hand, by Cor. 13.50, we may use $\mathbf{C}$ directly, $\mathbf{R}^2$ being identified with $\mathbf{C}$. Then the rotation of $\mathbf{R}^2$ represented by g is the map

$$y \rightsquigarrow g\,y\,\hat{g}^{-1} = g\,y\,g = g^2 y.$$

One can transfer from the one model to the other simply by setting $x = y\mathrm{j}$.

$\mathbf{R}^3$: The universal Clifford algebra $\mathbf{R}_{0,3}$ is ${}^2\mathbf{H}$, while the universal Clifford algebra $\mathbf{R}_{3,0}$ is $\mathbf{C}(2)$, the even Clifford algebras $\mathbf{R}_{0,3}^0$ and $\mathbf{R}_{3,0}^0$ each being isomorphic to $\mathbf{H}$. Besides these, there are the non-universal algebras $\mathbf{R}_{0,3}(1,0)$ and $\mathbf{R}_{0,3}(0,1)$, also isomorphic to $\mathbf{H}$. Any of these may be used to represent the rotations of $\mathbf{R}^3$.

The simplest to use is $\mathbf{R}_{0,3}(1,0) \cong \mathbf{H}$, $\mathbf{R}^3$ being identified with the linear subspace of pure quaternions. An alternative is to use $\mathbf{R}_{0,3}^0 \cong \mathbf{H}$, in which case $\mathbf{R}^3$ may be identified, by Prop. 13.49, with the linear subspace im$\{1,\mathrm{i},\mathrm{k}\}$. In either case Spin (3) $= Sp(1) = S^3$.

In the first of these two cases the rotation of $\mathbf{R}^3$ represented by $g \in$ Spin (3) is the map

$$x \rightsquigarrow g\,x\,g^{-1} = g\,x\,\bar{g},$$

while in the second case the rotation is the map

$$y \rightsquigarrow g\,y\,\hat{g}^{-1} = g\,y\,\tilde{g}.$$

One can transfer from the one model to the other by setting $x = y\mathrm{j}$, compatibility being guaranteed by the equation

$$g\,y\,\mathrm{j}\bar{g} = g\,y\,\tilde{g}\mathrm{j}.$$

$\mathbf{R}^4$: The universal Clifford algebras $\mathbf{R}_{0,4}$ and $\mathbf{R}_{4,0}$ are each isomorphic to $\mathbf{H}(2)$, the even Clifford algebra in either case being isomorphic to ${}^2\mathbf{H}$. There are various identifications of $\mathbf{R}^3$ with a linear subspace of ${}^2\mathbf{H}$ such that, for any $x \in \mathbf{R}^3$, $x^{(2)} = -x^2 = \bar{x}x$. Once one is chosen, $\mathbf{R}^4$ may be identified with $\mathbf{R} \oplus \mathbf{R}^3$, with $y^{(2)} = \bar{y}y$, for any $y \in \mathbf{R}^4$.

One method is to identify $\mathbf{R}^4$ with the linear subspace

$$\left\{\begin{pmatrix} y & 0 \\ 0 & \bar{y} \end{pmatrix} : y \in \mathbf{H}\right\}$$

of ${}^2\mathbf{H}$, $\mathbf{R}^3$ being identified with im $\left\{\begin{pmatrix} \mathrm{i} & 0 \\ 0 & -\mathrm{i} \end{pmatrix}, \begin{pmatrix} \mathrm{j} & 0 \\ 0 & -\mathrm{j} \end{pmatrix}, \begin{pmatrix} \mathrm{k} & 0 \\ 0 & -\mathrm{k} \end{pmatrix}\right\}$.

Then, for any $\begin{pmatrix} q & 0 \\ 0 & r \end{pmatrix} \in {}^2\mathbf{H}$,

$$\widehat{\begin{pmatrix} q & 0 \\ 0 & r \end{pmatrix}} = \begin{pmatrix} r & 0 \\ 0 & q \end{pmatrix},$$

while Spin 4 $= \left\{\begin{pmatrix} q & 0 \\ 0 & r \end{pmatrix} \in {}^2\mathbf{H} : |q| = |r| = 1\right\}$. The rotation of $\mathbf{R}^4$ represented by $\begin{pmatrix} q & 0 \\ 0 & r \end{pmatrix} \in$ Spin 4 is then, by Prop. 13.49, the map

$$\begin{pmatrix} y & 0 \\ 0 & \bar{y} \end{pmatrix} \rightsquigarrow \begin{pmatrix} q & 0 \\ 0 & r \end{pmatrix}\begin{pmatrix} y & 0 \\ 0 & \bar{y} \end{pmatrix}\widehat{\begin{pmatrix} q & 0 \\ 0 & r \end{pmatrix}}^{-1} = \begin{pmatrix} qy\bar{r} & 0 \\ 0 & r\bar{y}\bar{q} \end{pmatrix}.$$

This is essentially the map

$$y \rightsquigarrow q\,y\,\bar{r},$$

which is what we had before, in Chapter 10.

An alternative is to identify $\mathbf{R}^4$ with the linear subspace

$$\left\{\begin{pmatrix} y & 0 \\ 0 & \tilde{y} \end{pmatrix} : y \in \mathbf{H}\right\}.$$

The rotation induced by $\begin{pmatrix} q & 0 \\ 0 & r \end{pmatrix} \in$ Spin 4 is then, by a similar argument, the map

$$\begin{pmatrix} y & 0 \\ 0 & \hat{y} \end{pmatrix} \rightsquigarrow \begin{pmatrix} qy\tilde{r} & 0 \\ 0 & r\hat{y}\tilde{q} \end{pmatrix}$$

and this reduces to the map $y \rightsquigarrow qy\tilde{r}$.

Prop. 13.61. Spin 6 $\cong SU(4)$.

A proof of this may be based on Exercise 11.65. One proves first that if Y is the image of the injective real linear map $\gamma : \mathbf{C}^3 \to \mathbf{C}(4)$ constructed in that exercise, then, for each $y \in Y$, $\bar{y}^\tau y \in \mathbf{R}$, and that if Y is assigned the quadratic form $Y \to \mathbf{R}$; $y \rightsquigarrow \bar{y}^\tau y$, then γ is an orthogonal map and T is the unit sphere in Y. The rest is then a straightforward checking of the things that have to be checked. (See page 258.) Note that, for all $t \in SU(4)$, $t^{\sim} = t^{-1}$. □

For any $g \in$ Spin (n), $N(g) = 1$. For $g \in$ Spin (p,q), on the other hand, with neither p nor q equal to zero, $N(g)$ can be equal either to 1 or to -1.

The subgroup $\{g \in \text{Spin}\,(p,q) : N(g) = 1\}$ will be denoted by Spin$^+$ (p,q). By Prop. 2.7, the image of Spin$^+$ (p,q) in $SO(p,q)$ by ρ is a subgroup of $SO(p,q)$. This subgroup, called the (*proper*) *Lorentz group* of $\mathbf{R}^{p,q}$ will be denoted by $SO^+(p,q)$. In Prop. 20.96 the Lorentz group of $\mathbf{R}^{p,q}$ is shown to be the set of rotations of $\mathbf{R}^{p,q}$ that preserve the semi-orientations of $\mathbf{R}^{p,q}$ (cf. page 161).

Exercise 13.62. Let $g \in$ Spin (1,1). Prove that the induced rotation ρ_g of $\mathbf{R}^{1,1}$ preserves the semi-orientations of $\mathbf{R}^{1,1}$ if, and only if, $N(g) = 1$, and reverses them if, and only if, $N(g) = -1$. □

The subgroup $\{g \in \text{Spin}\,(p,q) : N(g) = 1\}$ of Spin (p,q) will be denoted by Spin$^+$ (p,q).

The next proposition covers the cases of interest in the theory of relativity.

Prop. 13.63.

$$\text{Spin}^+\,(1,1) \cong \left\{\begin{pmatrix} a & 0 \\ 0 & d \end{pmatrix} \in {}^2\mathbf{R} : ad = 1\right\} \cong \mathbf{R}^* \cong GL(1;\mathbf{R})$$

$$\text{Spin}^+\,(1,2) \cong \left\{\begin{pmatrix} a & c \\ b & d \end{pmatrix} \in \mathbf{R}(2) : \det \begin{pmatrix} a & c \\ b & d \end{pmatrix} = 1\right\} = SL(2;\mathbf{R})$$

$$\text{and}\quad \text{Spin}^+\,(1,3) \cong \left\{\begin{pmatrix} a & c \\ b & d \end{pmatrix} \in \mathbf{C}(2) : \det \begin{pmatrix} a & c \\ b & d \end{pmatrix} = 1\right\} = SL(2;\mathbf{C}).$$

Proof It is enough to give the proof for Spin$^+$ (1,3), which may be regarded as a subgroup of $\mathbf{R}_{1,2} = \mathbf{C}(2)$, since $\mathbf{R}_{1,3}^{\,0} = \mathbf{R}_{1,2}$. Now, by Prop. 13.58 and Prop. 13.33,

$$\text{Spin}^+\,(1,3) = \{g \in \mathbf{R}_{1,2} : g^{-}g = 1\},$$

so that the problem is reduced to determining the conjugation anti-involution on $\mathbf{R}_{1,2}$. To do so we have just to select a suitable copy of $\mathbf{R}^{1,2}$ in $\mathbf{R}_{1,2}$. Our choice is to represent e_0, e_1 and e_2 in $\mathbf{R}^{1,2}$ by $\begin{pmatrix} 1 & 0 \\ 0 & -1 \end{pmatrix}$, $\begin{pmatrix} 0 & -1 \\ 1 & 0 \end{pmatrix}$ and $\begin{pmatrix} 0 & i \\ i & 0 \end{pmatrix}$, respectively, in $\mathbf{C}(2)$, these matrices being mutually anticommutative and satisfying the equations

$$\begin{pmatrix} 1 & 0 \\ 0 & -1 \end{pmatrix}^2 = 1 = -e_0{}^2, \quad \begin{pmatrix} 0 & -1 \\ 1 & 0 \end{pmatrix}^2 = -1 = -e_1{}^2$$

and $\begin{pmatrix} 0 & i \\ i & 0 \end{pmatrix}^2 = -1 = -e_2{}^2$, as is necessary. Now the anti-involution $\begin{pmatrix} a & c \\ b & d \end{pmatrix} \rightsquigarrow \begin{pmatrix} d & -c \\ -b & a \end{pmatrix}$ sends each of these three matrices to its negative. This, therefore, by Prop. 6.38, is the conjugation anti-involution. Since, for any $\begin{pmatrix} a & c \\ b & d \end{pmatrix} \in \mathbf{C}(2)$, $\begin{pmatrix} d & -c \\ -b & a \end{pmatrix}\begin{pmatrix} a & c \\ b & d \end{pmatrix} = \det \begin{pmatrix} a & c \\ b & d \end{pmatrix}$, the proposition is proved. □

It is natural, therefore, to identify the spinor space $\mathbf{C}^2$ for $\mathbf{R}_{1,2}$ with the complex symplectic plane $\mathbf{C}_{sp}$ and, similarly, to identify the spinor space $\mathbf{R}^2$ for $\mathbf{R}_{1,0}$ with hb $\mathbf{R}$ and the spinor space $\mathbf{R}^2$ for $\mathbf{R}_{1,1}$ with $\mathbf{R}_{sp}$. When this is done, the induced adjoint anti-involution on the real algebra of endomorphisms of the spinor space coincides with the conjugation anti-involution on the Clifford algebra.

Note, incidentally, the algebra injections

$$\text{Spin}\,(2) \to \text{Spin}^+\,(1,2)$$

and

$$\text{Spin}\,(3) \to \text{Spin}^+\,(1,3)$$

induced by the standard (real orthogonal) injections

$$\mathbf{R}^{0,2} \to \mathbf{R}^{1,2} \quad \text{and} \quad \mathbf{R}^{0,3} \to \mathbf{R}^{1,3},$$

the image of Spin (2) $= U(1)$ in Spin$^+$ (1,2) being $SO(2)$ and the image of Spin (3) $= Sp(1)$ in Spin$^+$ (1,3) being $SU(2)$.

The isomorphisms $U(1) \cong SO(2)$ and $Sp(1) \cong SU(2)$ fit nicely, therefore, into the general scheme of things.

Proposition 13.64 is a step towards the determination and classification of the conjugation anti-involutions for the universal Clifford algebras $\mathbf{R}_{p,q}$ other than those already considered.

Prop. 13.64. Let V be the spinor space for the orthogonal space $\mathbf{R}^{p,q}$, with $\mathbf{R}_{p,q} = \text{End } V$. Then if $p > 0$ and if $(p,q) \neq (1,0)$, the conjugation anti-involution on $\mathbf{R}_{p,q}$ coincides with the adjoint anti-involution on End V induced by a neutral semi-linear correlation on V.

Proof By Theorem 11.32 there is a reflexive non-degenerate $\mathbf{A}^{\psi}$-linear correlation on the right $\mathbf{A}$-linear space V producing the conjugation anti-involution on $\mathbf{R}_{p,q}$ as its adjoint. What we prove is that this correlation must be neutral. This follows at once from the even-dimensionality of V over $\mathbf{A}$ unless $\mathbf{A}^{\psi} = \mathbf{R}, \mathbf{C}, \bar{\mathbf{H}}, {}^2\mathbf{R}$ or ${}^2\bar{\mathbf{H}}$. However, since $p > 0$, there exists in every case $t \in \operatorname{End} V$ such that $t^{-}t = -1$, namely $t = e_0$; for $e_0{}^{-}e_0 = -e_0{}^2 = e_0{}^{(2)} = -1$. The existence of such an element guarantees neutrality when $\mathbf{A}^{\psi} = \mathbf{R}$, by Prop. 9.55. The obvious analogue of Prop. 9.55 guarantees neutrality in each of the other exceptional cases. □

An analogous result holds for the algebras $\bar{\mathbf{C}}_{p,q}$.

Prop. 13.65. Conjugation on $\bar{\mathbf{C}}_{0,n}$ is the adjoint anti-involution induced by the standard positive-definite correlation on the spinor space. Conjugation on $\bar{\mathbf{C}}_{p,q}$, where $p > 0$ and $(p,q) \neq (1,0)$, is the adjoint anti-involution induced by a neutral semi-linear correlation on the spinor space. □

The classification of the conjugation anti-involutions for each of the algebras $\mathbf{R}_{p,q}$, $\bar{\mathbf{C}}_{p,q}$, $(\text{hb }\mathbf{R})_{p,q}$, $\mathbf{C}_n$ and $(\text{hb }\mathbf{C})_n$ is completed if we know to which of the ten types listed in Chapter 11 each belongs. In the tables which follow we use the following code:

0 =	$\mathbf{R}$,	symmetric
1 =	hb $\mathbf{R}$,	symmetric or skew
2 =	$\mathbf{R}$,	skew
3 =	$\mathbf{C}$,	skew
4 =	$\tilde{\mathbf{H}}$,	skew or $\bar{\mathbf{H}}$, symmetric
5 =	hb $\tilde{\mathbf{H}}$ or hb $\bar{\mathbf{H}}$,	symmetric or skew
6 =	$\tilde{\mathbf{H}}$,	symmetric or $\bar{\mathbf{H}}$, skew
7 =	$\mathbf{C}$,	symmetric
8 =	$\bar{\mathbf{C}}$,	symmetric or skew
9 =	hb $\mathbf{C}$ or hb $\bar{\mathbf{C}}$,	symmetric or skew.

k, k indicates that the algebra is of the form $A \times A$ with A of type k.

The verification of the tables is left as a hard exercise.

Tables 13.66.

The following are the types to which the various $\mathbf{L}^{\alpha}$-Clifford algebras belong, as classified by their conjugation anti-involution. The tables for $\mathbf{R}_{p,q}$, $(\text{hb }\mathbf{R})_{p,q}$ and $\mathbf{C}_n$ have periodicity 8, while those for $\bar{\mathbf{C}}_{p,q}$ and $(\text{hb }\mathbf{C})_n$ have periodicity 2.

$\mathbf{R}_{p,q}$ ($q \rightarrow$, $p \downarrow$)

0	8	4	4,4	4	8	0	0,0	. . .
1	2	3	4	5	6	7	0	
2	2,2	2	8	6	6,6	6	8	
3	2	1	0	7	6	5	4	
4	8	0	0,0	0	8	4	4,4	
5	6	7	0	1	2	3	4	
6	6,6	6	8	2	2,2	2	8	
7	6	5	4	3	2	1	0	
⋮								⋱

$\bar{\mathbf{C}}_{p,q}$ ($q \rightarrow$, $p \downarrow$)

8	8,8	. . .
9	8	
⋮	⋱	

$(\text{hb }\mathbf{R})_{p,q}$ ($q \rightarrow$, $p \downarrow$)

1	9	5	5,5	5	9	1	1,1	. . .
1,1	1	9	5	5,5	5	9	1	
1	1,1	1	9	5	5,5	5	9	
9	1	1,1	1	9	5	5,5	5	
5	9	1	1,1	1	9	5	5,5	
5,5	5	9	1	1,1	1	9	5	
5	5,5	5	9	1	1,1	1	9	
9	5	5,5	5	9	1	1,1	1	
⋮								⋱

$\mathbf{C}_n$ ($n \rightarrow$)

7	9	3	3,3	3	9	7	7,7	. . .

$(\text{hb }\mathbf{C})_n$ ($n \rightarrow$)

9	9,9	. . .

□

Prof. C. T. C. Wall has commented that these five tables may also be set out as follows (cf. [57]).

	$p - q + 2$ (mod 8)								(mod 2)	
↓ $p + q + 2$ (mod 8)	0		0		6		6		7	
		0,0		7		6,6		7		7,7
	0		0		6		6		7	
		1		8		5		8		9
	2		2		4		4		3	
		2,2		3		4,4		3		3,3
	2		2		4		4		3	
		1		8		5		8		9
↓ (mod 2)	1		1		5		5		9	
		1,1		9		5,5		9		9,9

and

$\longrightarrow$ $p - q + 2 \pmod 4$

$\downarrow$ $p + q + 2 \pmod 4$

8		8	
	8,8		9
8		8	
	9		8,8

where the horizontal projections from the $\mathbf{R}_{p,q}$ table to the $\mathbf{C}_n$ table and from the hb $\mathbf{R}_{p,q}$ table to the hb $\mathbf{C}_n$ table are induced by tensoring with $\mathbf{C}$ and the vertical projections from the $\mathbf{R}_{p,q}$ table to the hb $\mathbf{R}_{p,q}$ table and from the $\mathbf{C}_n$ table to the hb $\mathbf{C}_n$ table are induced by tensoring with hb $\mathbf{R}$. There is an alternative route from the $\mathbf{R}_{p,q}$ table to the hb $\mathbf{C}_n$ table via the $\bar{\mathbf{C}}_{p,q}$ table by tensoring first with $\bar{\mathbf{C}}$ and then with hb $\mathbf{R}$.

The Radon–Hurwitz numbers

An important application of the Clifford algebras for positive-definite finite-dimensional orthogonal spaces, involving the non-universal algebras in an essential way, is to the construction of linear subspaces of the groups $GL(s;\mathbf{R})$, for finite s, a *linear subspace of* $GL(s;\mathbf{R})$ being, by definition, a linear subspace of $\mathbf{R}(s)$ all of whose elements, with the exception of the origin, are invertible.

For example, the standard copy of $\mathbf{C}$ in $\mathbf{R}(2)$ is a linear subspace of $GL(2;\mathbf{R})$ of dimension 2, while either of the standard copies of $\mathbf{H}$ in $\mathbf{R}(4)$ is a linear subspace of $GL(4;\mathbf{R})$ of dimension 4. On the other hand, when s is odd, there is no linear subspace of $GL(s;\mathbf{R})$ of dimension greater than 1. For if this were so, there would exist linearly independent elements a and b of $GL(s;\mathbf{R})$, such that, for all $\lambda \in \mathbf{R}$, $a + \lambda b \in GL(s;\mathbf{R})$ and therefore such that $c + \lambda 1 \in GL(s;\mathbf{R})$, where $c = b^{-1}a$. However, as we prove later in Cor. 19.25, there is a real number λ such that $\det(c + \lambda 1) = 0$, the map $\mathbf{R} \longrightarrow \mathbf{R};\ \lambda \rightsquigarrow \det(c + \lambda 1)$ being a polynomial map of odd degree. This provides a contradiction.

Proposition 13.67 provides a method of constructing linear subspaces of $GL(s;\mathbf{R})$.

Prop. 13.67. Let End $\mathbf{K}^m$ be a possibly non-universal Clifford algebra with conjugation for the positive-definite orthogonal space $\mathbf{R}^n$, for any $n \in \omega$. Then $\mathbf{R} \oplus \mathbf{R}^n$ is a linear subspace of Aut $\mathbf{K}^m = GL(m;\mathbf{K})$ and therefore of $GL(m;\mathbf{R})$, $GL(2m;\mathbf{R})$ or $GL(4m;\mathbf{R})$, according as $\mathbf{K} = \mathbf{R}$, $\mathbf{C}$ or $\mathbf{H}$. Moreover, the conjugate of any element of $\mathbf{R} \oplus \mathbf{R}^n$ is the conjugate transpose of its representative in $GL(m;\mathbf{K})$ or, equivalently, the transpose of its representative in $GL(m;\mathbf{R})$, $GL(2m;\mathbf{R})$ or $GL(4m;\mathbf{R})$.

Proof Let $y = \lambda + x \in \mathbf{R} \oplus \mathbf{R}^n$, where $\lambda \in \mathbf{R}$ and $x \in \mathbf{R}^n$. Then $y^- y = (\lambda - x)(\lambda + x) = \lambda^2 + x^{(2)}$ is real, and is zero if, and only if, $y = 0$. Therefore y is invertible if, and only if, $y \neq 0$.

The last statement of the proposition follows at once from Prop. 13.59. □

The following theorem is an immediate corollary of Prop. 13.67 coupled with the explicit information concerning the Clifford algebras $\mathbf{R}_{0,n}$ contained in Table 13.26 and its extension by Cor. 13.25.

Theorem 13.68. Let $\chi : \omega \to \omega;\ k \rightsquigarrow \chi(k)$ be the sequence of numbers defined by the formula

$$\chi(8p+q) = \begin{array}{ll} 4p, & \text{for } q = 0, \\ 4p+1, & \text{for } q = 1, \\ 4p+2, & \text{for } q = 2 \text{ or } 3 \\ 4p+3, & \text{for } q = 4, 5, 6 \text{ or } 7. \end{array}$$

Then, if $2^{\chi(k)}$ divides s, there exists a k-dimensional linear subspace X of $GL(s;\mathbf{R})$ such that

(i) for each $x \in X$, $x^\tau = -x$, $x^\tau x = -x^2$ being a non-negative real multiple of ${}^s1$, zero only if $x = 0$, and

(ii) $\mathbf{R} \oplus X$ is a $(k+1)$-dimensional linear subspace of $GL(s;\mathbf{R})$. □

The sequence χ is called the *Radon–Hurwitz sequence*. It can be proved that there is no linear subspace of $GL(s;\mathbf{R})$ of dimension greater than that asserted by Theorem 13.68(ii). There is a close relationship between Theorem 13.68 and the problem of tangent vector fields on spheres discussed in Chapter 20. References to the literature will be given there, on page 420.

As a particular case of Prop. 13.67, there is an eight-dimensional linear subspace of $GL(8;\mathbf{R})$, since $\mathbf{R}(8)$ is a (non-universal) Clifford algebra for $\mathbf{R}^7$. This fact will be used in Chapter 14.

Further Exercises

13.69. Show how, for any finite n, the Clifford algebra $\mathbf{C}_n$ may be applied to the description of the orthogonal automorphisms of $\mathbf{C}^n$, and define the Clifford, Pin and Spin groups in this case. □

13.70. Discuss the Pfaffians of complex skew-symmetric matrices (elements of $\mathrm{End}_-(\mathbf{C}^n)$, for any finite n). Show, in particular, that, for a complex skew-symmetric matrix s, $(\text{pf } s)^2 = \det s$. (Cf. Exercise 12.24.) □

13.71. Let A be a Clifford algebra for a finite-dimensional isotropic real, or complex, orthogonal space X such that, for some basic frame $(e_i : i \in n)$ on X, $\prod e_n \neq 0$. Prove that A is a universal Clifford algebra for X. (Try first the case where $n = \dim X = 3$.)

(The universal Clifford algebra, $\bigwedge X$, for a finite-dimensional linear space X, regarded as an isotropic orthogonal space by having assigned to it the zero quadratic form, is called the *exterior* or *Grassmann* algebra for X, Grassmann's term being the *extensive* algebra for X [19]. The square of any element of X in $\bigwedge X$ is 0 and any two elements of X anticommute.

The notation $\prod e_n$ is a shorthand for $\prod_{i \in n} e_i$. Cf. page 243.) □

13.72. Let X be a real or complex n-dimensional linear space and let a be an element of $\bigwedge X$ expressible in terms of some basis $\{e_i : i \in n\}$ for X as a linear combination of k-fold products of the e_i's for some finite k. Show that if $\{f_i : i \in n\}$ is any other basis for X, then a is a linear combination of k-fold products of the f_i's. Show by an example that the analogous proposition is false for an element of a universal Clifford algebra of a non-degenerate real or complex orthogonal space. □

13.73. Let X be as in 13.72. Verify that the set of elements of $\bigwedge X$ expressible in terms of a basis $\{e_i : i \in n\}$ for X as a linear combination of k-fold products of the e_i's is a linear space of dimension $\binom{n}{k}$, where $\binom{n}{k}$ is the coefficient of x^k in the polynomial $(1 + x)^n$. (This linear space, which is defined by 13.72 independently of the choice of basis for X, is denoted by $\bigwedge^k X$.) □

13.74. Let X be as in 13.72, let $(a_i : i \in n)$ be an n-tuple of elements of X, let $(e_i : i \in n)$ be a basic frame on X and let $t : X \to X$ be the linear map sending e_i to a_i, for all $i \in n$. Prove that, in $\bigwedge X$,

$$\prod a_n = (\det t) \prod e_n.$$ □

13.75. Let X be as in 13.72 and let $(a_i : i \in k)$ be a k-tuple of elements of X. Prove that $(a_i : i \in k)$ is a k-frame on X if, and only if, in $\bigwedge X$, $\prod a_k \neq 0$. □

13.76. Let X be as in 13.72 and let $(a_i : i \in k)$ and $(b_i : i \in k)$ be k-frames on X. Prove that the k-dimensional linear subspaces im a and im b of X coincide if, and only if, $\prod b_k$ is a (non-zero) scalar multiple of $\prod a_k$ in $\bigwedge^k X$.

(Consider first the case $k = 2$. In this case

$$a_0 a_1 = b_0 b_1 \Rightarrow a_0 a_1 b_0 = 0 \Rightarrow b_0 \in \operatorname{im} a,$$

since $b_1 b_0 = -b_0 b_1$ and $b_0{}^2 = 0$. It should now be easy to complete the argument, not only in this case, but in the general case.) □

13.77. Construct an injective map $\mathscr{G}_k(X) \to \mathscr{G}_1(\bigwedge^k X)$, where X is as in 13.72.

(Use Exercise 13.76. This is the link between Grassmannians and Grassmann algebras.) □

13.78. Let ρ_g be the rotation of $\mathbf{R}^4$ induced by an element g of $\Gamma^0(\mathbf{R}^4)$ with real part equal to 1. Prove that ρ_g is expressible as the composite of *two* hyperplane reflections (cf. Theorem 9.41) if, and only if, g is of the form

$$1 + s_{01}e_0e_1 + s_{02}e_0e_2 + s_{03}e_0e_3 + s_{12}e_1e_2 + s_{13}e_1e_3 + s_{23}e_2e_3$$

where (e_0,e_1,e_2,e_3) is the standard basic frame on $\mathbf{R}^4$. Deduce that

$$1 + s_{01}e_0e_1 + s_{02}e_0e_2 + s_{03}e_0e_3 + s_{12}e_1e_2 + s_{13}e_1e_3 + s_{23}e_2e_3$$

is the product in the Clifford algebra $\mathbf{R}_{0,4}$ of two elements of $\mathbf{R}^4$ if, and only if,

$$\operatorname{pf} s = s_{01}s_{23} - s_{02}s_{13} + s_{03}s_{12} = 0. \quad \square$$

13.79. Prove that an invertible element

$$1 + s_{01}e_0e_1 + s_{02}e_0e_2 + s_{03}e_0e_3 + s_{12}e_1e_2 + s_{13}e_1e_3 + s_{23}e_2e_3$$

of the Clifford algebra $\mathbf{C}_4$ is the product of two elements of $\mathbf{C}^4$ if, and only if, $\operatorname{pf} s = 0$. □

13.80. Prove that an element

$$s_{01}e_0e_1 + s_{02}e_0e_2 + s_{03}e_0e_3 + s_{12}e_1e_2 + s_{13}e_1e_3 + s_{23}e_2e_3$$

of $\bigwedge^2(\mathbf{K}^4)$, where $\mathbf{K} = \mathbf{R}$ or $\mathbf{C}$, is the product of two elements of $\mathbf{K}^4$ if, and only if, $\operatorname{pf} s = 0$. Deduce that the image constructed in Exercise 13.77 of the Grassmannian $\mathscr{G}_2(\mathbf{R}^4)$ in the projective space $\mathscr{G}_1(\bigwedge^2(\mathbf{R}^4))$ is the projective quadric with equation

$$s_{01}s_{23} - s_{02}s_{13} + s_{03}s_{12} = 0. \quad \square$$

13.81. Let X be a four-dimensional real or complex linear space, let $Q = \mathscr{G}_2(X)$ be regarded as a projective quadric in $\mathscr{G}_1(\bigwedge^2 X)$ as in Exercise 13.80, let L be a line through 0 in X and let M be a three-dimensional linear subspace of X. Prove that the set of planes containing L as a linear subspace is a projective plane lying on Q and that the set of planes that are linear subspaces of M also is a projective plane lying on Q. Show also that these projective planes belong one to each of the two families of planes on Q. (Cf. Exercise 12.26.) (Consider first the case where $X = L \oplus M$. A suitable basis for X may then be chosen. Separate L, L' may be compared via some common linear complement M.) □

13.82. Determine $\mathrm{Spin}^+(2,2)$. □

13.83. The definitions of 'spinor space' on pages 249 and 251 are slightly dishonest. Why? □

CHAPTER 14

THE CAYLEY ALGEBRA

In this chapter we take a brief look at a non-associative algebra over **R** that nevertheless shares many of the most useful properties of **R**, **C** and **H**. Though it is rather esoteric, it often makes its presence felt in classification theorems and can ultimately be held 'responsible' for a rich variety of exceptional cases. Most of these lie beyond our scope, but the existence of the algebra and its main properties are readily deducible from our work on Clifford algebras in the previous chapter.

Real division algebras

A *division algebra* over **R** or *real division algebra* is, by definition, a finite-dimensional real linear space X with a bilinear product $X^2 \to X$; $(a,b) \rightsquigarrow ab$ such that, for all $a, b \in X$, the product $ab = 0$ if, and only if, $a = 0$ or $b = 0$ or, equivalently, if, and only if, the linear maps

$$X \to X; \quad x \rightsquigarrow xb \quad \text{and} \quad x \rightsquigarrow ax$$

are injective when a and b are non-zero, and therefore bijective.

We are already familiar with three associative real division algebras, namely **R** itself, **C**, the field of complex numbers, representable as a two-dimensional subalgebra of **R**(2), and **H**, the non-commutative field of quaternions, representable as a four-dimensional subalgebra of **R**(4). Each has unity and for each there is an anti-involution, namely conjugation, which may be made to correspond to transposition in the matrix algebra representation, such that the map of the algebra to **R**,

$$N; \quad a \rightsquigarrow N(a) = \bar{a}\, a,$$

is a real-valued positive-definite quadratic form that respects the algebra product, that is, is such that, for each a, b in the algebra,

$$N(ab) = N(a)\, N(b).$$

A division algebra X with a positive-definite quadratic form $N : X \to \mathbf{R}$ such that, for all $a, b \in X$, $N(ab) = N(a)\, N(b)$, is said to be a *normed* division algebra.

Alternative division algebras

An algebra X such that, for all $a, b \in X$, $a(ab) = a^2b$ and $(ab)b = ab^2$ is said to be an *alternative* algebra. For example, any associative algebra is an alternative algebra.

Prop. 14.1. Let X be an alternative algebra. Then for all $a, b \in X$, $(ab)a = a(ba)$.

Proof For all $a, b \in X$,

$$(a + b)^2a = (a + b)((a + b)a)$$
$$\Rightarrow (a^2 + ab + ba + b^2)a = (a + b)(a^2 + ba)$$
$$\Rightarrow a^2a + (ab)a + (ba)a + b^2a = aa^2 + a(ba) + ba^2 + b(ba)$$
$$\Rightarrow (ab)a = a(ba). \qquad \square$$

Prop. 14.2. Let X be an alternative division algebra. Then X has unity and each non-zero $a \in X$ has an inverse.

Proof If X has a single element, there is nothing to be proved. So suppose it has more than one element. Then there is an element $a \in X$, with $a \neq 0$. Let e be the unique element such that $ea = a$. This exists, since the map $x \rightsquigarrow xa$ is bijective. Then $e^2a = e(ea) = ea$. So $e^2 = e$. Therefore, for all $x \in X$, $e(ex) = e^2x = ex$ and $(xe)e = xe^2 = xe$. So $ex = x$ and $xe = x$. That is, e is unity.

Again let $a \neq 0$ and let b be such that $ab = e$. Then $a(ba) = (ab)a = ea = ae$. So $ba = e$. That is, b is inverse to a. $\square$

The Cayley algebra

There are many non-associative division algebras over $\mathbf{R}$. Such an algebra may fail even to be power-associative, that is, it may contain an element a such that, for example, $(a^2)a \neq a(a^2)$. A less exotic example is given in Exercise 14.13. However, only one of the non-associative division algebras is of serious interest. This is the alternative eight-dimensional *Cayley algebra* or *algebra of Cayley numbers* [9] (also known as the algebra of *octaves* or *octonions*). Despite the lack of associativity and commutativity there is unity, the subalgebra generated by any two of its elements is isomorphic to $\mathbf{R}$, $\mathbf{C}$ or $\mathbf{H}$ and so is associative, and there is a conjugation anti-involution sharing the same properties as conjugation for $\mathbf{R}$, $\mathbf{C}$ or $\mathbf{H}$.

The existence of the Cayley algebra depends on the fact that the matrix algebra $\mathbf{R}(8)$ may be regarded as a (non-universal) Clifford algebra for the positive-definite orthogonal space $\mathbf{R}^7$ in such a way that conjugation of the Clifford algebra corresponds to transposition in $\mathbf{R}(8)$.

For then, as was noted on page 273, the images of $\mathbf{R}$ and $\mathbf{R}^7$ in $\mathbf{R}(8)$ together span an eight-dimensional linear subspace T, passing through 1, and such that each non-zero element $t \in T$ is invertible.

This leads to the following.

Prop. 14.3. For any linear isomorphism $\mu : \mathbf{R}^8 \to T$ the map $\mathbf{R}^8 \times \mathbf{R}^8 \to \mathbf{R}^8$; $(a,b) \rightsquigarrow ab = (\mu(a))(b)$ is a bilinear product on $\mathbf{R}^8$ such that, for all $a, b \in \mathbf{R}^8$, $ab = 0$ if, and only if, $a = 0$ or $b = 0$. Moreover, any non-zero element $e \in \mathbf{R}^8$ can be made unity for such a product by choosing μ to be the inverse of the isomorphism $T \to \mathbf{R}^8$; $t \rightsquigarrow t(e)$. □

The division algebra with unity e introduced in Prop. 14.3 is called the *Cayley algebra* on $\mathbf{R}^8$ with unity e. It is rather easy to see that any two such algebras are isomorphic. We shall therefore speak simply of *the* Cayley algebra, denoting it by $\mathbf{O}$ (for octonions). Though the choice of e is essentially unimportant, it will be convenient to select an element of length 1 in $\mathbf{R}^8$, say the element e_0 of its standard basis.

Here we have implicitly assigned to $\mathbf{R}^8$ its standard positive-definite orthogonal structure. The space T also has an orthogonal structure induced by conjugation on the Clifford algebra $\mathbf{R}(8)$. The quadratic form on T is the map

$$T \to \operatorname{im} \alpha; \quad t \rightsquigarrow N(t) = t^\tau t,$$

where $\alpha : \mathbf{R} \to \mathbf{R}(8)$ is the natural injection, and we shall denote the quadratic form on $\mathbf{R}^8$, similarly, by

$$\mathbf{R}^8 \to \mathbf{R}; \quad a \rightsquigarrow N(a) = a \cdot a = a^\tau a.$$

The Cayley algebra $\mathbf{O}$ inherits both structures, the one on $\mathbf{R}^8$ directly and the one on T by μ and α. As the next proposition shows, the choice of e as an element in $\mathbf{R}^8$ of length 1 guarantees that these two structures on $\mathbf{O}$ coincide.

Prop. 14.4. For all $a \in \mathbf{R}^8$, $N(\mu(a)) = \alpha(N(a))$.

Proof For all $a \in \mathbf{R}^8$,

$$\begin{aligned} N(a) &= N(\mu(a)(e)), \quad \text{by the definition of } \mu, \\ &= e^\tau(\mu(a))^\tau \mu(a)e \\ &= e^\tau N(\mu(a))e. \end{aligned}$$

Moreover, $e^\tau e = 1$. Since $N(t) \in \operatorname{im} \alpha$, for all $t \in T$, it follows at once that $N(\mu(a)) = \alpha(N(a))$. □

Conjugation on $\mathbf{R}(8)$ induces a *linear* involution

$$\mathbf{O} \to \mathbf{O}; \quad a \rightsquigarrow \bar{a} = \mu^{-1}(\mu(a))^\tau,$$

which we shall call *conjugation on* $\mathbf{O}$. This involution induces a direct sum decomposition $\mathbf{O} = (\operatorname{im}\{e\}) \oplus \mathbf{O}'$ in which $\mathbf{O}' = \{b \in \mathbf{O} : \bar{b} = -b\}$.

The following proposition lists some important properties both of the quadratic form and of conjugation on **O**. The product on **R**(8) and the product on **O** will both be denoted by juxtaposition, as will be the action of **R**(8) on **O**. It is important to remember, throughout the discussion, that though the product on **R**(8) is associative, the product on **O** need not be.

Prop. 14.5. For all $a, b \in \mathbf{O}$, $N(ab) = N(a)N(b)$, implying that **O** is normed, $(a \cdot b)e = \frac{1}{2}(\bar{a}b + \bar{b}a)$, implying that $\mathbf{O}' = (\operatorname{im}\{e\})^{\perp}$, $(N(a))e = \bar{a}a = a\bar{a}$, and $\overline{ab} = \bar{b}\bar{a}$, this last property implying that conjugation is an algebra anti-involution.

Proof For all $a, b \in \mathbf{O}$,

$$N(ab) = N(\mu(a)b) = b^{\tau}\mu(a)^{\tau}\mu(a)b$$
$$= b^{\tau}\alpha(N(a))b = N(a)N(b).$$

Also

$$\bar{a}b + \bar{b}a = \bar{a}(be) + \bar{b}(ae)$$
$$= (\mu(a)^{\tau}\mu(b) + \mu(b)^{\tau}\mu(a))e$$
$$= 2(a \cdot b)e,$$

implying that if $a \in \operatorname{im}\{e\}$ and if $b \in \mathbf{O}'$, then $2(a \cdot b)e = ab - ba = 0$, since e, and therefore any real multiple of e, commutes with any element of **O**. It implies, secondly, since $N(a) = a \cdot a$, that $N(a) = \bar{a}a$ and, since $\mu(a)\mu(a)^{\tau} = \mu(a)^{\tau}\mu(a)$, that $a\bar{a} = N(a)$.

The last part is harder to prove. Note first that for all $a, b, x \in \mathbf{O}$,

$$((ab)\cdot x)e = (ab)^{\tau}x = (\mu(a)b)^{\tau}x = b^{\tau}\mu(a)^{\tau}x = b^{\tau}(\bar{a}x) = (b\cdot(\bar{a}x))e.$$

So $(ab)\cdot x = b \cdot (\bar{a}x)$.

Also $(ab)\cdot x = a\cdot(x\bar{b})$, for this is so for all $b \in \operatorname{im}\{e\}$, while, if $b \in \mathbf{O}'$,

$$((ab)\cdot x - a\cdot(x\bar{b}))e = ((ab)\cdot x + (xb)\cdot a)e$$
$$= (ab)^{\tau}x + (xb)^{\tau}a$$
$$= b^{\tau}\mu(a)^{\tau}x + b^{\tau}\mu(x)^{\tau}a$$
$$= b^{\tau}(\mu(a)^{\tau}\mu(x) + \mu(x)^{\tau}\mu(a))e$$
$$= (a\cdot x)(b^{\tau}e)$$
$$= 0, \text{ since } b^{\tau}e = 0.$$

Therefore, for all $a, b, x \in \mathbf{O}$,

$$x\cdot ab = ((ab)x)\cdot e = (ab)\cdot\bar{x} = b\cdot(\bar{a}\bar{x}) = (bx)\cdot\bar{a} = x\cdot(\bar{b}\bar{a}).$$

Since the scalar product on **O** is non-degenerate, it follows that $\overline{ab} = \bar{b}\bar{a}$. □

Note that, for $a, b \in \mathbf{O}'$, $(a\cdot b)e = -\frac{1}{2}(ab + ba)$, since $\bar{a} = -a$ and $\bar{b} = -b$. So $a \cdot b = 0 \Leftrightarrow ab + ba = 0$, implying that **O** is non-commutative, since $\dim \mathbf{O}' > 1$.

As we shall see later, **O** is not associative either. Nevertheless **O** is alternative, this being a corollary of the following proposition.

Prop. 14.6. For all $a, b \in \mathbf{O}$,

$$\bar{a}(ab) = (\bar{a}a)b = N(a)b = (ba)\bar{a}.$$

Proof The first part is straightforward, with

$$\bar{a}(ab) = \mu(\bar{a})\mu(a)b = \mu(a)^\tau\mu(a)b = N(a)b.$$

For the other part, let $c = (ba)\bar{a}$. Then, by conjugation, $\bar{c} = a(\bar{a}\bar{b})$, and this, by the first part, is equal to $N(a)\bar{b}$. Therefore $c = N(a)b$. □

To prove the alternativity of **O** from this, use the fact that, for all $a \in \mathbf{O}$, $a + \bar{a} \in \operatorname{im}\{e\}$.

Throughout the remainder of this chapter we shall identify **R** with $\operatorname{im}\{e\}$. In particular, we shall write 1 in place of e to denote unity in **O**.

Hamilton triangles

It has been remarked that two elements $a, b \in \mathbf{O}'$ are orthogonal if, and only if, they anticommute. An orthonormal 3-frame (i,j,k) in $\mathbf{O}'$, with i = jk, j = ki and k = ij, therefore spans, with 1, a subalgebra of **O** isomorphic with the quaternion algebra **H**. Such a 3-frame will be said to be a *Hamilton triangle* in **O** and will also be denoted by the diagram

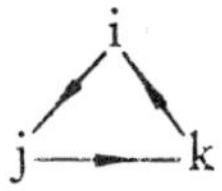

in which each vertex is the product of the other two in the order indicated by the arrows.

Prop. 14.7. Let a and b be mutually orthogonal elements of $\mathbf{O}'$ and let $c = ab$. Then $c \in \mathbf{O}'$ and is orthogonal both to a and to b.

Proof First,

$$\begin{aligned} a \cdot b = 0 &\Rightarrow ab + ba = 0 \\ &\Rightarrow \bar{c} = \overline{ab} = \bar{b}\bar{a} = (-b)(-a) = -c \\ &\Rightarrow c \in \mathbf{O}'. \end{aligned}$$

Also
$$\begin{aligned} a \cdot c &= \tfrac{1}{2}(\bar{a}(ab) + (\overline{ab})a) \\ &= \tfrac{1}{2}(N(a)b + \bar{b}N(a)), \quad \text{by Prop. 14.6,} \\ &= 0, \quad \text{since } b + \bar{b} = 0. \end{aligned}$$

Similarly, $b \cdot c = 0$. □

Cor. 14.8. Let (i,j) be an orthonormal 2-frame in $\mathbf{O}'$ and let k = ij. Then (i,j,k) is a Hamilton triangle in $\mathbf{O}'$. □

From this follows the assertion made earlier that the subalgebra generated by any two elements of **O** is isomorphic to **R**, **C** or **H** and so is, in particular, associative.

Cayley triangles

Finally, any Hamilton triangle in $\mathbf{O}'$ may be extended to a useful orthonormal basis for $\mathbf{O}'$. We begin by defining a *Cayley triangle* in $\mathbf{O}'$ to be an orthonormal 3-frame (a,b,c) in $\mathbf{O}'$ such that c is also orthogonal to ab.

Prop. 14.9. Let (a,b,c) be a Cayley triangle in $\mathbf{O}'$. Then

(i) $a(bc) + (ab)c = 0$, exhibiting the non-associativity of **O**,

(ii) $a \cdot (bc) = 0$, implying that the elements a, b, c form a Cayley triangle in whatever order they are listed,

(iii) $ab \cdot bc = 0$, implying that (a,b,bc) is a Cayley triangle,

and (iv) $(ab)(bc) = ac$, implying that (ab,bc,ac) is a Hamilton triangle.

Proof

(i) Since (a,b,c) is a Cayley triangle,
$ab + ba = ac + ca = bc + cb = (ab)c + c(ab) = 0.$
So
$$\begin{aligned} a(bc) + (ab)c &= -a(cb) - c(ab) \\ &= (a^2 + c^2)b - (a + c)(ab + cb) \\ &= (a + c)^2 b - (a + c)((a + c)b) = 0. \end{aligned}$$

(ii) From (i) it follows by conjugation that $(\bar{c}\bar{b})\bar{a} + \bar{c}(\bar{b}\bar{a}) = 0$ and therefore that $(bc)a + c(ab) = 0$. Since $(ab)c + c(ab) = 0$, it follows that $a(bc) + (bc)a = 0$, implying that $a \cdot (bc) = 0$.

(iii)
$$\begin{aligned} 2ab \cdot bc &= (ba)(bc) + (bc)(ba) \\ &= (ba)^2 + (bc)^2 - (b(a - c))^2 \\ &= -b^2a^2 - b^2c^2 + b^2(a - c)^2 \\ &= -b^2(ac + ca) \\ &= 2b^2 a \cdot c = 0. \end{aligned}$$

(iv) Apply (i) to the Cayley triangle (a,b,bc).
Then $(ab)(bc) = -a(b(bc)) = ac$, since $b^2 = 1$. □

We can reformulate this as follows.

Prop. 14.10. Let (i,j,l) be a Cayley triangle in $\mathbf{O}'$ and let k = ij. Then {i,j,k,l,il,jl,kl} is an orthonormal basis for $\mathbf{O}'$, and if these seven elements are arranged in a regular heptagon as follows:

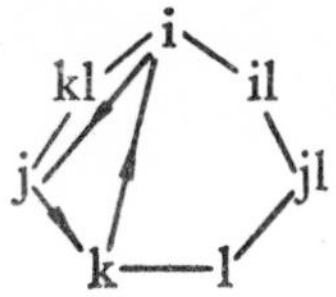

then each of the seven triangles obtained by rotating the triangle

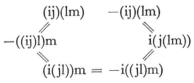

through an integral multiple of $2\pi/7$ is a Hamilton triangle, that is, each vertex is the product of the other two vertices in the appropriate order. □

This heptagon is essentially the multiplication table for the Cayley algebra **O**.

From this it is easy to deduce that there cannot be any division algebra over **R** of dimension greater than 8 such that the subalgebra generated by any three elements is isomorphic to **R**, **C**, **H** or **O**. Such an algebra A, if it exists, has a conjugation anti-involution, inducing a direct sum decomposition $\mathbf{R} \oplus A'$ of A in which A' consists of all the elements of A which equal the negative of their conjugate. Further details are sketched in Exercise 14.15. The following proposition then settles the matter.

Prop. 14.11. Let (i,j,l) be any Cayley triangle in A', let k = ij and let m be an element orthogonal to each of the seven elements i, j, k, l, il, jl and kl of the Cayley heptagon. Then m = 0.

Proof We remark first that parts (i) and (ii) of Prop. 14.9 hold for any $a, b, c \in A'$ such that $a \cdot b = a \cdot c = b \cdot c = ab \cdot c = 0$. Using this several times, we find, on making a circuit of the 'rebracketing pentagon', that

$$\begin{array}{ccc} & (ij)(lm) \qquad -(ij)(lm) & \\ & /\!/ \qquad\qquad \backslash\!\backslash & \\ -((ij)l)m & & i(j(lm)) \\ & \backslash\!\backslash \qquad\qquad /\!/ & \\ & (i(jl))m = -i((jl)m) & \end{array}$$

So (ij)(lm) = 0. But ij $\neq$ 0; so lm = 0, and therefore, since l $\neq$ 0, m = 0. □

Further results

There are various stronger results, for example

(i) Frobenius' theorem (1878) that any associative division algebra over **R** is isomorphic to **R**, **C** or **H**;
(ii) Hurwitz' theorem (1898) that any normed division algebra over **R**, with unity, is isomorphic to **R**, **C**, **H** or **O**;
(iii) the theorem of Skornyakov (1950) and Bruck-Kleinfeld (1951) that any alternative division algebra over **R** is isomorphic to **R**, **C**, **H** or **O**; and
(iv) the theorem of Kervaire, Bott-Milnor, and Adams (1958), that any division algebra over **R** has dimension 1, 2, 4 or 8.

The first two of these are little more difficult to prove than what we have proved here and can be left as exercises. The starting point in the proof of (i) is the remark that any element of an associative n-dimensional division algebra must be a root of a polynomial over **R** of degree at most n and therefore, by the fundamental theorem of algebra, proved later in Chapter 19, must be the solution of a quadratic equation. From this it is not difficult to define the conjugation map and to prove its linearity. Result (iii) is harder to prove. The discussion culminates in the following.

Theorem 14.12. Any real non-associative alternative division algebra is a Cayley algebra.

Indication of proof Let A be a real non-associative alternative division algebra, and, for any $x, y, z \in A$ let

$$[x,y] = xy - yx$$

and $[x,y,z] = (xy)z - x(yz)$. It can be shown that if x and y are such that $u = [x,y] \neq 0$, then there exists z such that $v = [x,y,z] \neq 0$. It can then be shown that $uv + vu = 0$ and therefore, by the previous remark, that there exists t such that $w = [u,v,t] \neq 0$. One can now verify that u^2, v^2 and w^2 are negative real numbers and that

$$\mathrm{i} = u/\sqrt{-u^2}, \quad \mathrm{j} = v/\sqrt{-v^2} \quad \text{and} \quad \mathrm{l} = w/\sqrt{-w^2}$$

form a Cayley triangle. Then A contains a Cayley algebra as a subalgebra. It follows, essentially by Prop. 14.11, that A coincides with this Cayley algebra.

The details are devious and technical, and the reader is referred to [36] for a full account. □

Finally, (iv) is very hard indeed. Its proof uses the full apparatus of algebraic topology. Cf. [1], [35], [45].

The Cayley projective line and plane

Most of the standard results of linear algebra do not generalize over the non-associative Cayley algebra, for the very definition of a linear space involves the associativity of the field. Nevertheless we can regard the map

$$\mathbf{O}^n \times \mathbf{O} \to \mathbf{O}^n; \quad ((y_i : i \in n), y) \rightsquigarrow (y_i y : i \in n)$$

as a quasi-linear structure for the additive group $\mathbf{O}^n$.

It is also possible to define a 'projective line' and a 'projective plane' over $\mathbf{O}$.

The *Cayley projective line* $\mathbf{O}P^1$ is constructed by fitting together two copies of $\mathbf{O}$ in the manner discussed on page 141. Any point is represented either by $[1,y]$ or by $[x,1]$, with $[1,y] = [x,1]$ if, and only if, $y = x^{-1}$, the square brackets here having their projective-geometry connotation. There is even a 'Hopf map' $h : \mathbf{O}^2 \rightarrowtail \mathbf{O}P^1$ defined by $h(y_0,y_1) = [y_0 y_1^{-1},1]$, whenever $y_1 \neq 0$, and by $h(y_0,y_1) = [1,y_1 y_0^{-1}]$, whenever $y_0 \neq 0$. Since any two elements of $\mathbf{O}$ (for example, y_0 and y_1) generate an associative subalgebra, it is true that $y_0 y_1^{-1} = (y_1 y_0^{-1})^{-1}$, and so the two definitions agree, whenever y_0 and y_1 are both non-zero.

The *Cayley projective plane* $\mathbf{O}P^2$ is similarly constructed by fitting together three copies of $\mathbf{O}^2$. Any point is represented in at least one of the forms $[1,y_0,z_0]$, $[x_1,1,z_1]$ or $[x_2,y_2,1]$. The obvious identifications are compatible, though this requires careful checking because of the general lack of associativity. What we require is that the equations

$$x_1 = y_0^{-1}, \quad z_1 = z_0 y_0^{-1} \quad \text{and} \quad x_2 = x_1 z_1^{-1}, \quad y_2 = z_1^{-1}$$

be compatible with the equations

$$x_2 = z_0^{-1}, \quad y_2 = y_0 z_0^{-1}.$$

But all is well, since

$$x_1 z_1^{-1} = y_0^{-1}(z_0 y_0^{-1})^{-1} = z_0^{-1}$$

and

$$z_1^{-1} = (z_0 y_0^{-1})^{-1} = y_0 z_0^{-1},$$

once again because the subalgebra generated by any two elements is associative.

The further study of the Cayley plane is beyond the scope of this book except for a few brief references later (cf. pages 401, 405 and 416).

Useful analogues over $\mathbf{O}$ of projective spaces of dimension greater than 2 do not exist. The reader is referred to [8] for a discussion.

FURTHER EXERCISES

14.13. Let X be a four-dimensional real linear space with basis elements denoted by 1, i, j and k, and let a product be defined on X by prescribing that

$$\mathrm{i}^2 = \mathrm{j}^2 = \mathrm{k}^2 = -1,$$
$$\mathrm{jk} + \mathrm{kj} = \mathrm{ki} + \mathrm{ik} = \mathrm{ij} + \mathrm{ji} = 0$$

and
$$\mathrm{jk} = \alpha\mathrm{i}, \ \mathrm{ki} = \beta\mathrm{j} \text{ and } \mathrm{ij} = \gamma\mathrm{k},$$

where α, β, γ are non-zero real numbers, all of the same sign. Prove that X, with this product, is a real division algebra and that X is associative if, and only if, $\alpha = \beta = \gamma = 1$ or $\alpha = \beta = \gamma = -1$. □

14.14. Prove that if $a, b \in \mathbf{O}'$ (cf. page 279) then $ab - ba \in \mathbf{O}'$. □

14.15. Let X be a division algebra over $\mathbf{R}$ such that for each $x \in X$ there exist $\alpha, \beta \in \mathbf{R}$ such that $x^2 - 2\alpha x + \beta = 0$ and let x' consist of all $x \in X$ for which there exists $\beta \in \mathbf{R}$ such that $x^2 + \beta = 0$, with $\beta \geqslant 0$.

Prove that X' is a linear subspace of X, that $X = \mathbf{R} \oplus X'$ and that the map

$$\mathbf{R} \oplus X' \to \mathbf{R} \oplus X'; \quad \lambda + x' \rightsquigarrow \lambda - x',$$

where $\lambda \in \mathbf{R}$, $x' \in X'$, is an anti-involution of X. □

14.16. Let A be a real alternative division algebra, and, for any $x, y, z \in A$, let

$$[x,y] = xy - yx$$

and
$$[x,y,z] = (xy)z - x(yz).$$

Prove that the interchange of any two letters in $[x,y,z]$ changes the sign, and that

$$[xy,z] - x[y,z] - [x,z]y = 3[x,y,z].$$

Hence show that if A is commutative, then A is also associative.

(Note: For all $x, y, z \in A$,

$$[x + y, x + y, z] = 0 = [x, y + z, y + z].) \quad □$$

14.17. Let A be a real alternative division algebra, and, for any $w, x, y, z \in A$, let

$$[w,x,y,z] = [wx,y,z] - x[w,y,z] - [x,y,z]w.$$

Prove that the interchange of any two letters in $[w,x,y,z]$ changes the sign.

(Note: For all $w, x, y, z \in A$,

$$w[x,y,z] - [wx,y,z] + [w,xy,z] - [w,x,yz] + [w,x,y]z = 0.) \quad □$$

14.18. Let A be a real alternative division algebra, let $x, y, z \in A$ and let $u = [x,y]$, $v = [x,y,z]$. Prove that $[v,x,y] = vu = -uv$. □

14.19. Prove that the real linear involution $\mathbf{O} \to \mathbf{O}$; $a \rightsquigarrow \tilde{a}$, sending j, l, jl to −j, −l, −jl, respectively, and leaving 1, i, k, il and kl fixed, is an algebra anti-involution of $\mathbf{O}$. □

14.20. Verify that the map $\beta : \mathbf{H}^2 \to \mathbf{O}$; $x \rightsquigarrow x_0 + \mathrm{l}x_1$ is a right $\mathbf{H}$-linear isomorphism and compute $\beta^{-1}(\beta(x)\,\beta(y))$, for any $x, y \in \mathbf{H}^2$. (Cf. Exercise 14.19.)

Let $Q = \{(x,y) \in (\mathbf{H}^2)^2 : \tilde{x}_0 y_0 + \tilde{x}_1 y_1 = 1\}$. Prove that for any $(a,b) \in \mathbf{O}^* \times \mathbf{H}$, $(\beta^{-1}(\tilde{a}), \beta^{-1}(a^{-1}(1 + \mathrm{l}b))) \in Q$ and that the map

$$\mathbf{O}^* \times \mathbf{H} \to Q; \quad (a,b) \rightsquigarrow (\beta^{-1}(\tilde{a}), \quad \beta^{-1}(a^{-1}(1 + \mathrm{l}b)))$$

is bijective. (Cf. Exercise 10.66.) □

14.21. Verify that the map $\gamma : \mathbf{C}^4 \to \mathbf{O}$; $x \rightsquigarrow x_0 + \mathrm{j}x_1 + \mathrm{l}x_2 + \mathrm{jl}x_3$ is a right $\mathbf{C}$-linear isomorphism and compute $\gamma^{-1}(\gamma(x)\gamma(y))$, for any $x, y \in \mathbf{C}^4$.

Let $Q = \{(x,y) \in (\mathbf{C}^4)^2 : \sum_{i \in 4} x_i y_i = 1\}$. Prove that, for any $(a,(b,c,d)) \in Y^* \times \mathbf{C}^3$, $(\gamma^{-1}(\tilde{a}), \gamma^{-1}(a^{-1}(1 + \mathrm{j}b + \mathrm{l}c + \mathrm{jl}d))) \in Q$ and that the map
$\mathbf{O}^* \times \mathbf{C}^3 \to Q; \quad (a,(b,c,d)) \rightsquigarrow (\gamma^{-1}(\tilde{a}), \gamma^{-1}(a^{-1}(1 + \mathrm{j}b + \mathrm{l}c + \mathrm{jl}d)))$
is bijective. □

14.22. Show that the fibres of the restriction of the Hopf map

$$\mathbf{O}^2 \to \mathbf{O}P^1; \quad (y_0,y) \rightsquigarrow [y_0,y_1]$$

to the sphere $S^{15} = \{(y_0,y_1) \in \mathbf{O}^2 : \bar{y}_0 y_0 + \bar{y}_1 y_1 = 1\}$ are 7-spheres, any two of which link. (See the comment following Exercise 10.69.) □

CHAPTER 15

NORMED LINEAR SPACES

Topics discussed in this chapter, which is independent of the four which immediately precede it, include norms on real affine spaces, subsets of such spaces open or closed with respect to a norm, continuity for maps between normed affine spaces, and completeness for normed affine spaces. These provide motivation for the study of topological spaces and the deeper properties of continuous maps in Chapter 16, and also provide concepts and techniques which will be extensively used in the theory of affine approximation in Chapters 18 and 19. The material is limited strictly to what is required in these chapters. For this reason such basic theorems as the Hahn–Banach theorem, the open mapping theorem and the closed graph theorem have been omitted. These theorems would be required if we wished to extend the section on smoothness in Chapter 19 to the non-finite-dimensional case. For them the reader is referred to [54] or [18], or any other introduction to functional analysis.

Norms

In Chapter 9 we defined the norm $|x|$ of an element x of a positive-definite real orthogonal space X to be $\sqrt{(|x^{(2)}|)}$, and in Prop. 9.58 we listed some of the properties of the map

$$X \longrightarrow \mathbf{R}; \quad x \rightsquigarrow |x|.$$

These included the following:

(i) for all $x \in X$, $|x| \geqslant 0$, with $|x| = 0$ if, and only if, $x = 0$;

(ii) for all $x \in X$ and all $\lambda \in \mathbf{R}$, $|\lambda x| = |\lambda|\,|x|$;

and (iii) for all $x, x' \in X$, $|x + x'| \leqslant |x| + |x'|$ (the *triangle inequality*), this last being equivalent, by (ii), with $\lambda = -1$, to

(iii)′ for all $x, x' \in X$, $||x| - |x'|| \leqslant |x - x'|$.

When X is any real linear space, any map $X \longrightarrow \mathbf{R}; x \rightsquigarrow |x|$ satisfying these three properties is said to be a *norm* on X, a norm being said to be *quadratic* if it is one that is induced by a positive-definite quadratic form on X.

The $| \ |$ notation is convenient, but we shall also sometimes use the double-line notation $|| \ ||$, especially when there is danger of confusion with the absolute value on **R**, or when two norms are under discussion at the same time.

Prop. 15.1. Any norm $|| \ ||$ on **R** is of the form $x \rightsquigarrow m\,|x|$ where $m = ||1|| > 0$, and conversely any such map is a norm on **R**. □

There is a greater choice of norms for $\mathbf{R}^2$ despite the fact that the restriction of such a norm to any line through 0 is essentially the absolute value on **R**, by Prop. 15.1. Examples include the *sum norm* $(x,y) \rightsquigarrow |x| + |y|$, the *product norm* $(x,y) \rightsquigarrow \sup\{|x|, |y|\}$ and the *quadratic norm* $(x,y) \rightsquigarrow \sqrt{(x^2 + y^2)}$, each of these being defined in terms of the standard basis for $\mathbf{R}^2$. A reason for the term 'product norm' will emerge presently. The check that the sum and product norms are norms is left as an easy exercise.

The following proposition provides an example of a norm on a possibly infinite-dimensional space. It may be regarded as a generalization of the norm referred to above as the product norm.

Prop. 15.2. Let $\mathscr{F}$ be the linear space of bounded real-valued functions on a set X and let $|| \ ||: \mathscr{F} \to \mathbf{R}$ be defined, for all $f \in \mathscr{F}$, by the formula $||f|| = \sup\{|f(x)| : x \in X\}$.

Then $|| \ ||$ is a norm on $\mathscr{F}$. □

A *normed linear space* $(X, | \ |)$ consists of a real linear space X and a norm $| \ |$ on X and a *normed affine space* $(X, | \ |)$ consists of a real affine space X and a norm $| \ |$ on the vector space X_*. In either case $(X, | \ |)$ is abbreviated to X wherever possible.

The restriction to a linear subspace W of the norm on a normed linear space X is a norm on W, and W is tacitly assigned this norm.

Prop. 15.3. Let X and Y be normed linear spaces. Then the map

$$X \times Y \to \mathbf{R}; \quad (x,y) \rightsquigarrow \sup\{|x|, |y|\}$$

is a norm on $X \times Y$. □

This norm is called the *product norm* on $X \times Y$. The definition generalizes in the obvious way to the product of any finite number of normed linear spaces.

Open and closed balls

The intuition surrounding the quadratic norms provides several descriptive terms which are also applied in using an arbitrary norm.

For example, let X be a normed affine space, subtraction being denoted simply by $-$ and the norm by $|\ \ |$, and let a and $b \in X$. The real number $|b - a|$ is then called the *distance* from a to b, and a subset A of X is said to be a *neighbourhood* of a in X if A contains all points of X *sufficiently close* to a; that is, if there exists a positive real number δ such that, for all $x \in X$,

$$|x - a| \leqslant \delta \Rightarrow x \in A,$$

or, equivalently, if there exists $\delta' > 0$ such that, for all $x \in X$,

$$|x - a| < \delta' \Rightarrow x \in A.$$

(The first statement clearly implies the second—we may take $\delta' = \delta$ —while the second implies the first on taking $\delta = \frac{1}{2}\delta'$.)

For any $a \in X$ and for any $\delta > 0$, the sets

$$\{x \in X : |x - a| < \delta\}, \quad \{x \in X : |x - a| \leqslant \delta\} \quad \text{and} \quad \{x \in X : |x - a| = \delta\}$$

are called, respectively, the *open ball*, the *closed ball* and the *sphere* in X with centre a and radius δ.

Thus a subset A of X is a neighbourhood of a in X if, and only if, there exists a ball (open or closed) B, with centre a, such that $B \subset A$.

A subset A of X is said to be *bounded* if there is a ball B in X such that $A \subset B$.

Prop. 15.4. Any ball, open or closed, in a normed affine space X is convex. □

Consider for example the three norms on $\mathbf{R}^2$ introduced above. A quadratic ball with centre 0 is a circular disc, centre 0, a product ball is a square disc with vertices the points $(\pm r, \pm r)$, r being the radius,

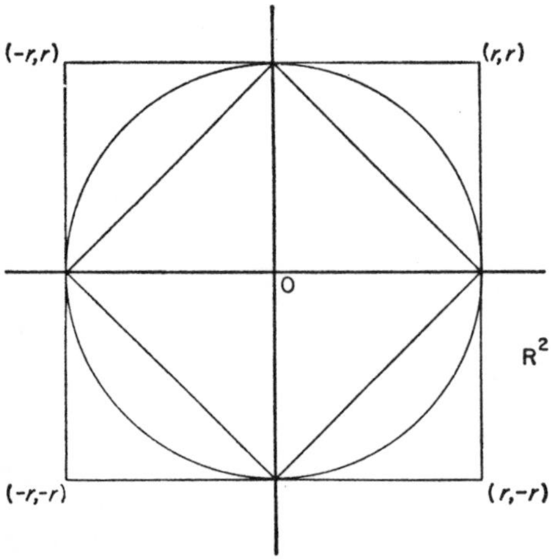

while a sum ball is also a square disc but with vertices the points $(\pm r, 0)$, $(0, \pm r)$, r again being the radius. The various balls are illustrated in the figure on page 290.

There may be some suspicion that any norm on a real linear space X is a quadratic norm, induced by some suitable quadratic form on X. That this is not so if $\dim X > 1$ follows from the following proposition.

Prop. 15.5. Let a and b be distinct points of a positive-definite real orthogonal space X, and suppose that $|a| = |b| = 1$, $|\ \ |$ denoting the quadratic norm. Then, for all $\lambda \in \mathbf{R}$,

$$|(1-\lambda)a + \lambda b| = 1 \quad \text{if, and only if,} \quad \lambda = 0 \text{ or } 1.$$

(Recall that $a \cdot b = |a||b| \Leftrightarrow |b|a = |a|b$.) □

It follows at once, for example, that the product norm on $\mathbf{R}^2$ is not a quadratic norm.

Open and closed sets

A subset A of a normed affine space X is said to be *open* in X if it is a neighbourhood of each of its points, and to be *closed* in X if its set complement in X, $X \backslash A$, is open.

Prop. 15.6. Any open ball in a normed affine space X is open in X, and any closed ball in X is closed in X. □

A subset of a normed affine space X need be neither open nor closed. For example, the interval $]-1,1]$ is neither open nor closed in $\mathbf{R}$ with respect to the absolute value norm. By contrast the null set and the whole space X are each both open and closed in X.

Prop. 15.7. Let X be a normed linear space. Then, if A and B are open subsets of X, $A \cap B$ is open in X while, if $\mathcal{S}$ is a set of open subsets of X, $\bigcup \mathcal{S}$ is open in X. □

It is a corollary of the first part of this proposition that the intersection of any non-null finite set of open subsets of X is open in X. However, the intersection of an infinite set of open subsets of X need not be open, an example being the set of all bounded open intervals in $\mathbf{R}$ with centre 0. By contrast, there is no requirement in the second part that $\mathcal{S}$ be finite, nor even countable.

It is natural to suppose that any affine subspace W of a normed affine space X is closed in X. This intuition is correct if X is finite-dimensional, by Theorem 15.26 and Cor. 15.24, though it is false in general.

Prop. 15.8. Let X be a real normed affine space and let W be a closed affine subspace in X of codimension 1. Then each side of W is open in X.

Proof Let A denote one of the sides of W in X and let $a \in A$. Since W is closed in X, $X \setminus W$ is open in X. So $X \setminus W$ contains a ball in X with centre a. Since the ball is convex, it cannot lie partly on one side

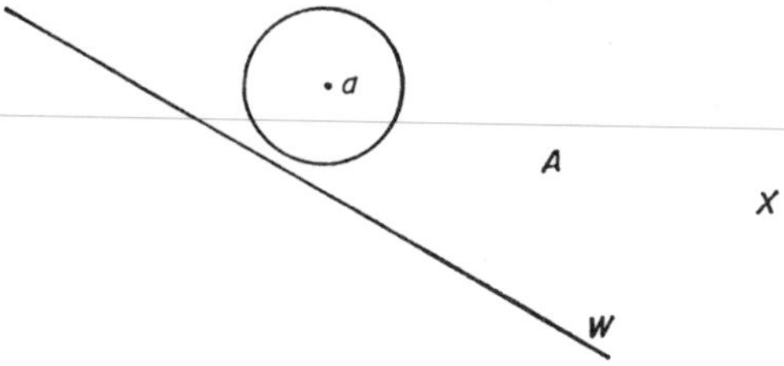

of W and partly on the other. So it lies entirely in A. Therefore A is open in X. □

Norms $|\ \ |$ and $||\ \ ||$ on an affine space X are said to be *equivalent* if they induce the same neighbourhoods or, equivalently, the same open sets on X. It follows at once that $|\ \ |$ and $||\ \ ||$ are equivalent if, and only if, each $|\ \ |$-ball in X contains as a subset a concentric $||\ \ ||$-ball and vice versa. For example, the standard product and quadratic norms on $\mathbf{R}^2$ are equivalent, since every square ball contains a concentric circular ball and vice versa.

An alternative criterion for the equivalence of norms will be given later (Prop. 15.18). It will also be proved later (Theorem 15.26) that any two norms on a *finite-dimensional* affine space X are equivalent.

Continuity

Let X and Y be normed affine spaces. A map $f: X \rightarrowtail Y$ (see page 39 for the notation) is said to be *continuous at* a point a of its domain if every neighbourhood B of $f(a)$ contains the image by f of some neighbourhood A of a.

In more intuitive language f is continuous at a if it sends points (sufficiently) close to a to points (as) close (as we please) to $f(a)$, the words in parentheses being strictly necessary if the statement is to be meaningful, though they are often omitted in practice. Note that, since the definition is in terms of neighbourhoods, the norm on either X or Y, or on both, may be replaced by an equivalent norm without affecting the continuity of f.

A map $f: X \rightarrowtail Y$ is said to be *continuous* if it is continuous at each point of its domain.

For example, any constant map on a normed affine space X is continuous. Also, the identity map 1_X is continuous.

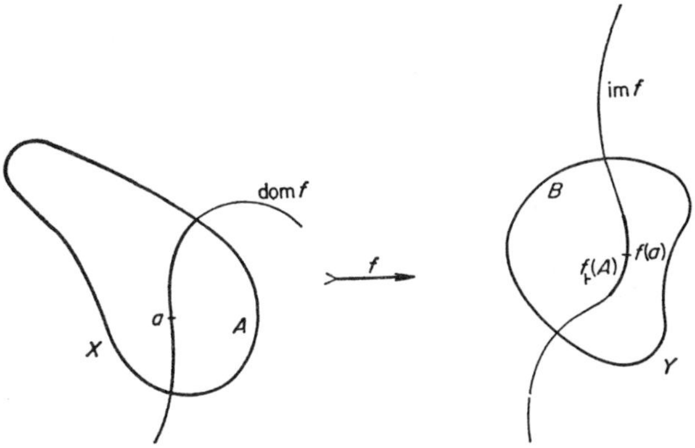

Prop. 15.9. Let X and Y be normed affine spaces. A map $f: X \rightarrow Y$ (with domain X) is continuous if, and only if, the inverse image $f^{-1}(B)$ of any open set B in Y is open in X. □

Exercise 15.10. Discuss possible extensions of the result of Prop. 15.9 to maps $f: X \rightarrowtail Y$ whose domain need not be the whole of X. □

The importance of Prop. 15.9 and Exercise 15.10 will become apparent in Chapter 16, where we develop the study of topological spaces and continuous maps between topological spaces. Meanwhile we establish several further forms of the definition of continuity at a point, each of which has certain technical advantages. We continue to assume that X and Y are normed affine spaces.

Prop. 15.11. The map $f: X \rightarrowtail Y$ is continuous at $a \in \operatorname{dom} f$ if, and only if, any ball B in Y with centre $f(a)$ contains the image by f of some ball A in X with centre a.

Proof $\Rightarrow$: Suppose f is continuous at a. Then, since any ball B in Y with centre $f(a)$ is a neighbourhood of $f(a)$, there exists a neighbourhood A' of a such that $f_\vdash(A') \subset B$. But since A' is a neighbourhood of a there exists a ball A, centre a, such that $A \subset A'$, that is, such that $f_\vdash(A) \subset B$.

$\Leftarrow$: Suppose that any ball B in Y with centre $f(a)$ contains the image by f of some ball A in X with centre a. Since any neighbourhood B' of $f(a)$ contains a ball B with centre $f(a)$ and since, by hypothesis,

B contains the image of a ball A in X with centre a, this ball being a neighbourhood of a, it follows that f is continuous at a. □

Prop. 15.12. The map $f: X \rightarrowtail Y$ is continuous at $a \in \operatorname{dom} f$ if, and only if, for every positive real number ε there is a positive real number δ such that

$$|x - a| < \delta \quad \text{and} \quad x \in \operatorname{dom} f \;\Rightarrow\; |f(x) - f(a)| < \varepsilon.$$

Proof The statement only differs from the statement of Prop 15.11 in that the balls B and A are required to be open. Since every open ball contains a concentric closed ball, and vice versa, the content of the two statements is the same. □

Clearly either or both of the symbols $<$ in the statement of Prop. 15.12 may be replaced by $\leqslant$.

There follow several routine elementary results which we shall later repeat in the wider context of topological spaces.

Prop. 15.13. Let W, X and Y be normed linear spaces and let $g: W \rightarrowtail X$ and $f: X \rightarrowtail Y$ be maps continuous at $a \in \operatorname{dom} g$ and $b = g(a) \in \operatorname{dom} f$, respectively. Then the map $fg: W \rightarrowtail Y$ is continuous at a. □

Prop. 15.14. Let $g: W \rightarrowtail X$ be a linear inclusion and let W have the norm induced by a norm on X. Then, with respect to these norms, g is continuous. □

Prop. 15.15. Let W, X and Y be normed linear spaces and let $X \times Y$ be assigned the product norm. Then a map

$$(f,g): W \rightarrowtail X \times Y$$

is continuous at a point $a \in \operatorname{dom} f \cap \operatorname{dom} g$ if, and only if $f: W \rightarrowtail X$ and $g: W \rightarrowtail Y$ are each continuous at a. □

Cor. 15.16. Let X and Y be normed linear spaces, let $i = (1_X, 0): X \rightarrow X \times Y$, $j = (0, 1_Y): Y \rightarrow X \times Y$ and let $(p,q) = 1_{X \times Y}$. Then i, j, p and q are continuous.

Proof The identity maps 1_X, 1_Y and $1_{X \times Y}$ and the constant maps $0: X \rightarrow Y$ and $0: Y \rightarrow X$ are all continuous. □

The next proposition is one which we shall frequently use and to which we shall return later in the chapter.

Prop. 15.17. A linear map $t: X \rightarrow Y$ is continuous with respect to norms $x \rightsquigarrow |x|$ on X and $y \rightsquigarrow |y|$ on Y if, and only if, for some real number $K > 0$,

$$|t(x)| \leqslant K|x|, \quad \text{for all } x \in X.$$

Proof $\Rightarrow$: Suppose t is continuous. Then by the continuity of t at 0 there exists a positive real number δ such that, for all $x \in X$,

$$|x| \leqslant \delta \Rightarrow |t(x)| \leqslant 1.$$

Now, if $x = 0$, $|t(0)| = 0 = \delta^{-1}|0|$ and, if $x \neq 0$, $|\delta|x|^{-1}x| = \delta$, implying that $\delta|x|^{-1}|t(x)| = |t(\delta|x|^{-1}x)| \leqslant 1$. So, for all $x \in X$, $|t(x)| \leqslant \delta^{-1}|x|$.

$\Leftarrow$: Suppose such a K exists. Then, for all $x, a \in X$,

$$|t(x) - t(a)| = |t(x - a)| \leqslant K|x - a|.$$

So, for any $\varepsilon > 0$,

$$|x - a| \leqslant K^{-1}\varepsilon \Rightarrow |t(x) - t(a)| \leqslant \varepsilon.$$

That is, t is continuous. □

As a corollary we have the following characterization of equivalent norms.

Prop. 15.18. Let $|\ \ |$ and $||\ \ ||$ be norms on a linear space X. Then $|\ \ |$ and $||\ \ ||$ are equivalent if, and only if, there exist positive real numbers H and K such that for all $x \in X$

$$||x|| \leqslant H|x| \quad \text{and} \quad |x| \leqslant K||x||.$$

Proof Let $X' = (X, |\ \ |)$ and let $X'' = (X, ||\ \ ||)$, and consider the identity maps $X' \rightarrow X''$ and $X'' \rightarrow X'$. In view of the definition of continuity by means of open sets, the norms $|\ \ |$ and $||\ \ ||$ will be equivalent if, and only if, each of these maps is continuous. The proposition follows, by Prop. 15.7, since the identity map on a linear space is linear. □

Complete normed affine spaces

Convergence has already been discussed, in Chapter 2, for sequences on $\mathbf{R}$. Recall that a sequence $\omega \rightarrow \mathbf{R}: n \rightsquigarrow x_n$ is said to be *convergent* with *limit* x if, and only if, for each $\varepsilon > 0$ there exists a number $n \in \omega$ such that, for all $p \in \omega$,

$$p \geqslant n \Rightarrow |x_p - x| \leqslant \varepsilon,$$

and to be *Cauchy* if, and only if, for each $\varepsilon > 0$, there exists a number n such that

$$p, q \geqslant n \Rightarrow |x_p - x_q| \leqslant \varepsilon.$$

These definitions remain meaningful if $\mathbf{R}$ is replaced by any normed affine space X. It can be proved, just as before, that a convergent sequence has a unique limit and that every convergent sequence is Cauchy.

The limit of a sequence $x \rightsquigarrow x_n$ on X is denoted by $\lim_{n\to\infty} x_n$, though $\lim_{n\to\omega} x_n$ would be more logical.

A normed affine space X such that every Cauchy sequence on X is convergent is said to be *complete*. For example, **R** is complete, by Prop. 2.63.

Prop. 15.19. The normed linear space $\mathbf{R}[x]$ of polynomials over **R** in x, with norm

$$\sum_{n \in \omega} a_n x^n \rightsquigarrow \sup \{ |a_n| : n \in \omega \},$$

is not complete.

(The norm exists, since all but a finite number of the a_n are zero.) □

A complete normed *linear* space is also called a *Banach space*.

Equivalence of norms

The following propositions are preparatory for Theorem 15.26.

Prop. 15.20. Let X be a real affine space, complete with respect to a norm $|\ \ |$ on X. Then X is complete with respect to any equivalent norm $||\ \ ||$ on X.

Proof The proof that follows is typical of many convergence and continuity arguments. Its logic should be carefully studied.

What has to be proved is that any sequence on X that is Cauchy with respect to $||\ \ ||$ converges with respect to $||\ \ ||$. So let $n \rightsquigarrow x_n$ be a sequence on X that is Cauchy with respect to $||\ \ ||$. The job is *to find* in X a limit for this sequence with respect to the norm $||\ \ ||$.

It is at this stage, *and not before*, that we turn to the data. What we are told is

(a) that any sequence on X, Cauchy with respect to $|\ \ |$, converges with respect to $|\ \ |$;
(b) that the norms $|\ \ |$ and $||\ \ ||$ on X are equivalent.

This suggests the following strategy:

(i) to prove (using (b)?) that our sequence $n \rightsquigarrow x_n$ is Cauchy, and therefore (by (a)) convergent, with respect to $|\ \ |$;
(ii) to guess that the limit in X of this sequence with respect to $|\ \ |$ is also the limit of the sequence with respect to $||\ \ ||$;
(iii) to verify our guess (using (b) again?).

Proof of (i) What has to be proved is that, for all $\varepsilon > 0$, there exists $n \in \omega$ such that, for all $p, q \in \omega$,

$$p, q \geqslant n \ \Rightarrow\ |x_p - x_q| \leqslant \varepsilon.$$

So let $\varepsilon > 0$. The job is *to find n*.

Look at the data. We have (b) and our original hypothesis that the sequence $n \rightsquigarrow x_n$ is Cauchy with respect to $|| \quad ||$.

Now, by (b) and by Prop. 15.18, there is a positive real number K such that, for all $x \in X$, $|x| \leqslant K ||x||$, and therefore such that for all $p, q \in \omega$

$$||x_p - x_q|| \leqslant K^{-1}\varepsilon \Rightarrow |x_p - x_q| \leqslant K||x_p - x_q|| \leqslant \varepsilon.$$

However, since $K^{-1}\varepsilon > 0$, there exists $n \in \omega$ such that, for all $p, q \in \omega$,

$$p, q \geqslant n \Rightarrow ||x_p - x_q|| \leqslant K^{-1}\varepsilon.$$

This number n is just what had to be found. So (i) is proved.

Proof of (iii) What has to be proved is that, for all $\varepsilon > 0$, there exists $n \in \omega$ such that, for all $p \in \omega$,

$$p \geqslant n \Rightarrow ||x_p - x|| \leqslant \varepsilon,$$

where x is the limit in X of the sequence $n \rightsquigarrow x_n$ with respect to $| \quad |$. So once again let $\varepsilon > 0$. The job is *to find n.*

What are we given? We have (b), as before, and the fact that x is the limit of the sequence with respect to $| \quad |$.

By (b) and by Prop. 15.18 once more, there is a positive real number H such that, for all $x \in X$, $||x|| \leqslant H|x|$, and therefore such that, for all $p \in \omega$,

$$|x_p - x| \leqslant H^{-1}\varepsilon \Rightarrow ||x_p - x|| \leqslant H|x_p - x| \leqslant \varepsilon.$$

However, since $H^{-1}\varepsilon > 0$, there exists $n \in \omega$ such that, for all $p \in \omega$,

$$p \geqslant n \Rightarrow |x_p - x| \leqslant H^{-1}\varepsilon.$$

So the required number n has been found.

This completes the proof of (iii) and therefore of the proposition. □

(*Please note* that again and again we have refrained from 'mucking around with the data' until we knew what had to be found!)

Prop. 15.21. A subset A of a normed affine space X is closed in X if, and only if, any sequence on A, convergent as a sequence on X, has its limit in A.

Proof $\Rightarrow$: Suppose that $X \setminus A$ is open and that $n \rightsquigarrow x_n$ is a sequence on A with limit x in $X \setminus A$. Since $X \setminus A$ is open, there exists a ball B with centre x such that $B \subset X \setminus A$, and since the sequence converges to x, we have $x_n \in B$ for n sufficiently large, a contradiction since $x_n \notin X \setminus A$, for any n.

$\Leftarrow$: Suppose that any sequence on A, convergent as a sequence on X, has its limit in A, let $x \in X \setminus A$ and let B_r denote the ball with centre x and radius r. Then there exists $\delta > 0$ such that $B_\delta \subset X \setminus A$. For, if not, we may choose, for each $n \in \omega$, an element x_n of $B_{2^{-n}} \cap A$. The sequence $n \rightsquigarrow x_n$ is then convergent with limit $x \in A$, contradicting the hypothesis that $x \notin A$.

It follows that $X \setminus A$ is open in X; that is, that A is closed in X. □

The theorem which follows is one of the most useful technical lemmas in the theory of complete normed linear spaces. It plays a vital role in the proof of Theorem 19.6, the inverse function theorem.

Theorem 15.22. (*The contraction lemma.*)

Let A be a closed subset of a complete normed linear space X, and suppose that $f: A \to A$ is a map such that, for some non-negative real number $M < 1$ and for all $a, b \in A$,

$$|f(b) - f(a)| \leqslant M|b - a|.$$

Then there is a unique point x of A such that $f(x) = x$.

Proof Let x_0 be any point of A, and consider the sequence $n \rightsquigarrow x_n = f^n(x_0)$, where $f^0(x_0) = x_0$. This sequence is Cauchy, since, for any $n > 1$,

$$x_{n+1} - x_n = f(x_n) - f(x_{n-1})$$

and so $\quad |x_{n+1} - x_n| \leqslant M|x_n - x_{n-1}| \leqslant M^n|x_1 - x_0|,$

from which it follows that, for all k,

$$\begin{aligned} |x_{n+k+1} - x_n| &\leqslant (\sum_{i \in k} M^{n+1})\,|x_1 - x_0| \\ &\leqslant (1 - M)^{-1} M^n |x_1 - x_0|. \end{aligned}$$

Let x be the limit of this sequence. This exists, since X is complete, and belongs to A, by Prop. 15.21. Also, for any $\varepsilon > 0$ and for n sufficiently large,

$$\begin{aligned} |f(x) - x| &\leqslant |f(x) - f(x_n)| + |x_{n+1} - x| \\ &\leqslant M|x - x_n| + |x_{n+1} - x| \\ &\leqslant (1 + M)((1 + M)^{-1}\varepsilon) = \varepsilon. \end{aligned}$$

So $f(x) = x$.

Finally, x is the only fixed point. For if $f(x') = x'$ then

$$|x' - x| \leqslant M|x' - x|,$$

implying that

$$(1 - M)|x' - x| \leqslant 0.$$

Therefore $|x' - x| = 0$ and $x' = x$. □

Cor. 15.23. Let X be a complete normed linear space and let $f: X \to X$ be a map such that for all $x, x' \in X$

$$|h(x) - h(x')| \leqslant \tfrac{1}{2}|x - x'|$$

where $h = f - 1_X$. Then f is bijective.

(Note that, for any $a, b \in X$, $f(a) = b \Leftrightarrow b - h(a) = a$.) □

Prop. 15.24. An affine subspace W of a normed affine space X is closed if W is complete, the condition being necessary as well as sufficient if X is complete. □

Prop. 15.25. Let X and Y be complete normed affine spaces. Then $X \times Y$ is complete.

Strategy of proof Let $n \rightsquigarrow (x_n, y_n)$ be a Cauchy sequence on $X \times Y$. Deduce that the sequences $n \rightsquigarrow x_n$ and $n \rightsquigarrow y_n$ are Cauchy and therefore convergent with limits x and y, say. Then prove that the sequence $n \rightsquigarrow (x_n, y_n)$ is convergent, with limit (x,y). □

This proposition extends in an obvious way to finite products of complete normed affine spaces. In particular, $\mathbf{R}^n$ is complete with respect to the product norm, for any finite n.

We are now in a position to prove the theorem on the equivalence of norms on a finite-dimensional space, referred to earlier.

Theorem 15.26.
(a) Any two norms on a finite-dimensional affine space are equivalent.
(b) Any finite-dimensional normed affine space is complete.

Proof The proof is by induction on dimension. Let $(a)_n$ and $(b)_n$ be the statements obtained from (a) and (b) by replacing the word 'finite' in each case by 'n', where n is any finite number. We prove

$$(a)_0, \quad (a)_n \Rightarrow (b)_n \quad \text{and} \quad (b)_n \Rightarrow (a)_{n+1}.$$

$(a)_0$: There is a unique norm on a zero-dimensional affine space.

$(a)_n \Rightarrow (b)_n$: Every n-dimensional affine space is isomorphic to $\mathbf{R}^n$, and $\mathbf{R}^n$ is complete with respect to the product norm and so, by $(a)_n$ and Prop. 15.20, with respect to any norm.

$(b)_n \Rightarrow (a)_{n+1}$: Since any $(n+1)$-dimensional affine space is isomorphic to $\mathbf{R}^{n+1}$ it is sufficient to prove that $(b)_n$ implies that every norm on $\mathbf{R}^{n+1}$ is equivalent to the standard product norm.

Let $|| \quad ||$ denote the product norm on $\mathbf{R}^{n+1}$ and $| \quad |$ some other norm. Then, for any $x = \sum_{i \in n+1} x_i e_i \in \mathbf{R}^{n+1}$, $|x| \leqslant L\,||x||$, where $L = (n+1) \sup \{|e_i| : i \in n+1\}$. From this it follows at once that any $| \quad |$-ball in $\mathbf{R}^{n+1}$ contains a product ball with the same centre.

Conversely, any open product ball in $\mathbf{R}^{n+1}$ is the intersection of a finite number of open half-spaces, these being open, by Prop. 15.8, with respect to any norm, since by $(b)_n$ the bounding affine hyperplanes are complete and therefore closed in $\mathbf{R}^{n+1}$, by Prop. 15.24. Each open product ball therefore contains a concentric $|\ \ |$-ball.

This completes the proof. □

An alternative proof of this theorem using compactness and the continuity of the map $\mathbf{R} \rightarrowtail \mathbf{R} : x \rightsquigarrow x^{-1}$, is presented in Chapter 16 (page 328).

The theorem has the following important corollary.

Prop. 15.27. Let X and Y be finite-dimensional affine spaces and let $t : X \to Y$ be affine. Then t is continuous with respect to any norm on X and any norm on Y.

Proof By Theorem 15.26 it is enough to show that t is continuous for some norm on X and some norm on Y.

Choose 0 in X and set $t(0) = 0$ in Y. Let $X_0 = \ker t$ and $Y_0 = \operatorname{im} t$ and let X_1 be a linear complement of X_0 in X and Y_1 a linear complement of Y_0 in Y. Choose any norms for X_0 and X_1, give Y_0 the norm induced by the bijection $X_1 \to Y_0$; $x_1 \rightsquigarrow t(x_1)$ and choose any norm for Y_1. Let X and Y have the sum norms with respect to these direct sum decompositions. Then for all $x = x_0 + x_1 \in X$, with $x_0 \in X_0$, $x_1 \in X_1$,

$$| t(x) | = | t(x_0 + x_1) | = | t(x_1) | = | x_1 | \leqslant | x_0 | + | x_1 | = | x | .$$

It follows, by Prop. 15.17, that t is continuous. □

The norm of a continuous linear map

In the proof of the last proposition we used the fact, proved in Prop. 15.17, that a linear map $t : X \to Y$ between normed linear spaces X and Y is continuous if, and only if, there is some real number K such that, for all $x \in X$, $| t(x) | \leqslant K | x |$. When such a number K exists the set $\{| t(x) | : | x | \leqslant 1\}$ is bounded above by K. This subset of $\mathbf{R}$ also is non-null, since it contains 0, and so it has a supremum. The supremum is denoted by $| t |$ and is called the *absolute gradient* or the *norm* of t.

Prop. 15.28. Let $t : X \to Y$ be a continuous map between normed linear spaces X and Y. Then, for all $x \in X$, $| t(x) | \leqslant | t | | x |$. Also, $| t |$ is the smallest real number K such that, for all $x \in X$, $| t(x) | \leqslant K | x |$.

Proof For all $x \in X$ such that $| x | \leqslant 1$,

$$| t(x) | \leqslant | t | \leqslant | t | | x |,$$

implying that, for all $x \in X$ such that $| x | \neq 0$,

$$| t(x) | = | t(| x |^{-1} x) | x | | = | t(| x |^{-1} x) | | x | \leqslant | t | | x |.$$

Also $| t(0) | \leqslant | t | | 0 |$. So, for all $x \in X$, $| t(x) | \leqslant | t | | x |$.

Finally, for any $K < | t |$ there exists $x \in X$ such that $| x | \leqslant 1$ and $| t(x) | > K \geqslant K | x |$, implying the last part of the proposition. □

In the sequel $| t |$ will often be thought of as 'the smallest K'.

The choice of the words 'absolute gradient' is motivated by the first two of the examples which follow. The choice of the word 'norm' is justified by Prop. 15.32 below.

Example 15.29. Let **R** be assigned the absolute value norm and let t be the map $\mathbf{R} \to \mathbf{R}$; $x \rightsquigarrow mx$, where $m \in \mathbf{R}$. Then

$$| t | = \sup \{| m | \, | x | : | x | \leqslant 1\} = | m |.$$

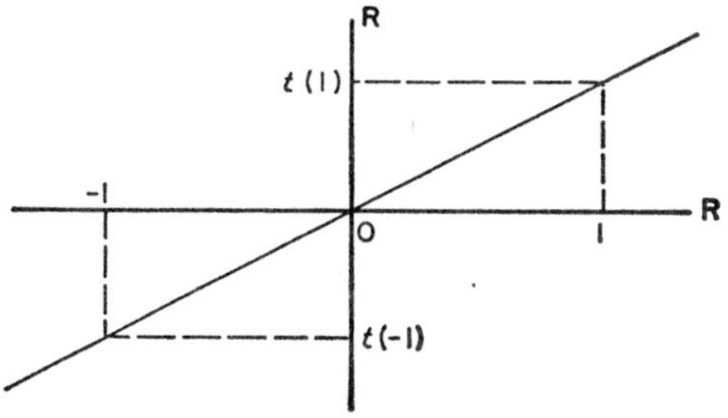

□

Example 15.30. Let **R** have the absolute value norm and $\mathbf{R}^2$ the standard quadratic norm and let t be the map $\mathbf{R}^2 \to \mathbf{R}; (x,y) \rightsquigarrow ax + by$ where $(a,b) \in \mathbf{R}^2$.

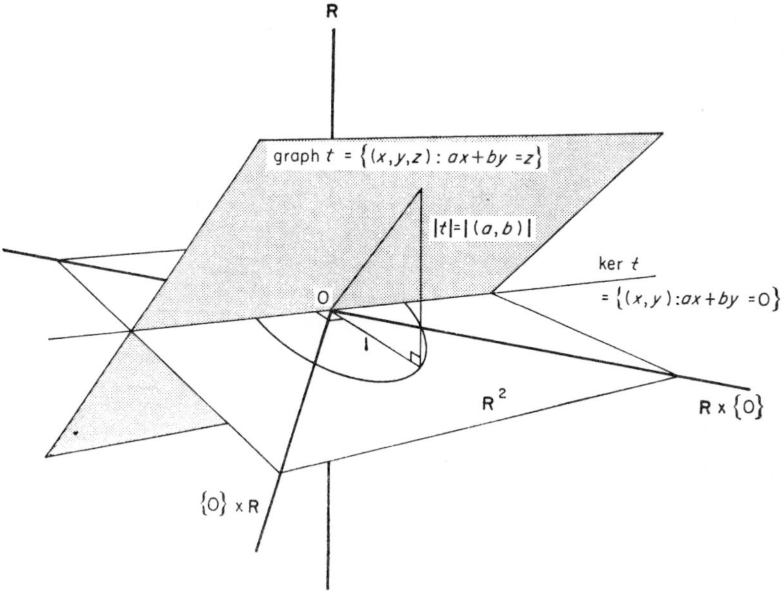

Since, by Cauchy–Schwarz (9.58),

$$|ax + by| = |(a,b)\cdot(x,y)| \leqslant |(a,b)|\,|(x,y)|$$

for any $(x,y) \in \mathbf{R}^2$, with equality for any (x,y) when $(a,b) = (0,0)$ and for $(x,y) = |(a,b)|^{-1}(a,b)$ when $(a,b) \neq (0,0)$, it follows that

$$|t| = \sup\{|ax + by| : |(x,y)| \leqslant 1\} = |(a,b)|. \qquad \square$$

Example 15.31. Let $\mathbf{R}[x]$ be the linear space of polynomials over $\mathbf{R}$ in x, with the norm $|| \quad ||$ defined in Prop. 15.19. Then, in particular, $||x^n|| = 1$, for any $n \in \omega$.

Now let $t : \mathbf{R}[x] \to \mathbf{R}[x]$ be the map (differentiation) defined for any $\sum_{i\in\omega} a_i x^i \in \mathbf{R}[x]$ by the formula $t(\sum_{i\in\omega} a_i x^i) = \sum_{i\in\omega^+} i a_i x^{i-1}$. Then t is linear, while, for any $n \in \omega^+$, $|t(x^n)| = |nx^{n-1}| = n$.

Since ω has no supremum in $\mathbf{R}$, it follows that t does not have a norm. $\square$

The set of continuous linear maps from a normed linear space X to a normed linear space Y is denoted by $L(X,Y)$.

Prop. 15.32. Let X and Y be normed linear spaces. Then $L(X,Y)$ is a linear subspace of $\mathscr{L}(X,Y)$ and $t \rightsquigarrow |t|$ is a norm on $L(X,Y)$.

Proof It is clear that $|t| \geqslant 0$ for all $t \in L(X,Y)$ and that $|t| = 0$ if, and only if, $t = 0$.

Moreover, for all $t, t' \in L(X,Y)$, all $\lambda \in \mathbf{R}$ and all $x \in X$,

$$|(t + t')(x)| \leqslant |t(x)| + |t'(x)| \leqslant (|t| + |t'|)\,|x|,$$

so that $t + t'$ is continuous and $|t + t'| \leqslant |t| + |t'|$, and

$$|(\lambda t)(x)| \leqslant |\lambda|\,|t(x)| \leqslant |\lambda|\,|t|\,|x|;$$

so λt is continuous and $|\lambda t| \leqslant |\lambda|\,|t|$. Also, $|0t| = |0|\,|t|$ and, if $\lambda \neq 0$,

$$|\lambda t| \leqslant |\lambda|\,|t| = |\lambda|\,|\lambda^{-1}|\,|\lambda t| = |\lambda t|,$$

so that in fact $|\lambda t| = |\lambda|\,|t|$. $\square$

For any normed linear space X the linear space $L(X,\mathbf{R})$ is called the *continuous linear dual* of X. We shall denote it by X^L.

When X and Y are finite-dimensional real linear spaces, it follows from Prop. 15.27 that $L(X,Y) = \mathscr{L}(X,Y)$ and that $X^L = X^{\mathscr{L}}$.

Prop. 15.33. Let $t : X \to Y$ and $u : W \to X$ be continuous linear maps, W, X and Y being normed linear spaces. Then $tu : W \to Y$ is continuous, with $|tu| \leqslant |t|\,|u|$.

Proof For all $w \in W$, $|tu(w)| \leqslant |t|\,|u(w)| \leqslant |t|\,|u|\,|w|$. So tu is continuous and $|tu| \leqslant |t|\,|u|$. $\square$

Cor. 15.34. For all $t \in L(X,Y)$, the map $L(W,X) \to L(W,Y)$; $u \rightsquigarrow tu$ is continuous. Also, for all $u \in L(W,X)$, the map $L(X,Y) \to L(W,Y)$; $t \rightsquigarrow tu$ is continuous. □

Cor. 15.35. Let $u \in L(X,X)$. Then, for all finite n, $|u^n| \leqslant |u|^n$. (By convention, $u^0 = 1_X$.) □

Cor. 15.36. Let $u \in L(X,X)$, with $|u| < 1$. Then the sequence $n \rightsquigarrow u^n$ on $L(X,X)$ is convergent, with limit 0. □

The inverse of a continuous linear map is not necessarily continuous. See Exercise 15.60.

The *continuous linear dual* of a continuous linear map $t : X \to Y$ is by definition the map

$$t^L : Y^L \to X^L; \quad \beta \rightsquigarrow \beta t.$$

Prop. 15.37. The continuous linear dual of a continuous linear map $t : X \to Y$ is continuous. □

Prop. 15.38. Let X and Y be normed linear spaces and let Y be complete. Then $L(X,Y)$ is complete.

Proof Let $n \rightsquigarrow t_n$ be a Cauchy sequence on $L(X,Y)$. Then, for any $x \in X$, the sequence $n \rightsquigarrow t_n(x)$ is a Cauchy sequence.

This is clear if $x = 0$, while if $x \neq 0$ the result follows at once from the implication

$$|t_n - t_p| \leqslant |x|^{-1}\varepsilon \Rightarrow |t_n(x) - t_p(x)| \leqslant \varepsilon,$$

for any $n, p \in \omega$, and any $\varepsilon > 0$.

For each $x \in X$, let $t(x) = \lim_{n\to\infty} t_n(x)$ and define $t : X \to Y$ to be the map $x \rightsquigarrow t(x)$. Various things now have to be checked. Firstly, t is linear, as is easily verified. Secondly, t is continuous. For let $\varepsilon > 0$. Then, for any $x \in X$,

$$\begin{aligned} |t(x)| &\leqslant |t_n(x)| + \varepsilon, \quad \text{for } n \text{ sufficiently large,} \\ &\leqslant |t_n|\,|x| + \varepsilon, \quad \text{since } t_n \text{ is continuous,} \\ &\leqslant \sup\{|t_n| : n \in \omega\}\,|x| + \varepsilon, \end{aligned}$$

the supremum existing by Prop. 2.63, since the sequence $n \rightsquigarrow |t_n|$ is Cauchy, by axiom (iii)′, page 288. Let $K = \sup\{|t_n| : n \in \omega\}$. Then, since $|t(x)| \leqslant K|x| + \varepsilon$ for all $\varepsilon > 0$, it follows, by Exercise 2.37, that $|t(x)| \leqslant K|x|$, for any $x \in X$. That is, t is continuous.

Finally, $\lim_{n\to\infty} t_n = t$, with respect to the norm on $L(X,Y)$. What has to be proved is that, for any $\varepsilon > 0$, $|t - t_n| \leqslant \varepsilon$, for n sufficiently large. Now $|t - t_n| = \sup\{|t(x) - t_n(x)| : |x| \leqslant 1\}$ and, for any x

and any p, $| t(x) - t_n(x) | \leqslant | t(x) - t_p(x) | + | t_p(x) - t_n(x) |$. However, for p sufficiently large, depending on x, we have

$$| t(x) - t_p(x) | \leqslant \tfrac{1}{2}\varepsilon,$$

and for p and n sufficiently large, independent of x, we have

$$| t_p(x) - t_n(x) | \leqslant | t_p - t_n | \, | x | \leqslant \tfrac{1}{2}\varepsilon \, | x |,$$

and so if $| x | \leqslant 1$ we have, for n sufficiently large,

$$| t(x) - t_n(x) | \leqslant \tfrac{1}{2}\varepsilon + (\tfrac{1}{2}\varepsilon)1 = \varepsilon.$$

Therefore, for n sufficiently large,

$$| t - t_n | \leqslant \varepsilon,$$

as had to be proved.

This completes the proof. □

Continuous bilinear maps

Prop. 15.39. Let X, Y and Z be normed linear spaces. A bilinear map $X \times Y \to Z$; $(x,y) \rightsquigarrow x \cdot y$ is continuous if, and only if, there exists a real number $K > 0$ such that, for all $(x,y) \in X \times Y$,

$$| x \cdot y | \leqslant K \, | x | \, | y |.$$

Proof $\Rightarrow$: Suppose the map is continuous. Then by its continuity at 0 there exists a positive real number δ such that

$$| (x,y) | \leqslant \delta \Rightarrow | x \cdot y | \leqslant 1.$$

Now, if either x or $y = 0$, then $| x \cdot y | = 0 = \delta^{-2} | 0 |$, while, if neither x nor $y = 0$, then $| \delta(| x |^{-1} x , | y |^{-1} y) | = \delta$, implying that

$$\delta^2 | x |^{-1} | y |^{-1} | x \cdot y | = | \delta | x |^{-1} x \cdot \delta | y |^{-1} y | \leqslant 1.$$

So, for all $(x,y) \in X \times Y$, $| x \cdot y | \leqslant \delta^{-2} | x | \, | y |$.

$\Leftarrow$: Suppose such a K exists. Then, for all $(x,y), (a,b) \in X \times Y$,

$$\begin{aligned} | x \cdot y - a \cdot b | &\leqslant | (x - a) \cdot (y - b) | + | a \cdot (y - b) | + | (x - a) \cdot b | \\ &\leqslant K(| x - a | + | a | + | b |) \, | (x,y) - (a,b) |, \end{aligned}$$

since $| (x,y) - (a,b) | = \sup \{ | x - a | , | y - b | \}$. So, for any $\varepsilon > 0$, $| (x,y) - (a,b) | \leqslant \inf\{1, K(1 + | a | + | b |)^{-1} \varepsilon\} \Rightarrow | x \cdot y - a \cdot b | \leqslant \varepsilon$. □

Examples 15.40. Let W, X and Y be normed linear spaces. Then

$$L(X,Y) \times X \to Y; \quad (t,x) \rightsquigarrow t(x) \text{ is continuous}$$

since, for all (t,x), $| t(x) | \leqslant | t | \, | x |$;

$$L(X,Y) \times L(W,X) \to L(W,Y); \quad (t,u) \rightsquigarrow tu \text{ is continuous}$$

since, for all (t,u), $| tu | \leqslant | t | \, | u |$; and

$$\mathbf{R} \times L(X,Y) \to L(X,Y); \quad (\lambda,t) \rightsquigarrow \lambda t \text{ is continuous}$$

since, for all (λ,t), $| \lambda t | = | \lambda | \, | t |$.

Finally, if X is a positive-definite real orthogonal space and if $| \ |$ denotes the induced norm, then

$$X \times X \longrightarrow \mathbf{R};\quad (x,x') \rightsquigarrow x \cdot x' \text{ is continuous}$$

since, for all (x,x'), $| x \cdot x' | \leqslant | x | \, | x' |$. □

Prop. 15.41. Let $X \times Y \longrightarrow Z$; $(x,y) \rightsquigarrow x \cdot y$ be a bilinear map, X, Y and Z being finite-dimensional normed linear spaces. Then the map is continuous.

Proof For all $(x,y) \in X \times Y$,

$$| x \cdot y | = | (x \cdot)(y) | \leqslant | (x \cdot) | \, | y | \leqslant K \, | x | \, | y |,$$

for some real K, since $(x \cdot) : Y \longrightarrow Z$ and $X \longrightarrow L(Y,Z)$; $x \rightsquigarrow (x \cdot)$ are linear maps between finite-dimensional normed linear spaces, and are therefore continuous. □

Analogous results hold for multilinear maps.

Prop. 15.42. Let $f : \bigtimes_{i \in n} X_i \longrightarrow Z$ be an n-linear map, the spaces X_i, for all $i \in n$, and Z being normed linear spaces. Then f is continuous if, and only if, there exists a real number $K > 0$ such that, for all $(x_i : i \in n) \in \bigtimes_{i \in n} X_i$,

$$| (x_i : i \in n) | \leqslant K \prod_{i \in n} | x_i |. \qquad \square$$

Prop. 15.43. Let $f : \bigtimes_{i \in n} X_i \longrightarrow Z$ be an n-linear map, the spaces X_i, for all $i \in n$, and Z being finite-dimensional normed linear spaces. Then f is continuous. □

Prop. 15.44. Let $X^2 \longrightarrow Z$; $(x,x') \rightsquigarrow x \cdot x'$ be a continuous bilinear map, X and Z being normed linear spaces. Then the map $X \longrightarrow Z$; $x \rightsquigarrow x \cdot x$ is continuous. □

Prop. 15.45. Let $\mathbf{K} = \mathbf{R}$ or the real algebra of complex numbers $\mathbf{C}$. Then any polynomial map $\mathbf{K} \longrightarrow \mathbf{K}$; $x \rightsquigarrow \sum_{i \in \omega} a_i x^i$ is continuous. (Since the map is a polynomial map, $a_i = 0$ for all sufficiently large $i \in \omega$.) □

Inversion

We have already remarked that the inverse of a bijective continuous linear map is not necessarily continuous. It can, however, be proved that if X and Y are complete then the inverse of a bijective continuous linear map $t : X \longrightarrow Y$ is also continuous. (See any of the references given on page 288.) The set of bijective continuous linear maps $t : X \longrightarrow Y$ with continuous inverse will be denoted by $GL(X,Y)$.

The next three propositions are concerned with the continuity of the inversion map $L(X,Y) \rightarrowtail L(Y,X);\ t \rightsquigarrow t^{-1}$ with domain $GL(X,Y)$. The spaces X and Y may on a first reading be taken to be normed real linear spaces. However, the propositions and their proofs all remain valid if the real field is replaced either by the real algebra of complex numbers **C** or by the real algebra of quaternions **H**, with $L(X,Y)$ and $GL(X,Y)$ denoting respectively the spaces of complex or quaternionic linear or invertible linear maps of X to Y. (In every case the linear spaces X and Y are normed as real linear spaces.)

Prop. 15.46. Let X be a complete normed linear space, let $u \in L(X,X)$ and suppose that $|u| < 1$. Then $1_X - u \in GL(X,X)$ with $(1_X - u)^{-1} = \sum_{k \in \omega} u^k$ and $|(1_X - u)^{-1}| \leqslant (1 - |u|)^{-1}$.

Proof Since $|u| < 1$ the sequence $n \rightsquigarrow \sum_{k \in n} |u|^k$ on **R** is convergent and therefore Cauchy.

Now, for all $p, q \in \omega$ with $p = q + r \geqslant q$,

$$\Big|\sum_{k \in p} u^k - \sum_{k \in q} u^k\Big| = \Big|\sum_{k \in r} u^{q+k}\Big| \leqslant \sum_{k \in r} |u|^{q+k} \text{ (by Cor. 15.35)}$$

and $\sum_{k \in r} |u|^{q+k} = \sum_{k \in p} |u|^k - \sum_{k \in q} |u|^k$. So the sequence $n \rightsquigarrow \sum_{k \in n} u^k$ on $L(X,X)$ is Cauchy and therefore convergent by Prop. 15.38.

Also, for any $n \in \omega$, $(1 - u)(\sum_{k \in n} u^k) = 1 - u^n$; so $(1 - u)(\sum_{k \in \omega} u^k) = 1$, if $|u| < 1$. Similarly, $(\sum_{k \in \omega} u^k)(1 - u) = 1$ if $|u| < 1$. That is, $(1 - u)^{-1}$ exists for $|u| < 1$ and $(1 - u)^{-1} = \sum_{k \in \omega} u^k \in L(X,X)$.

Finally, since $|\sum_{k \in n} u^k| \leqslant \sum_{k \in n} |u|^k$ for all $n \in \omega$,

$$|(1 - u)^{-1}| \leqslant \sum_{k \in \omega} |u|^k = (1 - |u|)^{-1}. \qquad \square$$

Prop. 15.47. When X is a complete normed linear space, inversion

$$\chi : L(X,X) \rightarrowtail L(X,X); \quad t \rightsquigarrow t^{-1}$$

is defined on a neighbourhood of 1 $(= 1_X)$ and is continuous at 1.

Proof Let $t \in L(X,X)$ and let $u = 1 - t$. Then, if $|u| < 1$, $t = 1 - u \in GL(X,X)$, by Prop. 15.46. That is, χ is defined on a neighbourhood of 1.

Let $\varepsilon > 0$. If $|u| < 1$,

$$\chi(t) - \chi(1) = (1 - u)^{-1} - (1 - u)(1 - u)^{-1} = u(1 - u)^{-1}$$

and, if $|u| \leqslant \frac{1}{2}$, by the estimate of Prop. 15.46,

$$|\chi(t) - \chi(1)| \leqslant |u|\,|(1 - u)^{-1}| \leqslant |u|(1 - |u|)^{-1} \leqslant 2|u|.$$

So

$$|t - 1| = |u| \leqslant \inf\{\tfrac{1}{2}, \tfrac{1}{2}\varepsilon\} \Rightarrow |\chi(t) - \chi(1)| \leqslant \varepsilon.$$

That is, χ is continuous at 1. □

Prop. 15.48. Let X and Y be complete normed linear spaces. Then the map $\psi : L(X,Y) \rightarrowtail L(Y,X); t \rightsquigarrow t^{-1}$ is continuous and $GL(X,Y)$, the domain of ψ, is open in $L(X,Y)$.

Proof For any t and $u \in GL(X,Y)$, $(u^{-1}t)^{-1}u^{-1} = t$. Therefore, for any $u \in GL(X,Y)$, the map ψ is the composite of the maps

$$L(X,Y) \to L(X,X); \quad t \rightsquigarrow u^{-1}t,$$

which by the inequality $|u^{-1}t| \leqslant |u^{-1}|\,|t|$ is continuous linear, since u^{-1} is continuous, and which sends u to 1_X;

$$\chi : L(X,X) \rightarrowtail L(X,X); \quad t \rightsquigarrow t^{-1},$$

which is defined in a neighbourhood of 1_X and is continuous at 1; and

$$L(X,X) \to L(Y,X); \quad t \rightsquigarrow tu^{-1},$$

which also is continuous linear. In diagram form ψ decomposes as follows:

$$\begin{array}{ccccccc} L(X,Y) & \to & L(X,X) & \overset{\chi}{\rightarrowtail} & L(X,X) & \to & L(Y,X) \\ t & \rightsquigarrow & u^{-1}t & \rightsquigarrow & t^{-1}u & \rightsquigarrow & t^{-1} \\ u & \rightsquigarrow & 1 & \rightsquigarrow & 1 & \rightsquigarrow & u^{-1} \end{array}$$

It follows that, for each $u \in GL(X,Y)$, the map ψ is defined on a neighbourhood of u and is continuous at u. So the domain of ψ is open in $L(X,Y)$, and ψ is continuous. □

In particular, by taking $X = Y = \mathbf{K}$, where $\mathbf{K} = \mathbf{R}$, $\mathbf{C}$ or $\mathbf{H}$, and by identifying $L(\mathbf{K},\mathbf{K})$ with $\mathbf{K}$, it follows that the map $\mathbf{K} \to \mathbf{K}; x \rightsquigarrow x^{-1}$ is continuous, though this can of course easily be proved directly.

The statement that $GL(X,Y)$ is open in $L(X,Y)$ means, in elementary terms, and in the particular case where $X = Y = \mathbf{K}^n$, that if we have a set of n linear equations in n variables with a unique solution, and we vary the coefficients, then, provided that we do not alter the coefficients too much, the new set of equations also will have a unique solution.

In the last proposition of the chapter the use of the notation $GL(X,Y)$ is extended in the case where X and Y are finite-dimensional, just as in Chapter 6. Here again, $\mathbf{K}$ may be $\mathbf{R}$, $\mathbf{C}$ or $\mathbf{H}$.

Prop. 15.49. Let X and Y be finite-dimensional $\mathbf{K}$-linear spaces, and let $GL(X,Y)$ denote the set of linear maps $t : X \to Y$ such that $\operatorname{rk} t = \inf\{\dim X, \dim Y\}$. Then $GL(X,Y)$ is an open subset of $L(X,Y)$. □

FURTHER EXERCISES

15.50. Let X be a normed linear space and let $a, b \in X$. Prove that, if $|a - b| \leqslant \frac{1}{2}|a|$, then $|a - b| \leqslant |b|$ and $|a| \leqslant 2|b|$. □

15.51. Let X be a real linear space, $|\ \ |$ and $||\ \ ||$ norms on X, and r and s real numbers such that, for all $x \in X$, $|x| = r \Rightarrow ||x|| \geqslant s$. Prove that $||x|| = s \Rightarrow |x| \leqslant r$.

Illustrate this by choosing $|\ \ |$ and $||\ \ ||$ to be two of the familiar norms on $\mathbf{R}^2$. □

15.52. Let $t : X \rightarrow Y$ be a linear map between normed linear spaces X and Y such that, for every $\varepsilon > 0$, there exists $\delta > 0$ such that, for all $x \in X$,

$$|x| \leqslant \delta \Rightarrow |t(x)| \leqslant \varepsilon |x|.$$

Prove that, for each $x \in X$ and for all $\varepsilon > 0$, $|t(x)| \leqslant \varepsilon |x|$. Hence prove that $t = 0$. □

15.53. Let $t : \mathbf{R}^2 \rightarrow \mathbf{R}^2$; $(x,y) \rightsquigarrow (u,v)$ be the linear map defined by the equations $u = ax + cy$, $v = bx + dy$, a, b, c and d being real, and let the domain of t be assigned the norm $(x,y) \rightsquigarrow |x| + |y|$ and the target of t the norm $(u,v) \rightsquigarrow \sup\{|u|, |v|\}$. Show that t is continuous with respect to these norms and that $|t| = \sup\{|a|, |b|, |c|, |d|\}$.

(Don't forget to show that the stated norm is the *smallest* K for the map t.) □

15.54. Why is it wrong, in the proof of Prop. 15.46, to deduce directly from the fact that, for any $u \in L(X,X)$ such that $|u| < 1$,

$$\lim_{n\to\infty} (1 - u) \sum_{k \in n} u^k = \lim_{n\to\infty} (\sum_{k \in n} u^k)(1 - u) = 1$$

the conclusion that $1 - u \in GL(X,X)$? Why was it necessary to establish first that $\lim_{n\to\infty} \sum_{k \in n} u^k \in L(X,X)$? □

15.55. Let X, Y and Z be normed linear spaces, and let $X \times Y \rightarrow Z$; $(x,y) \rightsquigarrow u(x) + v(y)$ be a continuous linear map such that $v : Y \rightarrow Z$ is a linear homeomorphism. Prove that the map

$$X \times Y \rightarrow X \times Z;\quad (x,y) \rightsquigarrow (x\ ,\ u(x) + v(y))$$

is a linear homeomorphism. □

15.56. Let X be a complete normed linear space and let $f : X \rightarrow X$ be a map such that, for all $x, x' \in X$, $|h(x) - h(x')| \leqslant \frac{1}{2}|x - x'|$, where $h = f - 1_X$. Prove that f is bijective.

(Note that, for all $x, y \in X$, $f(x) = y \Leftrightarrow y - h(x) = x$.) □

15.57. Let X be a normed linear space and let $u, v \in L(X,X)$. Prove that $uv - vu$ cannot be equal to 1_X.

(Prove that, if $uv - vu = 1_X$, then $uv^{n+1} - v^{n+1}u = (n+1)v^n$, for all $n \geqslant 1$. Use this to show that $|v^n| = 0$, for n sufficiently large. Hence show that $v = 0$, a contradiction. For the history of this exercise and its relation to the Heisenberg uncertainty principle, see [22].) □

15.58. Let Y be a normed real linear space. Prove that, for any $n \in \omega$, the map

$$L(\mathbf{R}^n, Y) \longrightarrow Y^n; \quad t \rightsquigarrow (t(e_i) : i \in n)$$

is a linear homeomorphism. □

15.59. Let W, X and Y be normed real linear spaces, X being finite-dimensional, and consider a map

$$f : W \rightarrowtail L(X, Y).$$

Prove that f is continuous if, and only if, for each $x \in X$, the map $f : W \rightarrowtail Y;\ w \rightsquigarrow f(w)(x)$ is continuous. (Use Exercise 15.58.) □

15.60. Let the linear space of polynomials $\mathbf{R}[x]$ have the norm assigned to it in Prop. 15.19. Prove that the map

$$u : \mathbf{R}[x] \longrightarrow \mathbf{R}[x]; \quad \sum_{n \in \omega} a_n x^n \rightsquigarrow \sum_{n \in \omega} \frac{a_n}{n+1} x^n$$

is linear and continuous, with gradient norm $|u| = 1$, that u is bijective, but that u^{-1} is not continuous. □

15.61. (Polya's *Peano curve.* Cf. page 384.)

For any right-angled triangle abc in $\mathbf{R}^2$, with right angle at b, let $p(abc)$ denote the base of the perpendicular from b to $[a,c]$.

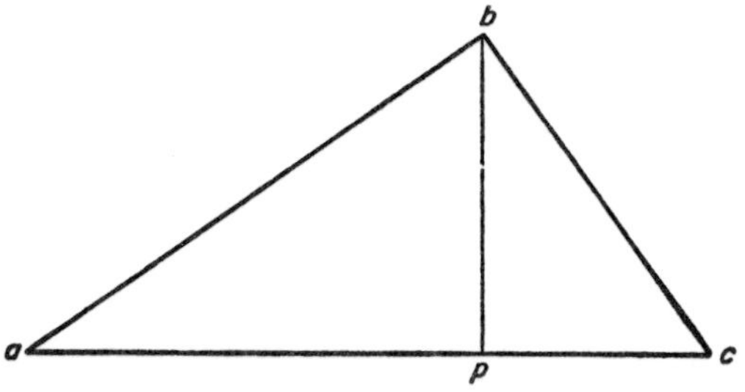

Suppose that $a_0b_0c_0$ is such a triangle and let $s \in 2^\omega$. Then we may construct recursively a sequence of right-angled triangles $n \rightsquigarrow a_nb_nc_n$ and a sequence of points $n \rightsquigarrow p_n$ by defining, for all $k \in n$, $p_k = p(a_kb_kc_k)$ and

$$a_{k+1} = \begin{cases} a_k \text{ if } s_k = 0 \\ b_k \text{ if } s_k = 1 \end{cases}, \quad b_{k+1} = p_k, \quad c_{k+1} = \begin{cases} b_k \text{ if } s_k = 0 \\ c_k \text{ if } s_k = 1 \end{cases}.$$

Prove that the sequence $n \rightsquigarrow p_n$ is convergent.

Hence, representing each number of the interval by its binary expansion or expansions (cf. Exercise 2.61), construct a *surjective continuous* map of the closed interval $[0,1]$ to the convex hull of the triangle $a_0b_0c_0$, sending 0 to a_0, $\frac{1}{2}$ to b_0 and 1 to c_0. □

15.62. Construct a continuous surjection $[0,1] \longrightarrow [0,1]^2$. Hence, construct a continuous surjection $[0,1] \longrightarrow [0,1]^n$, for any $n \in \omega$. (Cf. Exercise 1.69.) □

15.63. Let $BL(X_0 \times X_1, Y)$ denote the linear space of continuous bilinear maps $X_0 \times X_1 \longrightarrow Y$, where X_0, X_1 and Y are normed linear spaces and, for each $t \in BL(X_0 \times X_1, Y)$, let

$$|\, t \,| = \sup \{t(x_0,x_1) : |\,(x_0,x_1)\,| \leqslant 1\}.$$

Prove that the map

$$BL(X_0 \times X_1, Y) \longrightarrow \mathbf{R}; \quad t \rightsquigarrow |\, t \,|$$

is a norm on $BL(X_0 \times X_1, Y)$, that, for all $(x_0,x_1) \in X_0 \times X_1$,

$$|\, t(x_0,x_1) \,| \leqslant |\, t \,|\,|\, x_0 \,|\,|\, x_1 \,|,$$

that $|\,t\,| = \inf \{K \in \mathbf{R}\colon \text{ for all } (x_0,x_1) \in X_0 \times X_1, |\,t(x_0,x_1)\,| \leqslant K\,|\,x_0\,|\,|\,x_1\,|\}$ and that the map

$$L(X_0,L(X_1,Y)) \longrightarrow BL(X_0 \times X_1, Y); \quad t \rightsquigarrow t'$$

defined, for all $(x_0,x_1) \in X_0 \times X_1$, by the formula

$$t'(x_0,x_1) = t(x_0)(x_1),$$

is a normed linear space isomorphism. □

CHAPTER 16

TOPOLOGICAL SPACES

In Chapter 15 the concept of continuity has been defined for maps between normed linear spaces. The purpose of this chapter is to deepen and widen the discussion of continuity, by showing that the case so far considered is a particular case of a much more general concept, that of continuity for maps between topological spaces. The initial definitions of a topology and of a topological space are strongly motivated by the properties of the set of open sets of a normed linear space, as listed, for example, in Props. 15.7 and 15.9.

The most important new concepts introduced in the chapter are compactness and connectedness.

This chapter contains all the topology necessary for the reading of Chapters 18 and 19, with the exception of one detail of the proof of Theorem 19.20, where the reference is to Chapter 17.

Topologies

Cohesion may be given to a set X by singling out a subset $\mathscr{T}$ of Sub X such that

(i) $\emptyset, X \in \mathscr{T}$;

(ii) for all $A, B \in \mathscr{T}$, $A \cap B \in \mathscr{T}$, that is, the intersection of any two and therefore of any non-null finite set of elements of $\mathscr{T}$ belongs to $\mathscr{T}$;

(iii) for all $\mathscr{S} \subset \mathscr{T}$, $\bigcup \mathscr{S} \in \mathscr{T}$, that is, the union of any set of elements of $\mathscr{T}$ belongs to $\mathscr{T}$.

The set T is said to be a *topology* for X, and the elements of T are called the *open sets* of the topology.

Proposition 15.7 states that for any normed affine space X the set of subsets of X open with respect to the norm is a topology for X. In particular, the absolute value on **R** induces a topology for **R** which we shall refer to as the *standard topology* for **R**.

A set X may have many topologies. Examples include the *trivial* topology $\{\emptyset, X\}$, the *cofinite* topology $\{A \subset X;\ A = \emptyset \text{ or } X \backslash A \text{ finite}\}$

and the *discrete* topology, Sub X itself. A reason for using the word 'discrete' in this context willl be given later, on page 329.

Prop. 16.1. The sets $0 = \{\emptyset\}$ and $1 = \{0\}$ each have a unique topology, the set $2 = \{0,1\}$ has four topologies, and the set $3 = \{0,1,2\}$ has twenty-nine topologies. □

A *topological space* $(X,\mathcal{T})$ consists of a set X and a topology $\mathcal{T}$ for X. When there is no danger of confusion it is usual to abbreviate $(X,\mathcal{T})$ to X and to speak, simply, of the topological space X. In the same spirit the open sets of $\mathcal{T}$ are then referred to as the open sets of X.

An *open neighbourhood* of a point x of a topological space X is, by definition, an open subset A of X such that $x \in A$. A *neighbourhood* of x is a subset of X with an open neighbourhood of x as a subset. An open set is a neighbourhood of each of its points.

Unless there is an explicit statement to the contrary, a normed affine or linear space will tacitly be assigned the topology induced by its norm. A finite-dimensional affine or linear space will be assigned the topology induced by any of its norms, this being independent of the choice of norm, by Theorem 15.26, while a finite set will normally be assigned the discrete topology. Each of these topologies will be referred to as the *standard topology* for the set in question. A further standard example is provided by the next proposition.

Prop. 16.2. Let $\bar{\omega} = \omega \cup \{\omega\}$, ω being as usual the set of natural numbers, and let a subset A of $\bar{\omega}$ be defined to be *open* if either $A \subset \omega$ or $\bar{\omega} \setminus A$ is finite. Then the set of open sets of $\bar{\omega}$ is a topology for $\bar{\omega}$. □

This topology will be called the *standard topology* for $\bar{\omega}$.

It is most important to note the distinction between axioms (ii) and (iii) for a topology. To prove (iii) in a particular case it is not enough to consider pairs of open sets and their unions and then to argue by induction, for this would yield only a statement about *finite* sets of open sets, whereas the axiom makes a statement about every set of open sets. A set of open sets may well be infinite and possibly not even countable. For a finite topological space X the distinction disappears. In this case it is, for example, true that the intersection of the set of open neighbourhoods of a point $x \in X$ is itself an open neighbourhood of x. The corresponding statement for an arbitrary topological space is false. For example, the intersection of all the open neighbourhoods of 0 in $\mathbf{R}$ is the set $\{0\}$, which is not open in $\mathbf{R}$.

Continuity

Let $(X,\mathscr{T})$ and $(Y,\mathscr{U})$ be topological spaces. By analogy with the case where X and Y are normed affine spaces a map $f: (X,\mathscr{T}) \to (Y,\mathscr{U})$ is said to be *continuous* if, and only if, $(f^{-1})_{\vdash}(\mathscr{U}) \subset \mathscr{T}$, that is, if, and only if, for each open subset B of Y, $f^{-1}(B)$ is open in X. (For the notations, see page 11.)

The proofs of the following elementary propositions are left as exercises.

Prop. 16.3. Let X and Y be topological spaces and let $f: X \to Y$ be constant. Then f is continuous. □

Prop. 16.4. Let X be a topological space. Then the map $1_X : X \to X$ is continuous. □

Prop. 16.5. Let W, X and Y be topological spaces and let $g : W \quad X$ and $f: X \to Y$ be continuous. Then $fg : W \to Y$ is continuous. □

The inverse of a bijective continuous map need not be continuous. For example, let X be any set with more than one element. Then the map

$$1_X : (X,\text{Sub } X) \to (X,\{\emptyset,X\})$$

is continuous, but its inverse is not continuous.

A bijective continuous map whose inverse is also continuous is said to be a *homeomorphism*. (The word 'homeomorphism' is essentially synonymous with 'isomorphism', the prefixes being the Greek adjectives 'homoios' = 'like' and 'isos' = 'equal', respectively.)

Two topological spaces X and Y are said to be *homeomorphic*, $X \cong Y$, and either is said to be a *homeomorphic* or *topological model* of the other, if there exists a homeomorphism $f: X \to Y$. The relation $\cong$ is an equivalence on any set of topological spaces.

Exercise 16.6. Put the four topologies on $\{0,1\}$ and the twenty-nine topologies on $\{0,1,2\}$ into homeomorphism classes. □

Subspaces and quotient spaces

Let X be a topological space and let $g : W \to X$ and $f: X \to Y$ be maps. The next proposition states that if a subset of W is defined to be *open* in W if it is of the form $g^{-1}(A)$, where A is open in X, and if a subset C of Y is defined to be *open* in Y if $f^{-1}(C)$ is open in X, then the sets of open sets so defined for W and Y are topologies for W and Y.

Prop. 16.7. Let $g : W \to X$ and $f: X \to Y$ be maps and let $\mathscr{T}$ be a topology for X. Then $(g^{\dashv})_{\vdash}(\mathscr{T})$ is a topology for W and $(f^{\dashv})^{\dashv}(\mathscr{T})$ is a topology for Y.

Proof To prove that $(g^{\dashv})_{\vdash}(\mathscr{T})$ is a topology for W it is enough to remark that

(i) $\emptyset = g^{\dashv}(\emptyset)$ and $W = g^{\dashv}(X)$,

(ii) for all $A, B \in \text{Sub } X$, $g^{\dashv}(A) \cap g^{\dashv}(B) = g^{\dashv}(A \cap B)$,

and (iii) for all $\mathscr{S} \subset \text{Sub } X$, $\bigcup (g^{\dashv})_{\vdash}(\mathscr{S}) = g^{\dashv}(\bigcup \mathscr{S})$,

while to prove that $(f^{\dashv})^{\dashv}(\mathscr{T})$ is a topology for Y it is enough to remark that

(i) $f^{\dashv}(\emptyset) = \emptyset$ and $f^{\dashv}Y = X$,

(ii) for all $C, D \in \text{Sub } Y$, $f^{\dashv}(C \cap D) = f^{\dashv}(C) \cap f^{\dashv}(D)$,

and (iii) for all $\mathscr{U} \subset \text{Sub } Y$, $f^{\dashv}(\bigcup \mathscr{U}) = \bigcup (f^{\dashv})_{\vdash}(\mathscr{U})$.

(Here, and elsewhere, the axiom of choice will be used without comment.) □

The topologies defined in this way are said to be *induced* from the topology $\mathscr{T}$ on X by the maps g and f respectively. The induced topology on W is the smallest topology for W such that g is continuous, while the induced topology on Y is the largest topology for Y such that f is continuous.

When g is an inclusion, the topology $(g^{\dashv})_{\vdash}(\mathscr{T})$ is said to be the *subspace topology* on W *relative* to $(X,\mathscr{T})$, and $(W,(g^{\dashv})_{\vdash}(\mathscr{T}))$ is said to be a (*topological*) *subspace* of $(X,\mathscr{T})$. A subset C of W is open with respect to the subspace topology for W if, and only if, there is some open subset A of X such that $C = A \cap W$. Any subset of a topological space is tacitly assigned the subspace topology unless there is explicit mention to the contrary.

When f is a *partition*, the topology $(f^{\dashv})^{\dashv}(\mathscr{T})$ is said to be the *quotient* (or *identification*) *topology* on Y *relative* to $(X,\mathscr{T})$ and $(Y,(f^{\dashv})^{\dashv}(\mathscr{T}))$ is said to be a (*topological*) *quotient space* of $(X,\mathscr{T})$. A subset B of Y is open with respect to the quotient topology for Y if, and only if, $f^{\dashv}(B)$ is open in X. Any quotient of a topological space is tacitly assigned the quotient topology unless there is explicit mention to the contrary.

As an example of the subspace topology consider $X = \mathbf{R}$ with its usual topology and let $W = [a,b]$, where $a, b \in \mathbf{R}$; that is, W is a bounded closed interval of $\mathbf{R}$. Any open set of X is the union of a set of open intervals of X. Any open set of W is therefore the union of a set of intervals, each of which is of one of the three types:

$$[a,c[, \quad]d,e[, \quad \text{or} \quad]f,b],$$

where c, d, e and $f \in \,]a,b[$ and where $d < e$. This example shows that a set which is open in a subspace W of a topological space X need not be open in X.

Prop. 16.8. Let W be an open subspace of a topological space X. Then a subset B of W is open in W if, and only if, B is open in X. □

Prop. 16.9. Let $f: X \rightarrow Y$ be a continuous map and let W be a subspace of X. Then the map $f \mid W: W \rightarrow Y$ is continuous. □

These induced topologies have particular relevance to the canonical decomposition

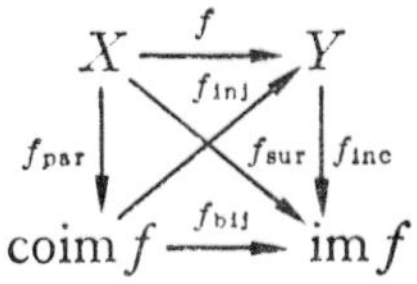

of a continuous map $f: X \rightarrow Y$, $\operatorname{im} f$ being assigned the subspace topology and $\operatorname{coim} f$ the quotient topology.

Prop. 16.10. Let $f: X \rightarrow Y$ be a continuous map, X and Y being topological spaces. Then f_{sur}, f_{inj} and f_{bij} are continuous. □

The map f_{bij} need not be a homeomorphism. It is at first sight tempting to single out for special study those continuous maps f for which f_{bij} is a homeomorphism. However, the composite of two such maps need not have this property. Consider, for example, the inclusion map

$$g: [0,2[\rightarrow \mathbf{R}$$

and the map

$$f: \mathbf{R} \rightarrow \mathbf{C}; \quad x \rightsquigarrow e^{\pi i x},$$

with image S^1, the unit circle. Clearly, g_{bij} is a homeomorphism, and it is easily verified that f_{bij} also is a homeomorphism. However, fg, though bijective, is not a homeomorphism, for $[0,1[$ is open in $[0,2[$, but $(fg)_\vdash([0,1[)$ is not open in S^1.

Later, in Cor. 16.44, we state sufficient, though not necessary, conditions for the map f_{bij}, induced by a continuous map f, to be a homeomorphism.

A continuous injection $f: X \rightarrow Y$ such that f_{bij}, or equivalently f_{sur}, is a homeomorphism is said to be a (*topological*) *embedding* of X in Y.

Prop. 16.11. Let $s: Y \rightarrow X$ be a continuous section of a continuous surjection $f: X \rightarrow Y$. Then s is a topological embedding. □

A continuous surjection $f: X \rightarrow Y$ such that f_{bij}, or equivalently f_{inj}, is a homeomorphism is said to be a (*topological*) *projection* of X on to Y.

Prop. 16.12. Let W, X and Y be topological spaces and let $g: W \rightarrow X$ and $f: X \rightarrow Y$ be maps, whose composite $fg: W \rightarrow Y$ is continuous.

Then, if f is an embedding, g is continuous and, if g is a projection, f is continuous.

Proof Suppose that f is an embedding and let A be any open set in X. Then, since f is an embedding, $A = f^{-1}(B)$ for some open subset B of Y. It follows that $g^{-1}(A) = g^{-1}f^{-1}(B) = (fg)^{-1}(B)$, which is open in W since fg is continuous. Therefore g is continuous.

The other part of the proposition is similarly proved. □

For the relationship between topological projections and product projections, see Cor. 16.53.

Closed sets

A subset B of a topological space X is said to be closed in X if its complement $X \setminus B$ is open. A point $x \in X$ is said to be *closed* if the subset $\{x\}$ is closed in X.

Prop. 16.13. A map $f: X \to Y$ between topological spaces X and Y is continuous if, and only if, for each closed subset B of Y, $f^{-1}(B)$ is closed in X. □

Examples 16.14. Any closed interval of $\mathbf{R}$ (with its standard topology) is closed.

Any finite subset of $\mathbf{R}$ is closed. In particular, any point of $\mathbf{R}$ is closed.

The set $f^{-1}\{0\}$ of zeros of a continuous map $f: X \to \mathbf{R}$ is closed in X.

Prop. 16.15. Let X be a topological space. Then $\emptyset$ and X are closed in X, the union of any finite set of closed sets is closed, and the intersection of any non-null set of closed sets is closed. □

Note that a subset of a topological space X may be both open and closed or neither open nor closed.

A point x of a topological space X is said to be in the *closure* $Cl_X A$ of a subset A of X if every open neighbourhood of x in X intersects A.

Prop. 16.16. The closure $Cl_X A$ of a subset A of a topological space X is closed in X, and if B is any closed subset of X with $A \subset B$, then $Cl_X A \subset B$. If V is an open subset of X such that $A \cap V$ is closed in V, then $A \cap V = Cl_X A \cap V$. □

Prop. 16.17. A map $f: X \to Y$ is continuous if, and only if, for each subset A of X, $f_{\vdash}(Cl_X A) \subset Cl_Y(f_{\vdash}(A))$, X and Y being topological spaces. □

A subset A of a topological space X is said to be *locally closed* in X

if, for every $a \in A$, there is an open neighbourhood V of a such that $A \cap V$ is closed in V. For example, the set $\{(x,0) \in \mathbf{R}^2 : -1 < x \leqslant 1\}$ is locally closed in $\mathbf{R}^2$.

Prop. 16.18. Any locally closed subset of a topological space X is of the form $B \cap C$, where B is open in X and C is closed in X, and any subset of this form is locally closed. □

Limits

Let $g : W \rightarrowtail X$ be a continuous map with domain a subset of the topological space W and let $a \in W \setminus \operatorname{dom} g$. Then, by definition, the map g has *a limit b at a* if the map $f : W \rightarrowtail X$ defined by $f(w) = g(w)$ for all $w \in \operatorname{dom} g$ and by $f(a) = b$ is continuous. If the limit is unique, then we write $\lim_a g = b$ or $\lim_{w \to a} g(w) = b$.

It is left to the reader to verify that this definition of limit agrees with the earlier definition in the case that $W = \bar{\omega}$, $a = \omega$ and X is a normed affine space, g being a sequence on X.

The uniqueness of the limit is discussed further in Prop. 16.36 below.

Covers

An *open cover* or *cover* for a topological space $(X,\mathscr{T})$ is, by definition, a subset $\mathscr{S}$ of $\mathscr{T}$ such that $\bigcup \mathscr{S} = X$.

Prop. 16.19. Let B be a subset of a topological space X and let $\mathscr{S}$ be a cover for X. Then B is open in X if, and only if, for each $A \in \mathscr{S}$, $B \cap A$ is open in A.

Proof $\Rightarrow$: by the definition of the induced topology;
$\Leftarrow$: by axiom (iii) for a topology, since

$$B = B \cap X = B \cap (\bigcup \mathscr{S}) = \bigcup \{B \cap A : A \in \mathscr{S}\},$$

$B \cap A$ being open in X as well as in A, for any $A \in \mathscr{S}$, by Prop. 16.8. □

Cor. 16.20. Let $f : X \to Y$ be a map between topological spaces X and Y and let $\mathscr{S}$ be a cover for X. Then f is continuous if, and only if, for each $A \in \mathscr{S}$, $f \mid A$ is continuous. □

Cor. 16.21. Two topologies on a set X are the same if, and only if, the induced topologies on each of the elements of some cover for X are the same. □

It follows that in studying a topological space X nothing is lost by choosing a cover for X and studying separately each element of the cover. This perhaps gives some insight into the way in which a topology gives cohesion to a set. Note, in particular, the role of axiom (iii).

Prop. 16.22. Let $f: X \rightarrow Y$ be a continuous injection and let $\mathscr{B}$ be a cover for Y. Then f is an embedding if, and only if, for each $B \in \mathscr{B}$ the map

$$f \mid f^{-1}B : f^{-1}(B) \rightarrow Y$$

is an embedding. □

Prop. 16.23. Let $f: X \rightarrow Y$ be a continuous surjection and let $\mathscr{B}$ be a cover for Y. Then f is a projection if, and only if, for each $B \in \mathscr{B}$ the map

$$(f \mid f^{-1}(B))_{\text{sur}} : f^{-1}(B) \rightarrow B$$

is a projection. □

Let W be a subspace of a topological space X. A set $\mathscr{S}$ of open sets of X such that $W \subset \bigcup \mathscr{S}$ will be called an *X-cover* for W. The set $\{A \cap W: A \in \mathscr{S}\}$ is then a cover for W, called the *induced* cover.

For example, the set $\{]-1,1[\,,\,]0,2[\}$ is an $\mathbf{R}$-cover for the closed interval $[0,1]$. The induced cover is the set $\{[0,1[\,,\,]0,1]\}$.

It follows from the definition of the induced topology that every cover for W is induced by some X-cover for W (generally not unique).

Prop. 16.24. Let W be a subspace of a topological space X, let $\mathscr{S}$ be an X-cover for W and let $\mathscr{P}$ be the induced cover for W. Then there is a finite subset $\mathscr{P}'$ of $\mathscr{P}$ covering W if, and only if, there is a finite subset $\mathscr{S}'$ of $\mathscr{S}$ covering W. □

Theorem 16.25. (*Heine-Borel.*)

Let $\mathscr{S}$ be an $\mathbf{R}$-cover of a bounded closed interval $[a,b] \subset \mathbf{R}$. Then a finite subset $\mathscr{S}'$ of $\mathscr{S}$ covers $[a,b]$.

Proof Let A be the set of points $x \in [a,b]$ such that a finite subset of $\mathscr{S}$ covers $[a,x]$. It has to be proved that $b \in A$.

Since $a \in A$, A is non-null. Also, A is bounded above by b. So, by the upper bound axiom, $s = \sup A$ exists.

Now $s \in A$. For there exists an open set $U \in \mathscr{S}$ such that $s \in U$, and therefore, since U is the union of a set of open intervals of $\mathbf{R}$ and since $s = \sup A$, there exists $r \leqslant s$ such that $r \in A$ and $[r,s] \subset U$. Let $\mathscr{R}$ be a finite subset of $\mathscr{S}$ covering $[a,r]$. Then $\mathscr{R} \cup \{U\}$, also finite, covers $[a,s]$. That is, $s \in A$.

Also, $s = b$. For suppose $s < b$. Then there exists t, $s < t \leqslant b$, such that $[s,t] \subset U$. So $\mathscr{R} \cup \{U\}$ covers $[a,t]$, contradicting the definition of s. That is, $s = b$.

Therefore $b \in A$. □

Cor. 16.26. Let $\mathscr{P}$ be any cover for $[a,b]$. Then there exists a finite subset $\mathscr{P}'$ of $\mathscr{P}$ covering $[a,b]$. □

Compact spaces

A topological space X is said to be *compact* if for *each* cover $\mathscr{S}$ for X a finite subset $\mathscr{S}'$ of $\mathscr{S}$ covers X. For example, any finite topological space is compact (the topology need not be discrete). The Heine-Borel theorem states that every bounded closed interval of **R** is compact. By contrast, the interval $]0,1]$ is not compact, since no finite subset of the cover $\{](n+1)^{-1},1];\ n \in \omega\}$ covers $]0,1]$.

We shall eventually prove that a subset A of a finite-dimensional normed affine space X is compact if, and only if, A is closed and bounded in X. The Heine-Borel theorem is the first stage in the proof. Propositions 16.27 and 16.37 are further stages, and the final stage is Theorem 16.60.

We recall that a subset A of a normed affine space X is *bounded* if there is a ball B in X such that $A \subset B$.

Prop. 16.27. A compact subspace A of a normed affine space X is bounded.

Proof Consider the set $\mathscr{S}$ of all balls of radius 1 with centre a point of A. Since A is compact, a finite subset $\mathscr{S}'$ of $\mathscr{S}$ covers A. It follows easily that A is bounded. □

Prop. 16.28. A closed subset A of a compact space X is compact.

Proof Let $\mathscr{S}$ be an X-cover for A. Since A is closed, $X \setminus A$ is open in X. So $\mathscr{S} \cup \{X \setminus A\}$ covers X. Since X is compact, a finite subset $\mathscr{S}' \cup \{X \setminus A\}$ of $\mathscr{S} \cup \{X \setminus A\}$ covers X, where $X \setminus A \notin \mathscr{S}'$. Discarding $X \setminus A$ again, we find that $\mathscr{S}'$ covers A; that is, a finite subset of $\mathscr{S}$ covers A. So A is compact. □

The next proposition relates compactness to continuity.

Prop. 16.29. Let $f: X \to Y$ be a continuous surjection and let X be compact. Then Y is compact.

Proof Let $\mathscr{B}$ be any cover for Y. Then $\mathscr{A} = (f^{\dashv})_{\vdash}(\mathscr{B})$ is a cover for X and, since X is compact, there is a finite subset $\mathscr{A}'$ of $\mathscr{A}$ covering X. Since f is surjective, $f_{\vdash}(f^{\dashv}(B)) = B$ for any $B \subset Y$, in particular for any $B \in \mathscr{B}$. It follows that $(f_{\vdash})_{\vdash}(\mathscr{A}')$ is a finite cover for Y contained in $\mathscr{B}$. So Y is compact. □

Cor. 16.30. Let $f: X \to Y$ be a continuous map and let A be any compact subset of X. Then $f_{\vdash}(A)$ is a compact subset of Y. □

Cor. 16.31. Let X be a compact space and let $f: X \to Y$ be a partition of X. Then the quotient Y is compact. □

Hausdorff spaces

A topological space X is said to be a *Hausdorff space* if, given any distinct points $a, b \in X$, there exist mutually *disjoint* open neighbour-

hoods A and B of a and b respectively in X. (The figure is due originally, we believe, to Professor M. F. Atiyah!)

The proofs of the following elementary propositions are left as exercises.

Prop. 16.32. Any normed affine space is a Hausdorff space. □

Prop. 16.33. The only Hausdorff topology for a finite set is the discrete topology. □

Prop. 16.34. Any subspace of a Hausdorff space is a Hausdorff space. □

By contrast, a quotient of a Hausdorff space need not be a Hausdorff space.

Consider, for example, the partition $\pi: X \to Y$ of the subspace $X = \{-1,1\} \times]-1,1[$ of $\mathbf{R}^2$ which identifies $(-1,x)$ with $(1,x)$, for all $x \in]-1,0[$.

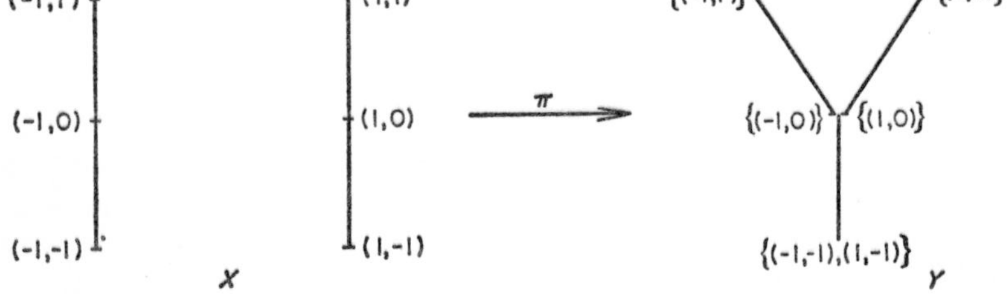

(The diagram is necessarily inadequate, since any subset of $\mathbf{R}^2$ is Hausdorff.)

The points $\{(-1,0)\}$ and $\{(1,0)\}$ of the quotient Y are then distinct but, since any open neighbourhood of 0 in $]-1,1[$ contains as a subset an open interval $]-\delta,\delta[$ where $0 < \delta \leqslant 1$, any open neighbourhoods of $\{(-1,0)\}$ and $\{(1,0)\}$ in Y intersect. That is, Y is not a Hausdorff space.

The space Y will be referred to in the sequel as *the Y space.*

Prop. 16.35. Let g and $h : W \to X$ be continuous maps, X being a Hausdorff space and let

$$M = \{w \in W : g(w) = h(w)\}.$$

Then M is closed in W. □

Prop. 16.36. Let $g : W \rightarrowtail X$ be a continuous map with domain a proper subset of the topological space W, X being a Hausdorff space, and let a be an element of the closure of W in X not belonging to W. Then if g has a limit b at a, b is unique. □

(This is in practice one of the most important features of a Hausdorff space.)

Prop. 16.37. Let W be a non-null compact subspace of a Hausdorff space X. Then W is closed in X.

(If $W = \emptyset$ the proposition is trivially true.)

Proof Let $\mathscr{T}$ be the topology on X and let $x \in X \setminus W$. Let $\mathscr{C} = \{(A,B) \in \mathscr{T}^2 : x \in A,\ A \cap B = \emptyset\}$ and let $\mathscr{B} = \{B \in \mathscr{T} :$ for some $A \in \mathscr{T}, (A,B) \in \mathscr{C}\}$.

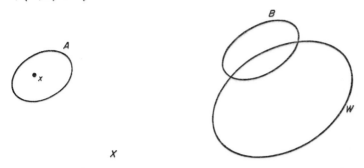

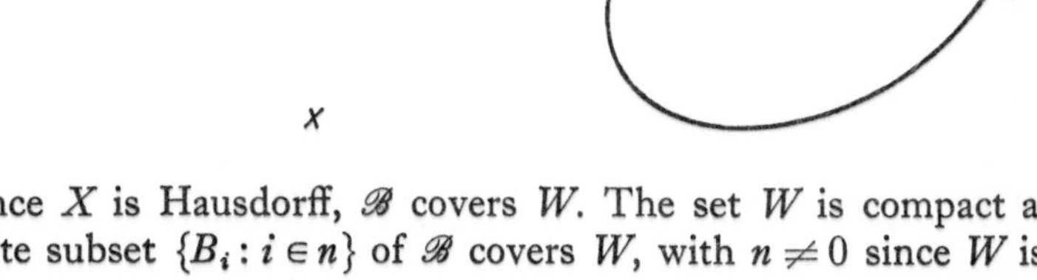

Since X is Hausdorff, $\mathscr{B}$ covers W. The set W is compact and so a finite subset $\{B_i : i \in n\}$ of $\mathscr{B}$ covers W, with $n \neq 0$ since W is non-null.

For each $i \in n$, choose $A_i \in \mathscr{T}$ such that $(A_i,B_i) \in \mathscr{C}$. Since $n \neq 0$ we may form $U = \bigcap_{i \in n} A_i$, this being an open neighbourhood of x in X, since n is finite. Also since, for each $i \in n$, $U \cap B_i = \emptyset$, $U \cap W = \emptyset$; that is, $U \subset X \setminus W$.

It follows that $X \setminus W$ is open in X; that is, W is closed in X. □

Cor. 16.38. A compact subset Y of a normed affine space X is closed. □

Putting together Cor. 16.26, Props. 16.27, 16.28 and Cor. 16.38, we obtain the following characterization of compact subsets of **R**.

Prop. 16.39. A subset A of **R** is compact if, and only if, it is closed and bounded.

Proof $\Rightarrow$: Let A be compact. Then, by Prop. 16.27, A is bounded and, by Cor. 16.38, A is closed.

$\Leftarrow$: Let A be closed and bounded. Since A is bounded, there exists a bounded closed interval $[a,b]$ such that $A \subset [a,b]$. By Cor. 16.26, $[a,b]$ is compact. Also, A is closed in $[a,b]$, since $[a,b]$ is closed in **R**. So, by Prop. 16.28, A is compact. □

Cor. 16.40. Let $f: X \to \mathbf{R}$ be a continuous map, and let A be a compact subspace of X. Then $f_{\vdash}(A)$ is closed and bounded in **R**. □

In particular, let $f: \mathbf{R} \rightarrowtail \mathbf{R}$ be a continuous map with domain a closed bounded interval $[a,b]$. Then f is bounded and 'attains its bounds'.

Open, closed and compact maps

Let $f: X \to Y$ be a continuous map. Then $f^{\dashv}$ sends open sets in Y to open sets in X and closed sets in Y to closed sets in X, while $f_{\vdash}$ sends compact sets in X to compact sets in Y. The map f is said to be

open if $f_{\vdash}$ sends open sets in X to open sets in Y

closed if $f_{\vdash}$ sends closed sets in X to closed sets in Y

and *compact* if $f^{\dashv}$ sends compact sets in Y to compact sets in X.

The map $p: \mathbf{R}^2 \to \mathbf{R}$; $(x,y) \rightsquigarrow x$ is open, since any open subset of $\mathbf{R}^2$ is the union of open squares, and the image by p of an open square is an open interval of **R**. On the other hand p is not closed, since the set $\{(x,y) \in \mathbf{R}^2 : xy = 1\}$ is closed in $\mathbf{R}^2$, it being the fibre over 1 of the map $\mathbf{R}^2 \to \mathbf{R}$; $(x,y) \rightsquigarrow xy$, but its image in **R** by p is $\mathbf{R} \setminus \{0\}$, which is not closed in **R**. (By Cor. 16.30 and Theorem 16.60 below, any closed subset of $\mathbf{R}^2$ with an image which is not closed must necessarily be unbounded.)

The restriction of p to the subset $(\mathbf{R} \times \{0\}) \cup (\{0\} \times \mathbf{R})$ of $\mathbf{R}^2$ is closed, but not open.

Prop. 16.41. Let $s: Y \to X$ be a continuous section of a continuous surjection $f: X \to Y$. Then s is open or closed if, and only if, im s is, respectively, open or closed in X. □

Exercise 16.42. Let $B = \{(x,y) \in \mathbf{R}^2 : x^2 - y^2 = -1 \text{ or } 0\}$. Show that $p: B \to \mathbf{R}$; $(x,y) \rightsquigarrow x$ has six continuous sections, all of which are closed, but only two of which are open.

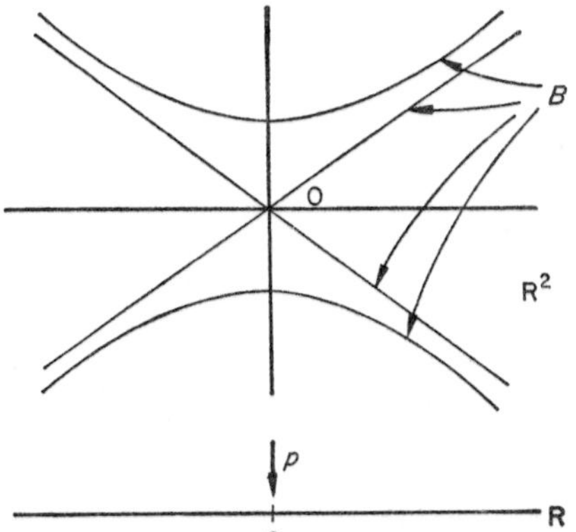

□

The following two propositions are frequently used in determining whether or not a continuous map is an embedding or a projection.

Prop. 16.43. Let $f: X \to Y$ be a continuous map. Then, if f is either open or closed, f_{bij} is a homeomorphism. □

Prop. 16.44. Let X be compact, Y Hausdorff and $f: X \to Y$ continuous. Then f is closed and compact.

Proof Let A be closed in X. Then A is compact, by Prop. 16.28, $f_{\vdash}(A)$ is compact, by Prop. 16.29, and $f_{\vdash}(A)$ is closed, by Prop. 16.37. That is, f is closed. (This implies, by Prop. 16.43, that f_{bij} is a homeomorphism.)

The proof of the compactness of f is similar. □

Theorem 16.45. A closed continuous map $f: X \to Y$ is compact if, and only if, each fibre of f is compact.

Proof $\Rightarrow$: For each $y \in Y$, $\{y\}$ is compact.

$\Leftarrow$: Let W be any compact subset of Y, let $V = f^{\dashv}(W)$ and let $\mathscr{A}$ be any X-cover for V. It has to be proved that if each fibre of f is compact a finite subset of $\mathscr{A}$ covers V.

Let $w \in W$. Then by hypothesis $f^{\dashv}\{w\}$ is compact and is therefore covered by a finite subset of $\mathscr{A}$, $\mathscr{A}'$, say. Let $A = \bigcup \mathscr{A}'$ and let $B = Y \setminus f_{\vdash}(X \setminus A)$.

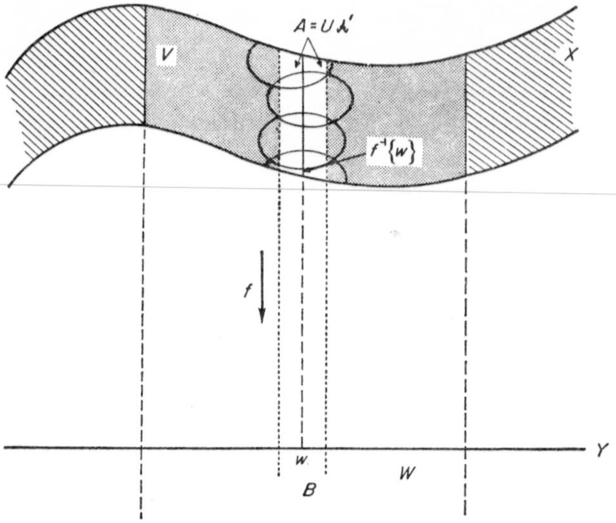

Since f is closed, B is open in Y and, since $f^{\dashv}(B) \subset A$, the set $f^{\dashv}(B)$ is covered by $\mathscr{A}'$.

Now let $\mathscr{B} = \{B \in \mathscr{T} : f^{\dashv}(B)$ covered by a finite subset of $\mathscr{A}\}$, where $\mathscr{T}$ is the topology on Y. By what we have just proved, $\mathscr{B}$ covers W. But W is compact, and so a finite subset of $\mathscr{B}$ covers W. It follows at once that a finite subset of $\mathscr{A}$ covers V. □

A closed compact map is called a *proper* map. For a full account of proper maps see [7].

Product topology

The following proposition generalizes the construction of the subspace topology.

Prop. 16.46. Let W be a set, X and Y topological spaces and $p : W \to X$ and $q : W \to Y$ maps. Define a subset C of W to be *open* in W if, and only if, C is the union of a set of subsets of W each of the form $p^{\dashv}A \cap q^{\dashv}B$, where A is open in X and B is open in Y. Then

(i) the set of open subsets of W is a topology for W;

(ii) this topology is the smallest topology for W such that both p and q are continuous. □

The topology so defined is said to be the topology for W *induced* by the maps p, q from the topologies for X and Y.

When $W = X \times Y$ and $(p,q) = 1_W$, the topology induced on W by p and q is called the *product topology* for W.

Prop. 16.47. Let X and Y be topological spaces and let $X \times Y$ have the product topology. Then a subset of $X \times Y$ is open if, and only if, it is the union of a set of subsets of $X \times Y$ each of the form $A \times B$ where A is open in X and B is open in Y.

Proof This proposition is just a reformulation of the definition of the product topology. For let $(p,q) = 1_{X \times Y}$. Then

$$\begin{aligned} A \times B &= (A \times Y) \cap (X \times B) \\ &= p^{\dashv}A \cap q^{\dashv}B. \qquad \square \end{aligned}$$

For example, let $\mathbf{R}^2 = \mathbf{R} \times \mathbf{R}$ have the product topology. Then a subset U of $\mathbf{R}^2$ is open if, and only if, it is the union of a set of subsets of $\mathbf{R}^2$ each of the form $A \times B$ where A is open in $\mathbf{R}$ and B is open in $\mathbf{R}$. Now a subset of $\mathbf{R}$ is open if, and only if, it is the union of bounded open intervals. It follows that $A \times B$ and therefore U is the union of bounded open rectangles, of the form $]a,b[\ \times\ ]c,d[$, where $a, b, c, d \in \mathbf{R}$.

From this last remark it follows that the product topology on $\mathbf{R}^2$ coincides with the topology induced by the product norm (the standard topology on $\mathbf{R}^2$). This is a special case of the following proposition.

Prop. 16.48. Let X and Y be normed affine spaces. Then the product norm on $X \times Y$ induces the product topology on $X \times Y$. $\square$

The product $X \times Y$ of two topological spaces will tacitly be assigned the product topology.

Prop. 16.49. A map $(f,g) : W \to X \times Y$ is continuous if, and only if, each of its components $f : W \to X$ and $g : W \to Y$ is continuous, W, X and Y being topological spaces.

Proof Let $(p,q) = 1_{X \times Y}$. Then $f = p(f,g)$, $g = q(f,g)$.

$\Rightarrow$: Let (f,g) be continuous. Since p is continuous and since $f = p(f,g)$, f is continuous. Similarly, g is continuous.

$\Leftarrow$: Any open set of $X \times Y$ is the union of sets of the form $A \times B = p^{\dashv}(A) \cap q^{\dashv}(B)$, where A is open in X and B is open in Y. Suppose f and g are continuous. Then

$$\begin{aligned} (f,g)^{\dashv}(A \times B) &= (f,g)^{\dashv}p^{\dashv}(A) \cap (f,g)^{\dashv}q^{\dashv}(B) \\ &= f^{\dashv}(A) \cap g^{\dashv}(B) \end{aligned}$$

which is open in W. It follows that (f,g) is continuous. $\square$

Prop. 16.50. Let X and Y be topological spaces, let A be a subspace of X and let B be a subspace of Y. Then $A \times B$ is a subspace of $X \times Y$.

Proof What has to be proved is that the product topology on $A \times B$ coincides with the topology on $A \times B$ induced by the inclusion $A \times B \to X \times Y$. Now, for any subset $C \subset X$ and any subset $D \subset Y$,

$$(A \times B) \cap (C \times D) = (A \cap C) \times (B \cap D).$$

The further details are left as an exercise. □

Cor. 16.51. Let X and Y be topological spaces and let $y \in Y$. Then the injection $X \to X \times Y$; $x \rightsquigarrow (x,y)$ is an embedding. □

Prop. 16.52. Let X and Y be topological spaces. Then the product projection $p : X \times Y \to X$; $(x,y) \rightsquigarrow x$ is an open map.

Proof Let A be any open subset of the topological space $X \times Y$. Then $A = \bigcup\{A \cap (X \times \{y\}) : y \in Y\}$. Since, for any $y \in Y$, $p \mid (X \times \{y\})$ is open, and since $p_{\vdash}A = \bigcup\{p_{\vdash}(A \cap (X \times \{y\})) : y \in Y\}$, the result follows. □

Cor. 16.53. If Y is non-null, the product projection $p : X \times Y \to X$ is a topological projection. □

A continuous surjection $f : X \to Y$ is said to be *trivial* if there is a topological space W and a homeomorphism $h : Y \times W \to X$ such that the map $fh : Y \times W \to Y$ is the product projection of $Y \times W$ on to Y. A continuous map $f : X \to Y$ is then said to be *locally trivial at* a point $y \in Y$ if there exists an open neighbourhood B of y in Y such that the map

$$(f \mid f^{\dashv}(B))_{\text{sur}} : f^{\dashv}(B) \to B$$

is trivial, and to be *locally trivial* if it is locally trivial at each $y \in Y$.

Prop. 16.54. A locally trivial continuous surjection $f : X \to Y$ is a topological projection. □

Prop. 16.55. Let $f : X \to Y$ be a continuous map of a Hausdorff space X to a topological space Y such that each fibre is finite and, for each $x \in X$, there is an open neighbourhood A of x in X such that $(f \mid A)_{\text{sur}}$ is a homeomorphism. Then f is locally trivial. □

Prop. 16.56. Let X and Y be topological spaces, Y being compact. Then the projection

$$p : X \times Y \to X; \quad (x,y) \rightsquigarrow x$$

is closed.

Proof If Y is null the map trivially is closed. Now suppose $Y \neq \emptyset$.

Let V be a closed subset of $X \times Y$ and let $W = p_{\vdash}(V)$. It has to be proved that W is closed in X.

Let $x \in X \setminus W$ and let $\mathscr{C} = \{A \times B : A$ an open neighbourhood of x in X, B open in Y and $(A \times B) \cap V = \emptyset\}$.

Since $p^{\dashv}\{x\}$ is a subset of the open set $(X \times Y) \setminus V$, $\mathscr{C}$ covers $p^{\dashv}\{x\}$ and so, by the compactness of $p^{\dashv}\{x\}$ (homeomorphic to Y), a finite subset $\{A_i \times B_i : i \in n\}$ of $\mathscr{C}$ covers $p^{\dashv}\{x\}$. Since Y is non-null, $n \neq 0$ and we may form $U = \bigcap_{i \in n} A_i$. This is an open neighbourhood of x in X, since n is finite.

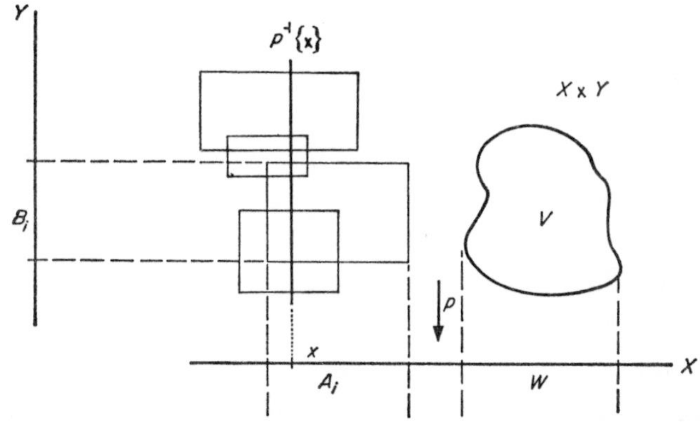

Also, $p^{\dashv}(U) \cap V = \emptyset$, so that $U \cap W = \emptyset$. That is, $U \subset X \setminus W$.

It follows that $X \setminus W$ is open in X and therefore that W is closed in X. □

Theorem 16.57. Let X and Y be non-null compact topological spaces. Then $X \times Y$ is compact.

Proof By Prop. 16.56 the projection $p : X \times Y \to X$; $(x,y) \rightsquigarrow x$ is closed, and therefore compact, by Theorem 16.45. But $X \times Y = p^{\dashv}(X)$ and X is compact. Therefore $X \times Y$ is compact. □

Cor. 16.58. Any finite product of compact topological spaces is compact. □

For example, a closed product ball in $\mathbf{R}^n$, being the product of a finite number of closed bounded intervals in $\mathbf{R}$ is compact.

(This last result can also be proved by constructing as in Exercise 15.62 a continuous 'Peano curve' of the interval $[0,1]$ with image the closed product ball in $\mathbf{R}^n$. Since $[0,1]$ is compact in $\mathbf{R}$ its image in $\mathbf{R}^n$ will be compact.)

Prop. 16.59. Any closed bounded subset of $\mathbf{R}^n$ is compact.

Proof Any bounded subset of $\mathbf{R}^n$ is a subset of some closed product ball in $\mathbf{R}^n$, and such a ball is compact, as has just been proved. Moreover, since the ball is closed, any subset of it that is closed in $\mathbf{R}^n$ is closed also in the ball. Since, by Prop. 16.28, any closed subset of a compact space is compact, the proposition follows. □

Theorem 16.60. A subset of $\mathbf{R}^n$ is compact if, and only if, it is closed and bounded.

Proof $\Rightarrow$: Prop. 16.27 and Cor. 16.38.
$\Leftarrow$: Prop. 16.59. □

The characterization of the compact sets of a finite-dimensional affine space to which we alluded earlier on page 319 is an immediate corollary.

Prop. 16.61. Let X be a finite-dimensional normed linear space, with norm $|| \quad ||$. Then the sphere $\{x \in X : ||x|| = 1\}$ is compact with respect to the topology induced by the norm.

Proof The map $x \rightsquigarrow ||x||$ is continuous with respect to $|| \quad ||$, implying that the sphere is closed, since $\{1\}$ is closed in $\mathbf{R}$. Also, the sphere is bounded. Hence the result. □

The notion of compactness provides an alternative proof of the equivalence of norms on a finite-dimensional linear space X, Theorem 15.26.

We suppose, as in Theorem 15.26, that we have two norms on X, denoted respectively by $| \quad |$ and $|| \quad ||$, the norm $|| \quad ||$ being the product norm induced by some basic framing $(e_i : i \in n)$ for X. Then, as before, it follows at once that, for all $x \in X$, $|x| \leqslant L\,||x||$, where $L = n \sup\{|e_i| : i \in n\}$, and therefore that the map $x \rightsquigarrow |x|$ is continuous, with respect to $|| \quad ||$. To obtain a similar inequality with the roles of $| \quad |$ and $|| \quad ||$ reversed, we remark first that the sphere $\{x \in X : ||x|| = 1\}$ is compact with respect to $|| \quad ||$, and the map $x \rightsquigarrow |x|^{-1}$, with domain the sphere, is continuous, since inversion is continuous. It follows, by Cor. 16.40, that there is a real number K such that $|x|^{-1} \leqslant K$ for all $x \in X$ such that $||x|| = 1$, and therefore such that $||x|| \leqslant K|x|$, for all $x \in X$.

The existence of K and L implies, as before, that the two norms are equivalent.

Prop. 16.62. Let X and Y be positive-definite finite-dimensional orthogonal spaces. Then $O(X, Y)$ is compact in $L(X, Y)$.

Proof By Prop. 9.14, $O(X, Y) = \{t \in L(X, Y) : t^*t = 1\}$ and the map $L(X, Y) \to L(X, Y);\ t \rightsquigarrow t^*t$ is continuous. So $O(X, Y)$ is closed in $L(X, Y)$.

Also, since $|t(x)| = |x|$, for all $x \in X$, $|\ \ |$ denoting in either case the quadratic norm, it follows that $|t| = 1$, for all $t \in O(X,Y)$. So $O(X,Y)$ is bounded, and therefore compact, in $L(X,Y)$. □

Exercise 16.63. Prove that $SL(2;\mathbf{R})$ is not compact in $\mathbf{R}(2)$. (Consider, for example, the map $\mathbf{R}(2) \to \mathbf{R}^2;\ t \rightsquigarrow t(0,1)$, that is the map $\begin{pmatrix} a & c \\ b & d \end{pmatrix} \rightsquigarrow \begin{pmatrix} c \\ d \end{pmatrix}$, and apply Cor. 16.30.) □

Connectedness

The simplest intuitive example of a disconnected set is the set $2 = \{0,1\}$, the standard set with two elements. Of the four topologies for 2 only the discrete topology is Hausdorff. Let 2 have this topology, its standard topology.

A non-null topological space X is said to be *disconnected* if there is a continuous surjection $f: X \to 2$, and to be *connected* if every continuous map $f: X \to 2$ is constant.

Any non-null topological space is easily seen to be either connected or disconnected, but not both. The null space is neither connected nor disconnected. Any set with at least two members is connected with respect to the trivial topology, disconnected with respect to the discrete topology, a reason for using the term 'discrete' (cf. Exercise 16.94).

Prop. 16.64. A topological space X is disconnected if, and only if, it is the union of two disjoint non-null open sets of X.

Proof $\Rightarrow$: Let X be disconnected. Then there exists a continuous surjection $f: X \to 2$. Now the sets $\{0\}$ and $\{1\}$ are open in 2. Since f is continuous, $f^{-1}\{0\}$ and $f^{-1}\{1\}$ are open in X and since f is surjective they are non-null. Also

$$f^{-1}\{0\} \cap f^{-1}\{1\} = \emptyset \quad \text{and} \quad f^{-1}\{0\} \cup f^{-1}\{1\} = X.$$

That is, X is the union of two disjoint non-null open sets.

$\Leftarrow$: Suppose A and B are non-null open sets of X such that $A \cap B = \emptyset$ and $A \cup B = X$. Then the map $f: X \to 2$ defined by $f(x) = 0$ for all $x \in A$ and by $f(x) = 1$ for all $x \in B$ is surjective and is continuous, for the inverse image of each of the four open sets of 2 is open in X. That is, X is disconnected. □

Prop. 16.65. Any bounded closed interval $[a,b]$ of $\mathbf{R}$ is connected.

Proof Suppose that $f : [a,b] \to 2$ is continuous and let C be the set $\{c \in [a,b] : f_{\vdash}[a,c] = \{f(a)\}\}$. Since $a \in C$ and since b is an upper bound for C, $s = \sup C$ exists.

Now $s \in C$. For since f is continuous there is an open neighbourhood of s on which f is constant. In particular, f is constant on an open interval around s. But since $s = \sup C$ there is a point of C in this open interval. So $f_{\vdash}[a,s] = f(a)$ and $s \in C$.

Also, $s = b$; for otherwise, by the same remark, there is a point $x \in]s,b]$ such that f is constant also on $[s,x]$, contradicting the definition of s.

So f is constant. That is, $[a,b]$ is connected. □

Theorem 16.66. A non-null subset C of $\mathbf{R}$ is connected if, and only if, it is an interval, that is, if, and only if, it is convex.

Proof $\Leftarrow$: Suppose C is convex and let $f: C \to 2$ be a continuous map. Then, for any $a, b \in C$, $[a,b] \subset C$ and $f \mid [a,b]: [a,b] \to 2$ is continuous. So $f(a) = f(b)$; that is, f is constant. So C is connected.

$\Rightarrow$: Suppose C is not convex. Then there exist $a, b \in C$ and $c \in \mathbf{R} \setminus C$ such that $a < c < b$. Let $A = C \cap]-\infty,c[$ and let $B = C \cap]c,\infty[$. Then A and B are open and non-null, $A \cap B = 0$ and $A \cup B = C$. That is, C is disconnected. □

In particular, $\mathbf{R}$ itself is connected.

Prop. 16.67. Let $f: X \to Y$ be a continuous surjection, and suppose that X is connected. Then Y is connected.

Proof Since X is non-null, Y is non-null. Also, if Y is disconnected there exists a continuous surjection $g: Y \to 2$ and hence a continuous surjection $gf: X \to 2$. So X is disconnected. Hence the result. □

Cor. 16.68. Let $f: X \to Y$ be a continuous map and let A be a connected subset of X. Then graph $(f \mid A)$ is a connected subset of $X \times Y$ and $f_{\vdash}(A)$ is a connected subset of Y. □

Cor. 16.69. Let $f: X \to \mathbf{R}$ be continuous, let X be connected and let $a, b \in f_{\vdash}(X)$. Then the interval $[a,b]$ is a subset of $f_{\vdash}(X)$. (This is sometimes called the *intermediate-value theorem*.) □

Prop. 16.70. Let X be a topological space such that for any $a, b \in X$ there exists a continuous map

$$f: [0,1] \to X$$

such that $f(0) = a$ and $f(1) = b$. Then X is connected. □

Prop. 16.71. For any finite $n > 0$ the unit sphere S^n is a connected subset of $\mathbf{R}^{n+1}$. □

Prop. 16.72. Let X and Y be non-null topological spaces. Then $X \times Y$ is connected if, and only if, X is connected and Y is connected.

Proof $\Rightarrow$: Let $X \times Y$ be connected and let $(p,q) = 1_{X \times Y}$. The map p is continuous and is surjective since Y is non-null. Therefore X is connected, by Prop. 16.67. Similarly, Y is connected.

$\Leftarrow$: Let X be connected, let Y be connected, let $f: X \times Y \to 2$ be continuous and let (x,y), (x',y') be any two points of $X \times Y$. Since $X \times \{y\}$ is homeomorphic to X, $X \times \{y\}$ is connected, and therefore $f(x,y) = f(x',y)$. Similarly $\{x'\} \times Y$ is connected, and $f(x',y) = f(x',y')$. So $f(x,y) = f(x',y')$. It follows that f is constant and that $X \times Y$ is connected.

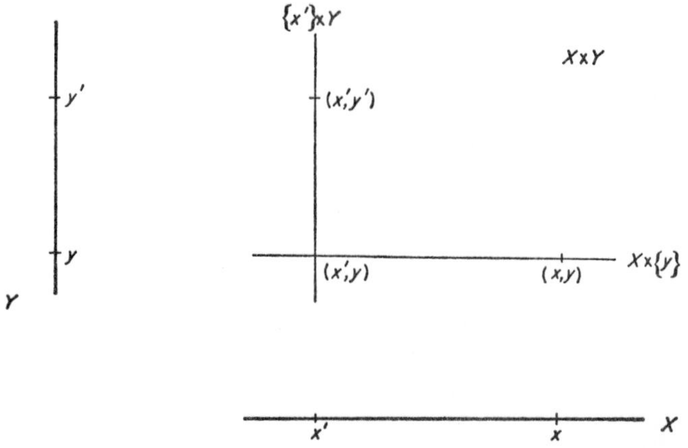

□

Proposition 16.72 may also be regarded as a particular case of the following proposition, whose proof is reminiscent of the proof of Prop. 5.15.

Prop. 16.73. Let $f: X \to Y$ be a topological projection of a topological space X on to a connected topological space Y, each of the fibres of f being connected. Then X is connected.

Proof Let $h: X \to 2$ be a continuous map.

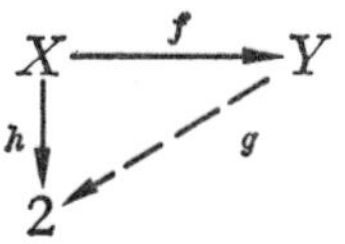

Since the fibres of f are connected, the restriction of h to any fibre is constant. So there exists a map $g: Y \to 2$ defined, for all $y \in Y$, by the formula $g(y) = h(x)$, for any $x \in f^{-1}\{y\}$, such that $h = gf$. Since h is continuous and since f is a projection, g is continuous, by Prop. 16.12, and therefore constant, since Y is connected. So h is constant.

Therefore X is connected. □

Prop. 16.74. Let A be a connected subset of a topological space X. Then $Cl_X A$ is connected. □

A *component* of a topological space X is defined to be a maximal connected subset of X, that is, a connected subset A of X such that any subset of X with A as a proper subset is disconnected.

Any component of a topological space X is closed in X, by Prop. 16.74. Surprisingly, a component need not be open. Consider for example $\bar{\omega} = \omega \cup \{\omega\}$ with its standard topology. The set $\{\omega\}$ is a component of $\bar{\omega}$ but is not open in $\bar{\omega}$.

A topological space for which each point is a component is said to be *totally disconnected.* The above example shows that a totally disconnected space need not be discrete.

Finally, we prove a uniqueness proposition, which will find application in Theorem 19.6.

Prop. 16.75. Let $f: X \to Y$ be a continuous surjection of a Hausdorff space X on to a connected space Y, let $g: Y \to X$ and $h: Y \to X$ be continuous sections of f, let g be an open map and let there be a point $y \in Y$ such that $h(y) = g(y)$. Then $h = g$.

Proof Let $B = \{y \in Y : h(y) = g(y)\}$. Since X is Hausdorff, B is a subset of Y, by Prop. 16.35. Now, since g and h are sections of f, $h(y) = g(y')$ only if $y = y'$. It follows from this that

$$B = \{y \in Y : h(y) = g(y') \text{ for some } y' \in Y\} = h^{\dashv}g_{\vdash}(Y).$$

Since g is an open map, $g_{\vdash}(Y)$ is open in X and so, by the continuity of h, B is open in Y. Finally, B is non-null. So $B = Y$, since Y is connected. That is, $h = g$.

(For an example of a map with continuous open sections, see Exercise 16.42.) □

FURTHER EXERCISES

16.76. In which of the twenty-nine topological spaces, with underlying set the set 3, is each subset of the space either open or closed (or both)? □

16.77. Let X be a topological space whose topology is the cofinite topology for the underlying set. Prove that every permutation of X is a homeomorphism. □

16.78. Sketch the subset $\{(x,x^{-1}) : x \in \mathbf{R}^+\}$ of $\mathbf{R}^2$. Prove that the map $\mathbf{R}^+ \to \mathbf{R};\ x \rightsquigarrow x - x^{-1}$ is a homeomorphism. □

16.79. Prove that if $x \in \mathbf{R}^+$ then $\dfrac{x-1}{x+1} \in]-1,1[$ and that the map $\mathbf{R}^+ \to]-1,1[;\ x \rightsquigarrow \dfrac{x-1}{x+1}$ is a homeomorphism. □

16.80. Prove that the map $]-1,1[\to \mathbf{R};\ x \rightsquigarrow \dfrac{x}{1-x^2}$ is a homeomorphism.

(This map may be visualized in terms of the 'stereographic' projection from $(0,1) \in \mathbf{R}^2$ of the piece of parabola $\{(x,y) \in \mathbf{R}^2 : x \in]-1,1[,\ y = x^2\}$ on to the line $\mathbf{R} \times \{0\}$.) □

16.81. Let A and B be disjoint compact convex subsets of $\mathbf{R}^2$. Prove that a line may be drawn between them. Is the corresponding statement true if the word 'compact' is replaced by (a) 'open' or (b) 'closed'? □

16.82. Let $f: X \to Y$ be a continuous map such that, for any space Z and any continuous maps $g, h: Y \to Z$, $gf = hf \Rightarrow g = h$. Prove that f is surjective.

(Let Z be the quotient of Y obtained by identifying all the points of $\operatorname{im} f$, let g be the partition of Y and let h be an appropriate constant map.) □

16.83. A subset A of a topological space X is said to be *dense* in X if $Cl_X A = X$.

Let $f, g: X \to Y$ be continuous maps that agree on some dense subset A of the topological space X, the topological space Y being Hausdorff. Prove that $f = g$. □

16.84. Let $f: X \rightarrowtail Y$ be a continuous map such that, for any Hausdorff space Z and any continuous maps $g, h: Y \to Z, gf = hf \Rightarrow g = h$. Show by an example that f need not be surjective. □

16.85. Let X and Y be non-null topological spaces. Prove that $X \times Y$ is Hausdorff if, and only if, X is Hausdorff and Y is Hausdorff. □

16.86. A map $f: X \to Y$ between topological spaces X and Y is said to be *locally a homeomorphism at* a point $a \in X$ if there is a neighbourhood A of a in X such that $B = f_{\vdash}(A)$ is a neighbourhood of $f(a)$ in Y and such that the map $(f \mid A)_{\text{sur}}: A \to B;\ x \rightsquigarrow f(x)$ is a homeomorphism.

Suppose that $b \in Y$ is such that $f^{\dashv}(\{b\})$ is finite and, for each $a \in f^{\dashv}(\{b\})$, f is locally a homeomorphism at a. Prove that f is locally trivial at b. □

(This will be required in the proof of Theorem 19.20.)

16.87. Let X and Y be topological spaces. Prove that if $X \times Y$ is compact, then both X and Y are compact, and that if either X or Y is non-compact, then $X \times Y$ is non-compact. □

16.88. Show, by an example, that the intersection of two compact sets need not be compact. □

16.89. Let X be a topological space, A a compact subset of X and S a *locally finite* cover for X, this meaning that each point $x \in X$ has an open neighbourhood intersecting only a finite number of the elements of S. Prove that A intersects only a finite number of elements of S. □

16.90. Let Y be a Hausdorff topological space such that each point of Y has a compact neighbourhood in Y, and let $f: X \rightarrow Y$ be a compact continuous map. Prove that f is closed, and therefore proper. □

16.91. Let $f: X \rightarrow Y$ be a map of a Hausdorff space X to a compact topological space Y. Prove that f is continuous if, and only if, graph f is closed in $X \times Y$. (Use Prop. 16.56.) □

16.92. Prove that an open continuous map $f: X \rightarrow Y$ is compact if, and only if, each fibre of f is compact. (This provides an alternative proof of Theorem 16.57.) □

16.93. Let X be a compact topological space. Prove that the number of components of X is finite. □

16.94. Let X be a topological space. Then a map $f: X \rightarrow Y$, where Y is a set, is said to be *locally constant* if, for every $x \in X$, there exists a neighbourhood N of x such that $f \mid N$ is constant. Prove that, for any topology on Y, a locally constant map $f: X \rightarrow Y$ is continuous. Prove also that the only topology for Y such that every continuous map $X \rightarrow Y$ is locally constant is the discrete topology. □

16.95. Let X be a non-null topological space. Prove that X is connected if, and only if, every locally constant map $X \rightarrow Y$ with domain X is constant. □

16.96. Rewrite the section of Chapter 16 on connectedness, basing connectedness on locally constant maps. □

16.97. Prove that the map $f: \mathbf{R} \rightarrow \mathbf{R}$, defined by $f(x) = 0$ when $x \leqslant 0$ and by $f(x) = \sin 1/x$ when $x > 0$, is discontinuous at 0, but that graph f is a connected subset of $\mathbf{R}^2$. Show, however, that there is no continuous map $g: [0,1] \rightarrow \text{graph} f$ with $(g(0))_0 < 0$ and with $(g(1))_0 > 0$. (This shows that the converse to Prop. 16.70 is false. A space X satisfying the hypothesis of Prop. 16.70 is said to be *path-connected*.) □

16.98. Determine whether or not the maps g and $h : \mathbf{R} \longrightarrow \mathbf{R}$ defined by

$$g(x) = \begin{cases} \frac{1}{2}, & x \leqslant 0 \\ -\frac{1}{2} + \sin\frac{1}{x}, & x > 0 \end{cases} \quad \text{and} \quad h(x) = \begin{cases} -\frac{1}{2}, & x \leqslant 0 \\ \frac{1}{2} + \sin\frac{1}{x}, & x > 0, \end{cases}$$

have connected graphs. Sketch both the graphs in different colours on the same diagram. □

16.99. Prove that the complement in $\mathbf{R}^2$ of a finite subset of points is connected. □

16.100. Let $[a,b]$ be a closed bounded interval of $\mathbf{R}$ and let $\mathscr{S}$ be a set of open intervals of $\mathbf{R}$ covering $[a,b]$. Prove that there exists a finite ordered set of elements of $\mathscr{S}$ covering $[a,b]$ such that two of the elements intersect if, and only if, they are adjacent in the ordering. (Cf. Exercise 2.86.) Hence, deduce, from the compactness of $[a,b]$, that $[a,b]$ also is connected.

(For any continuous map $f: [a,b] \longrightarrow 2$, construct a cover of $[a,b]$ by open intervals on each of which f is constant.) □

16.101. Prove that the intervals $]-1,1[$ and $[-1,1]$ are not homeomorphic.

(There are various proofs. One uses compactness. Another, which considers the complements of points of the space, uses connectedness.) □

16.102. Are $\mathbf{R}$ and $\mathbf{R}^2$ homeomorphic, or not? (One of the hints to Exercise 16.101 is relevant here also.) □

16.103. Are S^1 and S^2 homeomorphic, or not? □

16.104. Let $E = S^1 \times \{0\}$ be the equator of S^2, the unit sphere in $\mathbf{R}^3 = \mathbf{R}^2 \times \mathbf{R}$, and let $f: [0,1] \longrightarrow S^2$ be a continuous map such that $f(0) = (0,0,1)$ and $f(1) = (0,0,-1)$. Prove that $f^{-1}(E) \neq \emptyset$. □

16.105. Suppose that $f: [0,1]^2 \longrightarrow \mathbf{R}P^2$ is a continuous map and let $\pi : S^2 \longrightarrow \mathbf{R}P^2$ be the standard projection. Prove that there exists a continous map $g: [0,1]^2 \longrightarrow S^2$ such that $f = \pi g$, but that π has no continuous section.

(To prove the last part, show that if there were such a section, then S^2 would be homeomorphic to $2 \times \mathbf{R}P^2$, a contradiction, by Prop. 16.71.) □

16.106. One of the most intuitive properties of a circle, one that we have already remarked in Chapter 0, is that it cannot be continuously deformed within itself to a point. More precisely, there is no continuous map of the unit disc $\{(x,y) \in \mathbf{R}^2 : x^2 + y^2 \leqslant 1\}$ to the unit circle whose restriction to the circle is the identity. Try to prove this! Then read Chapter 6 of [7]. □

CHAPTER 17

TOPOLOGICAL GROUPS AND MANIFOLDS

As we have seen, there is an 'obvious' topology for a finite-dimensional real linear space X, the standard topology induced by any norm on X. It is a fair supposition that there should be more or less obvious topologies also for the general linear groups, groups of automorphisms of correlated spaces, Spin groups, Grassmannians and quadric Grassmannians, all of which are closely related to finite-dimensional linear spaces. In this chapter these examples are discussed in some detail. They provide good exercise material on the propositions and theorems of Chapter 16.

There are two new concepts of importance, the concept of a *topological group* and of a *topological manifold*.

Topological groups

A *topological group* consists of a group G and a topology for G such that the maps

$$G \times G \to G; \quad (a,b) \rightsquigarrow ab \quad \text{and} \quad G \to G; \quad a \rightsquigarrow a^{-1}$$

are continuous. An equivalent condition is that the map $G \times G \to G$; $(a,b) \rightsquigarrow a^{-1}b$ is continuous.

Example 17.1. Any finite group, assigned the discrete topology, is a topological group. □

Example 17.2. Any normed linear space, with addition as the group product, is a topological group. □

Example 17.3. Let X be a complete normed real linear space. Then the group $GL(X)$, regarded as a subspace of the topological space $L(X)$, is a topological group. This follows, by Props. 16.9 and 16.10, from Props. 15.33 and 15.48 which assert the continuity of the maps $L(X) \times L(X) \to L(X)$; $(t,u) \rightsquigarrow tu$ and $L(X) \rightarrowtail L(X)$; $t \rightsquigarrow t^{-1}$.

In particular, for each $n \in \omega$, the general linear group of degree n over $\mathbf{R}$, $GL(n;\mathbf{R})$, is a topological group. □

Topological group maps, isomorphisms and embeddings and topological subgroups are defined in the obvious ways. Suppose that G and

H are topological groups. Then a map $t: G \to H$ is a *topological group map* if it is both a group map and a continuous map, it is a *topological group isomorphism* if it is both a group isomorphism and a topological isomorphism (or homeomorphism), and it is a *topological group embedding* if it is both an injective group map and a topological embedding. A subset F of G is a *topological subgroup* of G if there is a topological group structure, necessarily unique, for F such that the inclusion $F \to G$ is a topological group embedding.

Prop. 17.4. Any subgroup of a topological group is a topological group. □

Cor. 17.5. For any $n, p, q \in \omega$ the groups listed in Table 11.53 are topological groups. In particular, $U(1) = S^1$ and $Sp(1) = S^3$ are topological groups. □

Prop. 17.6. For any $p, q \in \omega$, the group Spin (p,q), regarded as a subgroup of the Clifford algebra $\mathbf{R}_{p,q}$, is a topological group and the map

$$\text{Spin}\,(p,q) \to SO(p,q); \quad g \rightsquigarrow \rho_g,$$

defined in Prop. 13.48 and Prop. 13.56, is a topological group map. □

Prop. 17.7. The map $\mathbf{R}^* \to SL(2,\mathbf{R});\ \lambda \rightsquigarrow \begin{pmatrix} \lambda & 0 \\ 0 & \lambda^{-1} \end{pmatrix}$ is a topological group embedding. □

The compactness, or otherwise, of the groups listed in Table 11.53 and of the Spin groups is easily settled.

Prop. 17.8. For any $n \in \omega$, the topological groups $O(n)$, $SO(n)$, $U(n)$, $SU(n)$ and $Sp(n)$ are compact.

Proof The compactness of $O(n)$ was proved in Prop. 16.62. Each of the other groups is isomorphic to a closed subgroup of $O(n)$, $O(2n)$ or $O(4n)$, and is therefore compact, by Prop. 16.28. □

Prop. 17.9. For any $n \in \omega$, the topological group Spin (n) is compact. □

Prop. 17.10. All the groups listed in Table 11.53, with the exception of those listed in Prop. 17.8, are non-compact (unless n or $p + q = 0$). (Show, for example, that each contains an unbounded copy of $\mathbf{R}^*$.) □

Cor. 17.11. For any $p, q \in \omega$, with $p + q > 0$, the group Spin (p,q) is non-compact. □

Homogeneous spaces

Closely related to the concept of a topological group is the concept of a *homogeneous space.*

A Hausdorff topological space X is said to be a *homogeneous space* for a topological group G if there is a transitive continuous action of G on X, that is, a continuous map $G \times X \to X$; $(g,x) \rightsquigarrow gx$, such that

(i) for all $g, g' \in G$ and all $x \in X$,

$$(g'g)x = g'(gx), \quad \text{with } 1_X = x,$$

and (ii) (*transitivity*) for each $a, b \in X$, there is some $g \in G$ such that $b = ga$.

Prop. 17.12. Let $G \times X \to X$; $(g,x) \rightsquigarrow gx$ be a continuous action of the topological group G on the topological space X. Then, for each $g \in G$, the map $X \to X$; $x \rightsquigarrow gx$ is a homeomorphism. □

Cor. 17.13. Let X be a homogeneous space for a topological group G and let $a, b \in X$. Then there is a homeomorphism $h : X \to X$ such that $h(a) = b$. □

Hence the use of the word 'homogeneous' in this context.

Example 17.14. For any $n \in \omega$, S^n is a homogeneous space for $O(n+1)$. In particular S^0 is a homogeneous space for $O(1)$ and S^1 is a homogeneous space for $O(2)$. The action one has in mind is the obvious one, the map

$$O(n+1) \times S^n \to S^n; \quad (t,x) \rightsquigarrow t(x),$$

which is well defined by Prop. 9.61. The continuity of the action follows, by Prop. 16.9 and Prop. 16.10, from the continuity of the bilinear map

$$\mathbf{R}(n+1) \times \mathbf{R}^{n+1} \to \mathbf{R}^{n+1}; \quad (t,x) \rightsquigarrow t(x).$$

Also, S^n is Hausdorff. Finally, (i) is clearly satisfied, while (ii) follows from Prop. 9.40. □

Example 17.15. For any $n \in \omega$, S^{2n+1} is a homogeneous space for $U(n+1)$ and S^{4n+3} is a homogeneous space for $Sp(n+1)$, while, for any $n \in \omega^+$, S^n is a homogeneous space for $SO(n+1)$ and for Spin $(n+1)$, while S^{2n+1} is a homogeneous space for $SU(n+1)$. The action in each case is the obvious analogue of the action of $O(n+1)$ on S^n described in Example 17.14. □

The next few propositions explore the relationships between homogeneous spaces and coset space representations.

Prop. 17.16. Let G be a topological group, let X be a homogeneous space for X and let $a \in X$. Then the map $a_{\mathbf{R}} : G \to X$; $g \rightsquigarrow ga$ is surjective, the set $G(a) = \{g \in G : ga = a\}$ is a closed subgroup of G and the fibres of a_{R} are the left cosets of $G(a)$ in G.

Proof The map is surjective, by axiom (ii) for a homogeneous space.

Secondly, since $G(a) = a_{\mathrm{R}}^{\dashv}(\{a\})$, and since $\{a\}$ is closed in X, X being Hausdorff, $G(a)$ is closed in G. Also, since $1a = a$, since $(g'g)a = g'(ga) = a$, for any $g,g' \in G(a)$, and since $ga = a$ only if $g^{-1}a = a$, for any $g \in G(a)$, $G(a)$ is a subgroup of G.

Finally, since a_{R} is surjective none of the fibres is null and, for any $g, g' \in G$,

$$ga = g'a \;\Leftrightarrow\; g^{-1}g' \in G(a) \;\Leftrightarrow\; g' \in gG(a).$$

It follows that the fibres of a_{R} are the left cosets of $G(a)$ in G. □

In the terminology of Chapter 5, page 97, the pair of maps

$$G(a) \xrightarrow{\text{inc}} G \xrightarrow{a_{\mathrm{R}}} X$$

is left-coset exact.

Prop. 17.17. Let F be a subgroup of a topological group G. Then the partition $\pi : G \to G/F; g \rightsquigarrow gF$ is open.

(Show first that, for any $A \subset G$, $\pi^{\dashv}\pi_{\vdash}(A) = \bigcup\{Af : f \in F\}$.) □

Prop. 17.18. Let F be a closed subgroup of a topological group G. Then the space of left cosets G/F is a homogeneous space of G with respect to the action

$$G \times (G/F) \to G/F; \quad (g,g'F) \rightsquigarrow gg'F.$$

Proof First, the space G/F is Hausdorff. For let gF, $g'F$ be distinct points of G/F, where $g, g' \in G$. Since F is closed and since $g^{-1}g' \notin F$ there exists an open neighbourhood A of $g^{-1}g'$ in the set complement $G \setminus F$. It then follows from the continuity of the map $G \times G \to G$; $(g,g') \rightsquigarrow g^{-1}g'$ that there exist open neighbourhoods B of g and C of g' in G such that, for all $b \in B$ and $c \in C$, $b^{-1}c \notin F$. Now define $U = \pi_{\vdash}(B)$ and $V = \pi_{\vdash}(C)$, where π is the partition $G \to G/F$. Then $U \cap V = \emptyset$, while, by Prop. 17.17, U is an open neighbourhood of gF and V is an open neighbourhood of $g'F$ in G/F.

Secondly, the action is continuous, for in the commutative diagram of maps

$$\begin{array}{ccc} G \times G & \xrightarrow[\text{product}]{\text{group}} & G \\ \Big\downarrow{\scriptstyle 1\times\pi} & & \Big\downarrow{\scriptstyle\pi} \\ G \times (G/F) & \xrightarrow{\text{action}} & G/F \end{array}$$

where, for each $(g,g') \in G \times G$, $(1 \times \pi)(g,g') = (g,\pi(g'))$, each of the maps denoted by an unbroken arrow is continuous, while π, and therefore also $1 \times \pi$, is a projection. The continuity of the action then follows by Prop. 16.12.

Finally (i) and (ii) are readily checked. □

Prop. 17.19. Let X be a homogeneous space for a *compact* topological group G. Then, for any $a \in X$, the map $(a_{\mathrm{R}})_{\mathrm{bij}} : G/G(a) \longrightarrow X$ is a homeomorphism. □

Examples 17.20. Let $\bar{\mathbf{K}}^{n+1}$ be identified with $\bar{\mathbf{K}}^n \times \bar{\mathbf{K}}$, where $\bar{\mathbf{K}} = \mathbf{R}$, $\bar{\mathbf{C}}$ or $\bar{\mathbf{H}}$. Then, for any $n \in \omega$, $O(n+1)/O(n)$, $U(n+1)/U(n)$ and $Sp(n+1)/Sp(n)$ are homeomorphic, respectively, to S^n, S^{2n+1} and S^{4n+3}, while, for any positive n, $SO(n+1)/SO(n)$, Spin $(n+1)$/Spin (n) and $SU(n+1)/SU(n)$ are homeomorphic, respectively, to S^n, S^n and S^{2n+1}. (Recall Prop. 11.55.) □

Prop. 17.21. Let F be a connected subgroup of a topological group G and suppose that G/F is connected. Then G is connected.

Proof Apply Prop. 16.73 to the partition $G \longrightarrow G/F$. □

Cor. 17.22. For each $n \in \omega$ the groups $SO(n)$, Spin (n), $U(n)$, $SU(n)$ and $Sp(n)$ are connected.

Proof By Prop. 16.71, S^n is connected, for any positive n. Now argue by induction, using Examples 17.20. □

Prop. 17.23. For each positive $n \in \omega$, the group $O(n)$ is disconnected, with two components, namely $SO(n)$, the group of rotations of $\mathbf{R}^n$, and its coset, the group of antirotations of $\mathbf{R}^n$.

Proof The map $O(n) \longrightarrow S^0$; $t \rightsquigarrow \det t$, being the restriction of a multilinear map, is continuous, and for $n > 0$ it is surjective. □

It is harder to discuss the connectedness or otherwise of the various non-compact groups. The difficulty is in proving the appropriate analogue of Prop. 17.19, Prop. 16.44 no longer being applicable. The problem will be solved in Chapter 20 (pages 424 and 425).

What we can discuss here, with a view to their application in Chapter 20, is the connectedness and compactness, or otherwise, of the various quasi-spheres (cf. page 217). By virtue of the following proposition, the ten cases reduce to four, namely $\mathscr{S}(\mathbf{R}^{p,q+1})$, $\mathscr{S}(\mathbf{C}^{n+1})$, $\mathscr{S}(\tilde{\mathbf{H}}^{n+1})$ and $\mathscr{S}(\mathrm{hb}\ \tilde{\mathbf{H}}^{n+1})$, for all p and q and all n. The symbol $\cong$ denotes homeomorphism.

Prop. 17.24. For any $n, p, q \in \omega$,

$$\begin{aligned}
\mathscr{S}(\text{hb}\,\mathbf{R}^{n+1}) &= \{(a,b) \in (\mathbf{R}^{n+1})^2 : a^\tau b = 1\} \cong \mathscr{S}(\mathbf{R}^{n+1}_{\text{hb}}) \\
&\cong \mathscr{S}(\mathbf{R}^{n+1,n+1}), \\
\mathscr{S}(\text{hb}\,\mathbf{C}^{n+1}) &= \{(a,b) \in (\mathbf{C}^{n+1})^2 : a^\tau b = 1\} \cong \mathscr{S}(\mathbf{C}^{n+1}_{\text{hb}}) \\
&\cong \mathscr{S}(\mathbf{C}^{2n+2}), \\
\mathscr{S}(\mathbf{R}^{n+1}_{\text{sp}}) &\cong \{(a,b) \in (\mathbf{R}^{2n+2})^2 : a^\tau b = 1\} \cong \mathscr{S}(\mathbf{R}^{2n+2}_{\text{hb}}) \\
&\cong \mathscr{S}(\mathbf{R}^{2n+2,2n+2}), \\
\mathscr{S}(\mathbf{C}^{n+1}_{\text{sp}}) &\cong \{(a,b) \in (\mathbf{C}^{2n+2})^2 : a^\tau b = 1\} \cong \mathscr{S}(\mathbf{C}^{2n+2}_{\text{hb}}) \\
&\cong \mathscr{S}(\mathbf{C}^{4n+4}) \\
\mathscr{S}(\bar{\mathbf{C}}^{p,q+1}) &\cong \mathscr{S}(\mathbf{R}^{2p,2q+2}),
\end{aligned}$$

and $\mathscr{S}(\bar{\mathbf{H}}^{p,q+1}) \cong \mathscr{S}(\mathbf{R}^{4p,4q+4})$. □

The next four propositions cover the four outstanding cases.

Prop. 17.25. For any $p, q \in \omega$, $\mathscr{S}(\mathbf{R}^{p,q+1}) \cong \mathbf{R}^p \times S^q$, and so is connected for any positive q, but disconnected for $q = 0$, and non-compact for any positive p, but compact for $p = 0$.

Proof Cf. Exercise 9.81. It is not difficult to show that the bijection constructed in that exercise is a homeomorphism, by verifying that the map and its inverse are each continuous. □

Prop. 17.26. The quasi-sphere $\mathscr{S}(\mathbf{C}^{n+1})$ is connected and non-compact, for any positive number n.

Proof By definition, $\mathscr{S}(\mathbf{C}^{n+1}) = \{z \in \mathbf{C}^{n+1} : z^\tau z = 1\}$. For any $z \in \mathbf{C}^{n+1}$, let $z = x + iy$, where x and $y \in \mathbf{R}^{n+1}$, and let $\mathbf{R}^{n+1}$ have its standard positive-definite orthogonal structure. Then, since

$$z^\tau z = (x + iy)^\tau(x + iy) = x^{(2)} - y^{(2)} + 2i\, x \cdot y,$$

it follows that $z \in \mathscr{S}(\mathbf{C}^{n+1})$ if, and only if, $x^{(2)} - y^{(2)} = 1$ and $x \cdot y = 0$. In particular, since $x^{(2)} = 1 + y^{(2)}$, $x \neq 0$.

Now S^n is a subset of $\mathscr{S}(\mathbf{C}^{n+1})$. Consider the continuous map $\pi : \mathscr{S}(\mathbf{C}^{n+1}) \to S^n;\ z \rightsquigarrow x/|x|$. It is surjective, with $\pi \mid S^n = 1_{S^n}$. For any $b \in S^n$, the fibre of π over b is the image of the continuous embedding

$$(\text{im}\{b\}^\perp \to \mathscr{S}(\mathbf{C}^{n+1});\quad y \rightsquigarrow (\surd(1 + y^{(2)})b\,, y)$$

where $(\text{im}\,\{b\})^\perp$ denotes the orthogonal annihilator of $\text{im}\,\{b\}$ in $\mathbf{R}^{n+1}$. This image is connected, since $(\text{im}\,\{b\})^\perp$ is connected. It is also non-compact, since $(\text{im}\,\{b\})^\perp$ is non-compact, n being positive. Since each fibre of π is connected and since S^n is connected, for $n > 0$, it follows at once that $\mathscr{S}(\mathbf{C}^{n+1})$ is connected. Finally, since any fibre of π is a closed subset and is non-compact, $\mathscr{S}(\mathbf{C}^{n+1})$ is non-compact. □

It is tempting to suppose that $\mathscr{S}(\mathbf{C}^{n+1})$ is homeomorphic, for any n, to $\mathbf{R}^n \times S^n$, but this is not so except in a few special cases. See Exercises 17.54 and 20.43 and the remarks on page 420.

Prop. 17.27. The quasi-sphere $\mathscr{S}(\tilde{\mathbf{H}}^{n+1})$ is connected and non-compact, for any number n.

Proof This follows the same pattern as the proof of Prop. 17.26. Here it is convenient to identify $\mathbf{C}^{n+1}$ with $\{a + jb \in \mathbf{H}^{n+1} : a,b \in \mathbf{R}^{n+1}\}$ and to assign $\mathbf{C}^{n+1}$ its standard orthogonal structure, just as $\mathbf{R}^{n+1}$ was assigned its standard positive-definite orthogonal structure in the proof of Prop. 17.26.

By definition, $\mathscr{S}(\tilde{\mathbf{H}}^{n+1}) = \{q \in \mathbf{H}^{n+1} : \tilde{q}^\tau q = 1\}$. For any $q \in \mathbf{H}^{n+1}$, let $q = x + \mathrm{i}y$, where $x, y \in \mathbf{C}^{n+1}$. Then, since

$$\begin{aligned}\tilde{q}^\tau q = (\bar{x} + \bar{y}\mathrm{i})^\tau(x + \mathrm{i}y) &= \bar{x}^\tau x - \bar{y}^\tau y + \mathrm{i}(y^\tau x + x^\tau y)\\ &= \bar{x}^\tau x - \bar{y}^\tau y + 2\mathrm{i}(x \cdot y),\end{aligned}$$

it follows that $q \in \mathscr{S}(\tilde{\mathbf{H}}^{n+1})$ if, and only if,

$$\bar{x}^\tau x - \bar{y}^\tau y = 1 \quad \text{and} \quad x \cdot y = 0.$$

The rest of the proof consists of a consideration of the map $\pi : \mathscr{S}(\tilde{\mathbf{H}}^{n+1}) \longrightarrow S^{2n+1} : q \rightsquigarrow x/\sqrt{(\bar{x}^\tau x)}$ closely analogous to that given for the corresponding map in Prop. 17.26, the sphere S^{2n+1} being identified with $\mathscr{S}(\bar{\mathbf{C}}^{n+1})$ in this case. □

The final case is slightly trickier.

Prop. 17.28. The quasi-sphere $\mathscr{S}(\mathrm{hb}\,\tilde{\mathbf{H}}^{n+1})$ is connected and non-compact, for any number n.

Proof By definition, $\mathscr{S}(\mathrm{hb}\,\tilde{\mathbf{H}}^{n+1}) = \{(q,r) \in (\mathbf{H}^{n+1})^2 : \tilde{q}^\tau r = 1\}$. Let $u = \hat{q} + r$, $v = \hat{q} - r$. Then it easily follows that $\mathscr{S}(\mathrm{hb}\,\tilde{\mathbf{H}}^{n+1})$ is homeomorphic to

$$\mathscr{S}' = \{(u,v) \in (\mathbf{H}^{n+1})^2 : \tilde{u}^\tau u - \tilde{v}^\tau v = 1,\ \tilde{v}^\tau u = \tilde{u}^\tau v\}.$$

Now consider the map

$$\pi : \mathscr{S}' \longrightarrow S^{4n+3}; \quad (u,v) \rightsquigarrow u/\sqrt{(\tilde{u}^\tau u)}.$$

This is handled just like the corresponding maps in Props. 17.26 and 17.27. □

The various cases may be summarized as follows.

Theorem 17.29. Let (X,ξ) be an irreducible, non-degenerate, symmetric or essentially skew, finite-dimensional correlated space over

K or ^{2}**K**, where **K** = **R**, **C** or **H**. Then, unless (X,ξ) is isomorphic to **R**, hb **R** or **C**, the quasi-sphere $\mathscr{S}(X,\xi)$ is connected and, unless (X,ξ) is isomorphic to $\mathbf{R}^n$, $\bar{\mathbf{C}}^n$ or $\bar{\bar{\mathbf{H}}}^n$, for any n, or to **C** or $\tilde{\mathbf{H}}$, $\mathscr{S}(X,\xi)$ is non-compact. □

Topological manifolds

A topological space X is said to be *locally euclidean* if there is a cover $\mathscr{S}$ for X such that each $A \in \mathscr{S}$ is homeomorphic to an open subset of a finite-dimensional real affine space.

This definition may be reformulated as follows. A pair (E,i), where E is a finite-dimensional real affine space, and $i: E \rightarrowtail X$ is an open embedding with open domain, will be called a *chart* on the topological space X, and a set $\mathscr{S}$ of charts whose images form a cover for X will be called an *atlas* for X. Clearly, the topological space X is locally euclidean if, and only if, there is an atlas for X.

A chart *at* a point $x \in X$ is a chart (E,i) on X such that $x \in \operatorname{im} i$.

A locally euclidean space need not be Hausdorff. For example the Y space (page 321) is locally euclidean, but not Hausdorff. A Hausdorff locally euclidean space is said to be a *topological manifold*.

A topological manifold is often constructed by piecing together finite-dimensional real linear or affine spaces or open subsets of such spaces. It may help in understanding this process to consider first a slightly more general construction.

Prop. 17.30. Let X be a set and let $\mathscr{S}$ be a set of topological spaces such that $X = \bigcup \mathscr{S}$. Then

$$\{U \in \operatorname{Sub} X : \text{for each } A \in \mathscr{S},\ U \cap A \text{ is open in } A\}$$

is a topology for X. □

The topology defined in Prop. 17.30 is said to be the topology *induced* on X by the set $\mathscr{S}$.

If X, in Prop. 17.30, is assigned the topology induced on it by $\mathscr{S}$, it does not follow that $\mathscr{S}$ is a cover of X. In fact, for some $A \in \mathscr{S}$, the inclusion $A \rightarrow X$ need not even be an embedding. The topologies on A in its own right or as a subspace of X may well differ. For example, let $\mathscr{S}$ consist simply of two spaces, the set X with the discrete topology and the set X with the trivial topology. Then the induced topology on their union, X, is the trivial topology and the inclusion

$$(X, \text{discrete}) \rightarrow (X, \text{trivial})$$

is not an embedding. The case where $\mathscr{S}$ is a cover for X is covered by the next proposition.

Prop. 17.31. Let X be a set, let $\mathscr{S}$ be a set of subsets of X, each assigned a topology, and let X be assigned the topology induced by $\mathscr{S}$. Then $\mathscr{S}$ is a cover for X if, and only if, for each $A, B \in \mathscr{S}$, the map $A \rightarrowtail B;\ x \rightsquigarrow x$ is continuous, with open domain. □

Cor. 17.32. Let X be a set and let $\mathscr{S}$ be a set of finite-dimensional affine spaces or open subsets of such spaces such that $X = \bigcup \mathscr{S}$ and such that, for each $A, B \in \mathscr{S}$, the map $A \rightarrowtail B;\ x \rightsquigarrow x$ is continuous, with open domain. Then the topology for X induced by $\mathscr{S}$ is locally euclidean, the inclusions $A \rightarrow X$, where $A \in \mathscr{S}$, being open embeddings. □

A variant of this construction involves the concept of an atlas for a *set*.

An *atlas* $\mathscr{S}$ for a set X is a set of pairs, each pair (E,i) consisting of a finite-dimensional affine space E and an injective map $i : E \rightarrowtail X$, with open domain, such that

(i) $X = \bigcup \{\operatorname{im} i : (E,i) \in \mathscr{S}\}$,

(ii) for each $(E,i), (F,j) \in \mathscr{S}$ the map

$$j_{\text{sur}}^{-1} i : E \rightarrowtail F; \quad a \rightsquigarrow j_{\text{sur}}^{-1} i(a)$$

is continuous with open domain.

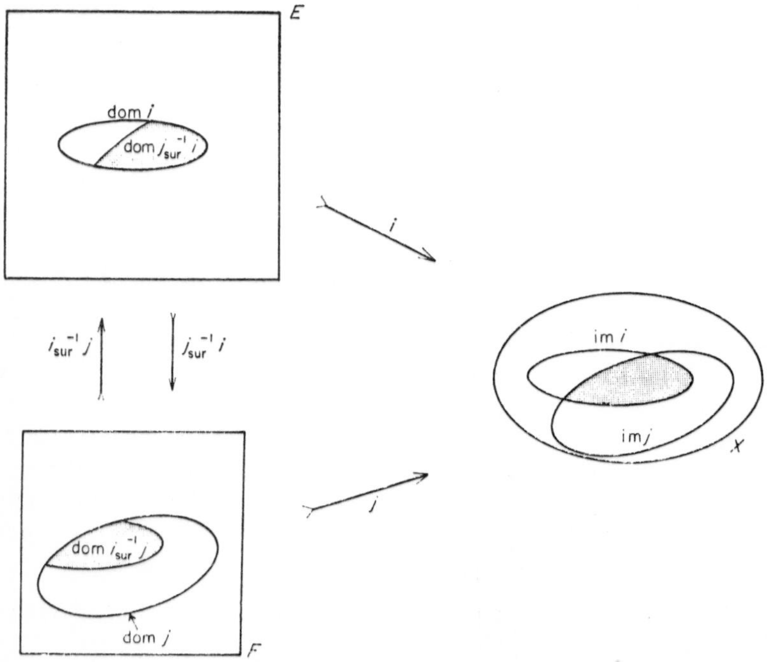

Prop. 17.33. Let $\mathscr{S}$ be an atlas for a set X, for each $(E,i) \in \mathscr{S}$ let $\operatorname{im} i$ be assigned the topology induced from E by the map i, and let X be assigned the topology induced by the set of topological spaces $\{\operatorname{im} i : (E,i) \in \mathscr{S}\}$. Then $\mathscr{S}$ is an atlas for the topological space X. □

The topology defined in Prop. 17.33 is said to be the topology *induced* on the set X by the atlas $\mathscr{S}$.

Two atlases on a set X are said to be *equivalent* if their union is also an atlas for X, or, equivalently, if they induce the same topology on X.

Grassmannians

A first application of Cor. 17.32 or of Prop. 17.33 is to the Grassmannians of finite-dimensional linear spaces. The natural charts on a Grassmannian were defined on page 223.

Prop. 17.34. Let X be a finite-dimensional linear space over $\mathbf{R}$, $\mathbf{C}$ or $\mathbf{H}$. Then, for any k, the set of natural charts for the Grassmannian $\mathscr{G}_k(X)$ of k-planes in X, is an atlas for $\mathscr{G}_k(X)$, and the topology on $\mathscr{G}_k(X)$ induced by this atlas is Hausdorff.

Proof Axiom (i) follows from Prop. 8.6 and axiom (ii) from the explicit form of the 'overlap maps' in Prop. 8.12. Finally, by Prop. 8.7 any two distinct points a and b of $\mathscr{G}_k(X)$ belong to the image of some common chart. Since a and b can be separated by open sets in this affine space and since the affine space is an open subset of $\mathscr{G}_k(X)$, they can be separated by open sets in $\mathscr{G}_k(X)$. □

Exercise 17.35. Extend Prop. 17.34 to the Grassmannians $\mathscr{G}_k^+(X)$, where X is a real linear space. □

In the following two propositions the Grassmannian $\mathscr{G}_k(V)$ of k-planes in a *real* finite-dimensional linear space V is related first to $GL(\mathbf{R}^k,V)$, the set of all *k-framings* on V, and then, for any choice of a positive-definite scalar product on V, to $O(\mathbf{R}^k,V)$, the set of all *orthonormal k-framings* on V. The set $GL(\mathbf{R}^k,V)$ is an open subset of $L(\mathbf{R}^k,V)$, while $O(\mathbf{R}^k,V)$ is a compact subset of $L(\mathbf{R}^k,V)$. Both are topological manifolds, $GL(\mathbf{R}^k,V)$ obviously, since it is an open subset of a finite-dimensional real linear space, and $O(\mathbf{R}^k,V)$ by an argument given in Chapter 20, and both are referred to as *Stiefel manifolds* for V.

Prop. 17.36. For any finite-dimensional real linear space V and any k, the map $\pi : GL(\mathbf{R}^k,V) \to \mathscr{G}_k(V);\ t \rightsquigarrow \operatorname{im} t$ is locally trivial.

Proof Let $V = X \oplus Y$, where $X \in \mathcal{G}_k(V)$, let $t \in L(X,Y)$ and let $u \in GL(\mathbf{R}^k,V)$ be such that $\operatorname{im} u = \operatorname{graph} t = \operatorname{im}(1,t)$, where $1 = 1_X$. Then, since u and $(1,t)$ are injective there exists a unique $s = (1,t)_{\text{sur}}^{-1}u_{\text{sur}} \in GL(\mathbf{R}^k,X)$ such that $u = (1,t)s$. Conversely, for any $s \in GL(\mathbf{R}^k,X)$, $\operatorname{im}(1,t)s = \operatorname{graph} t$.

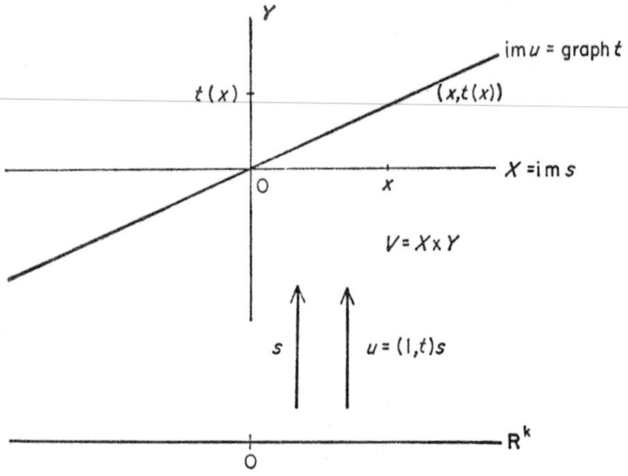

Now consider the commutative diagram of maps

$$\begin{array}{ccc} GL(\mathbf{R}^k,X) \times L(X,Y) & \xrightarrow{\alpha} & GL(\mathbf{R}^k,X \oplus Y) \\ \downarrow q & & \downarrow \pi \\ L(X,Y) & \longrightarrow & \mathcal{G}_k(V) \end{array}$$

where, for all $(s,t) \in GL(\mathbf{R}^k,X) \times L(X,Y)$, $q(s,t) = t$, $\gamma(t) = \operatorname{graph} t$ and $\alpha(s,t) = (s,ts) = (1,t)s$.

The chart γ is an open embedding, the map q is a projection and, by what has just been proved, $\pi^{-1}(\operatorname{im}\gamma) = \operatorname{im}\alpha$.

Finally, the continuous injection $\alpha;\ (s,t) \rightsquigarrow (s,ts)$ is an open embedding, since the map $\alpha_{\text{sur}}^{-1};\ (u,v) \rightsquigarrow (u,vu^{-1})$ is continuous, with domain open in $GL(\mathbf{R}^k,V)$, $GL(\mathbf{R}^k,V)$ itself being open in $L(\mathbf{R}^k,V)$.

The assertion follows. □

The simplest case of this proposition is for $k = 1$, when $GL(\mathbf{R},V)$ may be identified with $V \setminus \{0\}$ and π is the map associating to each non-zero point of V the one-dimensional linear subspace of V which it spans.

(On setting $X = \mathbf{R}$ and $Y = \mathbf{R}^n$ the commutative diagram reduces to

$$\begin{array}{ccc} (\mathbf{R}\setminus\{0\}) \times \mathbf{R}^n & \longrightarrow & \mathbf{R}^{n+1}\setminus\{0\} \\ (x,y) & & (x,xy) \\ \downarrow & & \downarrow \\ \mathbf{R}^n & \longrightarrow & \mathscr{G}_1(\mathbf{R}^{n+1}) \\ y & & \{(x,xy) : x \in \mathbf{R}\}. \end{array}$$

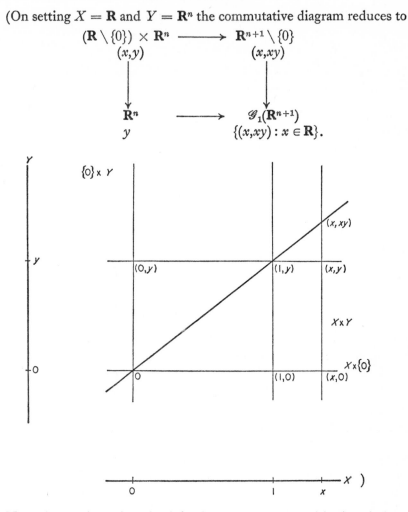

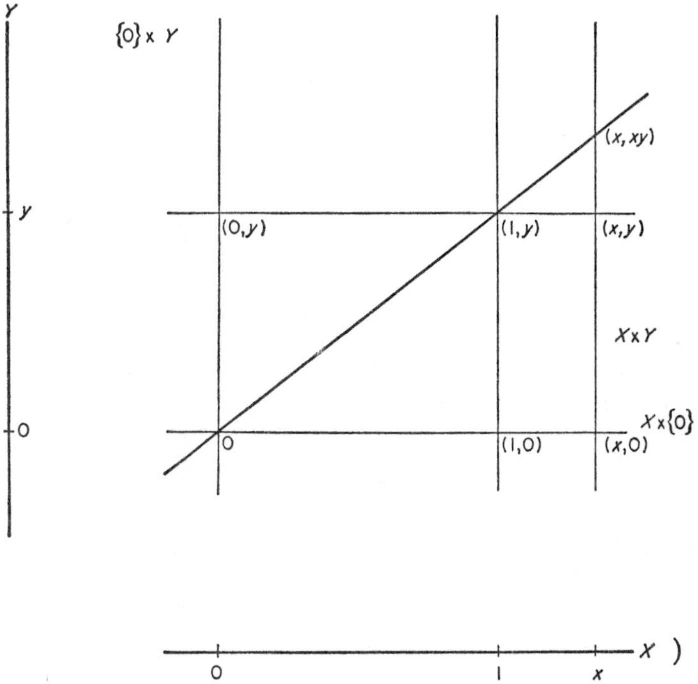

)

Note, in passing, that the injection α commutes with the obvious action of $GL(k)$ on the domain and target of α. For let $g \in GL(k)$. Then

$$\alpha((s,t)g) = \alpha(sg,t) = (1,t)(sg) = (\alpha(s,t))g.$$

(The map π is an example of a *principal fibre bundle* [27].)

Now choose a positive-definite scalar product on V.

Prop. 17.37. For any finite-dimensional non-degenerate real orthogonal space V the map

$$\pi' : O(\mathbf{R}^k, V) \longrightarrow \mathscr{G}_k(V); \quad t \rightsquigarrow \operatorname{im} t$$

is a projection.

Proof The map π' is the restriction to the compact subset $O(\mathbf{R}^k, V)$ of the locally trivial map

$$\pi : GL(\mathbf{R}^k, V) \to \mathscr{G}_k(V); \quad t \rightsquigarrow \operatorname{im} t.$$

It is surjective, by Theorem 9.32, and $\mathscr{G}_k(V)$ is Hausdorff, by Prop. 17.34. The result follows by Prop. 16.44. □

Cor. 17.38. The space $\mathscr{G}_k(V)$ is compact.

Proof The space $O(\mathbf{R}^k, V)$ is compact and π is a continuous surjection. □

The simplest case of Prop. 17.37 is for $k = 1$, when $O(\mathbf{R}, V)$ may be identified with the unit sphere in V, and $\mathscr{G}_1(V)$ is the projective space of V.

The next proposition presents $\mathscr{G}_k(\mathbf{R}^n)$ as a homogeneous space.

Prop. 17.39. The map $f: O(n) \to \mathscr{G}_k(\mathbf{R}^n)$ of Prop. 12.9 is a continuous surjection and the map $f_{\mathrm{bij}} : O(n)/(O(k) \times O(n-k)) \to \mathscr{G}_k(\mathbf{R}^n)$ is a homeomorphism.

Proof The map f admits the decomposition

$$O(n) \to O(\mathbf{R}^k, \mathbf{R}^n) \xrightarrow{\pi'} \mathscr{G}_k(\mathbf{R}^n)$$

where the first map is restriction to $\mathbf{R}^k$. It is therefore continuous, and we know already that it is surjective. Finally, since $O(n)$ is compact and since $\mathscr{G}_k(\mathbf{R}^n)$ is Hausdorff, f_{bij} is a homeomorphism. □

The natural topology on $\mathscr{G}_k(\mathbf{R}^n)$ is frequently defined to be that induced on $\mathscr{G}_k(\mathbf{R}^n)$ by the surjection f or, equivalently, the bijection f_{bij} of Prop. 17.39. This is, however, open to the objection that a particular orthogonal structure for $\mathbf{R}^n$ has first to be chosen. The atlas topology seems a much more natural starting point. There is further propaganda for this point of view in Chapter 20 where the smooth structure for $\mathscr{G}_k(\mathbf{R}^n)$ is introduced.

There are entirely parallel treatments of the complex and quaternionic Grassmannians obtained simply by replacing $\mathbf{R}$ by $\mathbf{C}$ and O by U in the former case and $\mathbf{R}$ by $\mathbf{H}$ and O by Sp in the latter case.

Prop. 17.40. For any $k, n \in \omega$, with $k \leqslant n$,

$$\mathscr{G}_k(\mathbf{C}^n) \cong U(n)/(U(k) \times U(n-k))$$

and

$$\mathscr{G}_k(\mathbf{H}^n) \cong Sp(n)/(Sp(k) \times Sp(n-k)),$$

$\cong$ denoting homeomorphism. □

In the real case there are also the Grassmannians of oriented k-planes.

Prop. 17.41. For any k, $n \in \omega$, with $k \leqslant n$,

$$\mathscr{G}_k^+(\mathbf{R}^n) \cong SO(n)/(SO(k) \times SO(n-k)),$$

the map $\mathscr{G}_k^+(\mathbf{R}^n) \to \mathscr{G}_k(\mathbf{R}^n)$ that forgets orientation being locally trivial. □

Quadric Grassmannians

The quadric Grassmannians of Chapter 12, being subsets of Grassmannians, are all Hausdorff topological spaces.

Prop. 17.42. Each of the parabolic charts on a quadric Grassmannian is an open embedding.

Proof In the notations of Prop. 12.7 the chart f and the map f_{sur}^{-} are each continuous. So f is an embedding. Finally, since any affine form of a quadric Grassmannian is an open subset of the quadric Grassmannian, f is an open embedding. □

Cor. 17.43. The quadric Grassmannians are topological manifolds. □

Cor. 17.44. For any $n \in \omega$, the groups $O(n)$, $U(n)$ and $Sp(n)$ are topological manifolds.

Proof The Cayley charts are open embeddings. □

Since $SO(n)$ is a component of $O(n)$, it follows at once that, for any n, the group $SO(n)$ is a topological manifold.

Next, Spin (n). As it will again be convenient to regard Spin (n) as a subgroup of the even Clifford algebra $\mathbf{R}_{0,n}^{0}$ rather than as a quotient group of $\Gamma^0(n)$, we begin by redefining the Pfaffian chart on Spin (n) at 1 (cf. page 263) to be the map

$$\operatorname{End}_-(\mathbf{R}^n) \to \operatorname{Spin}(n); \quad s \rightsquigarrow \operatorname{Pf} s/\sqrt{(N(\operatorname{Pf} s))}.$$

For any $g \in$ Spin (n), the *Pfaffian chart* on Spin (n) *at* g is then defined to be the Pfaffian chart at 1 composed with left multiplication by g.

Prop. 17.45. For any finite n, the group Spin (n) is a topological manifold and the group surjection ρ: Spin $(n) \to SO(n)$ is locally trivial.

Proof The Pfaffian charts are open embeddings. For example, since the components of the map $s \rightsquigarrow \operatorname{Pf} s$ are polynomial maps and since $N : \Gamma^0(n) \to$ Spin (n) is continuous, the Pfaffian chart on Spin (n) at 1 is continuous, while its 'inverse', the map

$$\operatorname{Spin}(n) \to \operatorname{End}_-(\mathbf{R}^n); \quad g \rightsquigarrow (g_\emptyset^{-1} g_{ij} : (i,j) \in n \times n),$$

where $g = g_\emptyset + \sum_{i<j} g_{ij}e_ie_j + \ldots$, with $g_{ji} = -g_{ij}$, is also continuous and has open domain. So this chart is an open embedding.

Moreover, for any $g \in \mathrm{Spin}\,(n)$, the diagram of maps

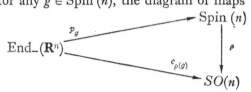

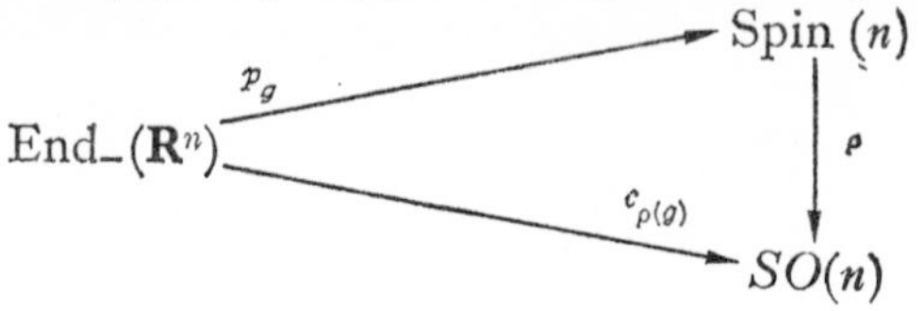

where p_g is the Pfaffian chart on Spin (n) at g, and $c_{\rho(g)}$ is the Cayley chart on $SO(n)$ at $\rho(g)$, is commutative, from which the second assertion readily follows. □

A direct proof that all the groups listed in Table 11.53, including the groups $SU(p,q)$, $SL(n;\mathbf{R})$ and $SL(n;\mathbf{C})$, are topological manifolds is given in Prop. 20.72, together with Cor. 20.76.

Prop. 17.46. Each of the coset space representations listed in Prop. 12.12 and in Theorem 12.19 is a homeomorphism. □

Particular cases of interest have already been considered in Chapter 12. Two of these are recalled in Prop. 17.47.

Prop. 17.47. For any positive p, q, the real projective quadric $\mathscr{I}_1(\mathbf{R}^{p,q})$ is homeomorphic to the set of antipodal pairs of points of $S^{p-1} \times S^{q-1}$, $(S^{p-1} \times S^{q-1})/\mathbf{Z}_2$, while, for any positive n, the complex projective quadric $\mathscr{I}_1(\mathbf{C}^n)$ is homeomorphic to the Grassmannian of oriented 2-planes in $\mathbf{R}^n$, $\mathscr{G}_2^+(\mathbf{R}^n)$. □

There are two interesting special cases:

Prop. 17.48. $\mathscr{I}_1(\mathbf{R}^{2,2}) = S^1 \times S^1$ and $\mathscr{I}_1(\mathbf{R}^{4,4}) = S^3 \times \mathbf{R}P^3$.

Proof The maps $S^1 \times S^1 \to S^1 \times S^1$; $(g,h) \rightsquigarrow (gh,h)$ and $S^3 \times S^3 \to S^3 \times S^3$; $(g,h) \rightsquigarrow (gh,h)$ are homeomorphisms, S^1 and S^3 being topological groups. Factorization by the actions of $\mathbf{Z}_2$ then produces the required homeomorphisms. □

The topological group $S^1 \times S^1$ is known as the *torus*. The projective quadric $\mathscr{I}(\mathbf{R}^{4,4})$ also features in Exercises 17.58 and 17.59.

Invariance of domain

We conclude by stating one of the fundamental theorems of topology. For the proof see, for example, [7] or [30].

Theorem 17.49. (Brouwer's 'invariance of domain'.)

If A and B are homeomorphic subsets of $\mathbf{R}^n$, and if A is open in $\mathbf{R}^n$, then B is open in $\mathbf{R}^n$. □

Cor. 17.50. For $m \neq n$, $\mathbf{R}^m$ is not homeomorphic to $\mathbf{R}^n$.

Proof Suppose $n = m + p$, where $p > 0$. Then $\mathbf{R}^n \cong \mathbf{R}^m \times \mathbf{R}^p$. Since $\mathbf{R}^m$ is homeomorphic to $\mathbf{R}^m \times \{0\}$, which is not open in $\mathbf{R}^m \times \mathbf{R}^p$, it follows that $\mathbf{R}^m$ is homeomorphic to a subset of $\mathbf{R}^n$ which is not open in $\mathbf{R}^n$. Hence the result. □

A direct proof of Cor. 17.50 is indicated in Exercise 16.102 in the particular case that $m = 1$ and $n = 2$.

By Theorem 17.49 one can define the dimension of a connected topological manifold.

Prop. 17.51. Let X be a connected topological manifold. Then the sources of the charts on X all have the same dimension. □

The common dimension of the sources of the charts on a connected topological manifold is said to be the *dimension* of the manifold. A manifold is said to be *n-dimensional* if each of its components has dimension n.

Further Exercises

17.52. Prove that the map $\mathbf{R}^n \to S^n$ 'inverse' to the stereographic projection of S^n on to $\mathbf{R}^n$ from its North pole is a topological embedding. □

17.53. Prove that the extension of a polynomial map $f: \mathbf{C} \to \mathbf{C}$ to a map $\bar{f}: \mathbf{C} \cup \{\infty\} \to \mathbf{C} \cup \{\infty\}$ as described in Example 8.14 is continuous. □

17.54. Construct the following homeomorphisms:

$O(1) \cong \mathscr{S}(\mathbf{R}^{0,1}) \cong S^0, \quad \mathscr{S}(\mathbf{R}^{0,2}) \cong S^1, \quad \mathscr{S}(\mathbf{R}^{1,1}) \cong \mathbf{R} \times S^0;$

$\mathbf{R}^* \cong \mathscr{S}(\text{hb}\,\mathbf{R}) \cong \mathbf{R} \times S^0, \quad \mathscr{S}(\text{hb}\,\mathbf{R})^2 \cong \mathbf{R}^2 \times S^1;$

$Sp(2,\mathbf{R}) \cong \mathscr{S}(\mathbf{R}_{\text{sp}}) \cong \mathbf{R}^2 \times S^1, \quad \mathscr{S}(\mathbf{R}^2_{\text{sp}}) \cong \mathbf{R}^4 \times S^3;$

$Sp(2,\mathbf{C}) \cong \mathscr{S}(\mathbf{C}_{\text{sp}}) \cong \mathbf{R}^3 \times S^3, \quad \mathscr{S}(\mathbf{C}^2_{\text{sp}}) \cong \mathbf{R}^7 \times S^7;$

$Sp(1) \cong \mathscr{S}(\bar{\mathbf{H}}^{0,1}) \cong S^3, \quad \mathscr{S}(\bar{\mathbf{H}}^{0,2}) \cong S^7, \quad \mathscr{S}(\bar{\mathbf{H}}^{1,1}) \cong \mathbf{R}^4 \times S^3;$

$\mathbf{H}^* \cong \mathscr{S}(\text{hb}\,\bar{\mathbf{H}}) \cong \mathbf{R} \times S^3, \quad \mathscr{S}(\text{hb}\,\bar{\mathbf{H}})^2 \cong \mathbf{R}^5 \times S^7;$

$O(1,\mathbf{H}) \cong \mathscr{S}(\tilde{\mathbf{H}}^1) \cong S^1, \quad \mathscr{S}(\tilde{\mathbf{H}}^2) \cong \mathbf{R}^2 \times S^3;$

$O(1,\mathbf{C}) \cong \mathscr{S}(\mathbf{C}^1) \cong S^0, \quad \mathscr{S}(\mathbf{C}^2) \cong \mathbf{R} \times S^1;$

$U(1) \cong \mathscr{S}(\bar{\mathbf{C}}^{0,1}) \cong S^1, \quad \mathscr{S}(\bar{\mathbf{C}}^{0,2}) \cong S^3, \quad \mathscr{S}(\bar{\mathbf{C}}^{1,1}) \cong \mathbf{R}^2 \times S^1;$

$\mathbf{C}^* \cong \mathscr{S}(\text{hb}\,\mathbf{C}) \cong \mathbf{R} \times S^1, \quad \mathscr{S}(\text{hb}\,\mathbf{C})^2 \cong \mathbf{R}^3 \times S^3.$

(Exercises 10.66, 14.20 and 14.21 may be of assistance in constructing several of the harder ones.) □

17.55. Verify that, for any $x \in \mathbf{R}^*$, $\frac{1}{2}\begin{pmatrix} x + x^{-1} & x - x^{-1} \\ x - x^{-1} & x + x^{-1} \end{pmatrix} \in SO(1,1)$ and that the map $\mathbf{R}^* \to SO(1,1);\ x \rightsquigarrow \frac{1}{2}\begin{pmatrix} x + x^{-1} & x - x^{-1} \\ x - x^{-1} & x + x^{-1} \end{pmatrix}$ is both a group isomorphism and a homeomorphism. □

17.56. Verify that, for any $x \in \mathbf{R}$, $\begin{pmatrix} \sqrt{(1 + x^2)} & x \\ x & \sqrt{(1 + x^2)} \end{pmatrix} \in SO^+(1,1)$ and that the map $\mathbf{R} \to SO^+(1,1);\ x \rightsquigarrow \begin{pmatrix} \sqrt{(1 + x^2)} & x \\ x & \sqrt{(1 + x^2)} \end{pmatrix}$ is a homeomorphism. □

17.57. Prove, in several ways, that $\mathbf{R}P^1$, $\mathbf{C}P^1$ and $\mathbf{H}P^1$ are homeomorphic, respectively, to S^1, S^2 and S^4.

Prove also that $\mathbf{O}P^1$ is homeomorphic to S^8. (Cf. page 285.) □

17.58. Prove that $SO(2) \cong \mathbf{R}P^1$, that $SO(3) \cong \mathbf{R}P^3$ and that $SO(4) \cong \mathscr{I}_1(\mathbf{R}^{4,4}) \cong \mathscr{I}_1(\mathbf{R}^4_{\text{hb}})$, the symbol $\cong$ denoting homeomorphism. □

17.59. Prove that $\mathscr{I}_n(\mathbf{R}^n_{\text{hb}})$ and $\mathscr{I}_n(\mathbf{C}^n_{\text{hb}})$ are each the union of two disjoint connected components, and that either component of $\mathscr{I}_4(\mathbf{R}^4_{\text{hb}})$ is homeomorphic to $\mathscr{I}_1(\mathbf{R}^4_{\text{hb}})$.

Is either component of $\mathscr{I}_4(\mathbf{C}^4_{\text{hb}})$ homeomorphic to $\mathscr{I}_1(\mathbf{C}^4_{\text{hb}})$? □

17.60. Prove that, for any $k, n \in \omega$, with $k \leqslant n$, each of the Grassmannians $\mathscr{G}_k(\mathbf{C}^n)$ and $\mathscr{G}_k(\mathbf{H}^n)$ is a compact Hausdorff topological space. □

17.61. Prove that, for any $k, n \in \omega$, with $k \leqslant n$, each of the Grassmannians $\mathscr{G}_k(\mathbf{R}^n)$, $\mathscr{G}_k^+(\mathbf{R}^n)$, $\mathscr{G}_k(\mathbf{C}^n)$ and $\mathscr{G}_k(\mathbf{H}^n)$ is connected, with one exception, namely $\mathscr{G}_1^+(\mathbf{R})$. □

CHAPTER 18

AFFINE APPROXIMATION

The maps studied in this chapter have normed affine spaces as source and target. The domain of a map may be a proper subset of the source, though when the map is affine the domain and source usually coincide. The vector space of an affine space X will be denoted, as in Chapter 4, by X_* and the linear part of an affine map $t: X \longrightarrow Y$ will be denoted by t_*. Subtraction in an affine space will be denoted simply by $-$.

The chapter falls naturally into two parts. The first part introduces the concept of *tangency* for pairs of maps from a normed affine space X to a normed affine space Y. The second part is concerned with the development of the concept of the *differential* of a map. The central theorem of the chapter is the chain rule, first proved as Theorem 18.7 and then reformulated, and extended, as Theorem 18.22. Deeper theorems on differentials are deferred to Chapter 19.

Tangency

Let $f: X \rightarrowtail Y$ and $t: X \rightarrowtail Y$ be maps between normed affine spaces X and Y, and let $a \in X$. We say that f is *tangent to t at a*, or that f and t are *mutually tangent at a* if

(i) $\operatorname{dom} f$ and $\operatorname{dom} t$ are neighbourhoods of a in X,

(ii) $f(a) = t(a)$,

and (iii) $\displaystyle\lim_{x \to a} \frac{|f(x) - t(x)|}{|x - a|} = 0$;

that is, in more technical language, but with all three axioms combined in one, if

(iv) for each $\varepsilon > 0$ there exists $\delta > 0$ such that (for all $x \in X$) $|x - a| \leqslant \delta \Rightarrow$ ($f(x)$ and $t(x)$ exist and) $|f(x) - t(x)| \leqslant \varepsilon |x - a|$,

the phrases in parentheses usually being omitted for brevity.

The inequality symbol $\leqslant$ before δ could be replaced by $<$ without changing the definition, but this is not the case with the symbol $\leqslant$ before ε, for when $x = a$ the right-hand side, and therefore also the left-hand side, of the inequality is equal to 0.

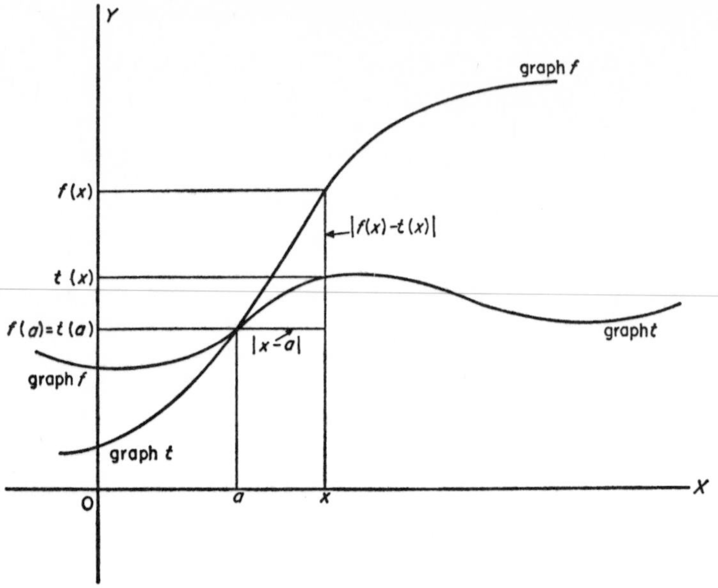

The diagram above illustrates the definition in the special case where $X = Y = \mathbf{R}$.

For example, the map $\mathbf{R} \longrightarrow \mathbf{R}$; $x \rightsquigarrow x^2$ is tangent at $a \in \mathbf{R}$ to the map $\mathbf{R} \longrightarrow \mathbf{R}$; $x \rightsquigarrow -a^2 + 2ax$, since, for any $\varepsilon > 0$,

$$|x - a| \leqslant \varepsilon \Rightarrow |x^2 - (-a^2 + 2ax)| = |(x - a)^2| \leqslant \varepsilon |x - a|.$$

In discussing the tangency of a pair of maps $f: X \rightarrowtail Y$ and $t: X \rightarrowtail Y$ at a particular point $a \in X$ it often simplifies notations to begin by setting $a = 0$ in X and $f(a) = t(a) = 0$ in Y. Then X and Y become linear while, if either of the maps, say t, is affine, it will, by this device, be identified with its linear part t_*. If either X or Y already has a linear structure, the procedure is equivalent to making a change of origin.

The next two propositions depend, for their proof, on the triangle inequality alone.

Prop. 18.1. Let f, g and h be maps from X to Y, and let f be tangent to g and g tangent to h at $a \in X$. Then f is tangent to h at a.

Proof By hypothesis dom f and dom h are neighbourhoods of a in X. Also $f(a) = g(a) = h(a)$. Set $a = 0$ in X and $f(a) = 0$ in Y. Then what remains to be proved is that for any $\varepsilon > 0$ there exists $\delta > 0$ such that

$$|x| \leqslant \delta \Rightarrow |f(x) - h(x)| \leqslant \varepsilon |x|.$$

Now $|f(x) - h(x)| \leqslant |f(x) - g(x)| + |g(x) - h(x)|$ and for any $\varepsilon > 0$ there exist $\delta', \delta'' > 0$ such that

$$|x| \leqslant \delta' \Rightarrow |f(x) - g(x)| \leqslant \tfrac{1}{2}\varepsilon |x|$$

and
$$|x| \leqslant \delta'' \Rightarrow |g(x) - h(x)| \leqslant \tfrac{1}{2}\varepsilon |x|.$$

On setting $\delta = \inf \{\delta', \delta''\}$ we obtain the required inequality. □

Prop. 18.2. Let f and t be maps from X to Y, tangent at $a \in X$. Then f is continuous at a if, and only if, t is continuous at a.

Proof It is enough to prove one of the implications, say $\Rightarrow$. Set $a = 0$ in X and $f(a) = t(a) = 0$ in Y. Suppose f is continuous at 0 and let $\varepsilon > 0$. Then there exists $\delta > 0$ such that $\delta \leqslant 1$ and such that

$$|x| \leqslant \delta \Rightarrow |f(x) - t(x)| \leqslant \tfrac{1}{2}\varepsilon |x| \quad \text{and} \quad |f(x)| \leqslant \tfrac{1}{2}\varepsilon.$$

Therefore

$$|x| \leqslant \delta \Rightarrow |t(x)| \leqslant \tfrac{1}{2}\varepsilon |x| + \tfrac{1}{2}\varepsilon \leqslant \varepsilon.$$

That is, t is continuous at 0. □

We next consider maps whose target or source is a product of normed affine spaces, the product in each case being assigned the product norm. In each case the proposition as stated involves a product with only two factors. Their generalization to products with any finite number of factors is easy and is left to the reader.

Prop. 18.3. Maps (f,g) and $(t,u): W \rightarrowtail X \times Y$ are tangent at $c \in W$ if, and only if, f and t are tangent at c and g and u are tangent at c.

Proof In either case $(f(c),g(c)) = (t(c),u(c))$. So set $c = 0$ in W, $f(c) = t(c) = 0$ in X and $g(c) = u(c) = 0$ in Y.

$\Leftarrow$: For any $\varepsilon > 0$ there exists $\delta > 0$ such that

$$|w| \leqslant \delta \Rightarrow |f(w) - t(w)| \leqslant \varepsilon |w|$$

and
$$|g(w) - u(w)| \leqslant \varepsilon |w|.$$

Therefore

$$|w| \leqslant \delta \Rightarrow |(f,g)(w) - (t,u)(w)| = \sup \{|f(w) - t(w)|, |g(w) - u(w)|\} \leqslant \varepsilon |w|.$$

That is, (f,g) and (t,u) are tangent at 0.

$\Rightarrow$: Reverse the above argument. □

In Prop. 18.4 it is convenient to introduce the notations $(-,b)$ and $(a,-)$ for the affine maps

$$X \to X \times Y; \quad x \rightsquigarrow (x,b) \quad \text{and} \quad Y \to X \times Y; \quad y \rightsquigarrow (a,y),$$

a being any point of X and b any point of Y.

Prop. 18.4. Let $f: X \times Y \rightarrowtail Z$ be tangent to $t: X \times Y \rightarrowtail Z$ at (a,b). Then

$$f(-,b) \text{ is tangent to } t(-,b) \text{ at } a$$

and

$$f(a,-) \text{ is tangent to } t(a,-) \text{ at } b.$$

Proof By hypothesis $f(a,b) = t(a,b)$. Set $(a,b) = (0,0)$ in $X \times Y$ and $f(a,b) = t(a,b) = 0$ in Z. Then, for any $\varepsilon > 0$, there exists $\delta > 0$ such that

$$|(x,y)| \leqslant \delta \Rightarrow |f(x,y) - t(x,y)| \leqslant \varepsilon\, |(x,y)|.$$

In particular,

$$|x| = |(x,0)| \leqslant \delta \Rightarrow |f(x,0) - t(x,0)| \leqslant \varepsilon\, |(x,0)| = \varepsilon\, |x|.$$

That is, $f(-,0)$ is tangent to $t(-,0)$ at 0.
Similarly $f(0,-)$ is tangent to $t(0,-)$ at 0. □

Proposition 18.4 may also be regarded as a special case of Prop. 18.6 below. Propositions 18.5 and 18.6 lead directly to the central theorem of the chapter, Theorem 18.7.

Prop. 18.5. Let f and $t: X \rightarrowtail Y$ be tangent at $a \in X$ and let $u: Y \rightarrow Z$ be continuous affine. Then uf is tangent to ut at a.

Proof Set $a = 0$ in X, $f(a) = t(a) = 0$ in Y and $uf(a) = 0$ in Z. Then u becomes linear. If $u = 0$ there is nothing to prove. So suppose $u \neq 0$ and let $\varepsilon > 0$.

Since u is linear, $uf(x) - ut(x) = u(f(x) - t(x))$, since u is continuous, $|u(f(x) - t(x))| \leqslant |u|\,|f(x) - t(x)|$ and, since f is tangent to t at 0, there exists $\delta > 0$ such that

$$|x| \leqslant \delta \Rightarrow |f(x) - t(x)| \leqslant \varepsilon\, |u|^{-1}\, |x|$$

(we assumed that $u \neq 0$), from which it follows that

$$|x| \leqslant \delta \Rightarrow |uf(x) - ut(x)| \leqslant \varepsilon\, |x|. \qquad \square$$

Prop. 18.6. Let $f: X \rightarrowtail Y$ be tangent to a continuous affine map $t: X \rightarrow Y$ at $a \in X$ and let g and $u: Y \rightarrowtail Z$ be tangent at $b = f(a)$. Then gf is tangent to uf at a.

Proof Set $a = 0$ in X, $b = f(a) = t(a) = 0$ in Y and $g(b) = u(b) = 0$ in Z. Then t becomes linear. Let ε, $K > 0$. Then $K\varepsilon > 0$ and, since g is tangent to u at 0, there exists $\eta > 0$ such that

$$|y| \leqslant \eta \Rightarrow |g(y) - u(y)| \leqslant K\varepsilon\, |y|.$$

Since t is continuous at 0, f is continuous at 0, by Prop. 18.2, and, since f is tangent to t at 0, f is defined on some neighbourhood of 0; so there exists $\delta' > 0$ such that

$$|x| \leqslant \delta' \Rightarrow |f(x)| \leqslant \eta.$$

Also, since f is tangent to t at 0 and since t is continuous, there exists $\delta'' > 0$ such that

$$\begin{aligned} |x| \leqslant \delta'' \quad &\Rightarrow \quad |f(x) - t(x)| \leqslant |x| \\ &\Rightarrow \quad |f(x)| - |t(x)| \leqslant |x| \\ &\Rightarrow \quad |f(x)| \leqslant (1 + |t|)\,|x|. \end{aligned}$$

In particular, such δ' and δ'' exist when $K = (1 + |t|)^{-1}$. Setting $\delta = \inf\{\delta', \delta''\}$, we obtain

$$|x| \leqslant \delta \quad \Rightarrow \quad |gf(x) - uf(x)| \leqslant \varepsilon\,|x|. \qquad \square$$

Theorem 18.7. (The *chain rule*.)

Let $f: X \rightarrowtail Y$ be tangent to the continuous affine map $t: X \to Y$ at $a \in X$ and let $g: Y \rightarrowtail Z$ be tangent to the continuous affine map $u: Y \to Z$ at $b = f(a)$.

Then gf is tangent to ut at a.

Proof By Prop. 18.6 gf is tangent to uf at a and by Prop. 18.5 uf is tangent to ut at a. Hence the result, by Prop. 18.1. □

An important special case of Prop. 18.6 is when $f = t$ and, in particular, when $f = t$ is an inclusion map, X being an affine subspace of Y with the induced norm. Then the conclusion is that $g|X$ is tangent to $u|X$ at a. The direct proof of this is very simple. By contrast, restriction of the target can be a tricky matter, as the remark after the next proposition indicates.

Prop. 18.8. Let $f: X \rightarrowtail Y$ be tangent to an affine map $t: X \to Y$ at a point a of X, X and Y being normed affine spaces, and suppose that W is a *closed* affine subspace of Y such that $\operatorname{im} f \subset W$. Then $\operatorname{im} t \subset W$ and the maps $X \rightarrowtail W;\ x \rightsquigarrow f(x)$ and $X \to W;\ x \rightsquigarrow t(x)$ are tangent to one another at a.

Proof Set $a = 0$ in X and $f(a) = t(a) = 0$ in Y and suppose that $x \in X$ is such that $t(x) \notin W$. Certainly $x \neq 0$. Since W is closed in Y, there exists $\varepsilon > 0$ such that the closed ball in Y with centre $t(x)$ and radius $\varepsilon\,|x|$ does not intersect W. On the other hand, since f is tangent to t at X, there exists a positive real number λ such that f is defined at λx and

$$|f(\lambda x) - t(\lambda x)| \leqslant \varepsilon\,|\lambda x|$$

and therefore such that

$$|(f(\lambda x)/\lambda) - t(x)| \leqslant \varepsilon\,|x|.$$

This implies that $f(\lambda x)/\lambda \notin W$, and therefore that $f(\lambda x) \notin W$, contrary to the hypothesis that $\operatorname{im} f \subset W$. □

One can give an example of a normed linear space Y, a linear subspace W of Y that is not closed in Y and a map $f : \mathbf{R} \to Y$ tangent at 0 to a linear map $t : \mathbf{R} \to Y$ such that $\operatorname{im} f \subset W$, but $\operatorname{im} t \not\subset W$. Such a phenomenon cannot occur if Y is finite-dimensional, since any affine subspace of a finite-dimensional affine space is closed.

Until now in this chapter we have supposed that the sources and targets of the maps under discussion are normed affine spaces. The next proposition shows that the concept of tangency depends only on the topologies induced by the norms and not on the particular norms themselves.

Prop. 18.9. Let X and Y be affine spaces, each assigned a pair of equivalent norms, denoted in either case by $|\ \ |$ and by $\|\ \ \|$, and let $f : X \rightarrowtail Y$ and $t : X \rightarrowtail Y$ be maps from X to Y. Then f and t are tangent at a point $a \in X$ with respect to the norms $\|\ \ \|$ if, and only if, they are tangent at a with respect to the norms $|\ \ |$.

Proof It is sufficient to prove the implication one way. Let X', X'' and Y', Y'' denote X and Y furnished with the norms $|\ \ |$, $\|\ \ \|$, respectively, and suppose that f and $t : X' \rightarrowtail Y'$ are tangent at a. Since $f : X'' \rightarrowtail Y''$ admits the decomposition

$$f : X'' \xrightarrow{1_X} X' \overset{f}{\rightarrowtail} Y' \xrightarrow{1_Y} Y''$$

and $t : X'' \rightarrowtail Y''$ the decomposition

$$t : X'' \xrightarrow{1_X} X' \overset{t}{\rightarrowtail} Y' \xrightarrow{1_Y} Y''$$

and since $1_X : X'' \to X'$ and $1_Y : Y' \to Y''$ are continuous affine, the norms on X and Y respectively being equivalent, it follows, by Prop. 18.5 and by Prop. 18.6, that f and $t : X'' \to Y''$ are tangent at a. □

From this it follows that, in discussing the tangency of maps between normed affine spaces, we are free at any stage to replace the given norms by equivalent ones. In the case of finite-dimensional affine spaces any norms will serve, since, by Theorem 15.26, any two norms on a finite-dimensional affine space are equivalent. There will always be a tacit assumption, in the finite-dimensional case, that some choice of norm has been made.

For an alternative definition of tangency depending only on the topological structure of the source and target, see Exercise 18.43.

Theorem 18.10 concerns an *injective* map $f : X \rightarrowtail Y$ between normed affine spaces X and Y, that is, a map $f : X \rightarrowtail Y$ such that the map $f_{\text{sur}} : \operatorname{dom} f \to \operatorname{im} f$ is bijective. In the present context it is convenient, to keep notations simple, to denote by $f^{-1} : Y \rightarrowtail X$ the map $g : Y \rightarrowtail X$ with $\operatorname{dom} g = \operatorname{im} f$ and $\operatorname{im} g = \operatorname{dom} f$ and with $g_{\text{sur}} = (f_{\text{sur}})^{-1}$.

Theorem 18.10. Let $f: X \rightarrowtail Y$ be an injective map between normed affine spaces X and Y, tangent at a point $a \in X$ to an affine homeomorphism $t: X \longrightarrow Y$, and let $f^{-1}: Y \rightarrowtail X$ be defined in a neighbourhood of $b = f(a) = t(a)$ and be continuous at b. Then f^{-1} is tangent to t^{-1} at b.

Proof Since $t: X \longrightarrow Y$ is an affine homeomorphism, the given norm on Y is equivalent to that induced on Y from X by t and there is no loss of generality in assuming that $Y = X$ and that $t = 1_X$. We may also set $a = 0$ and $b = 0$. What then has to be proved is that $f^{-1}: X \rightarrowtail X$ is tangent to $1_X^{-1} = 1_X$ at 0.

Let $\varepsilon > 0$. It has to be proved that there exists $\delta > 0$ such that

$$|y| \leqslant \delta \Rightarrow |f^{-1}(y) - y| \leqslant \varepsilon |y|.$$

First, since f is tangent to 1_X at 0, there exists $\eta > 0$ such that

$$|x| \leqslant \eta \Rightarrow |x - f(x)| \leqslant \tfrac{1}{2}\varepsilon |x|$$

and

$$|x - f(x)| \leqslant \tfrac{1}{2} |x|.$$

But then $|x| - |f(x)| \leqslant \frac{1}{2} |x|$, implying that $|x| \leqslant 2 |f(x)|$, and so

$$|x| \leqslant \eta \Rightarrow |x - f(x)| \leqslant \varepsilon |f(x)|.$$

Finally, since f^{-1} is defined in a neighbourhood of 0 and is continuous at 0, there exists $\delta > 0$ such that

$$\begin{aligned} |y| \leqslant \delta &\Rightarrow |f^{-1}(y)| \leqslant \eta \\ &\Rightarrow |f^{-1}(y) - y| \leqslant \varepsilon |y|. \qquad \square \end{aligned}$$

Theorems 18.7 and 18.10 indicate the special role played by continuous affine maps in the theory of tangency. This role is further clarified by the following intuitively obvious proposition. We isolate part of the proof as a lemma.

Lemma 18.11. Let $t: X \longrightarrow Y$ be a linear map between normed linear spaces X and Y and suppose that, for each $\varepsilon > 0$, there exists $\delta > 0$ such that

$$|x| \leqslant \delta \Rightarrow |t(x)| \leqslant \varepsilon |x|.$$

Then $t = 0$.

Proof Let $\varepsilon > 0$ and let δ be such that $|x| \leqslant \delta \Rightarrow |t(x)| \leqslant \varepsilon |x|$. For any x such that $|x| > \delta$, there exists a positive number λ, namely $\delta/|x|$, such that $|\lambda x| = \delta$ and therefore such that $|t(\lambda x)| \leqslant \varepsilon |\lambda x|$. But this inequality is equivalent to $|t(x)| \leqslant \varepsilon |x|$, since positive reals commute with linear maps and with norms. Therefore $|t(x)| \leqslant \varepsilon |x|$ for all $\varepsilon > 0$, without any restriction on $|x|$. It follows by Prop. 2.36 that, for each $x \in X$, $|t(x)| = 0$, and therefore $t(x) = 0$. So $t = 0$. $\square$

Prop. 18.12. Let t and $u : X \longrightarrow Y$ be affine maps, mutually tangent at a point a of X, X and Y being normed affine spaces. Then $t = u$.

Proof Set $a = 0$ in X and $t(a) = u(a) = 0$ in Y. Then t and u become linear.

Now apply the lemma to the map $t - u$. □

Cor. 18.13. A map $f : X \rightarrowtail Y$ is tangent at a point a to at most one affine map $t : X \longrightarrow Y$, this map being uniquely determined by its linear part. □

It may seem from this that Theorem 18.10 is nothing more than a corollary to Theorem 18.7. For if $f : X \rightarrowtail Y$ is an injective map, tangent at $a \in X$ to the continuous affine map $t : X \longrightarrow Y$, and if $f^{-1} : Y \rightarrowtail X$ is tangent at $b = f(a)$ to the continuous affine map $u : Y \longrightarrow X$, it follows, by Theorem 18.7, that $f^{-1}f$ is tangent to ut at a and ff^{-1} is tangent to tu at b. Now $f^{-1}f$ is also tangent to 1_X at a, and ff^{-1} is tangent to 1_Y at b, and therefore, by the above corollary, $ut = 1_X$ and $tu = 1_Y$. That is, $u = t^{-1}$. However, Theorem 18.7 does not prove the *existence* of an affine map u tangent to f^{-1} but only determines it if it does exist.

Differentiable maps

It has just been shown that a map $f : X \rightarrowtail Y$ between normed affine spaces X and Y is tangent at any given point $a \in X$ to *at most one continuous affine* map $t : X \longrightarrow Y$, this map, if it exists, being uniquely determined by its linear part by the condition that $t(a) = f(a)$. This linear part is called the *differential*, or more strictly the *value of the differential* of f *at* a, and is denoted by the symbol dfa, the map f then being said to be *differentiable at* a. For example, the differential at a of the map $\mathbf{R} \longrightarrow \mathbf{R}$; $x \rightsquigarrow x^2$ is the linear map $\mathbf{R} \longrightarrow \mathbf{R}$; $x \rightsquigarrow 2ax$. The *differential*, df, of f is the map

$$df : X \rightarrowtail L(X_*, Y_*); \quad x \rightsquigarrow dfx,$$

the map f being said to be *differentiable* if $\operatorname{dom}(df) = \operatorname{dom} f$, that is, if f is differentiable at every point of its domain.

In some applications, especially those considered in Chapter 19, maps are required to be not only differentiable but also smooth, that is, continuously differentiable. To be precise, a map $f : X \rightarrowtail Y$ between normed affine spaces X and Y is said to be *smooth at* $a \in X$ if df is defined on some neighbourhood of a and is continuous at a, the norm on $L(X_*, Y_*)$ being the gradient norm induced by the given norms on

X_* and Y_*, the map f being said to be *smooth* if it is differentiable and if df is continuous everywhere.

A smooth map is also said to be C^1. The explanation of this notation will be given at the end of Chapter 19, where differentials of higher order are briefly discussed.

The notations and the terminology are not quite standard. What we have called the *differential*, df, is called by some authors the *derivative* and denoted by Df, and what we have denoted by dfa is denoted by others by Dfa or df_a. The word *smooth* is often reserved to describe a map of class C^∞, this being one of the concepts discussed in Chapter 19.

In order to relate the definition of differentiability given here to one which may be more familiar to the reader, let us consider in more detail the special case where $X = Y = \mathbf{R}$. The affine map t is then of the form $x \rightsquigarrow mx + c$, where m and $c \in \mathbf{R}$ and, since $f(a) = t(a)$, $f(a) = ma + c$. Also in this special case,

$$\frac{|f(x) - t(x)|}{|x - a|} = \left|\frac{f(x) - t(x)}{x - a}\right|,$$

for any $x \in \operatorname{dom} f$ except a. Therefore f is differentiable at a if, and only if, f is defined on a neighbourhood of a and there exists a real number m such that the limit at a of

$$\left|\frac{f(x) - f(a) - m(x - a)}{x - a}\right| = \left|\frac{f(x) - f(a)}{x - a} - m\right|$$

exists and is equal to zero; that is, if, and only if, f is defined on a neighbourhood of a and the limit at a of $\dfrac{f(x) - f(a)}{x - a}$ exists. This number, usually denoted $f'(a)$, is called the *differential coefficient* of f at a, the differential at a, dfa, being the map $x \rightsquigarrow f'(a)x$. The map f'; $x \rightsquigarrow f'(x)$ is called the *derivative* of f, there being, in this case at least, no difference of opinion on the terminology.

The sets graph f and graph t are subsets of $\mathbf{R}^2$, graph t being a line, since t is affine. This line is defined to be the *tangent* to graph f at (a,b), where $b = f(a)$. In making a sketch, we may identify $\mathbf{R}^2{}_*$ with $\mathbf{R}^2$ either by the identity map or by the map

$$(x,y) \rightsquigarrow (a + x\,,\, b + y)$$

sending 0 to $0' = (a,b)$. (See the figure on page 362.)

In the first case graph dfa is identified with the line through the origin parallel to graph t, while in the second case graph dfa coincides with graph t. The former identification is the standard one, the second one being appropriate when we are particularly interested in the behaviour of f in the neighbourhood of a.

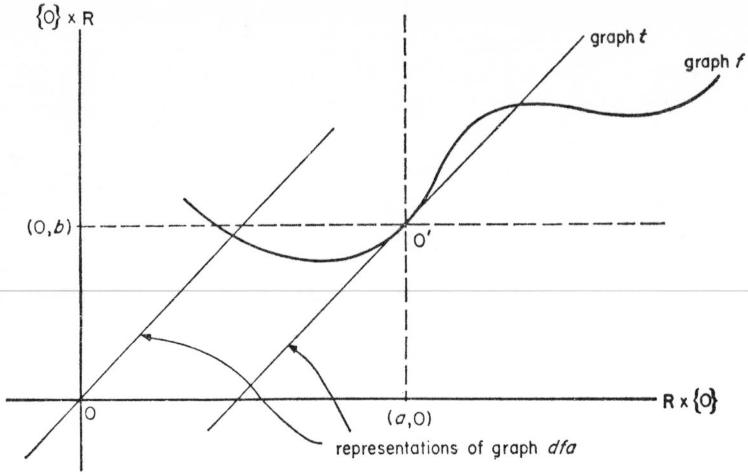

Since our general theory is dimension-free, the above picture is a useful illustration even in the general case.

In computational work the equation

$$y' = (dfx)(x'), \quad \text{where } x \in X, \quad x' \in X_* \quad \text{and} \quad y' \in Y_*,$$

is often written as $dy = \dfrac{dy}{dx}dx$, dx, dy and $\dfrac{dy}{dx}$ being alternative notations for x', y' and dfx respectively. These notations have a certain mnemonic value, as we shall see later in the discussion following Theorem 18.22.

In examples we often have the special case where $X = \mathbf{R}^n$ and $Y = \mathbf{R}^p$, for some finite n and p. In this case the linear map dfx may be represented by its matrix. This will be a $p \times n$ matrix over $\mathbf{R}$ known as the *Jacobian matrix* of f at x. The entries in this matrix are called the *partial differential coefficients* of f at x. The (i,j)th entry is usually denoted $\partial y_i/\partial x_j$, the virtue of this notation again being mnemonic, as we shall see.

Complex differentiable maps

The concept of differentiability can easily be extended to maps $f: X \rightarrowtail Y$, where X and Y are normed complex linear spaces (normed as real linear spaces). For such a map to be *complex differentiable* all that is required is that it should be real differentiable and that the differential at each point should be a complex linear map. For example, a map $f: \mathbf{C} \rightarrowtail \mathbf{C}$ is complex differentiable if, and only if, it is real differentiable and the differential at each point is complex linear, that is, is multiplication by an element of $\mathbf{C}$. If we identify $\mathbf{C}$ with $\mathbf{R}^2$ and let $(u,v) = f(x,y)$, then it at once follows, by Prop. 3.31, that the differential of f at (x,y) is complex linear if, and only if, the Jacobian matrix of f at (x,y),

$$\begin{pmatrix} \dfrac{\partial u}{\partial x} & \dfrac{\partial u}{\partial y} \\ \dfrac{\partial v}{\partial x} & \dfrac{\partial v}{\partial y} \end{pmatrix}, \quad \text{is of the form} \quad \begin{pmatrix} \alpha & -\beta \\ \beta & \alpha \end{pmatrix},$$

that is if, and only if $\partial u/\partial x = \partial v/\partial y$ and $\partial u/\partial y = -\partial v/\partial x$, equations known as the *Cauchy–Riemann equations.*

Properties of differentials

Numerous properties of differentials follow from the propositions and theorems already proved. In stating them it will be assumed, unless there is explicit mention to the contrary, that the letters W, X, Y and Z denote normed affine spaces.

Prop. 18.14. A continuous affine map $t : X \to Y$ is smooth, its differential $dt : X \to L(X_*, Y_*)$ being constant, with constant value the linear part of t, $t_* : X_* \to Y_*$.

In particular, any constant map is smooth, with differential zero, and the differential at any $x \in X$ of a continuous linear map $t : X \to Y$ is the map t itself, X and Y in this case being linear spaces. □

The next proposition is just a restatement of Prop. 18.3 in the case where t and u are continuous affine maps, the extension to smooth maps following at once from Prop. 15.15.

Prop. 18.15. A map $(f,g) : W \rightarrowtail X \times Y$ is differentiable, or smooth, at a point $w \in W$ if, and only if, each of the maps $f : W \rightarrowtail X$ and $g : W \rightarrowtail Y$ is, respectively, differentiable, or smooth, at w. In either case

$$d(f,g)w = (dfw, dgw). \qquad \square$$

Next, a restatement of Prop. 18.4.

Prop. 18.16. Let $f : X \times Y \rightarrowtail Z$ be differentiable, or smooth, at a point $(a,b) \in X \times Y$. Then the map $f(-,b) : X \rightarrowtail Z$ is, respectively, differentiable, or smooth, at a and the map $f(a,-) : Y \rightarrowtail Z$ is, respectively, differentiable, or smooth, at b. In either case, for all $x \in X_*$, $y \in Y_*$,

$$df(a,b)(x,y) = u(x) + v(y),$$

where $u = d(f(-,b))a$ and $v = d(f(a,-))b$.

(Note that if $t = df(a,b)$ then

$$\begin{aligned} t(x,y) &= t(x,0) + t(0,y) \\ &= (t(-,0))(x) + (t(0,-))(y).) \end{aligned} \qquad \square$$

The linear maps u and v are called the *partial differentials* of f *at* (a,b). There is, regrettably, no completely satisfactory notation for them. We shall denote them, for the moment, by $d_0f(a,b)$ and $d_1f(a,b)$. The *partial differentials* of f are then the maps

$$d_0f: X \times Y \rightarrowtail L(X_*,Z_*); \quad (x,y) \rightsquigarrow d_0 f(x,y)$$

and

$$d_1f: X \times Y \rightarrowtail L(Y_*,Z_*); \quad (x,y) \rightsquigarrow d_1 f(x,y).$$

The equation

$$(df(x,y))(x',y') = (d_0f(x,y))(x') + (d_1f(x,y))(y')$$

may be abbreviated to

$$df(x,y) = (d_0 f(x,y))p_* + (d_1 f(x,y))q_*,$$

where $(p,q) = 1_{X \times Y}$, and may then be abbreviated still further, since $p_* = dp(x,y)$ and $q_* = dq(x,y)$, to

$$df = (d_0f) \circ (dp) + (d_1f) \circ (dq),$$

where $\circ$ denotes composition of values. Traditionally, this last equation is often written

$$df = \frac{\partial f}{\partial x}dx + \frac{\partial f}{\partial y}dy,$$

at least in the particular case that $X = Y = Z = \mathbf{R}$, the letters x and y, doing double duty by denoting not only individual points of X and Y but also the projection maps $X \times Y \longrightarrow X$ and $X \times Y \longrightarrow Y$.

An alternative practice is to write the equation

$$dz = df(x,y)(dx,dy),$$

where $z = f(x,y)$ and dx, dy and dz are elements of X_*, Y_* and Z_* respectively, in the form

$$dz = \frac{\partial z}{\partial x}dx + \frac{\partial z}{\partial y}dy,$$

where $\partial z/\partial x = d_0 f(x,y)$ and $\partial z/\partial y = d_1 f(x,y)$.

In this case the letter f is no longer present in the formula, but this need not be a disadvantage, since we know (or should know!) in any computation which map we are at any instant working with.

The existence of two different interpretations of the symbols dx and dy is a constant source of confusion. One must, however, learn to live with both, since each has a sufficient number of advantages to justify its retention.

Both the last two propositions have obvious generalizations to the case where the product of two affine spaces is replaced by a product of n affine spaces, for any positive number n.

An immediate corollary of Prop. 18.16 is the usual practical rule for computing the partial differential coefficients of a differentiable map $f: \mathbf{R}^n \rightarrowtail \mathbf{R}$, namely, to differentiate with respect to each of the variables in turn, holding the remainder fixed. Propositions 18.15 and 18.16 together therefore provide a method for computing the Jacobian matrix at a point x of a differentiable map $f: \mathbf{R}^n \rightarrowtail \mathbf{R}^p$. Of the two interpretations of dx which have just been under discussion, it is the second which is closest to common usage in computations involving Jacobian matrices.

The matrix notation is also valuable and appropriate in discussing the differentiability of any map of the form

$$f: X_0 \times X_1 \rightarrowtail Y_0 \times Y_1,$$

X_0, X_1, Y_0 and Y_1 being normed linear spaces, the differential of this map at a point $(x_0,x_1) \in X_0 \times X_1$ taking the form

$$\begin{pmatrix} \dfrac{\partial y_0}{\partial x_0} & \dfrac{\partial y_0}{\partial x_1} \\ \\ \dfrac{\partial y_1}{\partial x_0} & \dfrac{\partial y_1}{\partial x_1} \end{pmatrix}$$

where $\partial y_i/\partial x_j$ is an abbreviation for the partial differential

$$d_j f_i(x_0,x_1): X_j \to Y_i$$

for all $i, j \in 2$. Again there is an obvious extension to products with any finite number of factors.

The converse of Prop. 18.16 is not true in the sense that one can have a map $f: X \times Y \rightarrowtail Z$ that is not differentiable at a point $(a,b) \in X \times Y$ even though the partial differentials $d_0 f(a,b)$ and $d_1 f(a,b)$ exist.

An example is the map $f: \mathbf{R}^2 \to \mathbf{R}$ defined by the formula

$$f(0,0) = 0 \quad \text{and} \quad f(x,y) = 2xy/(x^2 + y^2), \quad \text{for } (x,y) \neq (0,0),$$

for the partial differentials of f exist at (0,0) although f is not differentiable there. In fact, f is discontinuous at 0, for if $x = y \neq 0$, $f(x,y) = 1$.

It will, however, be proved later, in Prop. 19.5, that if in addition either $d_0 f$ or $d_1 f$ is defined on a neighbourhood of (a,b), and is continuous there, then f is differentiable at (a,b), while if both $d_0 f$ and $d_1 f$ have these properties, then f is smooth at (a,b).

Meanwhile, though Prop. 18.16 may help us to formulate the next proposition, it is of no help in its proof.

Prop. 18.17. Let $\beta: X \times Y \to Z$ be a continuous bilinear map,

X, Y and Z being normed linear spaces. Then, for any $(a,b) \in X \times Y$, β is tangent at (a,b) to the continuous affine map

$$X \times Y \to Z; \quad (x,y) \rightsquigarrow \beta(x,b) + \beta(a,y) - \beta(a,b),$$

that is, β is differentiable, and for all (a,b), $(x,y) \in X \times Y$,

$$d\beta(a,b)(x,y) = \beta(x,b) + \beta(a,y).$$

Also, $d\beta$ is a continuous linear map. In particular, β is smooth.

Proof Since β is continuous there exists a positive real number K such that, for all $(x,y) \in X \times Y$, $|\beta(x,y)| \leqslant K|x||y|$. Therefore, for all (a,b) and (x,y) in $X \times Y$,

$$|\beta(x,y) - \beta(x,b) - \beta(a,y) + \beta(a,b)|$$
$$= |\beta(x-a,y-b)| \leqslant K|x-a||y-b|.$$

From this it follows that, for any $\varepsilon > 0$,

$$|(x-a,y-b)| \leqslant K^{-1}\varepsilon \Rightarrow$$
$$|\beta(x,y) - \beta(x,b) - \beta(a,y) + \beta(a,b)| \leqslant \varepsilon|(x-a,y-b)|,$$

where $|(x-a,y-b)| = \sup\{|x-a|, |y-b|\}$.

This proves the first part. Also, since

$$d\beta(a,b)(x,y) = \beta(x,b) + \beta(a,y),$$
$$d\beta(a,b) = \beta(-,b)p + \beta(a,-)q, \quad \text{where } (p,q) = 1_{X \times Y},$$

implying that $d\beta$ is linear.

(Don't confuse the linearity of $d\beta$ with the linearity of $d\beta(a,b)$.)

Finally, since

$$|d\beta(a,b)(x,y)| \leqslant K|x||b| + K|a||y| \leqslant 2K|(a,b)||(x,y)|,$$
$$|d\beta(a,b)| \leqslant 2K|(a,b)|, \quad \text{for all } (a,b),$$

from which it follows at once, by Prop. 15.17, that $d\beta$ is continuous. So β is smooth. □

This result looks less formidable if $x \cdot y$ is written for $\beta(x,y)$. What it then states is that

$$d\beta(a,b)(x,y) = x \cdot b + a \cdot y,$$

or, by an inevitable abuse of notation,

$$d(a \cdot b)(x,y) = x \cdot b + a \cdot y.$$

It may also be written, in the differential notation, as

$$d(x \cdot y) = dx \cdot y + x \cdot dy.$$

Since the map $X \to X \times X$; $x \rightsquigarrow (x,x)$ is continuous affine, Prop. 18.17 has, by Prop. 18.6, the following corollary.

Cor. 18.18. Let $\beta : X \times X \to Z$ be a continuous bilinear map. Then, for any $a \in X$, the induced quadratic map $\eta : X \to Z;\ x \rightsquigarrow \beta(x,x)$ is tangent at a to the continuous affine map

$$X \to Z; \quad x \rightsquigarrow \beta(x,a) + \beta(a,x) - \beta(a,a),$$

that is, η is differentiable, with, for all $a, x \in X$,

$$d\eta a(x) = \beta(x,a) + \beta(a,x).$$

Also, $d\eta$ is a continuous linear map. In particular, η is smooth. □

A particular case of Cor. 18.18 is the example with which we opened the chapter, the map $\mathbf{R} \to \mathbf{R};\ x \rightsquigarrow x^2$, whose differential at any $a \in \mathbf{R}$ is the map $x \rightsquigarrow 2ax$.

There is a similar formula for the differential of a continuous multilinear map.

Prop. 18.19. Let $\beta : \bigtimes_{i \in k} \{X_i\} \to Y$ be a continuous k-linear map, k being some finite number. Then β is smooth and, for all $(x_i : i \in k)$ and $(x_i' : i \in k) \in \bigtimes_{i \in k} \{X_i\}$,

$$\begin{aligned}(d\beta(x_i : i \in k))(x_i' : i \in k) \\ = \beta(x_0', x_1, \ldots, x_{k-1}) + \beta(x_0, x_1', x_2, \ldots, x_{k-1}) + \ldots \\ + \beta(x_0, \ldots, x_{k-2}, x'_{k-1}).\end{aligned}$$ □

Immediate applications include the following.

Prop. 18.20. Let $\mathbf{K} = \mathbf{R}$ or $\mathbf{C}$. Then the differential at any point $a \in \mathbf{K}$ of the map $\mathbf{K} \to \mathbf{K};\ x \rightsquigarrow x^n$, for any $n \in \omega$, is the linear map $\mathbf{K} \to \mathbf{K};\ x \rightsquigarrow na^{n-1}x$, this being the zero map when $n = 0$. □

Prop. 18.21. Let X be a finite-dimensional $\mathbf{K}$-linear space, where $\mathbf{K} = \mathbf{R}$ or $\mathbf{C}$. Then the map

$$\det : L(X,X) \to \mathbf{K}; \quad t \rightsquigarrow \det t$$

is, respectively, real or complex differentiable, $d(\det)\,t$ being surjective if, and only if, $\operatorname{rk} t \geqslant \dim X - 1$. □

For any $t \in L(X,X)$ the field element $(d(\det)\,1_X)(t)$ is called the *trace* of t. With respect to any basis for X

$$\operatorname{trace} t = \sum_{i \in n} t_{ii},$$

where $n = \dim X$.

The chain rule, Theorem 18.7, may be restated in terms of differentials and extended as follows.

Theorem 18.22. Let $f : X \succ\!\to Y$ be differentiable, or smooth, at $a \in X$ and let $g : Y \succ\!\to Z$ be differentiable, or smooth, at $f(a)$. Then

$gf: X \rightarrowtail Z$ is, respectively, differentiable or smooth at a, with

$$d(gf)a = (dg(f(a))(dfa).$$

Proof The part of the theorem that concerns the differentiability of gf is just a restatement of Theorem 18.7. The smoothness of gf follows from Props. 15.13 and 15.15, since the restriction of $d(gf)$ to $(\text{dom } df) \cap f^{-1}(\text{dom } (dg))$ decomposes as follows:

$$\begin{array}{lcl} X \overset{df}{\rightarrowtail} L(X_*, Y_*) & & \\ \downarrow f \qquad\quad \times & \xrightarrow{\text{composition}} & L(X_*, Z_*), \\ Y \overset{dg}{\rightarrowtail} L(Y_*, Z_*) & & \end{array}$$

composition being, by 15.40, a continuous bilinear map. □

The formula in Theorem 18.22 may be abbreviated to

$$d(gf) = ((dg)f) \circ df,$$

denoting composition of values.

In terms of the abbreviated notation introduced on page 362 and developed on pages 364 and 365, Theorem 18.22 states that if f and g are differentiable maps and if $y = f(x)$ and $z = g(y)$, then

$$\frac{dz}{dx} = \frac{dz}{dy}\frac{dy}{dx}.$$

The whole purpose of the notation is to make this formula memorable and to 'mechanize' the matrix multiplication which arises when X, Y and Z are expressed as products, so that the computations can be performed without knowledge of matrices. For example, if $X = X_0 \times X_1$, $Y = Y_0 \times Y_1$ and $Z = Z_0 \times Z_1$ and if $(y_0, y_1) = f(x_0, x_1)$ and $(z_0, z_1) = g(y_0, y_1)$ the formula becomes

$$\begin{pmatrix} \dfrac{\partial z_0}{\partial x_0} & \dfrac{\partial z_0}{\partial x_1} \\ \\ \dfrac{\partial z_1}{\partial x_0} & \dfrac{\partial z_1}{\partial x_1} \end{pmatrix} = \begin{pmatrix} \dfrac{\partial z_0}{\partial y_0} & \dfrac{\partial z_0}{\partial y_1} \\ \\ \dfrac{\partial z_1}{\partial y_0} & \dfrac{\partial z_1}{\partial y_1} \end{pmatrix} \begin{pmatrix} \dfrac{\partial y_0}{\partial x_0} & \dfrac{\partial y_0}{\partial x_1} \\ \\ \dfrac{\partial y_1}{\partial x_0} & \dfrac{\partial y_1}{\partial x_1} \end{pmatrix}$$

and

$$\frac{\partial z_h}{\partial x_j} = \sum_{i \in 2} \frac{\partial z_h}{\partial y_i}\frac{\partial y_i}{\partial x_j}, \quad \text{for all } h, j \in 2.$$

The extension of these notations and formulae to products with any finite number of factors is easy and is left to the reader. When $X = \mathbf{R}^n$, $Y = \mathbf{R}^p$ and $Z = \mathbf{R}^q$ the matrices may be taken to be the appropriate Jacobian matrices, with entries in $\mathbf{R}$, rather than in $L(\mathbf{R}, \mathbf{R})$.

The next two propositions are complementary to Props. 15.47 and 15.48.

Prop. 18.23. Let X be a complete normed $\mathbf{K}$-linear space, where $\mathbf{K} = \mathbf{R}$, $\mathbf{C}$ or $\mathbf{H}$. Then the map $\chi : L(X,X) \rightarrowtail L(X,X);\ t \rightsquigarrow t^{-1}$ is differentiable at 1 $(= 1_X)$ and $d\chi 1 = -1\ (= -1_{L(X,X)})$.

Proof For any $u \in L(X,X)$ such that $|\,u\,| < 1$,

$$|\,\chi(1-u) - \chi(1) - (-1)(-u)\,| = |\,(1-u)^{-1} - 1 - u\,|$$
$$= |\,u^2(1-u)^{-1}\,| \leqslant |\,u\,|^2(1 - |\,u\,|)^{-1},$$

by the estimate of Prop. 15.46, and, if $|\,u\,| \leqslant \frac{1}{2}$, $(1 - |\,u\,|)^{-1} \leqslant 2$. Therefore

$$|\,u\,| \leqslant \inf\{\tfrac{1}{2}, \tfrac{1}{2}\varepsilon\} \Rightarrow |\,\chi(1-u) - \chi(1) - (-1)(-u)\,| \leqslant \varepsilon\,|\,u\,|.$$

That is, χ is differentiable at 1 and $d\chi 1 = -1$. □

Prop. 18.24. Let X and Y be complete normed $\mathbf{K}$-linear spaces; where $\mathbf{K} = \mathbf{R}$, $\mathbf{C}$ or $\mathbf{H}$. Then the map $\psi : L(X,Y) \rightarrowtail L(Y,X)$, $t \rightsquigarrow t^{-1}$ is smooth, and, for all $u \in GL(X,Y)$, and all $t \in L(X,Y)$,

$$d\psi u(t) = du^{-1}(t) = -u^{-1}\,t\,u^{-1}.$$

Proof Since, for any $u \in GL(X,Y)$, ψ admits the decomposition

$$\begin{array}{ccccccc} L(X,Y) & \longrightarrow & L(X,X) & \xrightarrow{\chi} & L(X,X) & \longrightarrow & L(Y,X) \\ t & \rightsquigarrow & u^{-1}t & \rightsquigarrow & t^{-1}u & \rightsquigarrow & t^{-1}, \end{array}$$

the first being left composition with u^{-1} and the third being right composition with u^{-1}, and since each factor is differentiable, the first and third being continuous linear, $d\psi u$ exists and admits, by Theorem 18.7, the decomposition

$$\begin{array}{ccccccc} L(X,Y) & \longrightarrow & L(X,X) & \xrightarrow{-1} & L(X,X) & \longrightarrow & L(Y,X) \\ t & \rightsquigarrow & u^{-1}t & \rightsquigarrow & -u^{-1}t & \rightsquigarrow & -u^{-1}tu^{-1}. \end{array}$$

That is, $d\psi u(t) = -u^{-1}tu^{-1}$.

The map $d\psi u$ is an element of the linear space $L(L(X,Y),L(Y,X))$. Now let

$$\eta : L(Y,X) \longrightarrow L(L(X,Y),L(Y,X))$$

be the continuous quadratic map defined by the formula

$$\eta(v)(t) = -vtv,$$

where $v \in L(Y,X)$ and $t \in L(X,Y)$. Then $d\psi = \eta\psi$, from which it follows at once that $d\psi$ is continuous and therefore that ψ is smooth. □

In the particular case where $X = Y = \mathbf{K}$, where $\mathbf{K} = \mathbf{R}$ or $\mathbf{C}$, and with $L(\mathbf{K},\mathbf{K})$ identified with $\mathbf{K}$, this proposition reduces to the statement that the map $\mathbf{K} \rightarrowtail \mathbf{K};\ x \rightsquigarrow x^{-1}$ is differentiable, with differential at $a \in \mathbf{K}^*$ the map $\mathbf{K} \longrightarrow \mathbf{K};\ x \rightsquigarrow -a^{-2}x$. From this and Prop. 18.20 it follows at once that Prop. 18.20 holds not only for all $n \in \omega$ but also for all $n \in \mathbf{Z}$.

Theorem 18.10 also may be restated and extended, with the same convention as before on the use of the notation f^{-1}.

Theorem 18.25. Let $f: X \rightarrowtail Y$ be an injective map between normed affine spaces X and Y, differentiable at $a \in X$, $dfa: X_* \rightarrow Y_*$ being a linear homeomorphism, and let $f^{-1}: Y \rightarrowtail X$ be defined in a neighbourhood of $f(a)$ in Y and continuous at $f(a)$. Then f^{-1} is differentiable at $f(a)$ and

$$d(f^{-1})(f(a)) = (dfa)^{-1}.$$

Moreover, if df is continuous, if f^{-1} is continuous with open domain, and if X (and therefore Y) is complete, then $d(f^{-1})$ is continuous.

Proof The first part is Theorem 18.10. The second part follows, by Prop. 15.48 and Prop. 15.13, from the following decomposition of $d(f^{-1})$:

$$Y \overset{f^{-1}}{\rightarrowtail} X \overset{df}{\rightarrowtail} L(X_*, Y_*) \overset{\text{inversion}}{\rightarrowtail} L(Y_*, X_*),$$

the completeness of X and Y being required in the proof that the inversion map in the decomposition is continuous. □

Note again that this theorem does not provide a criterion for a differentiable function to be invertible. The provision of such a criterion, based on the invertibility of the differential at a point, is one of the main purposes of the next chapter.

The differential of a more complicated map can often be computed by decomposing the map in some manner and then applying several of the above propositions and theorems.

Example 18.26.

Let W, X, Y and Z be normed linear spaces, let $(f,g): W \rightarrowtail X \times Y$ be differentiable and let $X \times Y \rightarrow Z$; $(x,y) \rightsquigarrow x \cdot y$ be a continuous bilinear map. Then the map $f \cdot g: W \rightarrowtail Z$; $w \rightsquigarrow f(w) \cdot g(w)$ is differentiable and, for all $w \in \operatorname{dom}(f \cdot g)$ and all $w' \in W$,

$$(d(f \cdot g)w)(w') = dfw(w') \cdot g(w) + f(w) \cdot dgw(w')$$

(a formula sometimes dangerously abbreviated to

$$d(f \cdot g) = df \cdot g + f \cdot dg).$$

The diagram of maps is

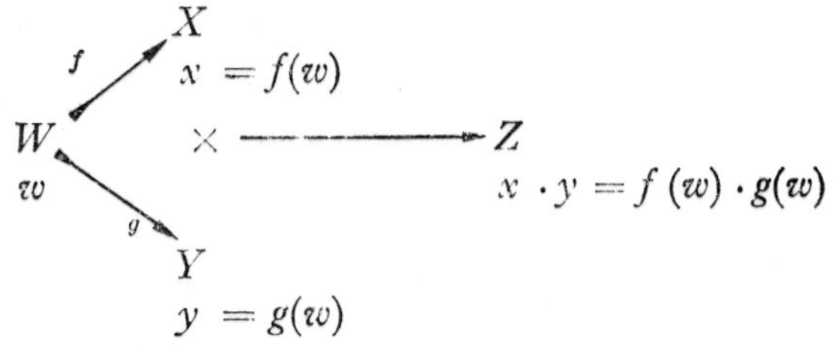

and the diagram of differentials at w and at (x,y) is

$$\begin{array}{ccccc} & & X & & \\ & \overset{dfw}{\nearrow} & x' = dfw(w') & & \\ W & & \times & \xrightarrow{d(x\cdot y)} & Z \\ w' & & & & x'\cdot y + x\cdot y' \\ & \underset{dgw}{\searrow} & & & \\ & & Y & & \\ & & y' = dgw(w') & & \end{array} .$$

The proof is a direct application of Props. 18.15, 18.17 and Theorem 18.22. □

Example 18.27. Let Y and Y' be linear complements of a linear subspace X of a finite-dimensional real, complex or right quaternionic linear space V, and let f be the map

$$L(X,Y) \rightarrowtail L(X,Y'); \quad t \rightsquigarrow q't(1_X + p't)^{-1},$$

where $(p',q') : Y \to V \cong X \times Y'$ is the inclusion and $L(X,Y)$ and $L(X,Y')$ are regarded as real linear spaces. (Cf. Prop. 8.12.) Then f is differentiable and, for $u \in \operatorname{dom} f$ and all $t \in L(X,Y)$,

$$dfu(t) = q'(1_Y - u(1_X + p'u)^{-1}p')t(1_X + p'u)^{-1}.$$

In particular, $df0$ is the map

$$L(X,Y) \to L(X,Y'); \quad t \rightsquigarrow q't. \qquad \square$$

Example 18.28. Let W, X and Y be finite-dimensional real, complex or right quaternionic linear spaces. Then the map

$$\alpha : L(W,X) \times L(X,Y) \to L(W, X \times Y); \quad (s,t) \rightsquigarrow (s,ts),$$

where $L(W,X \times Y)$ is identified with $L(W,X) \times L(W,Y)$, is smooth, and its differential at a point (s,t) is injective if, and only if, s is surjective. (Cf. Prop. 17.36.)

Proof The first component is linear and the second bilinear and both are continuous; so α is smooth and, for all $(s',t') \in \operatorname{dom} \alpha$,

$$(d\alpha(s,t))(s',t') = (s', ts' + t's).$$

Clearly $(s', ts' + t's) = 0 \Leftrightarrow s' = 0$ and $t's = 0$ and, by Prop. 3.8 and Exercise 5.26, $t's = 0 \Rightarrow t' = 0$, for all $t' \in L(X,Y)$, if, and only if, s is surjective, implying that $df(s,t)$ is injective if, and only if, s is surjective. □

Singularities of a map

A differentiable map $f : X \rightarrowtail Y$ is said to be *singular* at a point $a \in \operatorname{dom} f$ if its differential at a, $dfa : X \to Y$, is not injective, the point

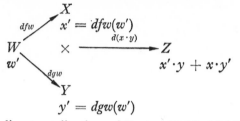

a being then a *singularity* of f. It is said to be *critical* at a if dfa is not surjective, the point a being then a *critical point* of f.

Example 18.29. The Jacobian matrix of the map

$$f: \mathbf{R}^2 \longrightarrow \mathbf{R}^2; \quad (x,y) \rightsquigarrow (\tfrac{1}{2}x^2, y)$$

at a point $(x,y) \in \mathbf{R}^2$ is $\begin{pmatrix} x & 0 \\ 0 & 1 \end{pmatrix}$. The set of singularities of f is therefore the line $\{0\} \times \mathbf{R}$, the rank of df at each singularity being 1. The image of the set of singularities is also the line $\{0\} \times \mathbf{R}$. □

Exercise 18.30. Show that the set of singularities of the map

$$f: \mathbf{R}^2 \longrightarrow \mathbf{R}^2; \quad (x,y) \rightsquigarrow (\tfrac{1}{3}x^3 - xy, y)$$

is the parabola $\{(x,y) \in \mathbf{R}^2 : y = x^2\}$, the rank of df at each singularity being 1. Sketch the image of the set of singularities. □

Exercise 18.31. Verify that the map

$$f: \mathbf{R} \longrightarrow \mathbf{R}^2; \quad t \rightsquigarrow (t^2 - 1, t(t^2 - 1))$$

has no singularities. Sketch the image of f. □

Prop. 18.32. The critical points of a differentiable map $f: X \rightarrowtail \mathbf{R}$ are just the zeros of df. □

A *local maximum* of a map $f: X \longrightarrow \mathbf{R}$ from a topological space X to $\mathbf{R}$ is a point $a \in X$ such that, for some neighbourhood A of a,

$$f(a) = \sup \{f(x) : x \in A\}.$$

A *local minimum* of f is a point $b \in X$ such that, for some neighbourhood B of b,

$$f(b) = \inf \{f(x) : x \in B\}.$$

Prop. 18.33. Let $f: X \rightarrowtail \mathbf{R}$ be differentiable at a, X being a normed affine space, and let a be either a local maximum or a local minimum of f. Then a is a critical point of f.

Proof, in the case that a is a local maximum of f.

Set $a = 0$ and $f(a) = 0$. Then, for all $\varepsilon > 0$, there exists $\delta > 0$ such that

$$|x| \leqslant \delta \Rightarrow f(x) \leqslant 0 \quad \text{and} \quad t(x) - f(x) \leqslant \varepsilon |x|,$$

where $t = dfa$. Therefore there exists $\delta > 0$ such that

$$|x| \leqslant \delta \Rightarrow t(x) \leqslant \varepsilon |x|$$

and so, by Lemma 18.11, $t = 0$.

The proof in the other case is similar. □

A critical point of f may of course be neither a local maximum nor a local minimum of f. For example, 0 is a critical point of the map $\mathbf{R} \to \mathbf{R}$; $x \rightsquigarrow x^3$ but is neither a local maximum nor a local minimum of the map.

Exercise 18.34.

Let X be a finite-dimensional real linear space, let $r \in \mathbf{R}^+$ and let $f: X \rightarrowtail \mathbf{R}$ be a map continuous on the set $\{x \in X : |x| \leqslant r\}$, differentiable on the set $\{x \in X : |x| < r\}$, and zero on the set $\{x \in X : |x| = r\}$. Show that there is a critical point a of f, with $|a| < r$. □

The study of local maxima and minima is continued in Chapter 19, page 386.

Further Exercises

18.35. Prove that the map $\mathbf{R} \to \mathbf{R}$; $x \rightsquigarrow x^2$ is not open and that the differential of this map at 0 is not surjective. □

18.36. Compute the differential at (x,y,z) of the map $f : \mathbf{R}^3 \to \mathbf{R}^3$; $(x,y,z) \rightsquigarrow (x^2 + y^2, y^2 + z^2, xz)$ and prove that $df(x,y,z)$ is invertible if, and only if, y is non-zero and either x or z is non-zero. □

18.37. Let $f : \mathbf{R}^3 \to \mathbf{R}^3$; $(x,y,z) \rightsquigarrow (u,v,w)$ be the map defined by the equations

$$\begin{aligned} u + v + w &= x \\ v + w &= xy \\ w &= xyz. \end{aligned}$$

Find S and $f_\vdash(S)$, S being the set of points in $\mathbf{R}^3$ at which the differential of f is not invertible.

(The neatest solution involves application of the chain rule.) □

18.38. Consider the map

$$f: L(X,X) \times L(X,Y) \rightarrowtail L(X,Y); \quad (a,b) \rightsquigarrow ba^{-1},$$

where X and Y are normed linear spaces. Prove that f is differentiable and that $df(1_X,0) = b$. □

18.39. Prove that the bijection

$$]-1,1[\to \mathbf{R}; \quad x \rightsquigarrow \frac{x}{1-x^2}$$

(cf. Exercise 16.80) is a smooth homeomorphism, that is, that it and its inverse are smooth. □

18.40. Let X and Y be finite-dimensional linear spaces over $\mathbf{R}$, let $g: X \rightarrow Y$ be a linear surjection and let $f: X \rightarrow \mathbf{R}$ be a map differentiable at 0. Prove that $f \mid W: W \rightarrow \mathbf{R}$ is differentiable at 0, where $W = \ker g$. Prove also that if $f \mid W$ has a local maximum at 0, then there exists a linear map $\lambda: Y \rightarrow \mathbf{R}$ such that $df0 = \lambda g$. □

18.41. Investigate the critical points of the map

$$\mathbf{R}^2 \rightarrow \mathbf{R}; \quad (x,y) \rightsquigarrow xy^2(x + y - 1).$$

(Start by making a rough contour map, then verify your conjectures by explicit computations.) □

18.42. Consider the map

$$f: \mathbf{R}^2 \rightarrow \mathbf{R}; \quad (x,y) \rightsquigarrow (y - x^2)(y - 2x^2).$$

Prove that (0,0) is *not* a local minimum of f but is a local minimum of the restriction of f to any line through (0,0). □

18.43. Let X and Y be two linear spaces, each assigned a norm topology, and let $f: X \rightarrow Y$ be a map. Prove that f is tangent to the zero map at 0 if, and only if, for any neighbourhood B of 0 in Y, there exists a neighbourhood A of 0 in X and a map $\phi: \mathbf{R} \rightarrowtail \mathbf{R}$ defined on some neighbourhood of 0, with $\phi(0) = 0$, such that

(i) $\lim_{t \to 0} \dfrac{\phi(t)}{t} = 0$

and (ii) for all $t \in \operatorname{dom} \phi$, $f_{\vdash}(tA) \subset \phi(t)B$. □

CHAPTER 19

THE INVERSE FUNCTION THEOREM

Let X and Y be normed affine spaces and let A be a subset of X and B a subset of Y. A map $f: A \rightarrow B$ is said to be a *smooth homeomorphism* if it is a homeomorphism and if each of the maps $X \rightarrowtail Y; x \rightsquigarrow f(x)$ and $Y \rightarrowtail X; y \rightsquigarrow f^{-1}(y)$ is smooth (C^1). A map $f : X \rightarrowtail Y$ is said to be *locally a smooth homeomorphism* at a point $a \in X$ if there are open neighbourhoods A of a in X and B of $f(a)$ in Y such that $f_{\vdash}(A) = B$ and the map $A \rightarrow B; x \rightsquigarrow f(x)$ is a smooth homeomorphism.

The main theorem of this chapter, the 'inverse function theorem', is a criterion for a map $f : X \rightarrowtail Y$ to be locally a smooth homeomorphism, when X and Y are *complete* normed affine spaces. Important corollaries include the 'implicit function theorem' and various propositions preliminary to the study of smooth submanifolds. Another corollary is the 'fundamental theorem of algebra'.

Higher differentials are considered briefly at the end of the chapter.

The increment formula

One of the main tools used in the proof of the inverse function theorem is the 'increment formula' ('la formule des accroissements finis'). This inequality replaces the 'mean value theorem', which occurs at this stage in many treatments of the calculus of real-valued functions of one real variable. The relation of the inequality to the mean value theorem is briefly discussed below.

Theorem 19.1. (The *increment formula*.)

Let a and b be points of the domain of a differentiable map $f : X \rightarrowtail Y$ such that the line-segment $[a,b]$ is a subset of dom f, X and Y being normed affine spaces, and suppose that M is a real number such that $| dfx | \leqslant M$ for all $x \in [a,b]$. Then

$$|f(b) - f(a)| \leqslant M \, | b - a |.$$

Proof Set $a = 0$ in X and $f(a) = 0$ in Y. What then has to be proved is that $|f(b)| \leqslant M \, | b |$. To prove this it is sufficient to prove that, for all $\varepsilon > 0$, $|f(b)| \leqslant (M + \varepsilon) \, | b |$.

Let $\varepsilon > 0$, let $A = \{\lambda \in [0,1] : |f(\lambda b)| \leqslant (M + \varepsilon)\lambda |b|\}$ and let $\sigma = \sup A$. The set A is non-null, since $0 \in A$, and it is bounded above by 1; so σ exists. Our task is to prove that $1 \in A$. To do so we prove first that $\sigma \in A$ and then that $\sigma = 1$.

Since f is differentiable at σb there exists $\delta > 0$ such that

$$\begin{aligned} |x - \sigma b| \leqslant \delta \;\Rightarrow\; & |f(x) - f(\sigma b) - df\sigma b(x - \sigma b)| \leqslant \varepsilon |x - \sigma b| \\ \Rightarrow\; & |f(x) - f(\sigma b)| \leqslant (M + \varepsilon) |x - \sigma b|. \end{aligned}$$

Also by the definition of σ there exists ρ, $0 \leqslant \rho \leqslant \sigma$, such that $|\rho b - \sigma b| \leqslant \delta$ and $\rho \in A$. Therefore

$$\begin{aligned} |f(\sigma b)| &\leqslant |f(\sigma b) - f(\rho b)| + |f(\rho b)| \\ &\leqslant (M + \varepsilon)(\sigma - \rho)|b| + (M + \varepsilon)\rho |b| = (M + \varepsilon)\sigma |b|. \end{aligned}$$

That is, $\sigma \in A$.

If $\sigma < 1$, there exists τ, $\sigma < \tau \leqslant 1$, such that $|\tau b - \sigma b| \leqslant \delta$, and

$$\begin{aligned} |f(\tau b)| &\leqslant |f(\tau b) - f(\sigma b)| + |f(\sigma b)| \\ &\leqslant (M + \varepsilon)(\tau - \sigma)|b| + (M + \varepsilon)\sigma |b| = (M + \varepsilon)\tau |b|. \end{aligned}$$

That is, $\tau \in A$, contradicting the definition of σ. So $\sigma = 1$. □

Cor. 19.2. Let $f : X \rightarrowtail Y$ be a differentiable map with convex domain, X and Y being normed affine spaces, and let M be a real number such that $|dfx| \leqslant M$, for all $x \in \operatorname{dom} f$. Then, for all $a, b \in \operatorname{dom} f$,

$$|f(b) - f(a)| \leqslant M |b - a|. \qquad \square$$

The following proposition is a refinement of Cor. 19.2.

Prop. 19.3. Let X and Y be normed affine spaces, let $\bar{A}$ denote the closure of an open convex subset A of X, let $f : \bar{A} \to Y$ be a continuous map and let $f|A$ be differentiable with $|dfa| \leqslant M$, for all $a \in A$, M being a real number. Then, for any $a, b \in \bar{A}$,

$$|f(b) - f(a)| \leqslant M |b - a|.$$

Proof Let $\varepsilon > 0$ and let a', $b' \in A$ be such that $M|a - a'|$, $M|b - b'|$, $|f(a) - f(a')|$ and $|f(b) - f(b')|$ are each $< \frac{1}{4}\varepsilon$. Then, by Cor. 19.2,

$$\begin{aligned} |f(b) - f(a)| &\leqslant |f(b) - f(b')| + |f(b') - f(a')| + |f(a') - f(a)| \\ &< \tfrac{1}{2}\varepsilon + M|b' - a'| \\ &< \varepsilon + M|b - a|. \end{aligned}$$

Since this inequality is true for each $\varepsilon > 0$, it follows that

$$|f(b) - f(a)| \leqslant M |b - a|.$$

(Note that it is possible for the segment $[a,b]$ to lie entirely in the boundary $\bar{A} \setminus A$ of A.) □

The classical 'mean value theorem' states that if $f : \mathbf{R} \rightarrowtail \mathbf{R}$ is a continuous map with domain a closed bounded interval $[a,b]$ and if f is differentiable on the open interval $]a,b[$, then, for some $\xi \in]a,b[$,

$$f(b) - f(a) = f'(\xi)(b - a).$$

It follows from this that, if $| f'(x) | \leqslant M$ for all $x \in]a,b[$, then

$$| f(b) - f(a) | \leqslant M \mid b - a \mid.$$

Theorem 19.1 and its corollaries are therefore generalizations of part of the classical theorem.

Next, a simple, but important, application of the increment formula.

Prop. 19.4. Let $f : X \rightarrowtail Y$ be a differentiable map with connected open domain, X and Y being normed affine spaces, and let $df = 0$. Then f is constant.

(Show, for some $y \in f_{\vdash}(X)$, that $f^{\dashv}\{y\}$ is both open, by the increment formula, and closed in dom f.) □

As a further application of the increment formula we have the following partial converse to Prop. 18.16.

Prop. 19.5. Let $f : X \times Y \rightarrowtail Z$ be a map, defined on some neighbourhood of $(a,b) \in X \times Y$, X, Y and Z being normed affine spaces, and suppose that of the two partial differentials of f at (a,b) the one, say

$$d_0 f : X \times Y \rightarrowtail L(X_*, Z_*),$$

exists on a neighbourhood of (a,b) and is continuous at (a,b) while the other,

$$d_1 f : X \times Y \rightarrowtail L(Y_*, Z_*),$$

exists at (a,b). Then f is differentiable at (a,b), while, if $d_1 f$ also exists on a neighbourhood of (a,b) and is continuous at (a,b), then f is smooth at (a,b). In either case,

$$df(a,b) = (d_0 f(a,b) \quad d_1 f(a,b)).$$

Proof Set $a = 0$ in X, $b = 0$ in Y and $f(a,b) = 0$ in Z, and let $\varepsilon > 0$. Since $d_0 f$ exists on a neighbourhood of $(0,0)$ and is continuous at $(0,0)$, there exists a real $\delta > 0$ such that

$$| (x,y) | < \delta \Rightarrow | d_0 f(x,y) - d_0 f(0,0) | \leqslant \tfrac{1}{2}\varepsilon.$$

The increment formula may then be applied, for any $y \in Y$, to the map

$$X \rightarrowtail Z; \quad x \rightsquigarrow f(x,y) - d_0 f(0,0)(x)$$

with domain the ball $\{x : | x | < \delta\}$ and differential at x the linear map $d_0 f(x,y) - d_0 f(0,0)$, the norm of this differential being bounded

by $\frac{1}{2}\varepsilon$ if $|y| < \delta$. This yields the inequality
$$|f(x,y) - f(0,y) - d_0 f(0,0)(x)| \leqslant \tfrac{1}{2}\varepsilon\,|x|,$$
whenever $|(x,y)| < \delta$.

By the existence of $d_1 f$ at $(0,0)$ we may suppose that δ is so small that
$$|y| < \delta \;\Rightarrow\; |f(0,y) - d_1 f(0,0)(y)| \leqslant \tfrac{1}{2}\varepsilon\,|y|.$$
It then follows that
$$|(x,y)| < \delta \;\Rightarrow\; |f(x,y) - d_0 f(0,0)(x) - d_1 f(0,0)(y)| \leqslant \varepsilon\,|(x,y)|,$$
that is, f is differentiable at $(0,0)$, with the stated map as differential there.

Since df admits the decomposition

$$
X \times Y
\begin{array}{c}
\xrightarrow{\;\;d_0 f\;\;} L(X_*,Z_*) \\[4pt]
\xrightarrow{\;\;d_1 f\;\;} L(Y_*,Z_*)
\end{array}
\;\times\; \xrightarrow{\;\;\sigma\;\;} L(X_* \times Y_*, Z_*)
$$

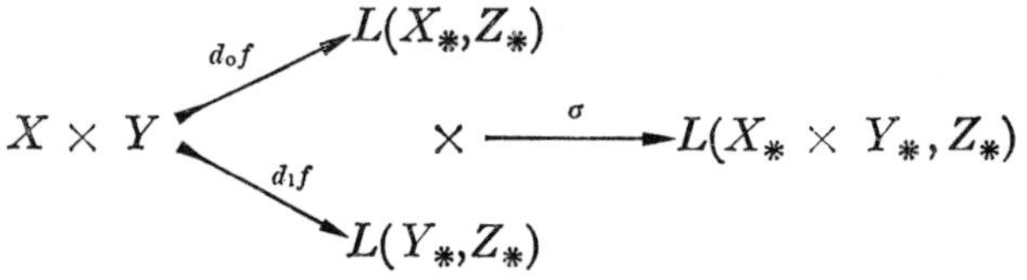

where σ, defined by the formula
$$\sigma(t,u)(x,y) = t(x) + u(y),$$
is continuous linear, the additional condition on $d_1 f$ at once implies the continuity of df, and hence the smoothness of f, at (a,b).   □

The inverse function theorem

The 'increment formula' is one of the main tools used in the proof of the inverse function theorem, which now follows. The other principal ingredient in the proof is the 'contraction lemma', Theorem 15.22. This requires that certain normed linear spaces are complete, this condition being automatically fulfilled whenever these spaces are finite-dimensional, as they will be in most of our applications.

Theorem 19.6.   (*The inverse function theorem.*)
Let X and Y be complete normed affine spaces, let $f : X \rightarrowtail Y$ be a smooth map and suppose that at some point $a \in X$, f is tangent to an affine homeomorphism. Then f is locally a smooth homeomorphism at a, that is, there exist open neighbourhoods A of a in X and B of $b = f(a)$ in Y and a smooth map $g : Y \rightarrowtail X$ with domain B such that
$$g_{\mathrm{sur}} = (f\,|\,A)_{\mathrm{sur}}^{-1}.$$

If, moreover, C is any connected subset of B containing b, there is a unique continuous map $g' : C \to X$ such that $(fg')_{\mathrm{sur}} = 1_C$, with $g'(b) = a$, namely $g' = g\,|\,C$.

Proof Set $a = 0$ in X and $f(a) = t(a) = 0$ in Y. Since the affine map t is a homeomorphism, there is no loss of generality in supposing also that $Y = X$ and that $t = 1_X$. (Strictly speaking we consider $t^{-1}f$ in place of f and $t^{-1}t = 1_X$ in place of t.) With these conventions $df0 = 1_X$. The argument is now based on the remark that, for all $x, y \in X$, $f(x) = y \Leftrightarrow x = y - h(x)$, where $h = f - 1_X$.

Since $df0 = 1_X$, $dh0 = 0$ and, by the continuity of df and therefore of dh at 0, there exists a positive real δ such that $|x| \leqslant \delta \Rightarrow |dhx| \leqslant \frac{1}{2}$. The ball $C_\delta = \{x \in X : |x| \leqslant \delta\}$ is convex. Therefore, by the increment formula applied to the restriction of h to C_δ,

$$|x|, |x'| \leqslant \delta \Rightarrow |h(x) - h(x')| \leqslant \tfrac{1}{2}|x - x'|.$$

In particular, since $h(0) = 0$,

$$|x| \leqslant \delta \Rightarrow |h(x)| \leqslant \tfrac{1}{2}|x| \leqslant \tfrac{1}{2}\delta.$$

Let B be the ball $\{y : |y| < \frac{1}{2}\delta\}$ and let $y \in B$. Since X is complete, since $|x| \leqslant \delta \Rightarrow |y - h(x)| \leqslant |y| + |h(x)| < \delta$ and since $|x|, |x'| \leqslant \delta \Rightarrow |(y - h(x)) - (y - h(x'))| \leqslant \frac{1}{2}|x - x'|$, the contraction lemma applied to the map $x \rightsquigarrow y - h(x)$ of C_δ to itself implies that there is a unique $x \in C_\delta$ such that $y - h(x) = x$ or, equivalently, such that $f(x) = y$. Indeed, $|x| < \delta$, for if $|x| \leqslant \delta$ and if $f(x) \in B$ then $|x| = |f(x) - h(x)| < \delta$. Now let $A = \{x : |x| < \delta\} \cap f^{\dashv}(B)$. Then $f_{\vdash}(A) = B$ and $(f \mid A)_{\text{sur}} : A \rightarrow B$ is bijective. Also, since f is continuous, A is open in X.

Next, $g = (f \mid A)_{\text{sur}}^{-1} : B \rightarrow A$ is continuous. For, since

$$\begin{aligned} |x|, |x'| \leqslant \delta \Rightarrow |f(x) - f(x')| &= |x + h(x) - x' - h(x')| \\ &\geqslant |x - x'| - |h(x) - h(x')| \\ &\geqslant \tfrac{1}{2}|x - x'|, \end{aligned}$$

it follows that

$$|y|, |y'| < \tfrac{1}{2}\delta \Rightarrow |g(y) - g(y')| \leqslant 2|y - y'|,$$

and so, for all $\varepsilon > 0$ and for all $y, y' \in B$,

$$|y - y'| \leqslant \tfrac{1}{2}\varepsilon \Rightarrow |g(y) - g(y')| \leqslant \varepsilon.$$

Since dom g is open in Y it follows, by Theorem 18.25, that $g : Y \rightarrowtail X$, with domain B, is smooth.

Finally, the uniqueness statement follows directly from Prop. 16.74, since X is Hausdorff and g is an open map. □

The implicit function theorem

The inverse function theorem, which we have just been considering, is concerned with the possibility of 'solving' the equation $f(x) = y$ for 'x in terms of y', near some point $a \in \operatorname{dom} f$. It states precise sufficient conditions for this to be possible in terms of the possibility of solving the affine equation $t(x) = y$, where the affine map t is tangent to f at a, there being a unique solution x of the latter equation, for each y, if, and only if, t is invertible.

The implicit function theorem is an apparently slightly more general theorem of the same type. The problem this time is to solve, near some point $(a,b) \in \operatorname{dom} f$, an equation of the form $f(x,y) = z$ for 'y in terms of x (and z)'. As in the case of the inverse function theorem, sufficient conditions for this to be possible are obtained in terms of the possibility of solving the affine, or linear, equation $t(x,y) = z$ for y in terms of x (and z), where t is tangent to f at (a,b). Such an equation has a unique solution of the required type if, and only if, the map $y \rightsquigarrow t(a,y)$ is invertible.

Theorem 19.7. (*The implicit function theorem.*)

Let X, Y and Z be complete normed affine spaces and suppose that $f : X \times Y \rightarrowtail Z$ is a smooth map tangent at a point $(a,b) \in X \times Y$ to a continuous affine map $t : X \times Y \to Z$ whose linear part is of the form

$$X_* \times Y_* \to Z_*; \quad (x,y) \rightsquigarrow u(x) + v(y),$$

where $u : X_* \to Z_*$ is a linear map and $v : Y_* \to Z_*$ is a linear homeomorphism.

Then there exists an open neighbourhood A of (a,b) in $X \times Y$ and a smooth map $h : X \rightarrowtail Y$ such that graph $h = A \cap f^{-1}\{f(a,b)\}$.

Moreover, if C is any connected subset of dom h containing a, there

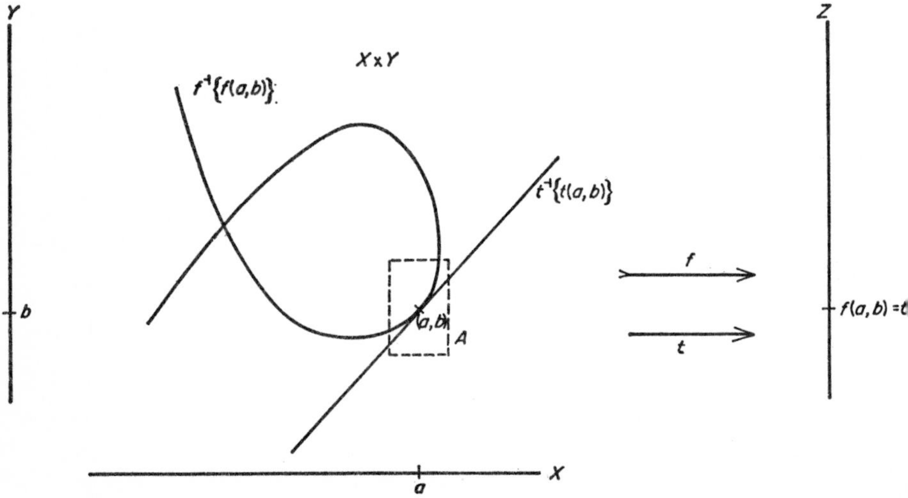

is a unique continuous map $h' : C \rightarrow Y$ such that for all $x \in C$, $f(x,h'(x)) = f(a,b)$ with $h'(a) = b$, namely $h' = h \mid C$.

Proof Set $a = 0$ in X, $b = 0$ in Y and $f(a,b) = t(a,b) = 0$ in Z. Then the map $F : X \times Y \rightarrowtail X \times Z$; $(x,y) \rightsquigarrow (x,f(x,y))$ is smooth, with $dF(0,0) : X \times Y \rightarrow X \times Z$; $(x,y) \rightsquigarrow (x,u(x) + v(y))$ a homeomorphism by Exercise 15.55, since v is a homeomorphism. So by Theorem 19.6 there exist open neighbourhoods A of 0 in $X \times Y$ and B of 0 in $X \times Z$ and a smooth map $G : X \times Z \rightarrowtail X \times Y$ with domain B such that $G_{\text{sur}} = (F \mid A)_{\text{sur}}^{-1}$.

Since $F(x,y) = (x,f(x,y))$, for all $(x,y) \in A$, $G(x,z)$ is of the form $(x,g(x,z))$, for all $(x,z) \in B$, where $g : X \times Z \rightarrowtail Y$ is smooth, with domain B, and $y = g(x,f(x,y))$, for all $(x,y) \in A$, and $z = f(x,g(x,z))$, for all $(x,z) \in B$.

Now let h be the map $X \rightarrowtail Y$; $x \rightsquigarrow g(x,0)$. Then h is smooth, since g is smooth, and, for all $(x,y) \in A$,

$$f(x,y) = 0 \;\Rightarrow\; y = h(x)$$

and

$$h(x) = g(x,0) = y \;\Rightarrow\; 0 = f(x,y);$$

that is, graph $h = A \cap f^{\dashv}\{0\}$.

The proof of the uniqueness statement is left as an exercise. □

The inverse function theorem may be regarded as a particular case of the implicit function theorem. The details are left to the reader.

Smooth subsets

The inverse function theorem is important for us since it provides us with several practical criteria for determining whether a subset of a normed affine space is 'smooth' in a sense that will shortly be defined. To avoid certain technical difficulties we shall restrict attention in this section to *finite-dimensional* affine spaces and subsets of such spaces. For extensions of the definition and theorems to the non-finite-dimensional case the reader is referred to [37].

Suppose, therefore, that X is a finite-dimensional affine space. A subset M of X is said to be *smooth* at $a \in M$ if there exist an affine subspace W of X passing through a, open neighbourhoods A and B of a in X and a smooth homeomorphism $h : A \rightarrow B$ tangent to 1_X at a, with $h_{\vdash}(A \cap W) = B \cap M$.

The affine subspace W is easily seen to be unique (by Prop. 18.8!). The tangent space to W at a, TW_a, is said to be the *tangent space* to M at a.

A subset M of X is said to be *smooth* if it is smooth at each of its

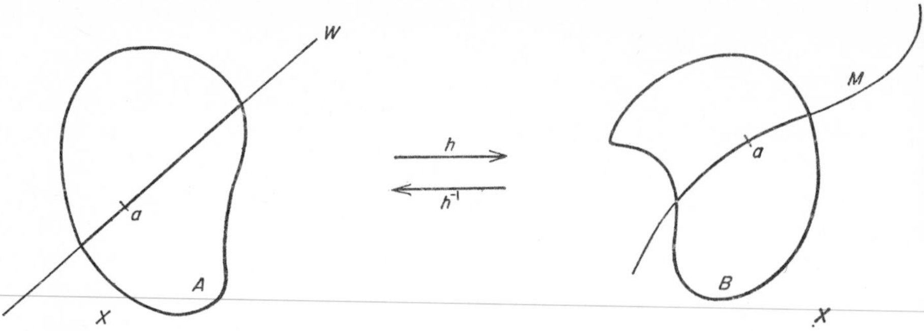

points. If each of the tangent spaces of a smooth subset M of X has the same finite dimension, m, say, we say that M is a *smooth m-dimensional submanifold* of X.

A one-dimensional smooth submanifold is also called a *smooth curve*, and a two-dimensional smooth submanifold a *smooth surface*.

In practice the subset M of X is often presented either as the image of a map or as a fibre of a map. For example, the unit circle in $\mathbf{R}^2$ is the image of the map

$$\mathbf{R} \longrightarrow \mathbf{R}^2; \quad \theta \rightsquigarrow e^{i\theta}$$

and is also a fibre of the map

$$\mathbf{R}^2 \longrightarrow \mathbf{R}; \quad (x,y) \rightsquigarrow x^2 + y^2.$$

Proposition 19.8 is concerned with the former possibility. It is of assistance in determining whether the image of a map is smooth.

Prop. 19.8. Let $f : W \rightarrowtail X$ be a smooth map, tangent at $c \in W$ to an injective affine map $t: W \longrightarrow X$, W and X being finite-dimensional affine spaces. Then there exists an open neighbourhood C of c in W such that the image of $f \mid C$, $f_\vdash(C)$, is smooth at $f(c)$, with tangent space the image of t, with $f(c)$ chosen as origin, $T(\operatorname{im} t)_{f(c)}$.

Proof Set $c = 0$ in W and $f(c) = 0$ in X, let $u: X \longrightarrow Y$ be a linear surjection with kernel im t, let $s: Y \longrightarrow X$ be a linear section of u and let

$$\{0\} \rightleftarrows W \underset{r}{\overset{t}{\rightleftarrows}} X \overset{u}{\rightleftarrows} Y \rightleftarrows \{0\}$$

be the induced split exact sequence associated with the direct sum decomposition $X = \operatorname{im} t \oplus \operatorname{im} s$.

Now define $h = fr + su : X \rightarrowtail X$. (If X is thought of as the product $X_0 \times X_1$, with $X_0 = \operatorname{im} t$, $X_1 = \operatorname{im} s$, then h is defined by $h(x_0,x_1) = f\, t_0^{-1}(x_0) + (0,x_1)$, for all $(x_0,x_1) \in X_0 \times X_1$.) Then $ht = f$,

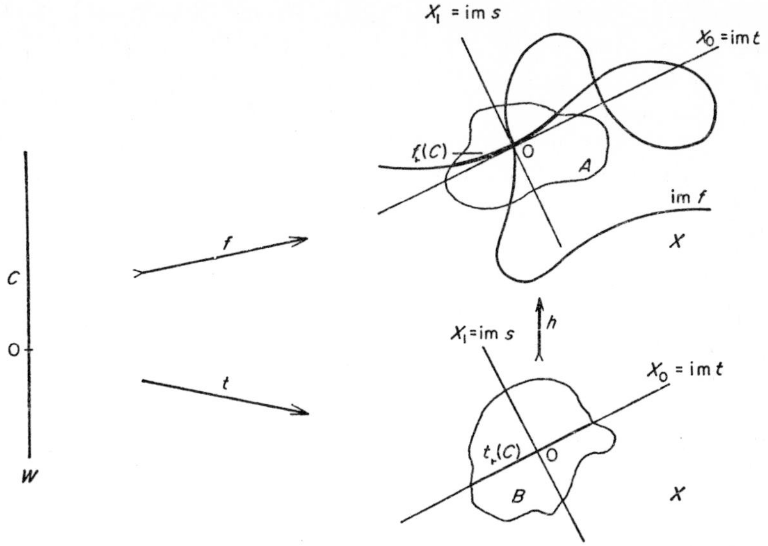

since $rt = 1_W$ and $ut = 0$; also, $h(0)$ and $dh0 = tr + su = 1_X$, that is, h is tangent to 1_X at 0.

From the inverse function theorem it follows that there are open neighbourhoods A and B of 0 in X such that $(h \mid B)_{\text{sur}} : B \longrightarrow A$ is a smooth homeomorphism. Let $C = t^{\dashv}(B)$. Then $f_{\vdash}(C) = h_{\vdash}(B \cap \text{im } t)$, with h tangent to 1_X at 0. That is, $f_{\vdash}(C)$ is smooth at 0, with tangent space $T(\text{im } t)_0$. □

Cor. 19.9. Let $f : W \rightarrowtail X$ be a smooth map, tangent at each point of its domain to an injective affine map, W and X being finite-dimensional affine spaces, and let f also be a topological embedding. Then the image of f is a smooth submanifold of X with dimension equal to $\dim W$. □

As the diagram suggests, the simplest cases occur when $W = \mathbf{R}$ and $X = \mathbf{R}^2$. The map f then has two components

$$f_0 : \mathbf{R} \rightarrowtail \mathbf{R} \quad \text{and} \quad f_1 : \mathbf{R} \rightarrowtail \mathbf{R}$$

and the map f has injective differential at a point $a \in \operatorname{dom} f$ unless the differential coefficients $(f_0)'(a)$ and $(f_1)'(a)$ are both zero.

In this context W is often called the *parameter space* and the map $f : W \rightarrowtail X$ a *parametric representation* of its image.

This is a suitable place to remark that the word 'curve' is widely used in two quite distinct senses, either to connote a one-dimensional subset of an affine space, as here in the phrase 'smooth curve', or to connote a continuous map of $\mathbf{R}$ or an interval of $\mathbf{R}$ to an affine space, as in the phrase 'Peano curve' (cf. Exercise 15.61). Peano's discovery of the first space-filling curve in 1890 caused a furore. It had been naively assumed until then that the image of an interval of **R** by a continuous map must be either a point or a one-dimensional subset of the target space. The whole study of dimension is a subtle one, with various candidates for the principal definitions. The classical work on the subject is [30]. For some further remarks on space-filling curves, and another example, see [54], page 341.

Example 19.10. Let $L_k(X,Y)$ denote the set of elements of $L(X,Y)$ of rank k, X and Y being finite-dimensional linear spaces. Then $L_k(X,Y)$ is a smooth submanifold of $L(X,Y)$, with dimension $k(n+p-k)$, where $\dim X = n$, $\dim Y = p$.

Proof Let $v \in L_k(X,Y)$ and let $X_0 = \ker v$, $Y_0 = \operatorname{im} v$. Then we may think of X as $X_0 \times X_1$ and Y as $Y_0 \times Y_1$, where X_1 is some linear complement of X_0 in X and Y_1 is some linear complement of Y_0 in Y. Moreover, $v_0 | X_1$ is bijective; so $u_0 | X_1$ is bijective for u sufficiently close to v, by Prop. 15.45.

Consider the map

$$f : L(X,Y_0) \times L(Y_0,Y_1) \rightarrowtail L(X,Y); \quad (s,t) \rightsquigarrow (s,ts),$$

with domain $GL(X,Y_0) \times L(Y_0,Y_1)$, $GL(X,Y_0)$ denoting the subset of surjective elements of $L(X,Y_0)$. This is injective, with image in $L_k(X,Y)$. It is also smooth, and the differential at any point of its domain is injective, by Example 18.28. Moreover, if u in $L_k(X,Y)$ is sufficiently near v, then u is in the image of f; for it is easily verified that in that case $u = (s,ts)$, where $s = u_0$ and $t = (u_1 | X_1)(u_0 | X_1)^{-1}$. From this formula it follows also that the map is an embedding. Finally

$$\dim (L(X,Y_0) \times L(Y_0,Y_1)) = nk + k(p-k).$$

So, by Cor. 19.9, $L_k(X,Y)$ is a smooth submanifold of $L(X,Y)$, with dimension $k(n+p+k)$. □

The next proposition and its corollary, which are complementary to Prop. 19.8, enable us to determine whether a fibre of a map is smooth.

Prop. 19.11. Let $f : X \rightarrowtail Y$ be a smooth map, tangent at $a \in X$ to a surjective affine map $t : X \to Y$, X and Y being finite-dimensional affine spaces. Then there exist open neighbourhoods A and B of a in X and a smooth homeomorphism $h : A \to B$, tangent to 1_X at a, with

$$h_\vdash(A \cap t^{\dashv}\{t(a)\}) = B \cap f^{\dashv}\{f(a)\}$$

and with $fh = t | A$.

Proof Set $a = 0$ in X and $f(a) = 0$ in Y, let $X_0 = t^{-1}\{0\}$, with inclusion map $i : X_0 \to X$, let $s : Y \to X$ be a linear section of t, with image X_1, and let

$$\{0\} \rightleftarrows X_0 \underset{p}{\overset{i}{\rightleftarrows}} X \underset{s}{\rightleftarrows} Y \rightleftarrows \{0\}$$

be the induced split exact sequence. The choice of s is equivalent to the choice of the linear complement X_1 of X_0 in X.

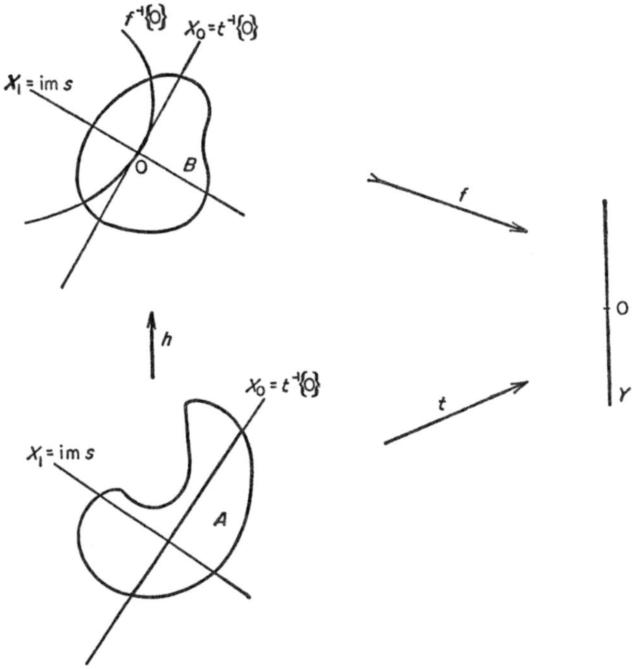

Define $g = ip + sf : X \rightarrowtail X$. (If X is identified with $X_0 \times X_1$ then, for all $x \in X$, $g(x) = (x_0, sf(x))$.) Since t is linear, $ti = 0$ and $ts = 1_Y$,

$$tg = tip + tsf = f.$$

Also, $g(0) = 0$ and $dg0 = ip + st = 1_X$; that is, g is tangent to 1_X at 0.

From the inverse function theorem it follows that there are open neighbourhoods B and A of 0 in X such that $(g \mid A)_{\text{sur}} : B \to A$ is a smooth homeomorphism, tangent to 1_X at 0. Define $h = (g \mid A)_{\text{sur}}^{-1}$. This map has the requisite properties. □

Cor. 19.12. Let $f : X \rightarrowtail Y$ be a smooth map, tangent at $a \in X$ to a surjective affine map $t : X \to Y$, X and Y being finite-dimensional affine spaces. Then the fibre of f through a, $f^{-1}\{f(a)\}$, is smooth at a, with tangent space the fibre of t through a, with a as origin, $T(t^{-1}\{t(a)\})_a$. □

Example 19.13. The sphere S^2 is a smooth two-dimensional submanifold of $\mathbf{R}^3$.

Proof Let $f : \mathbf{R}^3 \to \mathbf{R}$ be the map given by $f(x,y,z) = x^2 + y^2 + z^2$. Then $S^2 = f^{-1}\{1\}$. Now $df(x,y,z) : \mathbf{R}^3 \to \mathbf{R}$ is the linear map with matrix $(2x, 2y, 2z)$, of rank 1 unless $x = y = z = 0$, and therefore of rank 1 at every point of S^2. Therefore S^2 is a smooth submanifold of $\mathbf{R}^3$. It has dimension 2 since the kernel rank of $df(x,y,z)$ is 2 for every point $(x,y,z) \in S^2$. □

It follows, by the same argument, that for any n, S^n is a smooth n-dimensional submanifold of $\mathbf{R}^{n+1}$.

Example 19.14. The group $O(n)$ is a smooth submanifold of $\mathbf{R}(n)$, of dimension $\frac{1}{2}n(n-1)$, for any n.

Proof Let $f : \mathbf{R}(n) \to \mathbf{R}_+(n)$ be the map defined by $f(t) = t^*t$, t^* being the transpose of t and $\mathbf{R}_+(n)$ being the subset $\{t \in \mathbf{R}(n) : t^* = t\}$ of symmetric elements of $\mathbf{R}(n)$. Then $O(n) = f^{-1}\{1\}$, where 1 is the identity on $\mathbf{R}^n$. Now, since the map $\mathbf{R}(n) \to \mathbf{R}(n);\ t \rightsquigarrow t^*$ is linear, by Prop. 9.12, it follows, by Prop. 18.17, that, for any $t, u \in \mathbf{R}(n)$,

$$dft(u) = t^*u + u^*t.$$

If $t^* = t^{-1}$, then dft is surjective. For let $v \in \mathbf{R}_+(n)$. Then

$$dft(\tfrac{1}{2}tv) = \tfrac{1}{2}t^*tv + \tfrac{1}{2}v^*t^*t = v.$$

Also, since $\dim \mathbf{R}_+(n) = \frac{1}{2}n(n+1)$, $\operatorname{kr} dft = \frac{1}{2}n(n-1)$. The assertion follows, by Prop. 19.11. □

Notice that the tangent space to $O(n)$ at 1 is the translate through 1 of $\ker df1 = \{t \in \mathbf{R}(n) : t^* + t = 0\}$, the subspace of skew-symmetric elements of $\mathbf{R}(n)$. Analogous examples culled from Chapters 11, 12 and 13 are reserved for study in Chapter 20.

Local maxima and minima

The following corollary of Prop. 19.11 can be of value in locating the local maxima and minima of a real-valued map whose domain is a smooth subset of a finite-dimensional affine space.

Prop. 19.15. Let $f : X \rightarrowtail Y$ be a smooth map, tangent at $a \in X$ to a surjective affine map $t : X \to Y$, X and Y being finite-dimensional affine spaces, and let $\phi : X \rightarrowtail \mathbf{R}$ be a map tangent at a to an affine map $\beta : X \to \mathbf{R}$. Then, if the restriction of ϕ to the fibre of f through a is locally a maximum or minimum at a, there exists an affine map $\gamma : Y \to \mathbf{R}$ such that $\beta = \gamma t$.

Proof Set $a = 0$ in X, $f(a) = 0$ in Y and $\phi(a) = 0$ in $\mathbf{R}$. Then, since t is surjective, there exists, by Prop. 19.11, a smooth map $g : X \rightarrowtail X$ tangent to 1_X at 0 such that fg is the restriction to dom g of t. Let $\phi' = \phi g$.

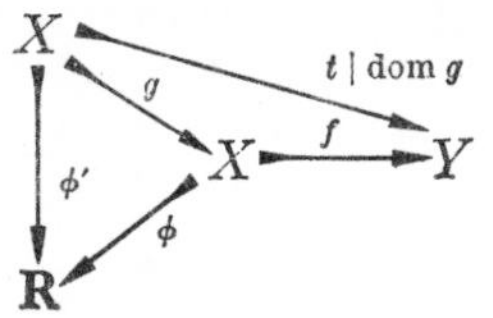

Then ϕ' is tangent to β at 0. Also the restriction of ϕ to $f^{-1}\{0\}$ has locally a maximum or minimum at 0 if, and only if, the restriction of ϕ' to $W = t^{-1}\{0\}$ has locally a maximum or minimum at 0. Since W is a linear space, a necessary condition for this is, by Prop. 18.33, that $d(\phi' \mid W)0 = (d\phi'0) \mid W = \beta \mid W$ shall be zero.

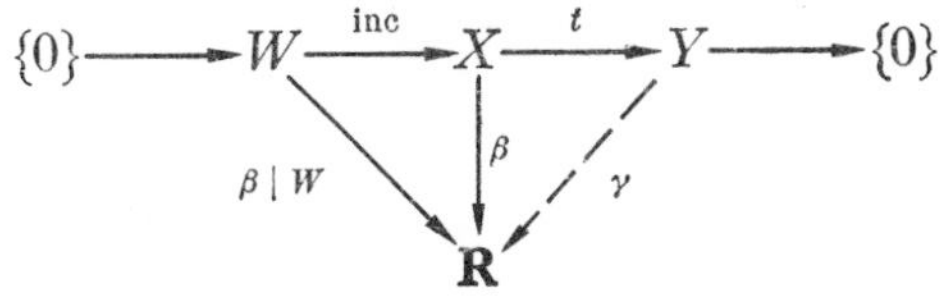

The existence of γ then follows from Prop. 5.15. □

An equivalent condition to that stated in Prop. 19.15 is that there exists a linear map $\gamma_* : Y \longrightarrow \mathbf{R}$ such that

$$d\phi a = \beta_* = \gamma_* t_* = \gamma_* dfa.$$

When $X = \mathbf{R}^n$ and $Y = \mathbf{R}^p$, the linear map $d\phi a$ is represented by the Jacobian matrix of ϕ at a, the linear map dfa is represented by the Jacobian matrix of f at a, and the linear map γ_* is represented by the row matrix of its coefficients. For example, when $n = 3$ and $p = 2$, with $(y_0, y_1) = f(x_0, x_1, x_2)$ and with $z = \phi(x_0, x_1, x_2)$, the condition is that there exist real numbers λ_0, λ_1 such that

$$\begin{pmatrix} \dfrac{\partial z}{\partial x_0} & \dfrac{\partial z}{\partial x_1} & \dfrac{\partial z}{\partial x_2} \end{pmatrix} = (\lambda_0 \;\; \lambda_1) \begin{pmatrix} \dfrac{\partial y_0}{\partial x_0} & \dfrac{\partial y_0}{\partial x_1} & \dfrac{\partial y_0}{\partial x_2} \\ \\ \dfrac{\partial y_1}{\partial x_0} & \dfrac{\partial y_1}{\partial x_1} & \dfrac{\partial y_1}{\partial x_2} \end{pmatrix}.$$

The matrix equation $d\phi a = \gamma_* dfa$ is called the *Lagrange equation* and the coefficients of γ_* are called the *Lagrange multipliers*.

Note that this method of locating possible local maxima and minima may fail if dfa is not surjective. For example, the minimum value of the restriction of the map $\mathbf{R}^2 \to \mathbf{R}$; $(x,y) \rightsquigarrow x$, to the subset $\{(x,y) \in \mathbf{R}^2 : y^2 = x^3\}$ is clearly 0, attained at (0,0). However, the *Lagrange* equation

$$(1 \quad 0) = \lambda(3x^2 \quad -2y),$$

that is, $3\lambda x^2 = 1$ and $-2y\lambda = 0$, admits no solution satisfying $y^2 = x^3$. Here f has been taken to be the map $\mathbf{R}^2 \to \mathbf{R}$; $(x,y) \rightsquigarrow x^3 - y^2$ and the method fails because $df0 = (0,0)$ does not have rank 1.

The Lagrange method of locating maxima and minima has heuristic value, but it is not the only, nor necessarily the best, method in practice.

Example 19.16. Find the maximum value of z, given that $(x,y,z) \in \mathbf{R}^3$ and that $x^2 + z^2 = 2y$ and $x - y + 4 = 0$.

Solution

(*a*) (Direct): From the equations $x^2 + z^2 = 2y$ and $x - y + 4 = 0$, we obtain $x^2 + z^2 = 2x + 8$. So $9 - z^2 = x^2 - 2x + 1 \geqslant 0$. Therefore $|z| \leqslant 3$, and $z = 3$ when $x = 1$ and $y = 5$.

(b) (Lagrange): Since all the relevant maps are smooth, the equation for possible local maxima and minima is

$$(0 \quad 0 \quad 1) = (\lambda \quad \mu) \begin{pmatrix} 2x & -2 & 2z \\ 1 & -1 & 0 \end{pmatrix}.$$

That is, $0 = 2\lambda x + \mu = -2\lambda - \mu$ and $1 = 2\lambda z$, implying that (1,5,3) and (1,5,−3) are possible candidates. The 2×3 matrix has rank <2 only when $x = 1$ and $z = 0$, and therefore has rank 2 for all (x,y,z) such that $x^2 + z^2 - 2y = x - y + 4 = 0$. So there are no other candidates. However, before we can conclude that 3 is the largest value attained by z we have to have some reason to suppose that on the set in question z is bounded and attains its bounds. We leave this to the reader and suspect that in doing so he will find himself rediscovering solution (a)! □

The rank theorem

The following proposition is the main ingredient in the proof of Theorem 19.19, the rank theorem.

Prop. 19.17. Let $f = (f_0,f_1) : X \rightarrowtail Y_0 \times Y_1$ be a smooth map tangent at 0 to a linear map $t = (t_0,t_1) : X \to Y_0 \times Y_1$, where t_0 is surjective and $t_1 = 0$, X, Y_0 and Y_1 being finite-dimensional linear spaces, and suppose that, for each $a \in \operatorname{dom} f$, im (dfa) is the graph of a linear map from Y_0 to Y_1. Then there exist an open neighbourhood A of 0 in dom f and a smooth map $\phi : Y_0 \rightarrowtail Y_1$, with $d\phi 0 = 0$, such that im $(f \,|\, A) =$ graph ϕ.

Proof The map f_0 is tangent to t_0 at 0, and since t_0 is surjective there exist, by Prop. 19.11, open neighbourhoods A and B of 0 in X and a smooth homeomorphism $g: B \to A$, tangent to 1_X at 0, such that $f_0 g = t_0 \mid B$. We may choose B so that it is of the form $B_0 + B_1$, where B_0 is a convex open neighbourhood of 0 in $X_0 = \ker t$, and B_1 is an open neighbourhood of 0 in some linear complement X_1 of X_0 in X.

Let $f' = fg$. Then $\operatorname{im}(f \mid A) = \operatorname{im} f'$. Also, since g is tangent to 1_X at 0, f' is tangent to t at 0. This implies, in particular, that $df_1'0 = 0$.

Let $b \in B$. Since $\operatorname{im}(dfa)$ is the graph of a linear map from Y_0 to Y_1 and since g is a smooth homeomorphism, $\operatorname{im}(df'b)$ is the graph of a linear map from Y_0 to Y_1. Since $df'_0 b = t_0$, it follows that $\ker(df'b) = \ker t_0 = X_0$. Since this is true for all $b \in B$, and since B_0 is convex, the restriction of f' to the intersection of B with any translate of X_0 in X has zero differential, and so is constant, by Prop. 19.4. So $\operatorname{im} f' = \operatorname{im}(f' \mid B_1)$.

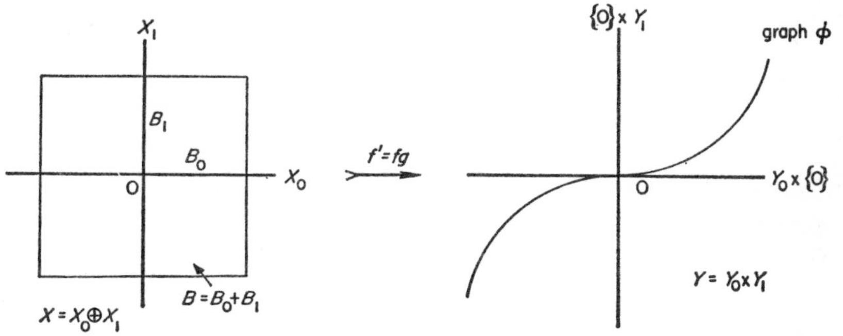

Now define $\phi = (f_1' \mid B_1)(t_0 \mid X_1)^{-1}$. Since $f_0' \mid B_1 = t_0 \mid B_1$, it at once follows that $\operatorname{im}(f' \mid B_1) = \operatorname{graph} \phi$. That is, $\operatorname{im}(f \mid A) = \operatorname{graph} \phi$. Finally, since $df_1'0 = 0$, $d\phi 0 = 0$. □

Cor. 19.18. Let $f: X \succ\!\!\rightarrow Y (= Y_0 \times Y_1)$ be a map satisfying the conditions of Prop. 19.17. Then there exist open neighbourhoods C and D of 0 in Y and a smooth homeomorphism $k: C \to D$, tangent to 1_Y at 0, such that $\operatorname{im}(kf) \subset Y_0$, the differential of the map

$$X \succ\!\!\rightarrow Y_0; \quad x \rightsquigarrow kf(x)$$

at 0 being surjective. □

In particular, it follows that there exists a smooth map $k_1: Y \succ\!\!\rightarrow Y_1$, defined on a neighbourhood of 0, and with surjective differential at 0, such that $k_1 f = 0$. Therefore, when $Y = \mathbf{R}^m$ and $\dim Y_0 = \operatorname{rk} df0 = r$, and given the conditions of Prop. 19.17, there exists, a smooth map $G: Y \succ\!\!\rightarrow R^{m-r}$, defined on a neighbourhood of 0, and with surjective differential at 0, such that $Gf = 0$. This is sometimes loosely referred to as the *functional dependence* of the components of f at 0. (Cf. [47], in particular Remark 2 on page 918.)

Theorem 19.19. (The *rank theorem*).
Let $f : X \rightarrowtail Y$ be a smooth map such that the restriction of the map $\operatorname{rk} f : X \rightarrowtail \omega$ to some neighbourhood of a point $a \in \operatorname{dom} f$ is constant, X and Y being finite-dimensional affine spaces, and let $t : X \longrightarrow Y$ be the affine map tangent to f at a. Then there exist open neighbourhoods A and B of a in X, open neighbourhoods C and D of $f(a)$ in Y, with $f_{\vdash}(B) \subset C$, a smooth homeomorphism $h : A \longrightarrow B$ tangent to 1_X at a, and a smooth homeomorphism $k : C \longrightarrow D$ tangent to 1_Y at $f(a)$ such that the map $kfh : X \rightarrowtail Y$ is the restriction to A of the affine map t.
(Apply Cor. 19.18 and Cor. 19.12.) □

The fundamental theorem of algebra

In Chapter 2 we sketched a proof of the fact that any polynomial map $\mathbf{C} \longrightarrow \mathbf{C}$; $z \rightsquigarrow \sum_{k \in n+1} a_k z^k$ has at most n zeros. What we could not then prove, except in trivial cases, was that if the polynomial has positive degree then it has at least one zero. This we are now able to do.

Theorem 19.20. (The *fundamental theorem of algebra.*)
Any polynomial map over $\mathbf{C}$

$$f : \mathbf{C} \longrightarrow \mathbf{C} : z \rightsquigarrow \sum_{k \in \omega} a_k z^k,$$

of positive degree, is surjective.

Proof By Example 8.14 and Exercise 17.53, the map f may be regarded as the restriction to $\mathbf{C}$ with target $\mathbf{C}$ of a continuous map

$$\bar{f} : \mathbf{C} \cup \{\infty\} = \mathbf{C}P^1 \longrightarrow \mathbf{C} \cup \{\infty\}$$

with $\bar{f}(\infty) = \infty$. Moreover, since $\mathbf{C}P^1$ is compact and Hausdorff, $\operatorname{im} \bar{f}$ is compact and therefore closed in $\mathbf{C} \cup \{\infty\}$, implying that the complement in $\mathbf{C}$ of $\operatorname{im} f$ is open in $\mathbf{C}$. Since $\mathbf{C}P^1$ is connected, $\operatorname{im} \bar{f}$ is connected, implying that f is not constant. Finally, since $\mathbf{C}$ is connected, $\operatorname{im} f$ is connected.

Now f is smooth and has bijective differential at all but a finite number of points. For the differential at any $z \in \mathbf{C}$ is multiplication by the complex number $\sum_{k \in \omega^+} k a_k z^{k-1}$ and is therefore bijective unless $\sum_{k \in \omega^+} k a_k z^{k-1} = 0$. However, since f has positive degree, the polynomial $\sum_{k \in \omega^+} k a_k z^{k-1}$ is not the zero polynomial and so, by Prop. 2.18, it is non-zero at all but a finite number of points.

By the inverse function theorem it follows that f is locally a homeomorphism at all but a finite number of points, and therefore at all but a finite number of values it is locally trivial, by Exercise 16.86, since each of its fibres is finite, again by Prop. 2.18.

Since also the complement in $\mathbf{C}$ of $\operatorname{im} f$ is open, the restriction of the map $\mathbf{C} \longrightarrow \omega;\ c \rightsquigarrow \#(f^{-1}\{c\})$ to the complement in $\mathbf{C}$ of the finite set of critical values of f is locally constant, that is, constant on some neighbourhood of each point of its domain. Since the complement in $\mathbf{C}$ of a finite subset is connected, by Exercise 16.99, this restriction is a constant map. The constant value cannot be zero, since f is not constant and $\operatorname{im} f$ is connected. Therefore each point of $\mathbf{C}$ is a value of f; that is, f is surjective. □

Cor. 19.21. Let $\sum_{k\in\omega} a_k z^k$ be a non-zero polynomial of degree n over $\mathbf{C}$. Then there exists $\alpha \in \mathbf{C}^n$ such that $\sum_{k\in\omega} a_k z^k = a_n \prod_{i\in n} (z - \alpha_i)$.

Proof Induction on the degree, with Theorem 19.20 as the inductive step, the cases $n = 0$ and $n = 1$ being trivial. □

A field $\mathbf{K}$ that satisfies Theorem 19.20 or, equivalently, Cor. 19.21, with $\mathbf{K}$ in place of $\mathbf{C}$, is said to be *algebraically closed*. For example, $\mathbf{C}$ is algebraically closed. The field $\mathbf{R}$ is not algebraically closed, since the polynomial map $\mathbf{R} \longrightarrow \mathbf{R};\ x \rightsquigarrow x^2$ is not surjective.

Lemma 19.22. Let $\sum_{k\in\omega} a_k z^k$ be a polynomial over $\mathbf{C}$ with real coefficients and suppose that, for some $\alpha \in \mathbf{C}$, $\sum_{k\in\omega} a_k \alpha^k = 0$. Then $\sum_{k\in\omega} a_k \bar{\alpha}^k = 0$.

Proof Since the coefficients are real, $\sum_{k\in\omega} a_k \bar{\alpha}^k = \overline{\sum_{k\in\omega} a_k \alpha^k}$. □

Cor. 19.23. Let $\sum_{k\in\omega} a_k z^k$ be a non-zero polynomial of degree n over $\mathbf{C}$, with real coefficients. Then there exists $m \in \omega$ such that $2m \leqslant n$ and $\alpha \in \mathbf{C}^m$ and $\beta \in \mathbf{R}^{n-2m}$ such that

$$\sum_{k\in\omega} a_k z^k = a_n \prod_{i\in m} (z - \alpha_i)(z - \bar{\alpha}_i) \prod_{j\in n-2m} (z - \beta_j). \qquad \square$$

Cor. 19.24. Let $\sum_{k\in\omega} a_k x^k$ be a polynomial of odd degree over $\mathbf{R}$. Then, for some $\beta \in \mathbf{R}$, $\sum_{k\in\omega} a_k \beta^k = 0$. □

Cor. 19.25. Any polynomial map of odd degree over $\mathbf{R}$ is surjective. □

Higher differentials

Until now the entire discussion in Chapters 18 and 19 has been of the first differential df of maps $f : X \rightarrowtail Y$, where X and Y are normed affine spaces. *Higher-order differentials* of f are defined recursively, by the formula

$$d^{n+1}f = d(d^n f), \quad \text{for all } n \in \omega,$$

where, by convention, $d^0 f = f$. In general, the targets of these differentials become progressively more complicated. For example, the first three differentials of the map $f : X \rightarrowtail Y$ are of the form

$$df : X \rightarrowtail L(X_*, Y_*)$$
$$d^2 f : X \rightarrowtail L(X_*, L(X_*, Y_*))$$

and

$$d^3 f : X \rightarrowtail L(X_*, L(X_*, L(X_*, Y_*))),$$

respectively, though, in the particular case that $X_* = \mathbf{R}$, each of the targets has a natural identification with Y_*, by Prop. 3.30. The map f is said to be ***k*-smooth**, or C^k, at a point $a \in X$, for some particular $k \in \omega$, if $d^k f$ is defined on a neighbourhood of a and is continuous at a, and to be ***infinitely smooth***, or C^∞, at a if, for each $k \in \omega$, $d^k f$ is defined on a neighbourhood of a. (Many writers use the term 'smooth' to mean 'infinitely smooth'.)

When f is C^∞ at a there is, for each $x \in X_*$, a sequence on Y_*

$$n \rightsquigarrow \sum_{m \in n} \frac{1}{m!}(d^m fa)(x) \underbrace{\ldots}_{(m \text{ arguments})} (x),$$

known as the *Taylor series* of f at a with increment x. The map f is said to be *analytic*, or C^ω, at a if, for some $\delta > 0$, this sequence is convergent whenever $|x| < \delta$, with limit $f(a + x)$.

The map f is said to be C^k, C^∞ or C^ω if, for each $a \in \operatorname{dom} f$, f is, respectively, C^k, C^∞ or C^ω at a. Examples exist of maps which are C^k, but not C^{k+1}, and C^∞, but not C^ω.

It is not possible here to prove any statements concerning analytic maps. It is, however, possible to prove some simple properties of C^k and C^∞ maps, and this we now do.

Prop. 19.26. Any continuous linear or bilinear map is C^∞.

Proof Let t be continuous linear. Then, by Prop. 18.14, dt is constant and $d^k t = 0$ for all $k > 1$.

Let β be continuous bilinear. Then, by Prop. 18.17, $d\beta$ is continuous linear. So $d^2\beta$ is constant and $d^k\beta = 0$ for all $k > 2$. □

Prop. 19.27. Let $(f,g) : W \rightarrowtail X \times Y$ be any map, where W, X and Y are normed affine spaces. Then (f,g) is C^k or C^∞ at a point $a \in W$ if, and only if, f and g are each, respectively, C^k or C^∞ at a. □

Prop. 19.28. Let $f : X \rightarrowtail Y$ be C^k at $a \in X$ and let $g : Y \rightarrowtail Z$ be C^k at $f(a)$, where $k \in \omega$ or $k = \infty$, X, Y and Z being normed affine spaces. Then $gf : X \rightarrowtail Z$ is C^k at a.

Proof The proof is by induction on k, the basis being the case $k = 1$, which is Theorem 18.22. Suppose the proposition true for $k = m$ and let f and g be C^{m+1} at a and $f(a)$, respectively. Then, since $d(gf)$ admits the decomposition

$$\begin{array}{ccc} X \xrightarrow{\;df\;} L(X_*, Y_*) & & \\ \downarrow \quad\quad\quad\quad\quad\quad & \times & \xrightarrow{\text{composition}} L(X_*, Z_*), \\ Y \xrightarrow{\;dg\;} L(Y_*, Z_*) & & \end{array}$$

and since f, df, dg and composition are C^m, composition being continuous bilinear, it follows by Prop. 19.27 and two applications of the inductive hypothesis, that $d(gf)$ is C^m. So gf is C^{m+1}.

(The proof of Prop. 19.27 uses a special case of Prop. 19.28 and conversely. Both inductions should therefore be carried out simultaneously.) □

Prop. 19.29. For any complete normed linear spaces X and Y the inversion map

$$\chi : L(X, Y) \rightarrowtail L(Y, X); \quad t \rightsquigarrow t^{-1}$$

is C^∞. □

Prop. 19.30. Let $f : X \rightarrowtail Y$ be a map satisfying at a point $a \in X$ the same conditions as in Theorem 18.25, with the same convention as before on the use of the notation f^{-1}, and suppose further that X (and therefore Y) is complete and that f is C^k at a, where $k \in \omega$ or $k = \infty$. Then f^{-1} is C^k at $f(a)$. □

Finally, the second differential of a map is symmetric, in the following sense.

Prop. 19.31. Let $f : X \rightarrowtail Y$ be a twice-differentiable map, X and Y being normed affine spaces. Then, for any $a \in \operatorname{dom} f$, and any $x, x' \in X_*$,

$$(d^2 fa(x'))(x) = (d^2 fa(x))(x').$$

Proof Set $a = 0$ in X, $f(a) = 0$ in Y and $dfa\ (= df0) = 0$ in $L(X, Y)$. This last we may do by replacing the map f by the map

$$x \rightsquigarrow f(x) - df0(x),$$

which has the same second differential as f. The result then follows from the *lemma*, that, for any $\varepsilon > 0$, there exists $\delta > 0$ such that $\sup\{|x|, |x'|\} < \frac{1}{2}\delta \Rightarrow$

$$|f(x + x') - f(x) - f(x') - (d^2f0(x'))(x)| \leqslant 3\varepsilon(|x| + |x'|)|x|.$$

Assuming its truth, we have at once that

$$|(d^2f0(x'))(x) - (d^2f0(x))(x')| \leqslant 3\varepsilon(|x| + |x'|)^2,$$

provided that $\sup\{|x|, |x'|\} < \frac{1}{2}\delta$. But this last condition can now be discarded, by the homogeneity of the previous inequality with respect to multiplication by positive reals (the argument is similar to that used in the proof of Lemma 18.11), from which it follows that

$$|(d^2f0(x'))(x) - (d^2f0(x))(x')| = 0$$

and therefore that $(d^2f0(x'))(x) = (d^2f0(x))(x')$.

Proof of the lemma Since df is differentiable at 0 it follows that, for all $\varepsilon > 0$, there exists $\delta > 0$ such that

$$|x| < \delta \Rightarrow |dfx - d^2f0x| \leqslant \varepsilon|x|.$$

Let $|x|, |x'| < \frac{1}{2}\delta$. Then $|x + x'| < \delta$. Now

$$\begin{aligned}&|f(x + x') - f(x) - f(x') - d^2f0(x')(x)|\\ &\leqslant |f(x + x') - f(x) - f(x') - dfx'(x)| + |dfx' - d^2f0(x')||x|.\end{aligned}$$

To estimate the first of the two terms on the right-hand side we apply the increment formula to the map

$$x \rightsquigarrow f(x + x') - f(x) - dfx'(x)$$

with domain the ball $\{x : |x| < \frac{1}{2}\delta\}$. The differential of this map at x, the linear map $df(x + x') - dfx - dfx'$, has norm

$$\begin{aligned}&\leqslant |df(x + x') - d^2f0(x + x')| + |dfx - d^2f0(x)| + |dfx' - d^2f0(x')|\\ &\leqslant 2\varepsilon(|x| + |x'|), \quad \text{since also } |x'| < \tfrac{1}{2}\delta.\end{aligned}$$

It follows from this that the left-hand side of the original inequality is

$$\begin{aligned}&\leqslant 2\varepsilon(|x| + |x'|)|x| + \varepsilon|x'||x|\\ &\leqslant 3\varepsilon(|x| + |x'|)|x|, \quad \text{as required.} \quad \square\end{aligned}$$

Cor. 19.32. Let $f : X \times Y \rightarrowtail Z$ be a twice-differentiable map, X, Y and Z being normed affine spaces. Then, for any $(a,b) \in \operatorname{dom} f$ and any $x \in X_*$, $y \in Y_*$,

$$d_1d_0f(a,b)(y)(x) = d_0d_1f(a,b)(x)(y).$$

Proof For any $(a,b) \in \operatorname{dom} f$ and any $x \in X_*$, $y \in Y_*$,

$$d_0d_1f(a,b)(x)(y) = d^2f(a,b)(x,0)(0,y)$$

and

$$d_1d_0f(a,b)(y)(x) = d^2f(a,b)(0,y)(x,0). \quad \square$$

FURTHER EXERCISES

19.33. 'The equation $f(x,y) = 0$ can be solved locally for y in terms of x if the partial differential of f with respect to y is invertible.' How accurate a version of the implicit function theorem is this? □

19.34. Deduce the inverse function theorem from the implicit function theorem. □

19.35. Let $f : \mathbf{R}^3 \rightarrowtail \mathbf{R}$ be a continuously differentiable map such that, at each point (x,y,z) of $f^{-1}(\{0\})$, each of its three partial differential coefficients is non-zero. State and prove a precise version of the loosely worded statement

$$\frac{\partial x}{\partial y}\frac{\partial y}{\partial z}\frac{\partial z}{\partial x} = -1 \quad \text{on } f^{-1}(\{0\}),$$

explaining carefully the meaning to be assigned to the symbols on the left-hand-side. □

19.36. What is the dimension of $\mathbf{R}(2)$, the real linear space of 2×2 real matrices?

Consider the map $f : \mathbf{R}(2) \longrightarrow \mathbf{R}$; $a \rightsquigarrow \det a$. Compute $dfa(b)$, for each $a, b \in \mathbf{R}(2)$, and show that, for each $b \in \mathbf{R}(2)$, $(df1)(b) = b_{11} + b_{22}$. Prove that $SL(2;\mathbf{R})$, the set of 2×2 real matrices with determinant 1, is a three-dimensional smooth submanifold of $\mathbf{R}(2)$. □

19.37. Consider the map $f : \mathbf{R}(2) \longrightarrow \mathbf{R}(2)$; $t \rightsquigarrow t^{-}t$, where, for all $t = \begin{pmatrix} a & c \\ b & d \end{pmatrix} \in \mathbf{R}(2)$, $t^{-} = \begin{pmatrix} d & c \\ b & a \end{pmatrix}$. Verify that, for every $u, t \in \mathbf{R}(2)$, $dfu(t) = u^{-}t + t^{-}u$. Describe the matrices in the kernel and image of $df1$ and prove that $f^{-1}(\{1\})$ is a smooth submanifold of $\mathbf{R}(2)$. □

19.38. Find the maximum value of $x - 2y - 2z$, given that $(x,y,z) \in \mathbf{R}^3$ and that $x^2 + y^2 + z^2 = 9$. □

19.39. Find the maximum and minimum values of

$$\text{(i) } 2x^2 - 3y^2 - 2x$$
$$\text{(ii) } 2x^2 + y^2 + 2x,$$

given that $(x,y) \in \mathbf{R}^2$ and that $x^2 + y^2 \leqslant 1$.

(The direct approach involves treating the interior and boundary of the circle separately in the search for candidate points, there being several ways of treating the boundary. An alternative is to regard the maps involved as maps from the unit sphere S^2 to $\mathbf{R}$.) □

19.40. Let $f : \mathbf{R}^2 \to \mathbf{R}^2$ be defined by

$$u = x^2 - y^2 + 2x,$$
$$v = 2xy - 2y,$$

where $(u,v) = f(x,y)$. Show that the subset S of points (x,y) for which $df(x,y)$ is non-invertible is the circle $\{(x,y) : x^2 + y^2 = 1\}$. Find the maximum and minimum values assumed by $u^2 + v^2$ on S. □

19.41. Prove that the map $(x,y,z,t) \rightsquigarrow \frac{1}{2}(x^2+y^2+z^2+t^2)$ restricted to the set $\{(x,y,z,t) \in \mathbf{R}^4 : xt - yz = 1\}$ attains a minimum value. Find this value and the set of points at which it is attained. □

19.42. Let X be a finite-dimensional real linear space. A critical point a of a twice-differentiable map $f : X \to \mathbf{R}$ is said to be *non-degenerate* if the linear map $d^2fa : X \to X^L$ is bijective. By applying the inverse function theorem to the map df, prove that if f is C^2 each non-degenerate critical point of f is *isolated*, that is, that there is some neighbourhood of the critical point containing no other critical points. □

19.43. Let $f : X \to \mathbf{R}$ be a C^2 map, where X is a finite-dimensional real linear space, let $a \in X$, let $X_0 = \ker d^2fa$ and let X_1 be a linear complement of X_0 in X. Prove that the map

$$X_0 \times X_1 (= X) \to X_0 \times X_1{}^L; \quad x \rightsquigarrow (x_0, d_1 fx)$$

has bijective differential at a and therefore that there are open neighbourhoods A of a in X and B of $(a_0, d_1 fa)$ in $X_0 \times X_1{}^L$ such that the map $h : A \to B;\ x \rightsquigarrow (x_0, d_1 fx)$ is a smooth homeomorphism.

Let $g = fh^{-1} : X_0 \times X_1{}^L \to \mathbf{R}$. Prove that

(i) the critical points of g all lie on $X_0 \times \{0\}$

(ii) $d_0 f = (d_0 g)h$

(iii) $g \mid (X_0 \times \{0\})$ is C^2. □

19.44. Let $f : X \to R$, $g : X \to R$, and $h : X \to X$ be twice-differentiable maps such that $g = hf$, each being defined on a neighbourhood of 0 and sending 0 to 0. Suppose, moreover, that $df0 = 0$ and $dg0 = 0$ and that $dh0$ is a linear homeomorphism. Prove that, for any $x' \in X$,

$$d^2g0(x')(x') = d^2f0(x)(x),$$

where $x = dh0(x')$. □

19.45. Give an example of a commutative diagram

$$\begin{array}{ccc} X & \xrightarrow{f} & Y \\ \uparrow h & & \downarrow k \\ X' & \xrightarrow{f'} & Y' \end{array}$$

of twice-differentiable maps between normed linear spaces, with h_{sur} and k_{sur} smooth homeomorphisms, such that, for some $a' \in X'$ and $a = h(a') \in X$,

$$d^2fa = 0 \quad \text{but} \quad d^2f'a' \neq 0. \qquad \square$$

19.46. Let

$$\begin{array}{ccc} X_0 \times X_1 & \overset{f}{\rightarrowtail} & Y_0 \times Y_1 \\ \uparrow h & & \downarrow \\ X_0' \times X_1' & \overset{f'}{\rightarrowtail} & Y_0' \times Y_1' \end{array}$$

be a commutative diagram of twice-differentiable maps between normed linear spaces, each a product of normed linear spaces as indicated, each map being defined on a neighbourhood of 0 and sending 0 to 0. Suppose, moreover, that

$$df0 = \begin{pmatrix} 0 & c \\ 0 & 0 \end{pmatrix} \quad \text{and} \quad df'0 = \begin{pmatrix} 0 & c' \\ 0 & 0 \end{pmatrix}$$

where $c : X_0 \longrightarrow Y_1$ and $c' : X' \longrightarrow Y_1'$ are linear homeomorphisms, and that $dh0$ and $dk0$ are linear homeomorphisms. Prove that $d_0h_10 = 0$ and $d_0k_10 = 0$, that $d_0h_00 : X_0' \longrightarrow X_0$ and $d_1k_10 : Y_1 \longrightarrow Y_1'$ are linear homeomorphisms and that, for any $x' \in X'$,

$$d(d_0f_1')0(x') = (d_1k_10)(d(d_0f_1)0(dh0(x')))(d_0h_00). \qquad \square$$

19.47. Let X and Y be normal linear spaces, let $f : X \rightarrowtail Y$ be twice differentiable at some point $a \in \operatorname{dom} f$, and let $\phi : X \rightarrowtail Y$ be defined on $\operatorname{dom} f$ by the formula

$$\phi(x) = f(x) - f(a) - dfa(x - a) - \tfrac{1}{2}d^2fa(x - a)(x - a).$$

Prove that, for any $\varepsilon > 0$, there exists $\delta > 0$ such that

$$|\, x - a \,| \leqslant \delta \;\Rightarrow\; |\, \phi(x) \,| \leqslant \varepsilon \,|\, x - a \,|^2.$$

(Since f is twice differentiable at a, df is differentiable at a and ϕ is C^1 at a. Show first that $|\, d\phi(x) \,| \leqslant \varepsilon \,|\, x - a \,|$, for all x sufficiently near a. Then apply the increment formula to ϕ near a.) $\square$

19.48. (*Taylor's theorem*—W. H. Young's form.) Let X and Y be normed linear spaces, let $f : X \rightarrowtail Y$ be n times differentiable at some point $a \in \operatorname{dom} f$, n being finite, and let $\phi : X \rightarrowtail Y$ be defined on $\operatorname{dom} f$ by the formula

$$\phi(x) = f(x) - \sum_{m\in} \frac{1}{m!}(d^mfa)(x - a)^m.$$

Prove that, for any $\varepsilon > 0$, there exists $\delta > 0$ such that

$$|\, x - a \,| \leqslant \delta \;\Rightarrow\; |\, \phi(x) \,| \leqslant \varepsilon \,|\, x - a \,|^n. \qquad \square$$

19.49. Let X and Y be normed linear spaces, let $f : X \rightarrowtail Y$ be C^n at some point $a \in \operatorname{dom} f$, n being finite, and let $\rho : X \times X \rightarrowtail Y$ be defined near (a,a) by the formula

$$\rho(x,b) = f(x) - \sum_{m \in n} \frac{1}{m!}(d^m f b)(x - b)^m.$$

Prove that, for any $\varepsilon > 0$, there exists $\delta > 0$ such that

$$|(x,b) - (a,a)| < \delta \;\Rightarrow\; |\rho(x,b)| \leqslant \varepsilon \, |x - b|^n. \qquad \square$$

19.50. Let X, Y and Z be finite-dimensional real linear spaces and let $f : X \times Y \rightarrow Z$ be a C^p map such that, for each $x \in X$, the map $f(x,-) : Y \rightarrow Z$ is continuous linear, p being any finite number greater than 0, or ∞. Prove that the map

$$X \rightarrow L(Y,Z); \quad x \rightsquigarrow f(x,-)$$

is C^p. (Recall Exercise 15.59.) $\square$

19.51. Consider what difficulties might arise in generalizing the section on smooth subsets to subsets of normed affine spaces which need not be finite-dimensional. $\square$

CHAPTER 20

SMOOTH MANIFOLDS

Consider again the definition on page 381 of a smooth submanifold M of a finite-dimensional real affine space X. The subset M of X is smooth (C^1) at a point $a \in M$ if there is an affine subspace W of X passing through a and there are open neighbourhoods A and B of a in X, and a smooth homeomorphism $h : A \rightarrow B$, tangent to 1_X at a, such that $h_{\vdash}(A \cap W) = B \cap M$.

Let i, in such a case, denote the map $W \rightarrowtail M$; $w \rightsquigarrow h(w)$. Its domain is $A \cap W$, which is open in W, and it is an open embedding, since h is a homeomorphism and $B \cap M$ is open in M. So (W,i) is a chart on M in the sense of Chapter 17. Such charts will be called the *standard charts* on M, as a smooth submanifold of X.

The following proposition follows at once from these remarks.

Prop. 20.1. Let M be a smooth submanifold of a finite-dimensional real affine space X. Then M is a topological manifold. □

Now consider two standard charts on a smooth submanifold.

Prop. 20.2. Let $i: V \rightarrowtail M$ and $j : W \rightarrowtail M$ be standard charts on a smooth submanifold M of a finite-dimensional real affine space X. Then the map $j_{\text{sur}}^{-1} i_{\text{sur}}$ is a smooth homeomorphism.

Proof From its construction, the map $j_{\text{sur}}^{-1} i_{\text{sur}}$ is the restriction to an open subset of the affine subspace V, with image an open subset of the affine subspace W, of a smooth homeomorphism whose domain and image are open subsets of X. □

These propositions provide the motivation for the following definitions and their subsequent development. The chapter is concerned mainly with the simplest properties of smooth manifolds and smooth submanifolds of smooth manifolds. Tangent spaces are discussed in detail, examples of smooth embeddings and smooth projections are given and the chapter concludes with the definition of a Lie group and the Lie algebra of a Lie group, and further examples.

Smooth manifolds and maps

Let X be a topological manifold. Then a *smooth* (C^1) *atlas* for a topological manifold X consists of an atlas $\mathscr{S}$ for X such that, for each (E,i), $(F,j) \in \mathscr{S}$, the map

$$j_{\text{sur}}^{-1}i : E \rightarrowtail F; \quad a \rightsquigarrow j_{\text{sur}}^{-1}i(a)$$

is smooth.

Since the map $i_{\text{sur}}^{-1}j$ must also be smooth, it is a corollary of the definition that $j_{\text{sur}}^{-1}i_{\text{sur}}$ is a smooth homeomorphism.

Example 20.3. Let X be a finite-dimensional real affine space. Then $\{X,1_X)\}$ is a smooth atlas for X. □

Example 20.4. Let A be an open subset of a finite-dimensional real affine space X. Then $\{(X,1_A)\}$ is a smooth atlas for A. □

Example 20.5. Let X be a finite-dimensional real affine space and let $\mathscr{S}$ be the set of maps $h : X \rightarrowtail X$ with open domain and image and with h_{sur} a smooth homeomorphism. Then $\mathscr{S}$ is a smooth atlas for X. □

Example 20.6. Let M be a smooth submanifold of a finite-dimensional real affine space X. Then the set of standard charts on M is a smooth atlas for M. □

Example 20.6 provides, in particular, a smooth atlas for the sphere S^n, for any $n \in \omega$, for, by Example 19.13, S^n is a smooth submanifold of $\mathbf{R}^{n+1}$. The next two examples also are of smooth atlases for the sphere S^n.

Example 20.7. For any $n \in \omega$, and for any $k \in n + 1$, let h_k be the map

$$\mathbf{R}^n \longrightarrow S^n; \quad x \rightsquigarrow (x_0,x_1, \ldots, x_{k-1},1,x_k, \ldots, x_{n-1})/\sqrt{(1 + x^{(2)})}.$$

Then the set of charts $\{(\mathbf{R}^n,h_k) : k \in n + 1\} \cup \{(\mathbf{R}^n,-h_k) : k \in n + 1\}$ is a smooth atlas for S^n.

For example, for any k, $l \in n + 1$ with $k < l$, $(h_l)_{\text{sur}}^{-1}h_k$ is the smooth map

$$\mathbf{R}^n \rightarrowtail \mathbf{R}^n; \quad x \rightsquigarrow (x_0,x_1, \ldots, x_{k-1},1,x_k, \ldots, x_{l-1},x_{l+1}, \ldots, x_{n-1})/x$$

with domain the half-space $\{x \in \mathbf{R}^n : x_l > 0\}$. □

Example 20.8. Let i and $j : \mathbf{R}^n \longrightarrow S^n$ be the inverses of stereographic projection onto the equatorial plane of the sphere S^n from the North and South poles, respectively. (Cf. Prop. 9.63 and Exercise 9.80.) Then $\{\mathbf{R}^n,i), (\mathbf{R}^n,j)\}$ is a smooth atlas for X; for the maps

$$i_{\text{sur}}^{-1}j = j_{\text{sur}}^{-1}i : \mathbf{R}^n \rightarrowtail \mathbf{R}^n; \quad u \rightsquigarrow u/u^{(2)}$$

are smooth. □

The Grassmannians and, in particular, the projective spaces also have smooth atlases.

Example 20.9. For any n-dimensional linear space V over $\mathbf{K} = \mathbf{R}$, $\mathbf{C}$ or $\mathbf{H}$ and for any $k \leqslant n$, the standard atlases for the Grassmannians $\mathscr{G}_k(V)$ and, in the real case, $\mathscr{G}_k^+(V)$ are smooth (see Example 18.27). In particular the standard atlases on the projective spaces $\mathbf{K}P^n$ are smooth. □

Example 20.10. For any $n \in \omega$, and for any $k \in n + 1$, let g_k be the map

$$\mathbf{K}^n \longrightarrow \mathbf{K}P^n; \quad x \rightsquigarrow [x_0, x_1, \ldots, x_{k-1}, 1, x_k, \ldots, x_{n-1}],$$

$\mathbf{K}$ being $\mathbf{R}$, $\mathbf{C}$ or $\mathbf{H}$ or, when $n = 1$ or 2, even being $\mathbf{O}$, the Cayley algebra. Then $\{(\mathbf{K}^n, g_k) : k \in n + 1\}$ is a smooth atlas for $\mathbf{K}P^n$, $\mathbf{K}^n$ being regarded as a *real* linear space. □

Two smooth atlases on a topological manifold X are said to be *equivalent*, or to define the same *smooth structure* on X if their union is smooth.

Example 20.11. The atlases for S_n given in Examples 20.7 and 20.8 are both equivalent to the standard atlas on S^n as a smooth submanifold of $\mathbf{R}^{n+1}$. □

A topological manifold with a smooth atlas is said to be a *smooth manifold*, smooth manifolds with the same underlying topological space and with equivalent atlases being said to be *equivalent*.

A chart (E,i) on a smooth manifold X, with atlas $\mathscr{S}$, is said to be *admissible* if $\mathscr{S} \cup \{(E,i)\}$ is a smooth atlas for X.

Infinitely smooth (C^∞) and *analytic* (C^ω) *atlases* and *manifolds* are defined by replacing the word 'smooth' in each of the above definitions by 'infinitely smooth' or by 'analytic' respectively. For example, the atlases for the spheres, the Grassmannians and the projective spaces in Examples 20.8, 20.9, 20.10 and 20.11 are C^∞ (and in fact C^ω).

For most purposes the distinction between different, but equivalent, manifolds is unimportant and may be ignored. It might seem to be more sensible to define a smooth manifold to be a set with a smooth structure rather than a set with a smooth atlas. The reason for choosing the second alternative is in order to sidestep certain logical difficulties concerned with the definition of the set of all admissible charts on a smooth manifold. There are various ways around the difficulty, and each author has his own preference. One place where it is logically important to have a particular atlas in mind is in the construction of the tangent bundle of a smooth manifold—see page 408 below—but even this turns out in the end to matter little since, by Cor. 20.41, the tangent bundles of equivalent smooth manifolds are naturally isomorphic.

Proposition 20.12 is of importance both in defining the dimension of a smooth manifold and in defining its tangent spaces. (Cf. Prop. 20.23 and page 407.)

Prop. 20.12. Let (E,i) and (F,j) be admissible charts on a smooth manifold X such that $\operatorname{im} i \cap \operatorname{im} i \neq \emptyset$. Then, for any $x \in \operatorname{im} i \cap \operatorname{im} i$, the map

$$d(j_{\text{sur}}^{-1}i)(i_{\text{sur}}^{-1}(x)) : E_* \to F_*$$

is a linear isomorphism.

Proof Since $i_{\text{sur}}^{-1}j_{\text{sur}} = (j_{\text{sur}}^{-1}i_{\text{sur}})^{-1}$, the given map is invertible, with inverse the map

$$d(i_{\text{sur}}^{-1}j)(j_{\text{sur}}^{-1}(x)) : F_* \to E_*. \qquad \square$$

The next proposition leads directly to the definition of smooth maps between smooth manifolds.

Prop. 20.13. Let $f : X \to Y$ be a map between smooth manifolds X and Y, let (E,i) and (E',i') be admissible charts on X, let (F,j) and (F',j') be admissible charts on Y, and suppose that x is a point of $\operatorname{im} i \cap \operatorname{im} i'$ such that $f(x) \in \operatorname{im} i \cap \operatorname{im} j'$.

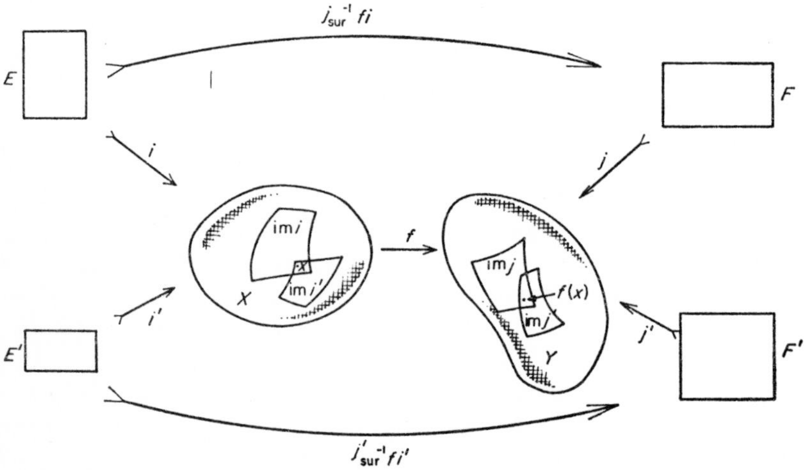

Then the map $j'^{-1}_{\text{sur}} f i' : E' \rightarrowtail F'$ is smooth at $i'^{-1}_{\text{sur}}(x)$ if, and only if, the map $j_{\text{sur}}^{-1} f i : E \rightarrowtail F$ is smooth at $i_{\text{sur}}^{-1}(x)$.

Proof Apply the chain rule (Theorem 18.22) to the equation

$$j'^{-1}_{\text{sur}} f i'(a) = (j'^{-1}_{\text{sur}} j)(j_{\text{sur}}^{-1} f i)(i_{\text{sur}}^{-1} i')(a)$$

for all $a \in E$ sufficiently close to $i_{\text{sur}}^{-1}(a)$. $\square$

A map $f: X \to Y$ between smooth manifolds X and Y is said to be *smooth at* a point x if the map $j_{sur}^{-1} f i: E \rightarrowtail F$ is smooth at $i_{sur}^{-1}(x)$ for some and therefore, by Prop. 20.12, for any admissible charts (E,i) at x and (F,j) at $f(x)$. The map is said to be *smooth* if it is smooth at each point of X.

Example 20.14. For any $n \in \omega$, let S^n and $\mathbf{R}P^n$ be assigned the smooth atlases given in Examples 20.7 and 20.10. Then the map

$$S^n \to \mathbf{R}P^n; \quad x \rightsquigarrow \operatorname{im}\{x\}$$

is smooth. □

Example 20.15. Let V be an n-dimensional real linear space. Then, for any $k \leqslant n$, the map

$$GL(\mathbf{R}^k, V) \to \mathscr{G}_k(V); \quad t \rightsquigarrow \operatorname{im} t$$

is smooth, $GL(\mathbf{R}^k,V)$, the Stiefel manifold of k-framings on V, being an open subset of $L(\mathbf{R}^k,V)$ by Prop. 15.49, and the Grassmannian $\mathscr{G}_k(V)$ being assigned its standard smooth structure. □

Notice that the definition of the smoothness of a map $f: X \to Y$ depends only on the smooth structures for X and Y and not on any particular choice of an atlas of admissible charts.

A bijective smooth map $f: X \to Y$ whose inverse f^{-1} also is smooth, is said to be a *smooth isomorphism* or a *smooth homeomorphism*.

Example 20.16. Let X' and X'' be equivalent smooth manifolds, X being the underlying topological space. Then the map $1_X: X' \to X''$ is a smooth isomorphism. □

Infinitely smooth and analytic maps and homeomorphisms are defined analogously in the obvious ways.

Inequivalent smooth atlases for a topological space X may yet be isomorphic. For example, let $h: \mathbf{R} \to \mathbf{R}$ be any homeomorphism of $\mathbf{R}$ on to itself that is not smooth. Then the atlases $\{(\mathbf{R},1_\mathbf{R})\}$ and $\{(\mathbf{R},h)\}$ for $\mathbf{R}$ are not equivalent, yet the map h from $\mathbf{R}$ with the atlas $\{(\mathbf{R},1_\mathbf{R})\}$ to $\mathbf{R}$ with the atlas $\{(\mathbf{R},h)\}$ is a smooth isomorphism, since $h^{-1} h\, 1_\mathbf{R} = 1_\mathbf{R}$, which is smooth. However, Milnor showed in 1956 [41] that there were atlases for S^7 which were not only not equivalent but not even isomorphic.

Infinitely smooth manifolds are also called *differentiable manifolds*, or, since Milnor's paper, *differential manifolds*. According to this new usage a *differentiable* or (*infinitely*) *smoothable* manifold is a topological manifold possessing at least one (infinitely) smooth atlas. A *differential* or (*infinitely*) *smooth* manifold is then a differentiable manifold with a particular choice of (infinitely) smooth atlas or (infinitely) smooth structure.

Submanifolds and products of manifolds

A subset W of a smooth manifold X is said to be *smooth at* a point $w \in W$ if there is an admissible chart (E,i) on X at w and an affine subspace D of E through $i_{sur}^{-1}(w)$ such that $i_\vdash(D) = W \cap i_\vdash(E)$, and to be a *smooth submanifold* of X if it is smooth at each point of W.

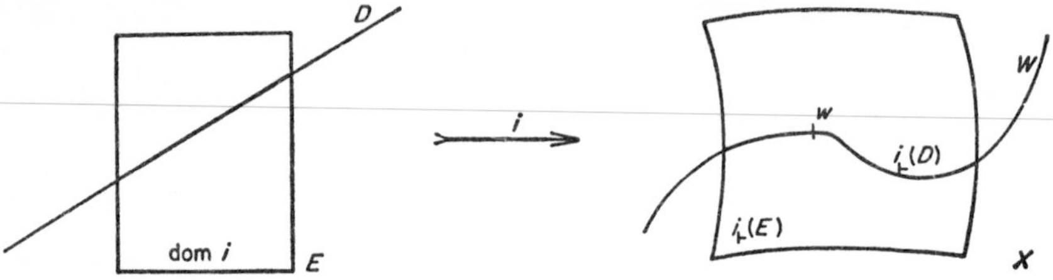

This definition generalizes that given of a smooth submanifold of a finite-dimensional real affine space in Chapter 19, page 381.

Prop. 20.17. Let W be a smooth submanifold of a smooth manifold X and let (E,i) be an admissible chart on X and D an affine subspace of E such that $i_\vdash(D) = W \cap i_\vdash(E)$. Then the restriction to D with target W is a chart on W. Moreover, any atlas for W formed from such charts is a smooth atlas for W, any two such atlases being equivalent. $\square$

That is, there is a well-determined smooth structure for each smooth submanifold of a smooth manifold. Any atlas of admissible charts for a smooth submanifold will be called an *admissible atlas* for the submanifold.

Prop. 20.18. Let $f : X \rightarrow Y$ be a smooth map, and let W be a smooth submanifold of X. Then the restriction $f \mid W : W \rightarrow Y$ is smooth. $\square$

The product of a pair (X,Y) of smooth manifolds is the smooth manifold consisting of the topological manifold $X \times Y$ together with the atlas consisting of all charts of the form $(E \times F, i \times j)$, where (E,i) and (F,j) are charts on X and Y respectively, and where $i \times j$ is the map $E \times F \rightarrowtail X \times Y;\ (a,b) \rightsquigarrow (i(a),j(b))$.

Many of the theorems of Chapter 18 have analogues for smooth maps between manifolds.

Prop. 20.19. Let W, X and Y be smooth manifolds. Then a map $(f,g) : W \rightarrow X \times Y$ is smooth if, and only if, its components f and g are smooth. $\square$

Prop. 20.20. Let X, Y and Z be smooth manifolds, and suppose that $f: X \times Y \to Z$ is a smooth map. Then, for any $(a,b) \in X \times Y$, the maps $f(-,b): X \to Z$ and $f(a,-): Y \to Z$ are smooth. □

Prop. 20.21. Let X, Y and Z be smooth manifolds and let $f: X \to Y$ and $g: Y \to Z$ be smooth maps. Then $gf: X \to Z$ is smooth. □

Prop. 20.22. Let V be a smooth submanifold of a smooth manifold X, let W be a smooth submanifold of a smooth manifold Y and let $f: X \to Y$ be a smooth map, with $f_{\vdash}(V) \subset W$. Then the restriction of f with domain V and target W is smooth. □

Dimension

As we saw at the end of Chapter 17, there are technical difficulties in the definition of the dimension of a topological manifold. There is no such difficulty for smooth manifolds, as Prop. 20.23 and Cor. 20.24 show.

Prop. 20.23. Let (E,i) and (F,j) be admissible charts on a smooth manifold X, such that $\operatorname{im} i \cap \operatorname{im} j \neq \emptyset$. Then $\dim E = \dim F$.

Proof Apply Prop. 20.12. □

Cor. 20.24. Let (E,i) and (F,j) be any admissible charts on a connected smooth manifold X. Then $\dim E = \dim F$. □

A smooth manifold X is said to have *dimension* n, $\dim X = n$, if the dimension of the source of every admissible chart on X is n.

Examples 20.25. Let $n, k \in \omega$ be such that $k \leqslant n$ and let $h = 1$, 2 or 4 according as $\mathbf{K} = \mathbf{R}$, $\mathbf{C}$ or $\mathbf{H}$. Then $\dim S^n = n$, $\dim \mathscr{G}_k(\mathbf{K}^n) = hk(n-k)$, $\dim \mathbf{K}P^n = hn$, $\dim \mathbf{O}P^2 = 16$.

The dimension of any open subset of a finite-dimensional real linear space X is equal to $\dim X$. In particular, for any n-dimensional right $\mathbf{K}$-linear space V, the Stiefel manifold of k-framings on V, $GL(\mathbf{K}^k, V)$, has real dimension hkn. □

Prop. 20.26. Let W be a connected smooth submanifold of a connected smooth manifold X. Then $\dim W \leqslant \dim X$. □

Prop. 20.27. Let X and Y be connected smooth manifolds. Then $\dim X \times Y = \dim X + \dim Y$. □

Tangent bundles and maps

The concept of the differential of a smooth map $f: X \rightarrowtail Y$, where X and Y are finite-dimensional real affine spaces, does not generalize directly to the case where X and Y are smooth manifolds. What does generalize is the concept of the *tangent map* of the map f, as defined below.

The *tangent bundle* of a finite-dimensional real affine space X is, by definition, the real affine space $TX = X \times X$, together with the projection $\pi_{TX}: TX \rightarrow X$; $(x,a) \rightsquigarrow a$, the fibre $\pi_{TX}^{-1}\{a\} = TX_a = X \times \{a\}$, for any $a \in X$, being assigned the linear structure with (a,a) as origin, as in Chapter 4. According to the definition of that chapter, $\pi_{TX}^{-1}\{a\}$ is the *tangent space* to X *at* a. The *tangent bundle space* TX may therefore be thought of as the union of all the tangent spaces to X, with the obvious topology, the product topology.

The *tangent bundle* of an open subset A of a finite-dimensional real affine space X is, by definition, the open subset $TA = X \times A$ of TX, together with the projection $\pi_{TA} = \pi_{TX} | TA$, the fibres of π_{TA} being regarded as linear spaces, as above.

Now suppose that a map $f: X \rightarrowtail Y$ is tangent at a point a of X to an affine map $t: X \rightarrow Y$, X and Y being finite-dimensional real affine spaces. Then, instead of representing the map t by its linear part $dfa: X_* \rightarrow Y_*$, we may equally well represent it by the linear map

$$Tf_a: TX_a \rightarrow TY_{f(a)}; \quad (x,a) \rightsquigarrow (t(x), f(a)),$$

the *tangent map* of f at a. Its domain is the tangent space to X at a and its target the tangent space to Y at $f(a)$. If the map f is differentiable everywhere, there is then a map

$$Tf: TX \rightarrow TY; \quad (x,a) \rightsquigarrow Tf_a(x,a),$$

with domain $T \operatorname{dom} f$, called, simply, the *tangent map* of f. Notice that, for any $(x,a) \in TX$, $Tf_a(x,a)$ may be abbreviated to $Tf(x,a)$. Notice also that the maps df and Tf are quite distinct. The maps dfa and Tf_a may be identified, for any $a \in X$, but not the maps df and Tf.

Prop. 20.28. Let $f: X \rightarrowtail Y$ be a smooth map, X and Y being finite-dimensional real affine spaces. Then the map Tf is continuous, with open domain.

Proof First of all, $\operatorname{dom} Tf = T \operatorname{dom} f$, which is open in TX, $\operatorname{dom} f$ being open in X. The continuity of Tf follows at once from the decomposition

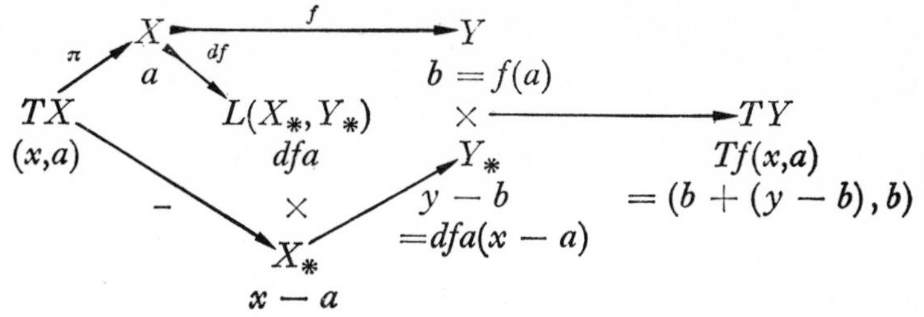

and the continuity of each of the component maps. □

Prop. 20.29. Let $f : X \rightarrowtail Y$ be a differentiable map for which Tf is continuous, X and Y being finite-dimensional real affine spaces. Then f is smooth. (Use Exercise 15.59.) □

Prop. 20.30. Let X be any finite-dimensional real affine space. Then $T1_X = 1_{TX}$. □

The following two propositions are corollaries of the chain rule.

Prop. 20.31. Let $f : X \rightarrowtail Y$ and $g : W \rightarrowtail X$ be smooth maps, W, X and Y being finite-dimensional real affine spaces. Then, for each $a \in \operatorname{dom}(fg)$ and each $w \in W$,

$$T(fg)(w,a) = Tf\,Tg(w,a). \qquad \square$$

Prop. 20.32. Let $f : X \rightarrowtail Y$ and $g : Y \rightarrowtail X$ be smooth maps, with $g_{\text{sur}} = f_{\text{sur}}^{-1}$, X and Y being finite-dimensional real affine spaces. Then, for any $a \in \operatorname{dom} f$ and any $x \in X$,

$$Tg\,Tf(x,a) = (x,a)$$

and, for any $b \in \operatorname{dom} g$ and any $y \in Y$,

$$Tf\,Tg(y,b) = (y,b). \qquad \square$$

Now consider a smooth manifold X with atlas $\mathscr{S}$, let $x \in X$, and let (E,i), (F,j) and (G,k) be charts at x with $x = i(a) = j(b) = k(c)$. Then, by Prop. 20.12, $d(j_{\text{sur}}^{-1}i)a : E \longrightarrow F$, or, equivalently, the tangent map $T(j_{\text{sur}}^{-1}i)_a : TE_a \longrightarrow TF_b$, is a linear isomorphism. Moreover

$$T(i_{\text{sur}}^{-1}i)_a = 1_{TE_a}$$

$$T(i_{\text{sur}}^{-1}j)_b = (T(j_{\text{sur}}^{-1}i)_a)^{-1}$$

and

$$T(k_{\text{sur}}^{-1}i)_a = (T(k_{\text{sur}}^{-1}j)_b)(T(j_{\text{sur}}^{-1}i)_a),$$

by Props. 20.30, 20.31 and 20.32. These remarks motivate and essentially prove the following proposition.

Prop. 20.33. Let X be a smooth manifold with atlas $\mathscr{S}$ and let $\mathscr{S} = \bigcup \{T(\operatorname{dom} i) \times \{i\} : (E,i) \in \mathscr{S}\}$ (to be thought of as the disjoint union of the $T(\text{dom } i)$). Then the relation $\sim$ on $\mathscr{S}$, given by the formula $((a',a),i) \sim ((b',b),j)$ if, and only if, $j(b) = i(a)$ and $d(j_{\text{sur}}^{-1}i)a(a') = b'$, or, equivalently, $T(j_{\text{sur}}^{-1}i)(a',a) = (b',b)$, is an equivalence. □

The *tangent bundle* of X, with atlas $\mathscr{S}$, is defined to be the quotient TX of the set $\mathscr{S}$ defined in Prop. 20.33 by the equivalence $\sim$, together with the surjection $\pi_{TX}: TX \to X$; $[((a',a),i)]_\sim \rightsquigarrow i(a)$.

Prop. 20.34. The set of maps $\{Ti: (E,i) \in \mathscr{S}\}$, where Ti is the map $T(\text{dom } i) \to TX$; $(a',a) \rightsquigarrow [((a',a),i)]_\sim$, is an atlas for the set TX. □

The set TX is assigned the topology induced by this atlas and called the *tangent bundle space* of X.

Prop. 20.35. The map π_{TX} is locally trivial.

Proof For any chart (E,i) the diagram

$$\begin{array}{ccc} T(\text{dom } i) & \xrightarrow{\ Ti\ } & TX \\ \Big\downarrow{\scriptstyle \pi_{T\,\text{dom}\,i}} & & \Big\downarrow{\scriptstyle \pi_{TX}} \\ \text{dom } i & \xrightarrow{\ i\ } & X \end{array}$$

is commutative, with $\text{im } Ti = \pi_{TX}^{\dashv}(\text{im } i)$, the maps i and Ti being topological embeddings. Since $\pi_{T\,\text{dom}\,i}$ is a product projection, it follows that π_{TX} is locally trivial. □

In particular, π_{TX} is a topological projection. It will be referred to as the *tangent projection* on X.

The next proposition examines the structure of the fibres of the tangent projection.

Prop. 20.36. Let X be a smooth manifold with atlas $\mathscr{S}$. Then, for any $x \in X$, the fibre $\pi_{TX}^{\dashv}\{x\}$ is the quotient of the set

$$\mathscr{S}_x = \bigcup \{TE_a \times \{i\} : (E,i) \in \mathscr{S} \text{ and } i(a) = x\}$$

by the restriction to this set of the equivalence $\sim$. Moreover there is a unique linear structure for the fibre such that each of the maps

$$TE_a \to \pi_{TX}^{\dashv}\{x\}; \quad (a',a) \rightsquigarrow [((a',a),i)]_\sim$$

is a linear isomorphism.

(The existence of the linear structure for the fibre follows directly from the remarks preceding Prop. 20.33.) □

The fibre $\pi_{TX}^{\dashv}\{x\}$ is assigned the linear structure defined in Prop. 20.36 and is called the *tangent space* to X at x. It will be denoted also by TX_x. Its elements are the *tangent vectors* to X at x.

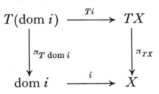

The projection π_{TX} always has at least one continuous section, the *zero section*, associating to any $x \in X$ the zero tangent vector to X at x. By Prop. 16.11 the zero section of π_{TX} is a topological embedding of X in TX.

As we remark in more detail later, the tangent bundle space TX of a smooth manifold X is not necessarily homeomorphic to the product of X with a linear space. (See page 420.)

Notice that the definitions of tangent bundle, tangent projection and tangent space for a smooth manifold agree with the corresponding definitions given earlier for a finite-dimensional real affine space X, or an open subset A of X, provided that X, or A, is assigned the single chart atlas of Example 20.3, or Example 20.4.

A smooth map $f: X \to Y$ induces in a natural way a continuous map $Tf: TX \to TY$, the *tangent (bundle) map* of f. It is defined in the next proposition.

Prop. 20.37. Let X and Y be smooth manifolds and let $f: X \to Y$ be a smooth map. Then there is a unique continuous map $Tf: TX \to TY$ such that, for any chart (E,i) on X and any chart (F,j) on Y,

$$(Tj_{\text{sur}})^{-1}(Tf)(Ti) = T(j_{\text{sur}}^{-1} fi),$$

where $T(j_{\text{sur}}^{-1} fi)$, Ti and Tj have the meanings already assigned to them. $\square$

It is readily verified that this definition of the tangent map of a smooth map where domain and target are smooth manifolds includes as special cases the tangent map of a smooth map with source and target finite-dimensional real affine spaces, and also the map Ti induced by an admissible chart (E,i) on a smooth manifold X, as previously defined.

The tangent map of f maps any tangent vector at a point x of X to a tangent vector at the point $f(x)$ in X, as the next proposition shows.

Prop. 20.38. Let $f: X \to Y$ be a smooth map. Then the diagram

$$\begin{array}{ccc} TX & \xrightarrow{Tf} & TY \\ \downarrow{\scriptstyle \pi_{TX}} & & \downarrow{\scriptstyle \pi_{TY}} \\ X & \xrightarrow{f} & Y \end{array}$$

commutes. That is, for any $x \in X$, $(Tf)_{\vdash}(TX_x) \subset TY_{f(x)}$. Moreover, for any $x \in X$, the map $Tf_x: TX_x \to TY_{f(x)};\ v \rightsquigarrow Tf(v)$ is linear. $\square$

The map Tf is easily computed in the following case.

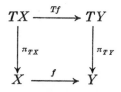

Prop. 20.39. Let X be a smooth submanifold of an affine space V, let Y be a smooth submanifold of an affine space W and let $g: V \rightarrowtail W$ be a smooth map, with $X \subset \operatorname{dom} g$, such that $g_{\vdash}(X) \subset Y$. Then the restriction $f: X \to Y$; $x \rightsquigarrow g(x)$ of g is smooth and, for any $a \in X$, $Tf_a: TX_a \to TY_{f(a)}$ is the restriction of Tg_a with domain TX_a and target $TY_{f(a)}$. □

In the application we make of this proposition, V, W and g are frequently linear.

Finally, Props. 20.30, 20.31 and 20.32 extend to smooth maps between smooth manifolds.

Prop. 20.40. Let W, X and Y be smooth manifolds. Then

$$T1_X = 1_{TX},$$

$$T(fg) = Tf\,Tg, \quad \text{for any smooth maps}$$

$$f: X \to Y \quad \text{and} \quad g: W \to X,$$

and

$$Tf^{-1} = (Tf)^{-1},$$

for any smooth homeomorphism $f: X \to Y$. □

Cor. 20.41. Let X' and X'' be equivalent smooth manifolds with underlying topological manifold X. Then $T1_X: TX' \to TX''$ is a 'tangent bundle isomorphism'. □

Cor. 20.42. Let W be a smooth submanifold of a smooth manifold X, let W be assigned any admissible atlas, and let $i: W \to X$ be the inclusion. Then the tangent map $Ti: TW \to TX$ is a topological embedding whose image is independent of the atlas chosen for W. □

The tangent bundle of W, in such a case, is normally identified with its image by Ti in TX.

For example, for any $n \in \omega$, the sphere S^n may be thought of as a smooth submanifold of $\mathbf{R}^{n+1}$, TS^n being identified with the subspace

$$\{(x,a) \in \mathbf{R}^{n+1} \times S^n : x \cdot a = 0\}$$

of $T\mathbf{R}^{n+1} = \mathbf{R}^{n+1} \times \mathbf{R}^{n+1}$.

Exercise 20.43. Prove that, for any $n \in \omega$, the complex quasi-sphere $\mathscr{S}(\mathbf{C}^{n+1}) = \{x \in \mathbf{C}^{n+1} : x^{(2)} = 1\}$ (cf. page 217 and Prop. 17.26) is homeomorphic to TS^n. □

Tangent bundles and maps are particular cases of *vector* (or *linear*) *bundles* and *maps*. See, for example, [27].

Particular tangent spaces

In many cases the tangent space to a smooth manifold at a point of the manifold may usefully be identified with, or represented by, some other linear space associated with the point. For example, as we have seen, the tangent space at a point w of a smooth submanifold W of a smooth manifold X may be identified with a subspace of the tangent space to X at w, while the tangent space at any point (x,y) of the product $X \times Y$ of smooth manifolds X and Y may be identified with the product $TX_x \times TY_y$ of the tangent spaces TX_x and TY_y. For any smooth manifold X, also, one can always choose some chart (A,i) on X at the point of interest x, A being a linear space and x being equal to $i(0)$. Then the tangent space TX_x may be identified with the linear space A by the linear isomorphism $Ti_0 : TA_0 (= A) \to TX_x$.

One important case where there is a natural candidate for the tangent space is the following.

Example 20.44. Let X be a point of the Grassmannian $\mathscr{G}_k(V)$ of k-planes in an n-dimensional real linear space V. Then the tangent space $T(\mathscr{G}_k(V))_X$ may naturally be identified with the linear space $L(X, V/X)$.

To see this, let Y and Y' be linear complements of X in V. Then the maps

$$L(X,Y) \to \mathscr{G}_k(V); \quad t \rightsquigarrow \text{graph}\, t$$

and

$$L(X,Y') \to \mathscr{G}_k(V); \quad t' \rightsquigarrow \text{graph}\, t'$$

are admissible charts on $\mathscr{G}_k(V)$ and each tangent vector at X has a unique representative both in $L(X,Y)$ and in $L(X,Y')$. By Prop. 8.12, the one is mapped to the other by the differential at zero of the map

$$L(X,Y) \rightarrowtail L(X,Y'); \quad t \rightsquigarrow q't(1_X + p't)^{-1}$$

where $(p',q') : Y \to V = X \times Y'$ is the inclusion. By Exercise 18.27 this differential is the map

$$L(X,Y) \to L(X,Y'); \quad t \rightsquigarrow q't.$$

Now, for any $y \in Y$, y and $q'(y)$ belong to the same coset of X in V, from which it follows at once that $t \in L(X,Y)$ composed with the natural isomorphism $Y \to V/X$ of Prop. 8.8 (with the roles of X and Y interchanged) is equal to $q't$ composed with the analogous natural isomorphism $Y' \to V/X$.

That is, each tangent vector at X corresponds in a natural way to an element of the linear space $L(X,V/X)$. □

When V has a prescribed positive-definite real orthogonal structure, an alternative candidate for the tangent space $T(\mathscr{G}_k(V))_X$ is the linear space $L(X,X^\perp)$.

There are analogous natural candidates for the tangent spaces of the other Grassmannians.

A definition of the tangent space at a point x of a smooth manifold X that is popular with differential geometers, and which has the technical, if not the intuitive, advantage that it is independent of the choice of smooth atlas defining the given smooth structure for the manifold, may be based on the following proposition, in which $F = F(X)$ denotes the linear space of smooth maps $X \to \mathbf{R}$, $\mathbf{R}^F$ denotes the linear space of maps $F \to \mathbf{R}$, and any tangent space $T\mathbf{R}_y$ of $\mathbf{R}$ is identified with $\mathbf{R}$ by the map $T\mathbf{R}_y \to \mathbf{R}$; $(y,b) \rightsquigarrow y - b$.

Prop. 20.45. For any $x \in X$ the map

$$TX_x \to \mathbf{R}^F; \quad v \rightsquigarrow \phi_v$$

is an injective map, where, for any $v \in TX_x$, ϕ_v is the map $F \to \mathbf{R}$; $f \rightsquigarrow Tf(v)$. □

By Prop. 20.45 the tangent vector v may be identified with the map ϕ_v. For details of this point of view see any modern book on differential geometry. For a discussion of certain technical points, see [56].

Smooth embeddings and projections

A smooth map $f: X \to Y$ between smooth manifolds X and Y is said to be a *smooth embedding* if $Tf: TX \to TY$ is a topological embedding, and to be a *smooth projection* if Tf is a topological projection.

Prop. 20.46. Let X and Y be finite-dimensional affine spaces. Then any affine injection $X \to Y$ is a smooth embedding and any affine surjection $X \to Y$ is a smooth projection. □.

Prop. 20.47. Let X and Y be smooth manifolds. Then a smooth map $f: X \to Y$ is a smooth embedding if, and only if, f is a topological embedding and, for each $x \in X$, Tf_x is injective.

Proof ⇒ : Suppose that Tf is a topological embedding. Then the restriction of Tf to the image in TX of the zero section of π_{TX}, with target the image in TY of the zero section of π_{TY}, is a topological embedding. But this is just f. Moreover, since Tf is injective, Tf_x is injective, for each $x \in X$.

⇐ : Let a be a point of X for which Tf_a is injective. By Prop. 19.8 there exists, for any chart $i: TX_a \rightarrowtail X$ sending 0 to a, a chart

$j: TY_{f(a)} \rightarrowtail Y$ sending 0 to $f(a)$, such that the diagram

$$\begin{array}{ccc} TX_a & \overset{i}{\rightarrowtail} & X \\ \Big\downarrow {\scriptstyle Tf_a} & & \Big\downarrow \\ TY_{f(a)} & \overset{j}{\rightarrowtail} & Y \end{array}$$

commutes. Therefore, since, by Prop. 20.46, Tf_a is a smooth embedding, $f \mid \operatorname{im} i$ is a smooth embedding. It follows, by Prop. 16.22, that, if Tf_x is injective for each $x \in X$, then f is a smooth embedding. □

Prop. 20.48. The image of a smooth embedding $f: X \to Y$ is a smooth submanifold of the smooth manifold Y, and the map $f_{\text{sur}}: X \to \operatorname{im} f$ is a smooth isomorphism. □

One commonly says that a smooth embedding $f: X \to Y$ *embeds* the manifold X *smoothly* in the manifold Y.

A smooth map $f: X \to Y$ is said to be an *immersion* if, for each $x \in X$, Tf_x is injective. An immersion need not be injective, nor need an injective immersion be a topological embedding.

Example 20.49. The map

$$f: \mathbf{R} \to \mathbf{R}^2; \quad x \rightsquigarrow (x^2 - 1, x^3 - x)$$

is an immersion that is not injective, and the restriction of this map to

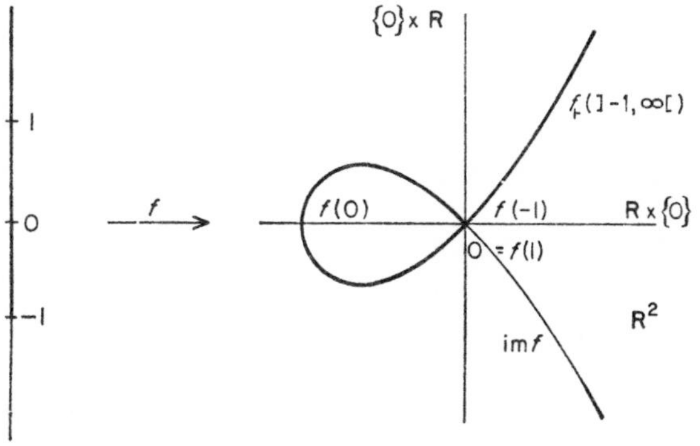

the interval $]-1, \infty[$ is an injective immersion that is not a topological embedding. □

Example 20.50. Let r be any irrational real number. Then the map $f: \mathbf{R} \to S^1 \times S^1$; $x \rightsquigarrow (e^{ix}, e^{irx})$ is an injective immersion that is not a topological embedding.

(To see that f is not a topological embedding, it is convenient first to represent the torus as the quotient of $\mathbf{R}^2$ by the equivalence $(x + 2m\pi, y + 2n\pi) \sim (x,y)$, for any $(x,y) \in \mathbf{R}^2$ and any $(m,n) \in \mathbf{Z}^2$. Then f is the composite of the map $\mathbf{R} \longrightarrow \mathbf{R}^2$; $x \rightsquigarrow (x,rx)$ with the partition induced by the equivalence.) □

By Cor. 16.44 the domain of any injective immersion that is not a topological embedding is necessarily non-compact.

Prop. 20.51. Let $f : X \longrightarrow Y$ be a smooth projection. Then f is a topological projection and, for each $x \in X$, Tf_x is surjective. □

A smooth map $f: X \longrightarrow Y$ is said to be a *submersion* if, for each $x \in X$, Tf_x is surjective.

Prop. 20.52. A submersion $f : X \longrightarrow Y$ is an open map. Its non-null fibres are smooth submanifolds of X, the tangent space at a point $x \in X$ to the fibre $f^{-1}\{f(x)\}$ through x being the kernel of Tf_x. A surjective submersion is a smooth projection.

Proof Let $f: X \longrightarrow Y$ be a submersion. Then, by Prop. 19.12, there exist, for any $x \in X$, admissible charts $i : TX_x \rightarrowtail X$, mapping zero to x and $j : TY_{f(x)} \rightarrowtail Y$, mapping 0 to $f(x)$, such that the diagram

$$\begin{array}{ccc} TX_x & \overset{i}{\rightarrowtail} & X \\ \downarrow{\scriptstyle Tf_x} & & \downarrow{\scriptstyle f} \\ TY_{f(x)} & \overset{j}{\rightarrowtail} & Y \end{array}$$

commutes. Since an affine surjection is open, it follows that x has a neighbourhood in X whose image in Y is open, from which it follows that f is an open map.

The statement concerning the fibres is an immediate corollary of Prop. 19.11.

The final statement is a corollary of Prop. 20.46 and Prop. 16.23. □

Example 20.53 below is an important example of a smooth projection. In this example, and in the section which follows, a tangent vector at any non-zero point of a real linear space X will be said to be *radial* if it is of the form $(x + \lambda x, x)$, for some $\lambda \in \mathbf{R}$, or, equivalently, if it is of the form λx, when TX_x has been identified in the standard way with X.

Example 20.53. For any finite n the map

$$\pi : \mathbf{R}^{n+1} \rightarrowtail S^n; \quad x \rightsquigarrow x/|x|$$

defined everywhere except 0, is a smooth projection, the kernel of the tangent map at any point consisting of the radial tangent vectors at that point.

Proof Let $g: \mathbf{R}^{n+1} \rightarrowtail \mathbf{R}^{n+1}$ be the composite of the map π with the inclusion of S^n in $\mathbf{R}^{n+1}$. For any non-zero a and any $x \in \mathbf{R}^{n+1}$,

$$dga(x) = |\,a\,|^{-1}\,x - |\,a\,|^{-3}(x \cdot a)a = |\,a\,|^{-1}\,x',$$

where $x' = x - (x \cdot g(a))g(a)$.

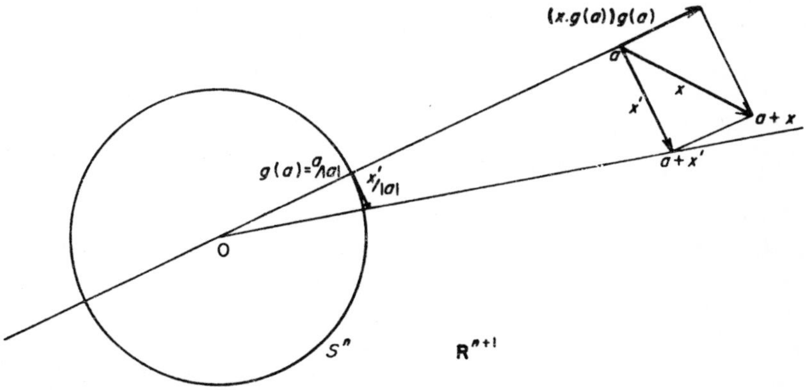

Moreover, for any non-zero $a \in \mathbf{R}^{n+1}$ and any $\lambda \in \mathbf{R}$,

$$x = \lambda a \quad \Rightarrow \quad x \cdot a = \lambda a \cdot a \quad \Rightarrow \quad x = (x \cdot g(a))g(a).$$

So $dga(x) = 0$ if, and only if, $x = \lambda a$, for some $\lambda \in \mathbf{R}$. That is, kr $(dga) = 1$, implying, by Prop. 6.32, that rk $(T\pi_a) =$ rk $(dga) = n$ and therefore that π is a submersion. Since π is surjective it follows, by the last part of Prop. 20.52, that π is a smooth projection. □

Embeddings of projective planes

It is a theorem of H. Whitney [59] and [60] that any compact smooth n-dimensional manifold may be embedded smoothly in $\mathbf{R}^{2n}$. The proof is hard, and even if one is content with an embedding in $\mathbf{R}^{2n+1}$, the proof, though much easier, is too long to be given here. If he can get hold of a copy, the reader should refer, for a proof of the simpler theorem, to Milnor's notes on Differential Topology [42]. The proof of the harder theorem may be extracted from [43].

It may be of interest to give an example of such an embedding for a manifold which is not normally presented as a submanifold of a linear space. The example chosen is the real projective plane $\mathbf{R}P^2 = \mathscr{G}_1(\mathbf{R}^3)$, which we shall embed in $\mathbf{R}^4$. Since it requires little extra effort to do so, we construct at the same time embeddings of $\mathbf{C}P^2$, of real dimension 4, in $\mathbf{R}^7$, $\mathbf{H}P^2$ of real dimension 8, in $\mathbf{R}^{13}$ and even the Cayley plane $\mathbf{O}P^2$ of real dimension 16, in $\mathbf{R}^{25}$.

Notice that in most of these cases the manifold is embedded in a linear space whose dimension is less than twice the dimension of the manifold. It can be shown that in each of these cases the dimension of the target space is the lowest possible. The problem for a given manifold X, and even for the projective spaces $\mathbf{R}P^n$, of determining the least number p for which X may be embedded in $\mathbf{R}^p$, is a hard one that has been the subject of many research papers in recent years. The present example is in a paper by I. James [32].

In the discussion which follows, $\mathbf{K}$ will denote $\mathbf{R}$, $\mathbf{C}$, $\mathbf{H}$ or $\mathbf{O}$ and conjugation will be the standard conjugation in each case, with $|x|^2 = \bar{x}x$, for each $x \in \mathbf{K}$, and with $\bar{x} = x$ if, and only if, $x \in \mathbf{R}$. The number $h = \dim_{\mathbf{R}} \mathbf{K}$.

To take account of the possibility that $\mathbf{K} = \mathbf{O}$, it is convenient to let $\mathbf{K}^3_{\mathrm{A}}$ denote the subset (not a linear subspace!) of $\mathbf{K}^3$ consisting of all $x \in \mathbf{K}^3$ such that the subalgebra $\mathbf{K}_x$ of $\mathbf{K}$ generated by the set $\{x_0, x_1, x_2\}$ is associative. Clearly $\mathbf{K}^3_{\mathrm{A}} = \mathbf{K}^3$ unless $\mathbf{K} = \mathbf{O}$. In every case $x \in \mathbf{K}^3_{\mathrm{A}}$, if one of the components of x is real. For each $i \in 3$, the subset of $\mathbf{K}^3_{\mathrm{A}}$ consisting of all $x \in \mathbf{K}^3$ such that x_i is real will be denoted by $\mathbf{K}^3_{i\mathrm{r}}$. This is a real linear space of dimension $(n-1)h + 1$. The open subset of $\mathbf{K}^3_{i\mathrm{r}}$ consisting of all $x \in \mathbf{K}^3$ such that x_i is real and positive will be denoted by $\mathbf{K}^3_{i+}$, and the affine subspace of $\mathbf{K}^3_{i\mathrm{r}}$ consisting of all $x \in \mathbf{K}^3$ such that $x_i = 1$ will be denoted by $\mathbf{K}^3_{i1}$.

The construction makes use of the map

$$f : \mathbf{K}^3 \to \mathbf{R} \times \mathbf{K}^3 \times \mathbf{R}; \quad x \rightsquigarrow (x_0\bar{x}_0, x_0\bar{x}_1, x_0\bar{x}_2 + x_1\bar{x}_1, x_1\bar{x}_2, x_2\bar{x}_2),$$

the smooth radial projection

$$\pi : \mathbf{R} \times \mathbf{K}^3 \times \mathbf{R} \rightarrowtail S^{3h+1}; \quad y \rightsquigarrow y/|y|$$

and the charts on $\mathbf{K}P^2$

$$h_i : \mathbf{K}^3_{i1} \to \mathbf{K}P^2; \quad x \rightsquigarrow [x], \quad \text{where } i \in 3.$$

The strategy is to embed $\mathbf{K}P^2$ in the first instance smoothly in the sphere S^{3h+1}. An embedding in $\mathbf{R}^{3h+1}$ is then readily obtained by composing the embedding in the sphere with the stereographic projection of the sphere to an equatorial hyperplane from some point of the sphere not in the image of the first embedding.

The steps in the construction of the smooth embedding of $\mathbf{K}P^2$ in S^{3h+1} are as follows:

Step 1 To show that, for all $i \in 3$, $f \mid \mathbf{K}^3_{i+}$ is injective.

Step 2 To show that, for all $i \in 3$, $f \mid \mathbf{K}^3_{i+}$ is an immersion, mapping radial tangent vectors to radial tangent vectors.

Step 3 To show that, for all $i \in 3$, $g_i = \pi(f \mid \mathbf{K}^3_{i1})$ is an injective immersion.

Step 4 To show that there is a unique map $g : \mathbf{K}P^2 \to S^{3h+1}$ such that, for all $i \in 3$, $gh_i = g_i$, and that g is an injective immersion.

Step 5 To show that g is a topological, and therefore a smooth embedding.

The following lemma is used in Steps 1, 2 and 3.

Lemma 20.54. For any $i \in 3$, any $x \in \mathbf{K}^3_{i+}$ and any $y \in \mathbf{K}^3_{ir}$, $f(y) = f(x)$ if, and only if, $y = \pm x$.

Proof $\Leftarrow$: Clear.

$\Rightarrow$: For all $x \in \mathbf{K}^3_i$, $f(x)$ determines x up to multiplication on the right by an element of $\mathbf{K}$ of modulus 1. For, if $x_0 \neq 0$, $x_0\bar{x}_0$ so determines x_0, and x_1 and x_2 are then uniquely determined by x_0, $x_0\bar{x}_1$ and $x_0\bar{x}_2 + x_1\bar{x}_1$; if $x_0 = 0$, but $x_1 \neq 0$, $x_0\bar{x}_2 + x_1\bar{x}_1 = x_1\bar{x}_1$ and x_1 is so determined, x_2 then being uniquely determined by $x_1\bar{x}_2$; finally, if $x_0 = x_1 = 0$, $x_2 \neq 0$ and is so determined by $\bar{x}_2x_2$. (There are no snags when $\mathbf{K} = \mathbf{O}$, since the computation takes place in the associative subalgebra $\mathbf{K}_x$ of $\mathbf{K}$.)

Now suppose that $x \in \mathbf{K}^3_{i+}$, $y \in \mathbf{K}^3_{ir}$ and $f(y) = f(x)$. By what has been proved, $y = xz$, where $z \in \mathbf{K}$, with $|z| = 1$. In particular, $y_i = x_iz$. But x_i and y_i are real, and $x_i \neq 0$. So z is real. Therefore $z = \pm 1$. □

Cor. 20.55. For each $i \in 3$, the map $f \mid \mathbf{K}^3_{ir}$ is injective. □

This completes Step 1.

Cor. 20.56. For any $i \in 3$, any $x \in \mathbf{K}^3_{i+}$, any $y \in \mathbf{K}^3_{ir}$ and any $\mu \in \mathbf{R}$,
$$f(y) = \mu f(x) \text{ if, and only if, } \mu \geqslant 0 \text{ and } y = \pm\sqrt{\mu}x. \quad \square$$

Cor. 20.57. For each $i \in 3$, the map
$$g_i : \mathbf{K}^3_{i1} \to S^{3h+1} \text{ is injective.} \quad \square$$

Corollary 20.57 will be used in Step 3.
In Step 2 the differential of f has to be computed.

Lemma 20.58. The map f is smooth and, for all $x, y \in \mathbf{K}^3$,
$$dfx(y) = \tfrac{1}{2}(f(x+y) - f(x-y)).$$

Proof Let $F : \mathbf{K}^3 \times \mathbf{K}^3 \to \mathbf{K}^5$ be defined, for all $(x,y) \in \mathbf{K}^3 \times \mathbf{K}^3$, by the formula
$$F(x,y) = (x_0\bar{y}_0, x_0\bar{y}_1, x_0\bar{y}_2 + x_1\bar{y}_1, x_1\bar{y}_2, x_2\bar{y}_2).$$

Then F is a real bilinear map such that, for all $x \in \mathbf{K}^3$, $f(x) = F(x,x)$. So f is smooth and, for all x, $y \in \mathbf{K}^3$,

$$\begin{aligned} dfx(y) &= F(x,y) + F(y,x) \\ &= \tfrac{1}{2}(F(x+y,x+y) - F(x-y,x-y)) \\ &= \tfrac{1}{2}(f(x+y) - f(x-y)). \quad \square \end{aligned}$$

Lemma 20.59. Let $f_{(i)} = f \mid \mathbf{K}^3_{i+}$. Then $f_{(i)}$ is an immersion.

Proof Since $\mathbf{K}^3_{i+}$ is an open subset of the real linear space $\mathbf{K}^3_{ir}$ of $\mathbf{K}^3$, $f_{(i)}$ is differentiable and, for all $x \in \mathbf{K}^3_{i+}$, $y \in \mathbf{K}^3_{ir}$,

$$df_{(i)}x(y) = \tfrac{1}{2}(f(x+y) - f(x-y)).$$

So, for any such x and y, $df_{(i)}x(y) = 0 \;\Rightarrow\; f(x+y) = f(x-y)$
$\Rightarrow\; x + y = \pm(x - y)$,

by 20.54, for, if $x_i + y_i = x_i - y_i = 0$, $x_i = 0$, contrary to hypothesis,
$\Rightarrow\; y = 0, \quad$ since $x \neq 0$.

Therefore, for each such x, $df_{(i)}x$ is injective. That is, $f_{(i)}$ is an immersion. $\square$

The next lemma completes Step 2 and leads on at once to Step 3.

Lemma 20.60. For any $x \in \mathbf{K}^3_{i+}$, $(Tf_{(i)})_x$ maps radial tangent vectors to radial tangent vectors.

Proof With F as in the proof of Lemma 20.58 we have, for all $x \in \mathbf{K}^3_{i+}$ and all $\lambda \in \mathbf{R}$,

$$\begin{aligned} df_{(i)}x\,(\lambda x) &= F(x,\lambda x) + F(\lambda x,x) \\ &= 2\lambda F(x,x) \\ &= 2\lambda f(x). \quad \square \end{aligned}$$

Cor. 20.61. For each $i \in 3$, the restriction of f to $\mathbf{K}^3_{i1}$ is an immersion, and in each case none of the tangent vectors of the image is radial. $\square$

Lemma 20.62. For any $i \in 3$, the map g_i is an injective immersion.

Proof The injectivity of g_i was Cor. 20.57. That g_i is an immersion follows at once from Cor. 20.61. $\square$

This completes Step 3.
Step 4 is an easy corollary of the following remark.

Lemma 20.63. For any $x \in \mathbf{K}^3_{\mathrm{A}}$, and all $\lambda \in \mathbf{K}_x$,

$$f(x\lambda) = |\lambda|^2 f(x). \quad \square$$

We therefore have a map $g : \mathbf{K}P^2 \longrightarrow S^{3h-1}$ which is an injective immersion and is, in particular, continuous. Now $\mathbf{K}P^2$ is compact and S^{3h-1} is Hausdorff. So, by Cor. 16.44, we finally have:

Lemma 20.64. The map g is a topological embedding. □

Cor. 20.65. The map g is a smooth embedding. □

This completes Step 5 and the construction is made.

Theorem 20.66. There exists a smooth embedding of the projective plane $\mathbf{K}P^2$ in $\mathbf{R}^{3h+1}$. □

Tangent vector fields

A *tangent vector field* on a smooth manifold X is a continuous section $X \to TX$ of the tangent bundle projection π_{TX}.

Example 20.67. The map

$$S^1 \to TS^1; \quad a \rightsquigarrow (a + ai, a)$$

is a nowhere-zero tangent vector field on the circle S^1.

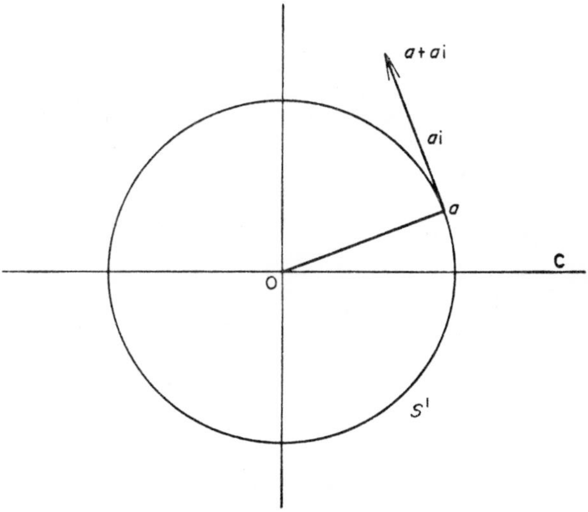

□

A set $\mathscr{S}$ of tangent vector fields on a smooth manifold X is said to be *free at* a point $a \in X$ if the set $\{s(a) : s \in \mathscr{S}\}$ is a free subset of the tangent space TX_a and is said to be *everywhere free* if it is free at each point.

The problem of determining the maximum number of everywhere free tangent vector fields on a smooth manifold X has played an important role in algebraic topology in recent years. Like the embedding problem referred to above, it has been a test bed for new techniques. For spheres the problem was solved by Adams in 1961 [2]. The easy half is to construct, for any finite n, a free set of tangent vector fields on S^n of the asserted maximum number. This we do in Theorem 20.68 below. The proof that this number can never be exceeded is the achievement of Adams. (See the comment following Exercise 12.28.) Particular cases can sometimes be dealt with more easily. For example, it is not so hard to prove that an even-dimensional sphere has no nowhere-zero vector field. (Cf. [55], pp. 201–203.) The sphere S^2 is an example of a smooth manifold whose tangent bundle is not homeomorphic to the product of the manifold with a linear space of the same dimension.

Theorem 20.68. Let $\chi : \omega \longrightarrow \omega$ be the Radon–Hurwitz sequence (cf. Theorem 13.68). For any $n \in \omega$, if $2^{\chi(k)}$ divides $n + 1$, there exists on S^n an everywhere free set of k tangent vector fields.

Proof By Theorem 13.68(i), if $2^{\chi(k)}$ divides $n + 1$, there exists a linear subspace X of $GL(n + 1; \mathbf{R})$ of dimension k such that, for each $x \in X$, $x^\tau = -x$. Then for any $a \in S^n$, the space $\{x(a) : x \in X\}$ is a k-dimensional linear subspace of $\mathbf{R}^{n+1}$, orthogonal to a; for, since $a \cdot x(a) = x^\tau(a) \cdot a = -x(a) \cdot a$, it follows that $a \cdot x(a) = 0$, for all $x \in X$.

Let $\{e_i : i \in k\}$ be any basis for X and for each $i \in k$ let E_i be the map:

$$S^n \longrightarrow TS^n; \quad a \rightsquigarrow (a + e_i(a), a).$$

Then E_i is a tangent vector field on S^n and the set $\{E_i : i \in k\}$ is free. □

It is easy to see in addition that if the basis $\{e_i : i \in k\}$ for X is chosen to be orthonormal with respect to the positive-definite quadratic form $x \rightsquigarrow x^\tau x$ (cf. Theorem 13.68(i) again), then the values of the tangent vector fields E_i at any point a of S^n form an orthonormal set of k tangent vectors to S^n at a.

Cor. 20.69. For $n = 0, 1, 3$ or 7, $TS^n \cong \mathbf{R}^n \times S^n$. □

It is a corollary of Adams's theorem that these are the only values of n for which TS^n is homeomorphic to $\mathbf{R}^n \times S^n$. This particular result—the 'parallelizability' of the spheres S^0, S^1, S^3 and S^7, and no others—was proved by several people round about 1958 [1], [35], [45]. The Radon–Hurwitz sequence dates from 1923 [51], [31].

Lie groups

Numerous examples of smooth manifolds and smooth maps are to be found in Chapters 11, 12 and 13. Those that are groups have already in Chapter 17 been shown to be, in a natural way, topological groups. What we now show is that these topological groups have a natural smooth structure.

A *Lie group* is a topological group G with a specified smooth (C^1) structure, such that the maps

$$G \times G \to G;\quad (a,b) \rightsquigarrow ab \quad \text{and} \quad G \to G;\quad a \rightsquigarrow a^{-1}$$

are smooth (C^1)[38]. For some purposes it is desirable to insist on a higher degree of smoothness than 1, but C^1 will do for the moment.

Elementary properties of Lie groups include the following.

Prop. 20.70. Let G be a Lie group. Then, for any $a, b \in G$ the maps

$$G \to G;\quad g \rightsquigarrow ag \quad \text{and} \quad g \rightsquigarrow gb$$

are smooth homeomorphisms.

Proof The map $g \rightsquigarrow ag$ is smooth, by Prop. 20.20, and its inverse, the map $G \to G$; $g' \rightsquigarrow a^{-1}g'$, is smooth.

Similarly for the other map. □

A *Lie group map* is a smooth group map $G \to H$, where G and H are Lie groups, and a *Lie group isomorphism* is a bijective Lie group map whose inverse also is a Lie group map.

Prop. 20.71. Let G be a Lie group. Then, for any $a \in G$, the map $G \to G$; $g \rightsquigarrow aga^{-1}$ is a Lie group isomorphism. □

Examples of Lie groups include all the groups in Table 11.53.

Prop. 20.72. Let (X,ξ) be a non-degenerate finite-dimensional irreducible $\mathbf{A}^{\varphi}$-correlated space. Then the group of correlated automorphisms $O(X,\xi)$ is a smooth submanifold of End X and is, with this smooth structure, a Lie group.

Proof By Cor. 11.38, $O(X,\xi) = \{t \in \operatorname{End} X : t^{\xi}t = 1_X\}$. Now, by Prop. 11.31, the map End $X \to$ End X; $t \rightsquigarrow t^{\xi}$ is real linear. It follows that the map

$$\pi : \operatorname{End} X \to \operatorname{End}_+(X,\xi);\quad t \rightsquigarrow t^{\xi}t$$

(cf. page 208) is smooth.

For any $u \in \operatorname{End} X$,

$$d\pi u(t) = ut + tu,$$

from which it follows, as in Example 19.14, that, for any $u \in O(X,\xi)$,

$d\pi u$ is surjective and, by Cor. 19.12, or by Prop. 20.52, that $O(X,\xi)$ is a smooth submanifold of End X of real dimension

$$\dim_{\mathbf{R}} \operatorname{End}_X - \dim_{\mathbf{R}} \operatorname{End}_+ (X,\xi) = \dim_{\mathbf{R}} \operatorname{End}_-(X,\xi). \quad \square$$

The tangent space to $O(X,\xi)$ at 1_X, being the affine subspace of End X through 1_X parallel to the real linear subspace $\operatorname{End}_-(X,\xi)$, with 1_X chosen as origin, is commonly and tacitly identified with this real linear space.

Cor. 20.73. For any finite p, q, n, with $p + q = n$,

$$\begin{aligned} &\dim O(p,q) = \tfrac{1}{2}n(n-1), && \dim O(n;\mathbf{C}) = n(n-1), \\ &\dim O(n;\mathbf{H}) = n(2n-1), && \dim U(p,q) = n^2 \\ &\dim Sp(2n;\mathbf{R}) = n(2n+1), && \dim Sp(2n;\mathbf{C}) = 2n(2n+1) \end{aligned}$$

and $\dim Sp(p,q) = n(2n+1)$. $\square$

The dimensions of the groups $GL(n;\mathbf{R})$, $GL(n;\mathbf{C})$ and $GL(n;\mathbf{H})$ could be computed similarly, but it is simpler to observe, as in Prop. 15.48, that these are open subsets of $\mathbf{R}(n)$, $\mathbf{C}(n)$ and $\mathbf{H}(n)$, respectively.

Cor. 20.74. For any finite n,

$$\dim GL(n;\mathbf{R}) = n^2, \quad \dim GL(n;\mathbf{C}) = 2n^2$$

and $\dim GL(n;\mathbf{H}) = 4n^2$. $\square$

Prop. 20.75. For any finite p, q, n the maps

$$O(p,q) \to S^0;\ t \rightsquigarrow \det t, \quad U(p,q) \to S^1;\ t \rightsquigarrow \det t$$
$$GL(n;\mathbf{R}) \to \mathbf{R}^*;\ t \rightsquigarrow \det t \quad \text{and} \quad GL(n;\mathbf{C}) \to \mathbf{C}^*;\ t \rightsquigarrow \det t$$

are smooth projections.

Proof Use Prop. 20.52, Prop. 20.39 and Prop. 18.21. $\square$

Cor. 20.76. For any p, q, n with $n = p + q$,

$SO(p,q)$ is a smooth submanifold of $O(p,q)$, of dimension $\frac{1}{2}(n-1)$,
$SU(p,q)$ is a smooth submanifold of $U(p,q)$, of dimension $n^2 - 1$,
$SL(n;\mathbf{R})$ is a smooth submanifold of $GL(n;\mathbf{R})$, of dimension $n^2 - 1$,
$SL(n;\mathbf{C})$ is a smooth submanifold of $GL(n;\mathbf{C})$, of dimension $2(n^2 - 1)$.
$\square$

There are many examples of a Lie group acting smoothly on a smooth manifold.

Prop. 20.77. Let G be a Lie group, X a smooth manifold and $G \times X \to X$; $(g,x) \rightsquigarrow gx$ a smooth action of G on X. Then, for any $a \in G$, the map $X \to X$; $x \rightsquigarrow ax$ is a smooth homeomorphism. $\square$

Prop. 20.78. Let G be a Lie group, X a smooth manifold and

$G \times X \to X$; $(g,x) \rightsquigarrow gx$ a smooth action of G on X. Then, for any $b \in X$, the map

$$\pi : G \to X; \quad g \rightsquigarrow gb$$

is a submersion if, and only if, $T\pi_1$ is surjective, where $1 = 1_{(G)}$.

Proof For any $a \in G$, the map π admits the decomposition

$$G \to G \xrightarrow{\pi} X \to X; \quad g \rightsquigarrow a^{-1}g \rightsquigarrow a^{-1}gb \rightsquigarrow gb.$$

From this, and Prop. 20.70 and Prop. 20.40, it follows that if $T\pi_1$ is surjective then $T\pi_a$ is surjective. This proves $\Leftarrow$. The proof of $\Rightarrow$ is trivial. □

The quasi-spheres (cf. page 217) are all smooth manifolds, and the appropriate correlated group for each quasi-sphere acts smoothly on it.

Prop. 20.79. Let (X,ξ) be a symmetric non-degenerate finite-dimensional irreducible $\mathbf{A}^\varphi$-correlated space. Then the quasi-sphere $\mathscr{S}((X,\xi) \times \mathbf{A}^\varphi)$ is a smooth submanifold of $X \times \mathbf{A}$ with tangent space at $(0,1)$ the linear subspace

$$\{(c,d) \in X \times \mathbf{A} : d^\varphi + d = 0\}$$

(or, more strictly, its parallel through $(0,1)$ with that point chosen as 0).

Proof The quasi-sphere is the fibre over 1 of the map

$$X \times \mathbf{A} \to \{\lambda \in \mathbf{A} : \lambda^\varphi = \lambda\}; \quad (c,d) \rightsquigarrow c^\xi c + d^\varphi d.$$

This map is easily proved to be a smooth submersion with tangent map at $(0,1)$ the map

$$X \times \mathbf{A} \to \{\lambda \in \mathbf{A}; \quad \lambda^\varphi = \lambda\}; \quad (c,d) \rightsquigarrow (0, d^\varphi + d). \qquad □$$

There is an analogue of Prop. 20.79 for the essentially skew cases. The reader is invited to formulate the analogue and prove it.

Prop. 20.80. Let (X,ξ) be a non-degenerate finite-dimensional symmetric irreducible $\mathbf{A}^\varphi$-correlated space, and let G and S be the group of correlated automorphisms and the unit quasi-sphere, respectively, of the $\mathbf{A}^\varphi$-correlated space $(X,\xi) \times \mathbf{A}^\varphi$. Then the map

$$G \times S \to S; \quad (g,x) \rightsquigarrow g(x)$$

is smooth.

Proof This map is a restriction of the linear map

$$\operatorname{End}(X \times \mathbf{A}) \times (X \times \mathbf{A}) \to X \times \mathbf{A}; \quad (t,x) \rightsquigarrow t(x). \qquad □$$

Prop. 20.81. Let G and S be as in Prop. 20.80. Then the map $\pi : G \to S$; $g \rightsquigarrow g(0,1)$ is a smooth projection.

Proof By Theorem 11.55, π is surjective, with fibres the left cosets in G of the group $O(X,\xi)$ regarded as a subgroup of G in the obvious way. By Prop. 20.80 and Prop. 20.20 the maps π and $S \longrightarrow S$; $x \rightsquigarrow u(x)$, for any $u \in G$, are smooth. By Prop. 20.78 it remains to prove that $T\pi_1$ is surjective.

Now $\begin{pmatrix} a & c \\ b & d \end{pmatrix} \in TG_1$ if, and only if, $a^\xi + a = 0$, $b = c^\xi$ and $d^\varphi + d = 0$, and $(c,d) \in TS_{(0,1)}$ if, and only if, $d^\varphi + d = 0$, from which the surjectivity of the linear map

$$T\pi_1 : TG_1 \longrightarrow TS_{(0,1)}; \quad g \rightsquigarrow g(0,1)$$

is evident. □

Cor. 20.82. For any finite p, q, n the maps

$$\begin{aligned}
O(p,q+1) &\longrightarrow \mathscr{S}(\mathbf{R}^{p,q+1}) \\
O(n+1;\mathbf{C}) &\longrightarrow \mathscr{S}(\mathbf{C}^{n+1}) \\
O(n+1;\mathbf{H}) &\longrightarrow \mathscr{S}(\mathbf{H}^{n+1}) \\
U(p,q+1) &\longrightarrow \mathscr{S}(\bar{\mathbf{C}}^{p,q+1}) \\
Sp(p,q+1) &\longrightarrow \mathscr{S}(\bar{\mathbf{H}}^{p,q+1}) \\
GL(n+1;\mathbf{R}) &\longrightarrow \mathscr{S}(\mathrm{hb}\mathbf{R}^{n+1}) \\
GL(n+1;\mathbf{C}) &\longrightarrow \mathscr{S}(\mathrm{hb}\mathbf{C}^{n+1}) \\
GL(n+1;\mathbf{H}) &\longrightarrow \mathscr{S}(\mathrm{hb}\tilde{\mathbf{H}}^{n+1})
\end{aligned}$$

defined in Thereom 11.55 are open continuous surjections. □

Cor. 20.83. For any finite p, q, n the groups

$O(n;\mathbf{H})$, $U(p,q)$, $Sp(p,q)$, $GL(n;\mathbf{R})$ $(n > 1)$, $GL(n;\mathbf{C})$ and $GL(n;\mathbf{H})$ are each connected.

Proof Add the information in Cor. 20.82 to Theorem 17.29 and apply Prop. 16.73. □

Similar methods prove the following.

Prop. 20.84. For any finite p, q, n the maps

$$\begin{aligned}
SO(p,q+1) &\longrightarrow \mathscr{S}(\mathbf{R}^{p,q+1}) \\
SO(n+1;\mathbf{C}) &\longrightarrow \mathscr{S}(\mathbf{C}^{n+1}) \\
SU(p,q+1) &\longrightarrow \mathscr{S}(\bar{\mathbf{C}}^{p,q+1})
\end{aligned}$$

defined in Theorem 11.55 are open continuous surjections. □

Cor. 20.85. For any finite p, q, n the groups

$$SO(n;\mathbf{C}) \quad \text{and} \quad SU(p,q)$$

are connected. □

The groups $SO(p,q)$, by contrast, are not connected unless p or $q = 0$. See Prop. 20.95 below.

Once again there is an analogue for the essentially skew cases. The conclusion is as follows.

Prop. 20.86. For any finite n the groups $Sp(2n;\mathbf{R})$ and $Sp(2n;\mathbf{C})$ are connected. □

Further examples of smooth manifolds and maps are provided by the quadric Grassmannians studied in Chapter 12.

Prop. 20.87. Let (X,ξ) be any non-degenerate finite-dimensional irreducible symmetric or skew $\mathbf{A}^{\varphi}$-correlated space. Then, for any $k \leqslant \dim X$, the quadric Grassmannian $\mathscr{I}_k(X,\xi)$ is a smooth submanifold of $\mathscr{G}_k(X)$. The parabolic atlas is a smooth atlas for $\mathscr{I}_k(X,\xi)$ and determines the same smooth structure. (Cf. pages 229–231.) □

Prop. 20.88. Let $G = \left\{\begin{pmatrix} a & \bar{b} \\ b & \bar{a}\end{pmatrix} \in \mathbf{C}(n) : \begin{pmatrix} a & \bar{b} \\ b & \bar{a}\end{pmatrix}^{\xi}\begin{pmatrix} a & \bar{b} \\ b & \bar{a}\end{pmatrix} = 1\right\}$, where, for any $\begin{pmatrix} a & \bar{b} \\ b & \bar{a}\end{pmatrix} \in \mathbf{C}(2n,)$ $\begin{pmatrix} a & \bar{b} \\ b & \bar{a}\end{pmatrix}^{\xi} = \begin{pmatrix} \bar{a}^{\tau} & \bar{b}^{\tau} \\ b^{\tau} & a^{\tau}\end{pmatrix}$. Then G is a Lie group, with tangent space at 1 the real linear subspace of $\mathbf{C}(2n)$

$$\left\{\begin{pmatrix} a & \bar{b} \\ b & \bar{a}\end{pmatrix} \in \mathbf{C}(2n) : a \in \operatorname{End}_-(\bar{\mathbf{C}}^n), \quad b \in \operatorname{End}_-(\mathbf{C}^n)\right\},$$

isomorphic in an obvious way with $\operatorname{End}_-(\bar{\mathbf{C}}^n) \times \operatorname{End}_-(\mathbf{C}^n)$. Moreover the map

$$f\colon O(2n) \to \mathbf{C}(2n); \quad t \rightsquigarrow c^{-1}tc,$$

with $c = \dfrac{1}{\sqrt{2}}\begin{pmatrix} 1 & \mathrm{i} \\ \mathrm{i} & 1\end{pmatrix}$, is a smooth embedding, with image G, and f_{sur} is a Lie group isomorphism. □

Prop. 20.89. For any n, the map

$$f\colon O(2n) \to \mathscr{I}_n(\mathbf{C}^n_{\mathrm{hb}}) : \begin{pmatrix} a & \bar{b} \\ b & \bar{a}\end{pmatrix} \rightsquigarrow \operatorname{im}\begin{pmatrix} a \\ b\end{pmatrix}$$

(cf. page 233) is a smooth projection.

Proof In this instance $O(2n)$ is embedded in $\mathbf{C}(2n)$, as in Prop. 20.89. It is enough to prove that the map is smooth at 1, with surjective differential there, the surjectivity of f having been already proved in Chapter 12.

The image of 1 by f is $\operatorname{im}\begin{pmatrix} 1 \\ 0\end{pmatrix}$ and near this point of $\mathscr{I}_n(\mathbf{C}^n_{\mathrm{hb}})$ one has the chart

$$\mathrm{End}_-(\mathbf{C}^n) \to \mathscr{I}_n(\mathbf{C}^n_{\mathrm{hb}}); \quad b' \rightsquigarrow \mathrm{im}\begin{pmatrix}1\\ b'\end{pmatrix},$$

with inverse

$$\mathscr{I}_n(\mathbf{C}^n_{\mathrm{hb}}) \to \mathrm{End}_-(\mathbf{C}^n); \quad \mathrm{im}\begin{pmatrix}a\\ b\end{pmatrix} \rightsquigarrow ba^{-1},$$

sending $\mathrm{im}\begin{pmatrix}1\\ 0\end{pmatrix}$, in particular, to 0.

Near 1, therefore, the map f is representable by the map

$$O(2n) \to \mathrm{End}_-(\mathbf{C}^n); \quad \begin{pmatrix}a & \bar{b}\\ b & \bar{a}\end{pmatrix} \rightsquigarrow ba^{-1},$$

which is smooth, with tangent map at 1

$$\mathrm{End}_-(\bar{\mathbf{C}}^n) \times \mathrm{End}_-(\mathbf{C}^n) \to \mathrm{End}_-(\mathbf{C}^n); \quad (a,b) \rightsquigarrow b.$$

(Cf. Exercise 18.38.) This tangent map is clearly surjective. □

There are nine other examples like this one and the reader is invited to formulate and to discuss them! (Cf. Prop. 12.12.)

It remains to consider several examples from Chapter 13.

The Pfaffian charts on Spin (n), regarded as a subgroup of the even Clifford algebra $\mathbf{R}_{0,n}^{0}$, were defined on page 349.

Prop. 20.90. The group Spin (n) is a smooth submanifold of $\mathbf{R}_{0,n}^{0}$ and is, with this structure, a Lie group.

Proof The Pfaffian charts on Spin (n) are open smooth embeddings. The group operations are restrictions of maps that are known to be smooth. □

The Cayley chart on $SO(p,q)$ at 1, for any finite p,q, was defined on page 236. For any $t \in SO(p,q)$ the *Cayley chart* on $SO(p,q)$ *at* t is defined to be the Cayley chart at 1 composed with left multiplication by t.

Prop. 20.91. The Cayley charts on $SO(p,q)$ are smooth. □

The section on the Pfaffian chart for $SO(n)$ in Chapter 13 extends to the indefinite case to the following extent.

Prop. 20.92. Let p, q be finite, let $s \in \mathrm{End}_-(\mathbf{R}^{p,q})$, and let $s' \in \mathbf{R}(p+q)$ be defined by the formula

$$s'_{ij} = -s_{ij} \quad \text{if } p \leqslant i \quad \text{and} \quad j > p$$

and

$$s'_{ij} = s_{ij} \quad \text{otherwise.}$$

Then $s' \in \mathrm{End}_-(\mathbf{R}^n)$.

If, moreover, $|s|$ is *sufficiently small,*

$$\text{Pf } s' \in \Gamma^0(p,q) \quad \text{and} \quad \rho_{\text{Pf}\,s'} = (1+s)(1-s)^{-1}. \qquad \square$$

The map

$$\text{End}_-(\mathbf{R}^{p,q}) \rightarrowtail \text{Spin}\,(p,q); \quad s \rightsquigarrow \text{Pf}s' / \,|\, N(\text{Pf}\, s') \,|$$

is called the *Pfaffian chart* on Spin (p,q) *at* 1, while, for any $g \in$ Spin (p,q), the *Pfaffian chart* on Spin (p,q) *at* g is the Pfaffian chart on Spin (p,q) at 1 composed with left multiplication by g.

Prop. 20.93. For any finite p and q the group Spin (p,q) is a smooth submanifold of $\mathbf{R}_{p,q}^{0}$ and is, with this structure, a Lie group. $\square$

Prop. 20.94. The group surjection ρ : Spin $(p,q) \rightarrow SO(p,q)$ is a smooth locally trivial projection.

(Use Pfaffian and Cayley charts, as in Prop. 17.45.) $\square$

Prop. 20.95. The groups Spin$^+$ (p,q) and SO^+ (p,q) (cf. page 268) are Lie groups. All are connected, with the exception of Spin$^+$ (0,0), Spin$^+$ (0,1), Spin$^+$ (1,0) and Spin$^+$ (1,1), homeomorphic to S^0, S^0, S^0 and $S^0 \times \mathbf{R}$, respectively. $\square$

Prop. 20.96. The Lorentz group SO^+ (p,q) consists of the rotations of $\mathbf{R}^{p,q}$ preserving the semi-orientations of $\mathbf{R}^{p,q}$. (Cf. page 161.)

Proof Since SO^+ (p,q) is connected, by Prop. 20.95, the continuous map $SO^+(p,q) \rightarrow \mathbf{R}^*$; $\begin{pmatrix} a & c \\ b & d \end{pmatrix} \rightsquigarrow \det a$ is of constant sign and, since its value at 1 is 1, it is always of positive sign. Similarly, det d is positive on SO^+ (p,q). By a similar argument det a and det d are negative on the coset SO^- (p,q) of $SO^+(p,q)$ in $SO(p,q)$. $\square$

Lie algebras

In all the examples of Lie groups given above the standard atlases or embeddings defining the smooth structure and the group operations have been not only C^1, but also C^2, C^∞ and even C^ω. In this final section all groups will be C^2 at least. This is no restriction, since it can be shown [49] that any C^1 Lie group admits a unique C^2, C^∞ or even C^ω Lie group structure compatible with the given C^1 Lie group structure. By a theorem of Gleason, Montgomery and Zippin [17], [46] (Hilbert's 5th problem [26]) it can even be shown that any topological group that is also a manifold has a unique C^1 Lie group structure compatible with the given structure.

Prop. 20.97. Let G be a C^2 Lie group. Then the map

$$G \times G \to G; \quad (a,g) \rightsquigarrow a\,g\,a^{-1}$$

is C^2. □

In particular, the group map

$$\rho_a : G \to G; \quad g \rightsquigarrow a\,g\,a^{-1}$$

is C^2, for any $a \in G$. The map

$$\mathrm{Ad}_G : G \to \mathrm{Aut}\, TG_1; \quad a \rightsquigarrow (T\rho_a)_1,$$

where $1 = 1_{(G)}$, is called the *adjoint representation* of the Lie group G.

Prop. 20.98. Let G be a C^2 Lie group. Then Ad_G is a C^1 group map.

Proof By Prop. 20.40, Ad_G is a group map. To prove that it is C^1 it is enough to prove that it is C^1 at 1.

Let $L = TG_1$ and let $h : L \rightarrowtail G$ be any C^2 chart on G with $h(0) = 1$ and $Th_0 = 1_L$, the identity map on L. Such a chart exists. Let $f : L \times L \rightarrowtail L$ be the map defined, for any $(x,y) \in L$ sufficiently near to 0, by the formula

$$h(f(x,y)) = h(x)\,h(y)\,(h(x))^{-1}.$$

Then, for any $a \in G$ sufficiently near 1,

$$(T\rho_a)_1 = d_1 f(x,0), \quad \text{where } h(x) = a,$$

so that $(\mathrm{Ad}_G)\,h = d_1 f(-,0)$, which is C^1 at 0. Therefore Ad_G is C^1 at 1. □

The adjoint representation of a Lie group need not be injective.

Example 20.99. Let $G = S^1$. Then Ad_G is the constant map with value 1. □

Example 20.100. Let G be any abelian Lie group. Then Ad_G is the constant map with value 1. □

Example 20.101. Let $G = S^3$. Then Ad_G has image $SO(3)$, while the map $(\mathrm{Ad}_G)_{\mathrm{sur}} : S^3 \to SO(3)$ is the familiar double covering of Chapter 10 or Chapter 13. □

The map

$$\mathrm{ad}_G = T(\mathrm{Ad}_G)_1 : TG_1 \to \mathrm{End}\, TG_1$$

is called the *adjoint representation* of TG_1.

Prop. 20.102. For any C^2 Lie group G, the map

$$TG_1 \times TG_1 \to TG_1; \quad (x,y) \rightsquigarrow [x,y] = \mathrm{ad}_G\,(x)(y)$$

is bilinear. □

The product defined in Prop. 20.102 is known as the *Lie bracket*, and the linear space TG_1 with this product is known as the *Lie algebra of* G. The Lie bracket is normally neither commutative nor associative. (Cf. Theorems 20.106 and 20.110 below.)

Prop. 20.103. Let G be a C^2 Lie group and let h and f be defined as in the proof of Prop. 20.98. Then, for any $x, y \in TG_1$,

$$[x,y] = d_0 d_1 f(0,0)(x)(y). \qquad \square$$

Theorem 20.104. Let $t : G \rightarrow H$ be a C^1 group map, where G and H are C^2 Lie groups. Then Tt_1 is a Lie algebra map; that is, for all $x, y \in TG_1$,

$$Tt_1([x,y]) = [Tt_1(x), Tt_1(y)].$$

Proof For any $a, g \in G$,

$$t(\rho_a(g)) = t(a\,g\,a^{-1}) = \rho_{t(a)})\,t(g),$$

and therefore, for any $a \in G$, the diagram of maps

$$\begin{array}{ccc} G & \xrightarrow{\rho_a} & G \\ \downarrow{\scriptstyle t} & & \downarrow \\ H & \xrightarrow{\rho_{t(a)}} & H \end{array}$$

is commutative. The induced diagram of tangent maps is

$$\begin{array}{ccc} TG_1 & \xrightarrow{\mathrm{Ad}_G\, a} & TG_1 \\ \downarrow{\scriptstyle Tt_1} & & \downarrow{\scriptstyle Tt_1} \\ TH_1 & \xrightarrow{\mathrm{Ad}_H\, t(a)} & TH_1 \end{array},$$

leading, for any $y \in TG_1$ to the commutative diagram

$$\begin{array}{ccccc} G & \xrightarrow{\mathrm{Ad}_G} & \mathrm{Aut}\, TG_1 & \xrightarrow[\text{at } y]{\text{evaluation}} & TG_1 \\ \downarrow{\scriptstyle t} & & & & \downarrow{\scriptstyle Tt_1} \\ H & \xrightarrow{\mathrm{Ad}_H} & \mathrm{Aut}\, TH_1 & \xrightarrow[\text{at } Tt_1(y)]{\text{evaluation}} & TH_1 \end{array}.$$

The induced tangent map diagram this time is

$$\begin{array}{ccccc} TG_1 & \xrightarrow{\mathrm{ad}_G} & \mathrm{End}\ TG_1 & \xrightarrow[\text{at } y]{\text{evaluation}} & TG_1 \\ \downarrow{\scriptstyle Tt_1} & & & & \downarrow{\scriptstyle Tt_1} \\ TH_1 & \xrightarrow{\mathrm{ad}_H} & \mathrm{End}\ TH_1 & \xrightarrow[\text{at } Tt_1(y)]{\text{evaluation}} & TH_1\ . \end{array}$$

This also is commutative. That is, for all $x,y \in TG_1$,

$$Tt_1[x,y] = [Tt_1(x),Tt_1(y)],$$

which is what had to be proved. □

Prop. 20.105. Let G be a C^2 Lie group and let L and h be defined as in the proof of Prop. 20.98. For all x, $y \in L$, let $\phi(x,y) = x \cdot y$ be defined by the formula

$$h(x \cdot y) = h(x)\, h(y)$$

whenever $h(x)\, h(y) \in \operatorname{im} h$, and let $\chi(x) = x^{(-1)}$ be defined by the formula

$$h(x^{(-1)}) = h(x)^{-1},$$

whenever $h(x)^{-1} \in \operatorname{im} h$. Then ϕ and χ are C^2 maps with non-null open domains,

$$d_0\phi(0,0) = 1_L, \quad d_1\phi(0,0) = 1_L$$

and $d\chi 0 = -1_L$.

(Note that, for any $x \in L$ sufficiently near 0, $\phi(x,0) = x$, $\phi(0,x) = x$ and $\phi(x,\chi(x)) = 0$.) □

Theorem 20.106. Let G be a C^2 Lie group and let $L = TG_1$. Then, for all x, $y \in L$,

$$[y,x] = -[x,y].$$

Proof Let h and f be defined as in the proof of Prop. 20.98 and let ϕ and χ be defined as in Prop. 20.105. Then, for any $x \in \operatorname{dom} \chi$, since the map $f(x,-)$ admits the decomposition

$$\begin{array}{ccccc} L & \rightarrowtail & L & \rightarrowtail & L \\ y & \rightsquigarrow & x \cdot y = w & \rightsquigarrow & w \cdot x^{(-1)}, \end{array}$$

it follows that

$$d_1 f(x,0) = d_0\phi(0,x^{(-1)})\, d_1\phi(x,0).$$

From this, and from Prop. 20.105, it follows that, for any $x \in L$,

$$\begin{aligned} d_0 d_1 f(0,0)(x) &= (d_1 d_0\phi(0,0))\,(d\chi 0(x))\, d_1\phi(0,0) \\ &\quad + (d_0\phi(0,0))\,(d_0 d_1\phi(0,0))(x) \\ &= d_0 d_1\phi(0,0)(x) - d_1 d_0\phi(0,0)(x), \end{aligned}$$

implying, by Prop. 20.103 and by Cor. 19.32 that, for any $x, y \in L$,

$$[x,y] = d_0 d_1 \phi(0,0)(x)(y) - d_0 d_1 \phi(0,0)(y)(x),$$

and therefore that $[y,x] = -[x,y]$. □

Note that, though $d_0 d_1 f(0,0)$ is independent of the choice of chart h, this is not so for $d_0 d_1 \phi(0,0)$. Consider, for example, $G = \mathbf{R}^*$. Then, if h is the chart $\mathbf{R} \rightarrowtail \mathbf{R}^*$; $x \rightsquigarrow 1 + x$, ϕ is a restriction of the map

$$\mathbf{R} \times \mathbf{R} \to \mathbf{R}; \quad (x,y) \rightsquigarrow x + y + xy$$

and $d_0 d_1 \phi(0,0)(x)(y) = xy$, while, if h is the chart $\mathbf{R} \to \mathbf{R}^*$; $x \rightsquigarrow e^x$, ϕ is the map

$$\mathbf{R} \times \mathbf{R} \to \mathbf{R}; \quad (x,y) \rightsquigarrow x + y$$

and $d_0 d_1 \phi(0,0)(x)(y) = 0$.

Cor. 20.107. Let X be a finite-dimensional real linear space. Then, for any $u, v \in T(\text{Aut } X)_1 = \text{End } X$,

$$[u,v] = uv - vu.$$

Proof Let h be the chart

$$\text{End } X \rightarrowtail \text{Aut } X; \quad t \rightsquigarrow t.$$

Then, for any $u,v \in \text{End } X$, since $\phi(u,v) = uv$,

$$d_0 d_1 \phi(0,0)(u)(v) = uv. \qquad \square$$

Cor. 20.108. Let G be a Lie subgroup of Aut X, where X is a finite-dimensional real linear space. Then, for any $u, v \in TG_1$,

$$[u,v] = uv - vu. \qquad \square$$

Example 20.109. For any $x, y \in (TS^3)_1$, the space of pure quaternions,

$$[x,y] = xy - yx = 2x \times y,$$

where $\times$ denotes the vector product. □

Theorem 20.110. Let G be a C^2 Lie group and let $L = TG_1$. Then, for all $x, y, z \in L$,

$$[[x,y],z] = [x,[y,z]] - [y,[x,z]].$$

Proof By Theorem 20.104 applied to the C^1 group map Ad_G, ad_G is a Lie algebra map. Therefore, for all $x, y \in L$, by Cor. 20.107,

$$\begin{aligned}\text{ad}_G [x,y] &= [\text{ad}_G x, \text{ad}_G y] \\ &= (\text{ad}_G x)(\text{ad}_G y) - (\text{ad}_G y)(\text{ad}_G x),\end{aligned}$$

and so, for all $x, y, z \in L$,

$$[[x,y],z] = [x,[y,z]] - [y,[x,z]]. \qquad \square$$

The equation proved in Theorem 20.110 is known as the *Jacobi identity* for the Lie algebra L. By Theorem 20.106 this can also be written in the more symmetrical form

$$[x,[y,z]] + [y,[z,x]] + [z,[x,y]] = 0.$$

A *Lie algebra* over a commutative field K is an algebra L over K such that, for any $x, y \in L$,

$$[y,x] = -[x,y]$$

and, for any $x, y, z \in L$,

$$[x,[y,z]] + [y,[z,x]] + [z,[x,y]] = 0,$$

where $L \times L \to L;\ (x,y) \rightsquigarrow [x,y]$ is the algebra product.

By Theorem 20.106 and Theorem 20.110 the Lie algebra of a Lie group is a Lie algebra in this more general sense.

For a good survey article on Lie algebras see [33].

The theory of Lie groups is developed in many books. See, for example, [58], [10], [49], [24], [25].

Further Exercises

20.111. A smooth atlas $\mathscr{S}$ for a smooth manifold X is said to be *orientable* if an orientation can be chosen for the source of each chart in such a way that, for any two charts $h_0 : V_0 \rightarrowtail X$, $h_1 : V_1 \rightarrowtail X$ of $\mathscr{S}$ and for each $v \in \operatorname{dom}((h_1)_{\text{sur}}^{-1}h_0)$, the differential at v of $(h_1)_{\text{sur}}^{-1}h_0$ respects the orientations chosen.

Let $\mathscr{S}$ be an orientable smooth atlas for a smooth manifold X and let $\mathscr{S}'$ be any equivalent smooth atlas for X, the domain of any chart in $\mathscr{S}'$ being connected. Show that $\mathscr{S}'$ also is an orientable smooth atlas for X. □

20.112. Suggest definitions for the terms *orientable* smooth manifold and *non-orientable* smooth manifold. □

20.113. Show that, for any odd $n \in \omega$, $\mathbf{R}P^n$ is orientable and that, for any even $n \geqslant 2$, $\mathbf{R}P^n$ is non-orientable. □

20.114. Show that, for any $n \in \omega$, $\mathbf{C}P^n$ is orientable. □

20.115. For which n is $\mathbf{H}P^n$ orientable? □

20.116. Show that any Lie group is orientable. □

20.117. A *complex smooth atlas* for an even-dimensional smooth manifold X consists of a smooth atlas for X, the sources of whose charts are complex linear spaces, the atlas being such that the overlap maps are not only smooth but also complex differentiable (cf. page 362). A manifold with such an atlas chosen is said to be a *complex manifold*. Prove that any complex manifold is orientable. □

20.118. Two smooth submanifolds U and V of a smooth manifold X are said to intersect *transversally at* a point w of their intersection W if $TX_w = TU_w + TV_w$, and to intersect *transversally* if they intersect transversally at each point of W.

Prove that, if U and V intersect transversally, then W is a smooth submanifold of U, V and X. (Recall Exercise 3.52.) □

20.119. Consider the map

$$f: \mathbf{R}^2 \longrightarrow \mathbf{R}^4; \quad x \rightsquigarrow (x_0, x_1 - 2x_1h(x), h(x), x_0x_1h(x)),$$

where $h(x) = ((1 + x_0^2)(1 + x_1^2))^{-1}$.

Prove that f is an immersion, injective except that $f(0,1) = f(0,-1)$, and show that any sufficiently small neighbourhood of 0 in $\operatorname{im} f$ consists of two two-dimensional submanifolds of $\mathbf{R}^4$ intersecting transversally at 0. Show also that, for any $\varepsilon > 0$, there exists $\delta > 0$ such that, for $|x| > \delta$, $|f(x) - (x_0, x_1, 0, 0)| < \varepsilon$. □

20.120. Verify that the restriction to any line through 0 of the map

$$\mathbf{R}^4 \longrightarrow \mathbf{R}^6; x \rightsquigarrow (x_0^2 - x_1^2, x_0x_1, x_0x_2 - x_1x_3, x_1x_2 + x_0x_3, x_2^2 - x_3^2, x_2x_3)$$

followed by the radial projection $\mathbf{R}^6 \setminus \{0\} \longrightarrow S^5$; $y \rightsquigarrow y/|y|$ is constant and prove that the induced map $\mathbf{R}P^3 \longrightarrow S^5$ is a smooth embedding. □

20.121. Try to immerse $\mathbf{R}P^2$ in $\mathbf{R}^3$. (This was first done by Werner Boy [5]. Such an immersion is essentially constructed in [48].) □

20.122. Let $f: X \longrightarrow Y$ be a smooth map, X and Y being smooth manifolds. Then a map $\phi: X \longrightarrow TY$ such that $\pi_{TY}\phi = f$ is said to be a *tangent vector field along f*. Verify that a tangent vector field along 1_X is the same thing as a tangent vector field on X.

Suppose that $F: \mathbf{R} \times X \longrightarrow Y$ is a smooth map such that $F(0,-) = f$. Verify that the map

$$\phi_F: X \longrightarrow TY; \quad x \rightsquigarrow T(F(-,x))_0(1)$$

is a continuous tangent vector field along f. □

20.123. Let X, Y and Z be smooth manifolds, let $F: \mathbf{R} \times X \longrightarrow Y$ and $G: \mathbf{R} \times Y \longrightarrow Z$ be smooth maps and let H be the map

$$\mathbf{R} \times X \longrightarrow Z; \quad (t,x) \rightsquigarrow G(t,F(t,x)).$$

Prove that

$$\phi_H = \phi_G f + (Tg)\phi_F,$$

where $f = F(0,-)$ and $g = G(0,-)$.

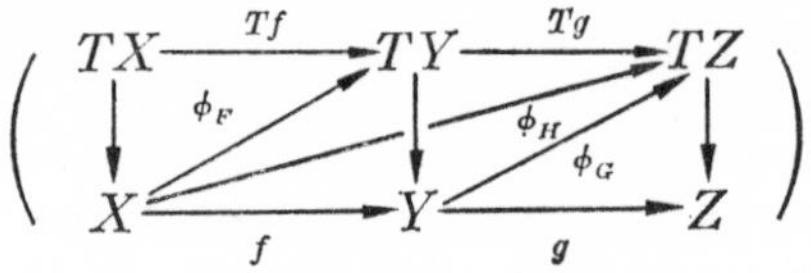

□

20.124. Let X and Y be smooth manifolds, and let $F : \mathbf{R} \times X \to Y$, $G : \mathbf{R} \times X \to X$ and $H : \mathbf{R} \times Y \to Y$ be smooth maps such that $G(0,-) = 1_X$, $H(0,-) = 1_Y$ and, for all $(t,x) \in \mathbf{R} \times X$,

$$F(t,x) = H(t, fG(t,x)),$$

or, equivalently, for all $t \in \mathbf{R}$,

$$F(t,-) = H(t,-)\, f\, G(t,-),$$

where $f = F(0,-)$. Prove that

$$\phi_F = \phi_H f + (Tf)\phi_G.$$

(This exercise is the key to recent work on the structural stability of maps by J. N. Mather [40].) □

20.125. Let G be a Lie group. Prove that TG is homeomorphic to $TG_1 \times G$. □

20.126. The *dual tangent bundle* of a finite-dimensional real affine space X consists of the space $T^L X = \bigcup_{a \in X} (X \times \{a\})^L$, with the topology induced by the obvious bijection $\bigcup_{a \in X} (X \times \{a\})^L \to X_*^{\,L} \times X$, together with the obvious projection $T^L X \to X$. The space $T^L X$ may be regarded as an affine space with vector space $X_*^{\,L} \times X_*$. Moreover the linear space $X_*^{\,L} \times X_*$ may be assigned the non-degenerate skew-symmetric real bilinear product

$$(X_*^{\,L} \times X_*)^2 \to \mathbf{R};\ ((\alpha, v), (\alpha', v')) \rightsquigarrow \alpha(v') - \alpha'(v).$$

Suggest a definition for the *dual tangent bundle* $T^L X$ of a finite-dimensional smooth manifold X. On the assumption that X is $\mathbf{C}^2$, show that there is a non-degenerate skew-symmetric product on each tangent space of $T^L X$, inducing an isomorphism of the tangent space with its dual, such that the induced bijection

$$T(T^L X) \to T^L(T^L X)$$

is a homeomorphism.

(The dual tangent bundle plays the role of the *phase space* in modern treatments of Hamiltonian dynamics. See, for example, [0].) □

BIBLIOGRAPHY

[0] ABRAHAM, R. *Foundations of Mechanics*, Benjamin, New York (1967).

[1] ADAMS, J. F. 'On the nonexistence of elements of Hopf invariant one, *Bull. Am. math Soc.* **64,** 279–282 (1958); also *Ann. Math.* **72,** 20–104 (1960).

[2] ADAMS, J. F. 'Vector fields on spheres' *Ann. Math.* **75**, 603–632 (1962); see also *Mathl. Rev.* **24,** # A3662 (1962).

[3] ARTIN, E. *Geometric Algebra,* Interscience, New York (1957).

[4] ATIYAH, M. F., BOTT, R. and SHAPIRO, A. 'Clifford modules' *Topology* **3** (Supp. 1), 3–38 (1964).

[5] BOY, W. 'Über die Curvatura integra und die Topologie geschlossener Flächen' *Math. Annln.* **57,** 151–184 (1903).

[6] BROUWER, L. E. J. 'On looping coefficients' *Proc. Akad. Wet., Amsterdam* **15,** 113–122 (1912).

[7] BROWN, R. *Elements of Modern Topology,* McGraw-Hill, London (1968).

[8] BRUCK, R. H. 'Recent advances in the foundations of Euclidean plane geometry' *Am. math. Mon.* **62,** 2–17 (1955).

[9] CAYLEY, A. 'On Jacobi's elliptic functions, in reply to Rev. Brice Brownin and on quaternions' *Phil. Mag.* **3,** 210–213 (1845).

[10] CHEVALLEY, C. *Theory of Lie Groups*, Princeton U.P., Princeton, N.J. (1946).

[11] CLIFFORD, W. K. *On the Classification of Geometric Algebras* (1876): published as Paper XLIII in *Mathematical Papers* (ed. R. Tucker), Macmillan, London (1882).

[12] COHEN, L. W. and EHRLICH, G. *The Structure of the Real Number System,* Van Nostrand, Princeton, N.J. (1963).

[13] CROWELL, R. H. and FOX, R. H. *Introduction to Knot Theory,* Ginn, Boston (1963).

[14] DIEUDONNÉ, J. *Foundations of Modern Analysis,* Academic Press, New York (1960).

[15] FUCHS, L. *Partially Ordered Algebraic Systems,* Pergamon, Oxford (1963).

[16] GAUSS, K. F. *Zur mathematischen Theorie der electrodynamischen Wirkungen* (1833): republished in *Werke*, Vol. 5, 605, Göttingen (1877).

[17] GLEASON, A. M. 'Groups without small subgroups' *Ann. Math.* **56,** 193–212 (1952).

[18] GOFFMAN, C. and PEDRICK, G. *First Course in Functional Analysis,* Prentice-Hall, Englewood Cliffs, N.J. (1965).

[19] GRASSMANN, H. *Die Wissenschaft der extensiven Grösse oder die Ausdehnungslehre, eine neue mathematischen Disclipin*, Leipzig (1844).

[20] GRUENBERG, K. W. and WEIR, A. J. *Linear Geometry,* Van Nostrand' Princeton, N.J. (1967).

[21] HALMOS, P. R. *Naive Set Theory*, Van Nostrand, Princeton, N.J. (1960).

[22] HALMOS, P. R. 'A glimpse into Hilbert space' *Lectures on Modern Mathematics* (ed. T. L. Saaty), Vol. I, Wiley, New York (1963).

[23] HAMILTON, W. R. *Lectures on Quaternions* (1853): republished in *The Mathematical Papers of Sir William Rowan Hamilton*, Vol. III, *Algebra,*

Cambridge University Press, London (1967). See also *Mathl. Rev.* **35,** # 4081 (1968).

[24] HELGASON, S. *Differential Geometry and Symmetric Spaces,* Academic Press, New York (1962).

[25] HERMANN, R. *Lie Groups for Physicists,* Benjamin, New York (1966).

[26] HILBERT, D. 'Mathematische Probleme' *Arch. Math. Phys.* (3) **1,** 44–63, 213–237 (1901).

[27] HIRZEBRUCH, F. *Topological Methods in Algebraic Geometry* (1956); 3rd edn, Springer, Berlin (1966).

[28] HOPF, H. Über die Abbildungen der dreidimensionalen Sphäre auf die Kugelfläche' *Mathl. Ann.* **104,** 637–665 (1931).

[29] HOPF, H. 'Über die Abbildungen von Sphären auf Sphären niedrigerer Dimension' *Fundam. Math.* **25,** 427–440 (1935).

[30] HUREWICZ, W. and WALLMAN, H. *Dimension Theory,* Princeton University Press, Princeton, N.J. (1941).

[31] HURWITZ, A. 'Über die Komposition der quadratischen Formen' *Mathl. Ann.* **88,** 1–25 (1923).

[32] JAMES, I. M. 'Some embeddings of projective spaces' *Proc. Camb. phil. Soc.* **55,** 294–298 (1959).

[33] KAPLANSKY, I. 'Lie algebras' *Lectures on Modern Mathematics* (ed. T. L. Saaty), Vol. I, Wiley, New York (1963).

[34] KELLEY, J. L. *General Topology,* Van Nostrand, New York (1955).

[35] KERVAIRE, M. 'Non-parallelizability of the n-sphere for $n > 7$' *Proc. natn. Acad. Sci. U.S.A.* **44,** 280–283 (1958).

[36] KLEINFELD, E. 'A characterization of the Cayley Numbers' *Studies in Modern Algebra* (ed. A. A. Albert), Vol. 2, Mathematical Association of America, Prentice-Hall, Englewood Cliffs, N.J. (1963).

[37] LANG, S. *Introduction to Differentiable Manifolds,* Interscience, New York (1962).

[38] LIE, S. *Theorie der Transformationsgruppen,* Leipzig (1888–89); *Vorlesungen über continuerliche Gruppen,* Leipzig (1893).

[39] MACLANE, S. *Homology,* Springer, Berlin (1963).

[40] MATHER, J. N. 'Stability of C^∞-Mappings: II, Infinitesimal stability implies stability' *Ann. Math.* **89,** 254–291 (1969).

[41] MILNOR, J. 'On manifolds homeomorphic to the 7-sphere' *Ann. Math.* **64,** 399–405 (1956).

[42] MILNOR, J. *Differential Topology*; notes by J. Munkres: mimeograph, Princeton University (1958).

[43] MILNOR, J. 'Lectures on the h-cobordism theorem' *Princeton math. Notes* (1965).

[44] MILNOR, J. W. *Topology from the Differentiable Viewpoint,* University Press of Virginia, Charlottesville, Va. (1965).

[45] MILNOR, J. and BOTT, R. 'On the parallelizability of the spheres' *Bull. Am. math. Soc.* **64,** 87–89 (1958).

[46] MONTGOMERY, D. and ZIPPIN, L. 'Small subgroups of finite-dimensional groups' *Ann. Math.* **56,** 213–241 (1952).

[47] NEWNS, W. F. 'Functional dependence' *Am. math. Mon.* **74,** 911–920 (1967).

[48] PHILLIPS, A. 'Turning a surface inside out' *Scient. Am.* (May 1966).

[49] PONTRJAGIN, L. S. *Topological Groups* (1946); 2nd edn, Gordon & Breach, New York (1966).

[50] Pontryagin, L. S. *Smooth Manifolds and Their Application in Homotopy Theory* (1955): *Amer. Math. Soc. Translations.* (2) **11,** 1–114 (1959).

[51] Radon, J. 'Lineare Scharen orthogonaler Matrizen' *Abh. math. Semin. Univ. Hamburg* **1,** 1–14 (1923).

[52] Russell, B. *My Philosophical Development*, Allen & Unwin, London (1959).

[53] Segre, B. 'Gli automorphismi del corpo complesso ed un problema di Corrado Segre' *Atti Accad. naz. Lincei Rc.* (8) **3,** 414–420 (1947).

[54] Simmons, G. F. *Introduction to Topology and Modern Analysis*, McGraw-Hill, New York (1963).

[55] Steenrod, N. *The Topology of Fibre Bundles*, Princeton University Press, Princeton, N.J. (1951).

[56] Walker, A. G. and Newns, W. F. 'Tangent planes to a differentiable manifold' *J. Lond. math. Soc.* **31,** 400–407 (1956).

[57] Wall, C. T. C. 'Graded algebras, antiinvolutions, simple groups and symmetric spaces' *Bull. Am. math. Soc.* **74,** 198–202 (1968).

[58] Weyl, H. *The Classical Groups*, Princeton University Press, Princeton, N.J. (1939).

[59] Whitney, H. 'Differentiable manifolds' *Ann. Math.* **37,** 645–680 (1936).

[60] Whitney, H. 'The self-intersections of a smooth n-manifold in $2n$-space' *Ann. Math.* **45,** 220–246 (1944).

[61] Witt, E. 'Theorie der quadratischen Formen in beliebigen Körpern' *J. de Crelle*, **176,** 31–44 (1937).

[62] 'Correspondence, from an ultramundane correspondent' *Ann. Math.* **69,** 247–251 (1959).

LIST OF SYMBOLS

INDEX